Roumen Kountchev and Kazumi Nakamatsu (Eds.)

Advances in Reasoning-Based Image Processing Intelligent Systems

Intelligent Systems Reference Library, Volume 29

Editors-in-Chief

Further volumes of this series can be found on our homepage: springer.com

Vol. 4. Lakhmi C. Jain and Chee Peng Lim (Eds.)
Handbook on Decision Making: Techniques and Applications, 2010
ISBN 978-3-642-13638-2

Vol. 5. George A. Anastassiou
Intelligent Mathematics: Computational Analysis, 2010
ISBN 978-3-642-17097-3

Vol. 6. Ludmila Dymowa
Soft Computing in Economics and Finance, 2011
ISBN 978-3-642-17718-7

Vol. 7. Gerasimos G. Rigatos
Modelling and Control for Intelligent Industrial Systems, 2011
ISBN 978-3-642-17874-0

Vol. 8. Edward H.Y. Lim, James N.K. Liu, and Raymond S.T. Lee
Knowledge Seeker – Ontology Modelling for Information Search and Management, 2011
ISBN 978-3-642-17915-0

Vol. 9. Menahem Friedman and Abraham Kandel
Calculus Light, 2011
ISBN 978-3-642-17847-4

Vol. 10. Andreas Tolk and Lakhmi C. Jain
Intelligence-Based Systems Engineering, 2011
ISBN 978-3-642-17930-3

Vol. 11. Samuli Niiranen and Andre Ribeiro (Eds.)
Information Processing and Biological Systems, 2011
ISBN 978-3-642-19620-1

Vol. 12. Florin Gorunescu
Data Mining, 2011
ISBN 978-3-642-19720-8

Vol. 13. Witold Pedrycz and Shyi-Ming Chen (Eds.)
Granular Computing and Intelligent Systems, 2011
ISBN 978-3-642-19819-9

Vol. 14. George A. Anastassiou and Oktay Duman
Towards Intelligent Modeling: Statistical Approximation Theory, 2011
ISBN 978-3-642-19825-0

Vol. 15. Antonino Freno and Edmondo Trentin
Hybrid Random Fields, 2011
ISBN 978-3-642-20307-7

Vol. 16. Alexiei Dingli
Knowledge Annotation: Making Implicit Knowledge Explicit, 2011
ISBN 978-3-642-20322-0

Vol. 17. Crina Grosan and Ajith Abraham
Intelligent Systems, 2011
ISBN 978-3-642-21003-7

Vol. 18. Achim Zielesny
From Curve Fitting to Machine Learning, 2011
ISBN 978-3-642-21279-6

Vol. 19. George A. Anastassiou
Intelligent Systems: Approximation by Artificial Neural Networks, 2011
ISBN 978-3-642-21430-1

Vol. 20. Lech Polkowski
Approximate Reasoning by Parts, 2011
ISBN 978-3-642-22278-8

Vol. 21. Igor Chikalov
Average Time Complexity of Decision Trees, 2011
ISBN 978-3-642-22660-1

Vol. 22. Przemysław Różewski, Emma Kusztina, Ryszard Tadeusiewicz, and Oleg Zaikin
Intelligent Open Learning Systems, 2011
ISBN 978-3-642-22666-3

Vol. 23. Dawn E. Holmes and Lakhmi C. Jain (Eds.)
Data Mining: Foundations and Intelligent Paradigms, 2011
ISBN 978-3-642-23165-0

Vol. 24. Dawn E. Holmes and Lakhmi C. Jain (Eds.)
Data Mining: Foundations and Intelligent Paradigms, 2011
ISBN 978-3-642-23240-4

Vol. 25. Dawn E. Holmes and Lakhmi C. Jain (Eds.)
Data Mining: Foundations and Intelligent Paradigms, 2011
ISBN 978-3-642-23150-6

Vol. 26. Tauseef Gulrez and Aboul Ella Hassanien (Eds.)
Advances in Robotics and Virtual Reality, 2011
ISBN 978-3-642-23362-3

Vol. 27. Cristina Urdiales
Collaborative Assistive Robot for Mobility Enhancement (CARMEN), 2011
ISBN 978-3-642-24901-3

Vol. 28. Tatiana Valentine Guy, Miroslav Kárný and David H. Wolpert (Eds.)
Decision Making with Imperfect Decision Makers, 2012
ISBN 978-3-642-24646-3

Vol. 29. Roumen Kountchev and Kazumi Nakamatsu (Eds.)
Advances in Reasoning-Based Image Processing Intelligent Systems, 2012
ISBN 978-3-642-24692-0

Roumen Kountchev and Kazumi Nakamatsu (Eds.)

Advances in Reasoning-Based Image Processing Intelligent Systems

Conventional and Intelligent Paradigms

Springer

Prof. Roumen Kountchev
Technical University of Sofia
Drujba 2
Bl. 404, Entr. 2, Ap. 54
Sofia 1582
Bulgaria
E-mail: rkountch@tu-sofia.bg

Prof. Kazumi Nakamatsu
University of Hyogo
Nakamachi-dori 3-1-3-901 Chuo-ku
Kobe 650-0027
Japan
E-mail: nakamatu@shse.u-hyogo.ac.jp

ISBN 978-3-642-24692-0 e-ISBN 978-3-642-24693-7

DOI 10.1007/978-3-642-24693-7

Intelligent Systems Reference Library ISSN 1868-4394

Library of Congress Control Number: 2011939334

Typeset by Scientific Publishing Services Pvt. Ltd., Chennai, India.

Printed on acid-free paper

9 8 7 6 5 4 3 2 1

springer.com

Contents

Part I: Intelligent Image Processing

1 Advances in Reasoning-Based Image Processing and Pattern Recognition: Conventional and Intelligent Paradigms 3
Roumen Kountchev, Kazumi Nakamatsu
1.1 Introduction 3
1.2 Performance Analysis and Comparison of the Dirac Video Codec with H.264/ MPEG-4, Part 10 4
1.3 Linear and Non-linear Inverse Pyramidal Image Representation: Algorithms and Applications 4
1.4 Preserving Data Integrity of Encoded Medical Images: The LAR Compression Framework 5
1.5 Image Processing in Medicine 5
1.6 Attention in Image Sequences: Biology, Computational Models, and Applications 6
1.7 Visual Perception for Mobile Robots Motion Control 6
1.8 Motion Estimation for Object Analysis and Detection in Videos 6
1.9 Shape-Based Invariant Features Extraction for Object Recognition 7
1.10 Object-Based Image Retrieval System Using Rough Set Approach 7
1.11 Paraconsistent Artificial Neural Networks and Delta, Theta, Alpha, and Gamma Band Detection 7
1.12 Paraconsistent Artificial Neural Networks and Pattern Recognition: Speech Production Recognition and Cephalometric Analysis 8
1.13 On Creativity and Intelligence in Computational Systems 8
1.14 Method for Intelligent Representation of Research Activities of an Organization over Taxonomy of Its Field 8

2 Performance Analysis and Comparison of the Dirac Video Codec with H.264/MPEG-4, Part 10 9
Aruna Ravi, K.R. Rao
2.1 Introduction 9
2.2 Dirac Architecture 10
2.2.1 Dirac Encoder 10
2.2.2 Dirac Decoder 11
2.3 Stages of Encoding and Decoding in Dirac 12
2.3.1 Wavelet Transform 12
2.3.2 Scaling and Quantization 15
2.3.3 Entropy Coding 16

2.3.4 Motion Estimation 16
2.3.5 Motion Compensation 18
2.3.6 Decoder 19
2.4 Implementation 20
2.4.1 Code Structure Overview 20
2.4.2 Simplicity and Relative Speed of Encoding 20
2.5 Results 22
2.5.1 Compression Ratio Test 22
2.5.2 SSIM Test 24
2.5.3 PSNR Test 26
2.6 Conclusions 31
2.7 Future Research 31
References 32
Abbreviations 34

3 Linear and Non-linear Inverse Pyramidal Image Representation: Algorithms and Applications 35
Roumen Kountchev, Vladimir Todorov, Roumiana Kountcheva
3.1 Basic Methods for Pyramidal Image Decomposition 35
3.2 Basic Principles of the Inverse Pyramid Decomposition 41
3.2.1 Inverse Pyramid Decomposition with Orthogonal Transforms 41
3.2.2 Comparison of the Inverse and the Laplacian Pyramid Decompositions 46
3.2.3 Reduced Inverse Pyramid Decomposition 50
3.2.4 Inverse Pyramid Decomposition with Non-linear Transforms Based on Neural Networks 58
3.3 Multi-view Image Representation Based on the Inverse Pyramidal Decomposition 67
3.3.1 Multi-view 3D Object Representation with Modified IPD 68
3.3.2 Experimental Results 73
3.4 Multispectral Images Representation with Modified IPD 78
3.4.1 Selection of the Reference Image in Multispectral Sequence 80
3.4.2 Experimental Results 81
3.5 Conclusions 84
References 84

4 Preserving Data Integrity of Encoded Medical Images: The LAR Compression Framework 91
Marie Babel, François Pasteau, Clément Strauss, Maxime Pelcat, Laurent Bédat, Médéric Blestel, Olivier Déforges
4.1 Introduction 91
4.2 How to Protect Content in an Image? 93
4.2.1 Cryptography 93
4.2.2 Data Hiding and Image Coding 96
4.3 Secure Transmission of Encoded Bitstreams 97
4.3.1 Error Resilience and Channel Coding 98

4.3.2 IP Packets Securization Processes ... 99
4.3.3 LTE Standard Application Case: Securization Process for Advanced Functionalities ... 100
4.4 Application Example: LAR Medical Framework ... 105
4.4.1 LAR Codec Overview ... 105
4.4.2 Principles and Properties ... 107
4.4.3 Content Protection Features ... 111
4.4.4 Transmission Error Protection - Error Resilience ... 118
4.5 Conclusion ... 121
References ... 121

5 Image Processing in Medicine ... 127
Baigalmaa Tsagaan, Hiromasa Nakatani
5.1 Introduction ... 127
5.2 Overview of Medical Imaging ... 128
5.2.1 Imaging Modality ... 128
5.2.2 Image Reconstruction ... 130
5.2.3 Image Format ... 131
5.2.4 Diagnostic Practice Using Medical Images ... 131
5.3 Conventional Approaches of Image Processing ... 132
5.3.1 Image Segmentation ... 132
5.3.2 Image Registration ... 135
5.3.3 Visualization ... 137
5.4 Application ... 137
5.4.1 CAD, CAS and Virtual Endoscopy ... 138
5.4.2 Image-Guided Navigation for Paranasal Sinus Surgery ... 139
5.5 Summary ... 142
References ... 143
List of Abbreviations ... 146

6 Attention in Image Sequences: Biology, Computational Models, and Applications ... 147
Mariofanna Milanova, Engin Mendi
6.1 Introduction ... 147
6.2 Computational Models of Visual Attention ... 149
6.2.1 A Taxonomy of Computational Model of Bottom-Up Visual Attention ... 149
6.2.2 Hybrid Computational Models of Visual Attention ... 155
6.3 Selected Datasets ... 160
6.3.1 LABELME ... 160
6.3.2 Amsterdam Library of Object Images (ALOI) ... 160
6.3.3 Spatially Independent, Variable Area, and Lighting (SIVAL) ... 161
6.3.4 MSRA ... 161
6.3.5 Caltech ... 161
6.3.6 PASCAL VOC ... 161

6.4 Software Implementations of Attention Modeling 161
6.4.1 Itti-Koch Model 162
6.4.2 Matlab Implementations 162
6.4.3 TarzaNN 163
6.4.4 Model Proposed by Matei Mancas 163
6.4.5 JAMF 163
6.4.6 LabVIEW 163
6.4.7 Attention Models Evaluation and Top-Down Models 163
6.5 Applications 163
6.6 Example 164
References 167

Part II: Pattern Recognition, Image Data Mining and Intelligent Systems

7 Visual Mobile Robots Perception for Motion Control 173
Alexander Bekiarski
7.1 The Principles and Basic Model of Mobile Robot Visual Perception 173
7.2 Log-Polar Visual Mobile Robot Perception Principles and Properties 177
7.2.1 Definition of Log-Polar Transformation for Mobile Robot Visual Perception 177
7.2.2 Log-Polar Transformation of Image Points in Visual Mobile Robot Perception System 178
7.2.3 The Properties of Log-Polar Transformation Suitable for Visual Mobiles Robot Perception System 182
7.2.4 Visual Perception of Objects Rotation in Log-Polar Mobile Robot Visual Perception Systems 182
7.2.5 Visual Perception of Objects Translation and Scaling in Log-Polar Mobile Robot Visual Perception Systems 188
7.3 Algorithm for Motion Control with Log-Polar Visual Mobile Robot Perception 194
7.3.1 The Basic Principles and Steps of the Algorithm for Motion Control with Log-Polar Visual Mobile Robot Perception 194
7.3.2 Simulation and Test Results for the Algorithm of Motion Control with Log-Polar Visual Mobile Robot Perception 200
7.4 Conclusion 206
References 207

8 Motion Estimation for Objects Analysis and Detection in Videos 211
Margarita Favorskaya
8.1 Introduction 211
8.2 Classification of Motion Estimation Methods 213
8.2.1 Comparative Motion Estimation Methods 214

8.2.2 Gradient Motion Estimation Methods 226
8.3 Local Motion Estimation Based on Tensor Approach 234
8.3.1 The Initiation Stage 235
8.3.2 Motion Estimation in Visual Imagery 236
8.3.3 Motion Estimation in Infrared Imagery 240
8.3.4 Elaboration of Boundaries of Moving Regions 242
8.3.5 Classification of Dynamic Regions 244
8.4 Experimental Researches 245
8.5 Tasks for Self-testing 248
8.6 Conclusion 250
References 250

9 Shape-Based Invariant Feature Extraction for Object Recognition 255
Mingqiang Yang, Kidiyo Kpalma, Joseph Ronsin
9.1 Introduction 255
9.2 One-Dimensional Function for Shape Representation 258
9.2.1 Complex Coordinates 258
9.2.2 Centroid Distance Function 259
9.2.3 Tangent Angle 259
9.2.4 Contour Curvature 260
9.2.5 Area Function 261
9.2.6 Triangle-Area Representation 261
9.2.7 Chord Length Function 262
9.2.8 Discussions 262
9.3 Polygonal Approximation 263
9.3.1 Merging Methods 263
9.3.2 Splitting Methods 265
9.3.3 Discussions 265
9.4 Spatial Interrelation Feature 266
9.4.1 Adaptive Grid Resolution 266
9.4.2 Bounding Box 267
9.4.3 Convex Hull 269
9.4.4 Chain Code 269
9.4.5 Smooth Curve Decomposition 271
9.4.6 Symbolic Representation Based on the Axis of Least Inertia 271
9.4.7 Beam Angle Statistics 272
9.4.8 Shape Matrix 273
9.4.9 Shape Context 275
9.4.10 Chord Distribution 276
9.4.11 Shock Graphs 277
9.4.12 Discussions 278
9.5 Moments 278
9.5.1 Boundary Moments 278
9.5.2 Region Moments 279
9.5.3 Discussions 282

9.6 Scale Space Approaches 282
9.6.1 Curvature Scale-Space 283
9.6.2 Intersection Points Map 284
9.6.3 Discussions 285
9.7 Shape Transform Domains 285
9.7.1 Fourier Descriptors 285
9.7.2 Wavelet Transform 288
9.7.3 Angular Radial Transformation 288
9.7.4 Shape Signature Harmonic Embedding 289
9.7.5 $\mathscr{R}$-Transform 290
9.7.6 Shapelet Descriptor 292
9.7.7 Discussions 293
9.8 Summary Table 293
9.9 Illustrative Example: A Contour-Based Shape Descriptor 295
9.9.1 Fundamental Concepts 295
9.9.2 Equal Area Normalization 296
9.9.3 Normalized Part Area Vector 299
9.9.4 Experimental Results 302
9.10 Conclusion 310
References 311

10 Object-Based Image Retrieval System Using Rough Set Approach 315
Neveen I. Ghali, Wafaa G. Abd-Elmonim, Aboul Ella Hassanien
10.1 Introduction 315
10.2 Basic Concepts 316
10.2.1 Rough Sets: Short Description 316
10.2.2 Rough Image Processing 317
10.2.3 Image Retrieval Systems: Problem Definition and Categories 320
10.3 Object-Based Image Retrieval System 321
10.3.1 Pre-processing: Segmentation and Feature Extraction 322
10.3.2 Similarity and Retrieval System 323
10.4 Experimental Results and Discussion 325
10.5 Conclusion 327
References 328

11 Paraconsistent Artificial Neural Networks and Delta, Theta, Alpha, and Beta Bands Detection 331
Jair Minoro Abe, Helder F.S. Lopes, Kazumi Nakamatsu
11.1 Introduction 331
11.2 Background 333
11.3 The Main Artificial Neural Cells 335
11.3.1 Paraconsistent Artificial Neural Cell of Analytic Connection – PANCac 336
11.3.2 Paraconsistent Artificial Neural Cell of Maximization– PANCmax 338

11.3.3 Paraconsistent Artificial Neural Cell of Minimization– PANCmin ... 339
11.3.4 Paraconsistent Artificial Neural Unit ... 340
11.3.5 Paraconsistent Artificial Neural System ... 340
11.4 PANN for Morphological Analysis ... 340
11.4.1 Data Preparation ... 340
11.4.2 The PANN Architecture ... 341
11.4.3 Expert System 1 – Checking the Number of Wave Peaks ... 344
11.4.4 Expert System 2 – Checking Similar Points ... 346
11.4.5 Expert System 3 – Checking Different Points ... 347
11.5 A Didactic Sample ... 348
11.6 Experimental Procedures – Attention-Deficit / Hyperactivity Disorder ... 351
11.7 Experimental Procedures – Applying in Alzheimer Disease ... 355
11.7.1 Expert system 1 – Detecting the Diminishing Average Frequency Level ... 358
11.7.2 Expert System 2 – High Frequency Band Concentration ... 359
11.7.3 Expert System 3 – Low Frequency Band Concentration ... 360
11.7.4 Results ... 360
11.8 Discussion ... 361
11.9 Conclusions ... 361
References ... 362

12 Paraconsistent Artificial Neural Networks and Pattern Recognition: Speech Production Recognition and Cephalometric Analysis ... 365
Jair Minoro Abe, Kazumi Nakamatsu
12.1 Introduction ... 365
12.2 Background ... 367
12.3 The Paraconsistent Artificial Neural Cells – PANC ... 369
12.4 The Paraconsistent Artificial Neural Cell of Learning - PANC-L ... 370
12.5 Unlearning of a PANC-l ... 371
12.6 Using PANN in Speech Production Recognition ... 373
12.7 Practical Results ... 374
12.8 Cephalometric Variables ... 376
12.9 Architecture of the Paraconsistent Artificial Neural Network ... 377
12.10 Results ... 379
12.11 Discussion ... 380
12.12 Conclusions ... 381
References ... 381

13 On Creativity and Intelligence in Computational Systems ... 383
Stuart H. Rubin
13.1 Introduction ... 383
13.2 On the Use of Ray Tracing for Visual Recognition ... 385
13.2.1 Case Generalization for Ray Tracing ... 385
13.2.2 The Case Generalization for Ray Tracing Algorithm ... 388

13.3 On Unmanned Autonomous Vehicles (UAVs)....395
13.4 Overview....396
13.5 Alternate Approaches....398
13.5.1 Theory....400
13.6 Algorithm for Image Randomization....407
13.7 A Theory for Machine Learning....411
13.7.1 Case vs. Rule-Based Learning....411
13.7.2 The Inference Engine....413
13.7.3 On Making Predictions....415
13.7.4 On Feature Induction....417
13.8 Conclusions and Outlook....420
References....420

14 Method for Intelligent Representation of Research Activities of an Organization over a Taxonomy of Its Field....423
Boris Mirkin, Susana Nascimento, Luís Moniz Pereira
14.1 Introduction....423
14.1.1 Motivation....423
14.1.2 Background....429
14.2 Taxonomy-Based Profiles....432
14.2.1 E-Screen Survey Tool....432
14.3 Representing Research Organization by Fuzzy Clusters of ACM-CCS Topics....433
14.3.1 Deriving Similarity between ACM-CCS Research Topics....433
14.3.2 Fuzzy Additive-Spectral Clustering....434
14.3.3 Experimental Verification of FADDI-S....437
14.4 Parsimonious Lifting Method....441
14.5 Case Study....444
14.6 Conclusion....451
References....452

Author Index....455

Part I
Intelligent Image Processing

Chapter 1
Advances in Reasoning-Based Image Processing and Pattern Recognition

Conventional and Intelligent Paradigms

The book puts special stress on the contemporary techniques for reasoning-based image processing and analysis: learning based image representation and advanced video coding; intelligent image processing and analysis in medical vision systems; similarity learning models for image reconstruction; visual perception for mobile robot motion control, simulation of human brain activity in the analysis of video sequences; shape-based invariant features extraction; essential of paraconsistent neural networks, creativity and intelligent representation in computational systems.

The book comprises 14 chapters. Each chapter is a small monograph, representing resent investigations of authors in the area. The topics of the chapters cover wide scientific and application areas and complement each-other very well. The chapters' content is based on fundamental theoretical presentations, followed by experimental results and comparison with similar techniques. In some chapters are included examples and tests, which facilitate the learning of the material and help the individual training of students and researchers. The size of the chapters is well-ballanced which permits a thorough presentation of the investigated problems. The authors are from universities and R&D institutions all over the world; some of the chapters are prepared by international teams. The book will be of use for university and PhD students, researchers and software developers working in the area of digital image and video processing and analysis.

Organization

The book is divided into 2 parts, as follows:

Part I: Intelligent Image Processing

1.1 Introduction

In the last decade significant developments have been made in intelligent image processing, based on the use of large image databases, and the rules for their classification, through analyzing visual apperception mechanisms. A large number of new approaches for computer intelligence have been created, such as: structure evaluation of the image quality, based on vision models, artificial neural networks, fuzzy logic, evolutionary computation, expert systems, etc.

R. Kountchev et al. (Eds.): Advances in Reasoning-Based Image Processing, ISRL 29, pp. 3–8.
springerlink.com

The basic trends in the image intelligent processing and analysis comprise:

- Data structures for image compression and analysis, based on various linear and non-linear models for image representation;
- Low level image processing: image acquisition by sensors;
- Preprocessing: noise suppression and enhancement of some object features, relevant to image understanding;
- Image restoration;
- Image segmentation: edge and region extraction to separate objects from the image background;
- Object description and classification: shape and texture representation and description;
- Motion analysis and 3D vision;
- Image and video retrieval;
- Intelligent data and video systems.

The chapters, included in this book depict the achievements of the authors in these scientific areas.

1.2 Performance Analysis and Comparison of the Dirac Video Codec with H.264/ MPEG-4, Part 10

The chapter presents the Dirac video codec, which is a hybrid motion-compensated state-of-the-art video codec that uses modern techniques such as wavelet transforms and arithmetic coding. It is an open technology designed to avoid patent infringement and can be used without the payment of license fees. It is well suited to the business model of public service broadcasters since it can be easily recreated for new platforms. Dirac is aimed at applications ranging from HDTV (high definition television) to web streaming. H.264, MPEG-4 part-10 or AVC, is the latest digital video codec standard which has proven to be superior to earlier standards in terms of compression ratio, quality, bit rates and error resilience. However unlike Dirac, it requires the payment of patent fees. The objective of this chapter is to analyze the Dirac video codec (encoder and decoder), based on several input test sequences, and to compare its performance with H.264/MPEG-4 Part 10 AVC. Analysis has been done on Dirac and H.264 using QCIF, CIF and SDTV video test sequences as input and the results recorded graphically for various parameters, including compression ratio, bit rate, PSNR, SSIM and MSE. In these tests, encoding and decoding has been performed for quality factor ranging from 0 - 10 and for lossless compression. Apart from this, comparison between Dirac and H.264's performance has been analyzed at various constant 'target' bit rates ranging from 10 KBps to 200 KBps. The test results indicate that Dirac's performance is comparable to that of H.264.

1.3 Linear and Non-linear Inverse Pyramidal Image Representation: Algorithms and Applications

In the chapter is presented one specific approach for image representation, known as Inverse Pyramid Decomposition (IPD), and its main applications. The chapter contains a review of the state of the art, aimed at the presentation of various pyramidal decompositions and outlines their advantages and demerits. In the next sections are considered in detail the principles of the IPD based on linear (DFT, DCT, WHT, KLT, etc.) and non-linear transforms: deterministic, based on oriented surfaces, and adaptive, based on pyramidal neural networks. Furthermore, the work introduces non-recursive and recursive implementations of the IPD. Special attention is paid to the main application areas of the IPD: image compression (lossless, visually lossless and lossy), multi-view and multispectral image representation. A significant part of the chapter is devoted to evaluation and comparison of the new representation with the well known compression standards JPEG and JPEG2000. In the conclusion are outlined the main advantages of IPD and the trends for future development and investigations of the new approach.

1.4 Preserving Data Integrity of Encoded Medical Images: The LAR Compression Framework

Through the development of medical imaging systems and their integration into a complete information system, the need for advanced joint coding and network services becomes predominant. PACS (Picture Archiving and Communication System) aims to acquire, store and compress, retrieve, present and distribute medical images. These systems also need to be accessible via the Internet or wireless channels. Thus protection processes against transmission errors have to be added to get a powerful joint source-channel coding tool. Moreover, such sensitive data requires confidentiality and privacy for archiving and transmission purposes, leading to the use of cryptography and data embedding solutions. This chapter presents dedicated tools of content protection and secure bitstream transmission for medical image purposes. In particular, the LAR image coding method is defined, together with advanced security-providing services.

1.5 Image Processing in Medicine

In this chapter the authors focus on image processing, pattern analysis and computer vision methods in medicine. The chapter comprises a brief overview of medical image acquisition systems and general approaches to image processing and vision applications in medicine. The first part reviews conventional issues of medical imaging: image modalities, image reconstruction, and use of medical imaging in diagnostic practice. The second part emphasizes those methods that are appropriate when medical images are the subjects of image processing and analysis. A brief overview of segmentation and registration algorithms is presented. The final section of the chapter presents a more detailed view of the recent practices incorporating interdisciplinary fields of computer aided diagnosis (CAD), computer-assisted surgery (CAS) systems and virtual endoscopy which encompass knowledge from medicine, image processing, pattern recognition and computer vision. Recent issues in development of medical imaging systems are summarized at the end of chapter.

1.6 Attention in Image Sequences: Biology, Computational Models, and Applications

Research in the area of visual attention modeling has grown since it was first introduced by Koch and Ullman in 1985. The chapter reviews different combined visual attention models. Concepts, such as the feature maps, the saliency map, the winner take-all (WTA) and the inhibition of return (IOR) were adopted from the Koch-Ulman model. To use only the visual input for a guided search is a bottom-up strategy. Such a strategy is not appropriate for search because different locations are fixated depending on the task. A strategy that involves combining the incoming image with information on the target or so called top-down information is also presented. The chapter also presents applications of visual attention models for adapting images on small displays and applications of the proposed models in Video Quality Assessment.

Part II: Pattern Recognition, image data mining and intelligent systems

1.7 Visual Perception for Mobile Robots Motion Control

Visual perception methods are developed mainly for human perception description and understanding. The results of such research are now very popular for robot visual perception modeling. In this chapter is presented a brief review of the basic visual perception methods suitable for intelligent mobile robot applications. Analysis of these methods is directed to mobile robot motion control, where visual perception is used for objects or human body localization, like: Bayesian visual perception methods for localization; log-polar visual perception; robot observation mapping using visual perception; landmark-based finding and localization with visual perception etc. The development of an algorithm for mobile robot visual perception is proposed, based on the features of log-polar transformation to represent some of the objects and scene fragments in the area of mobile robot observation in a more simple form for image processing. The features and advantages of the proposed algorithm are demonstrated by way of mobile robots visual perception situation of motion control in a road or corridor with outdoor road edges, painted lane separation lines or indoor two side existing room or corridor lines. The proposed algorithm is tested with suitable simulations and experiments with real mobile robots like the Pioneer 3-DX (Mobil Robots INC), WiFiBot and Lego Robot Mindstorms NXT. The results are summarized and presented in graphical form, and as test images and comparative tables in the conclusion.

1.8 Motion Estimation for Object Analysis and Detection in Videos

Motion estimation methods are used for the modeling of various physical processes, object behavior, and event prediction. In this chapter moving objects in videos are generally considered. Motion estimation methods are classified as comparative and gradient. Comparative motion estimation methods are usually used in real-time applications. Many aspects of block-matching modifications are discussed including the Gaussian mixture model, Lie operators, bilinear deformations, the multi-level motion model, etc. Gradient motion estimation methods assist in the realization of motion segmentation in complex dynamic scenes because only they provide the required accuracy. The application of the 2D tensors (in spatial domain) or the 3D tensors (in spatio-temporal domain) depends on the problem under study.

1.9 Shape-Based Invariant Features Extraction for Object Recognition

In this study, a shape descriptor is proposed for two-dimensional object retrieval, which in theory remains invariant under affine transforms. These transforms are main part of generally observed deformations. The proposed descriptor operates on the affine enclosed area. After a normalization, the number of points on a contour between two appointed positions doesn't change with affine transforms. This work proves that for any linearly filtered contour, the area of a triangle whose vertices are the centroid of the contour and a pair of successive points on the normalized contour remains linear under affine transforms. Experimental results indicate that the proposed method is invariant to boundary starting point variation, affine transforms (even in the case of high deformations), and also resistant to noise on the shapes.

1.10 Object-Based Image Retrieval System Using Rough Set Approach

In this chapter is presented an object-based image retrieval system using rough set theory. The system incorporates two major modules: Preprocessing and Object-based image retrieval. In preprocessing, an image-based object segmentation algorithm in the context of rough set theory is used to segment the images into meaningful semantic regions. A new object similarity measure is proposed for image retrieval. Performance is evaluated on an image database, and the effectiveness of the proposed image retrieval system is demonstrated. The experimental results show that the proposed system performs well in terms of speed and accuracy.

1.11 Paraconsistent Artificial Neural Networks and Delta, Theta, Alpha, and Gamma Band Detection

In this work is presented a study of brain EEG waves - delta, theta, alpha, and gamma bands - employing a new ANN based on Paraconsistent Annotated Evidential Logic, which is capable of manipulating concepts like impreciseness, inconsistency, and paracompleteness in a nontrivial manner. The Paraconsistent Artificial Neural Network is presented in some detail, and some specific applications also discussed.

1.12 Paraconsistent Artificial Neural Networks and Pattern Recognition: Speech Production Recognition and Cephalometric Analysis

In this expository work is sketched a theory of artificial neural network, based on a paraconsistent annotated evidential logic. Such theory, called Paraconsistent

Artificial Neural Network, is built from the Para-analyzer algorithm and has as characteristics the capability of manipulating uncertainty, inconsistent and paracomplete concepts. Some applications are presented in speech production analysis and cephalometrich variable analysis.

1.13 On Creativity and Intelligence in Computational Systems

The chapter presents an investigation of the potential for creative and intelligent computing in the domain of machine vision. It addresses such interrelated issues as randomization, dimensionality reduction, incompleteness, heuristics, as well as various representational paradigms. In particular, randomization is shown to underpin creativity, and heuristics are shown to serve as the basis for intelligence, and incompleteness implies the need for heuristics in any non trivial machine vision application, among others. Furthermore, the evolution of machine vision is seen to imply the evolution of heuristics, which follows from the examples supplied herein.

1.14 Method for Intelligent Representation of Research Activities of an Organization over Taxonomy of Its Field

In the chapter is presented a novel method for the analysis of research activities of an organization by mapping that to a taxonomy tree of the field. The method constructs fuzzy membership profiles of the organization members or teams in terms of the taxonomy's leaves (research topics), and then it generalizes them in two steps: the fuzzy clustering research topics according to their thematic similarities in the department, ignoring the topology of the taxonomy, and the optimally lifting clusters mapped to the taxonomy tree to higher ranked categories by ignoring "small" discrepancies. The method is illustrated by applying it to data collected using an in-house e-survey tool from a university department and from a university research center. The method can be considered for knowledge generalization over any taxonomy tree.

Acknowledgments. The book editors express their special thanks to the excellent scientists and book chapter reviewers Adel Elmaghraby, Alexander Bekiarsky, Benjamin Gadat, Chris Hinde, Demin Wang, Dominik Slezak, Fabio Romeu de Carvalho, Gordon Lee, Janne Nappi, João Mexia, Kidiyo Kpalma, Marie Babel, Pavel Babayan, Pooja Agawane, Robert Cierniak, Roumiana Kountcheva, Shinichi Tamura, Soumya Banerjee, Tim Borer, Tomasz Smolinski, and Witold Pedrycz (in alphabetical order) for their efforts and good will to help for the successful preparation of the book.

Roumen Kountchev
Kazumi Nakamatsu

Chapter 2
Performance Analysis and Comparison of the Dirac Video Codec with H.264/MPEG-4, Part 10

Aruna Ravi[1] and K.R. Rao[2]

[1] Department of Electrical Engineering, University of Texas at Arlington, Arlington, Texas 76019, USA
aruna.ravi@mavs.uta.edu
[2] Department of Electrical Engineering, University of Texas at Arlington, Box 19016 Arlington, Texas 76019, USA
rao@uta.edu

Abstract. Dirac is a hybrid motion-compensated state-of-the-art video codec that can be used without the payment of license fees. It can be easily adapted for new platforms and is aimed at applications ranging from HDTV to web streaming. In this chapter we analyze the Dirac video codec [1] based on several input test sequences, and compare its performance with H.264 / MPEG-4 Part 10 AVC [11-14]. Both Dirac and H.264 are implemented using different video test sequences at various constant 'target' bit rates ranging from 10KBps to 200KBps at image resolutions from QCIF to SD. The results have been recorded graphically and we arrive at a conclusion whether Dirac's performance is comparable to H.264. We also research whether Dirac outperforms H.264 / MPEG-4 Part 10 in terms of computational speed and efficiency.

2.1 Introduction

Video compression is used to exploit limited storage and transmission capacity as efficiently as possible which is important for the internet and high definition media. Dirac is an open and royalty-free video codec developed by the BBC [1] [2]. It aims to provide high-quality video compression from web video up to HD, [4] and as such competes with existing formats such as H.264 [11 - 14] and SMPTE VC-1 [17]. Dirac can compress any size of picture from low-resolution QCIF (176x144 pixels) to HDTV (1920x1080) and beyond, similar to common video codecs such as the ISO/IEC Moving Picture Experts Group (MPEG)'s MPEG-4 Part 2 [18][27] and Microsoft's SMPTE VC-1 [17].

R. Kountchev et al. (Eds.): Advances in Reasoning-Based Image Processing, ISRL 29, pp. 9–34.
springerlink.com

Dirac employs wavelet compression, instead of the discrete cosine transforms used in most other codecs. The Dirac software is a prototype implementation that can freely be modified and deployed. Dirac's decoder in particular is designed to be fast and more agile than other conventional decoders. The resulting specification is simple and straightforward to implement and is optimized for real-time performance. [1] Open source software such as the VLC [54] player can decode and display Dirac wrapped in MPEG-2 transport stream or in mp4 (".mov") files.

In addition to the C++ Dirac reference code, there is also a high speed open source ANSI C implementation called Schrödinger [4] under active development. Schrödinger is a cross-platform implementation of the Dirac video compression specification as a C library. Many media frameworks such as GStreamer [52] and ffmpeg [53] and applications such as VLC use Schrödinger to encode and decode video. Schrödinger is more optimized than Dirac reference code and performs better in most encoding situations, both in terms of encoding speed and visual quality. [19] Current development of Dirac implementations is hosted at diracvideo.org.

Substantial parts of the Dirac codec relating to intra coding have been ratified as an international standard in SMPTE 2042 (VC-2). This intra-frame version of Dirac is called DiracPro, [45] with emphasis on quality and low latency. It is optimized for professional production, archiving applications and not for end user distribution. [44]

2.2 Dirac Architecture

In the Dirac codec, image motion is tracked and the motion information is used to make a prediction of a later frame. A transform is applied to the prediction error between the current frame and the previous frame aided by motion compensation and the transform coefficients are quantized and entropy coded. [1] Temporal and spatial redundancies are removed by motion estimation, motion compensation and discrete wavelet transform respectively. Dirac uses a flexible and efficient form of entropy coding called arithmetic coding which packs the bits efficiently into the bit stream. [1]

2.2.1 Dirac Encoder [1] [21]

Video encoding is the process of preparing the video for output, where the digital video is encoded to meet proper formats and specifications for recording and playback through the use of video encoder software. [21] Streaming video quality is partly dependent upon the video encoding process and the amount of bandwidth required for it to be viewed properly. While encoding a video, a high degree of compression is applied to both the video and audio tracks so that it will stream at this speed.

In the Dirac encoder (Fig. 2.1) the entire compressed data is packaged in a simple byte stream. This has synchronization, permitting access to any frame quickly and efficiently - making editing simple. The structure is such that the entire byte

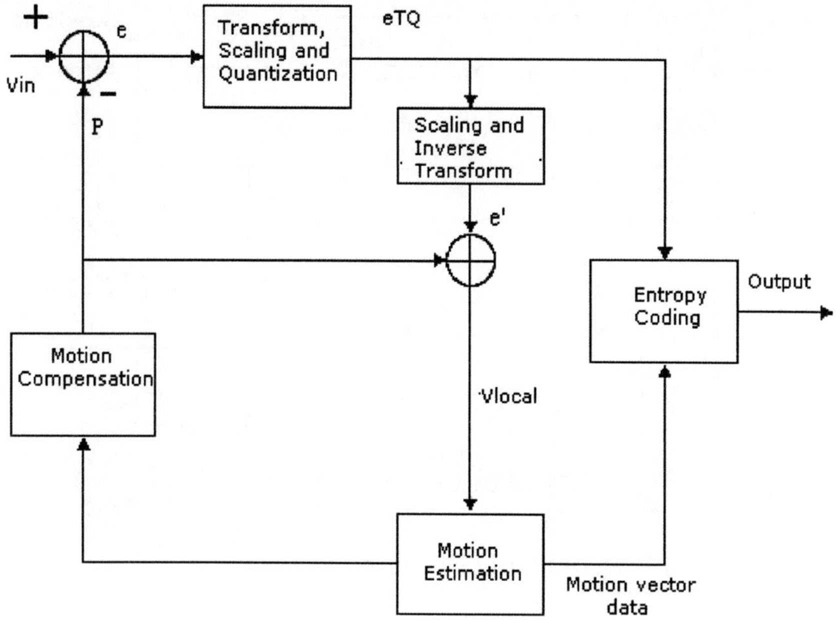

Fig. 2.1 Dirac encoder architecture [1] [2]

stream can be packaged in many of the existing transport streams. This feature allows a wide range of coding options, as well as easy access to all the other data transport systems required for production or broadcast metadata.

In the above figure, each input video frame *Vin* is compared with the previous motion compensated reference frame *P* to obtain *e*, the motion compensated prediction error (MCPE). *eTQ* is the MCPE after application of wavelet transform, scaling and quantization and it aids entropy coding. *e'* is the MCPE after scaling and inverse transform. This is combined with *P* to get *Vlocal* which is useful during the motion estimation stage to generate motion vector data. *P* is updated each time after motion compensation.

2.2.2 Dirac Decoder [1] [21]

The Dirac decoder (Fig. 2.2) performs the inverse operations of the encoder. The Dirac's decoder implementation is designed to provide fast decoding whilst remaining portable across various software platforms.

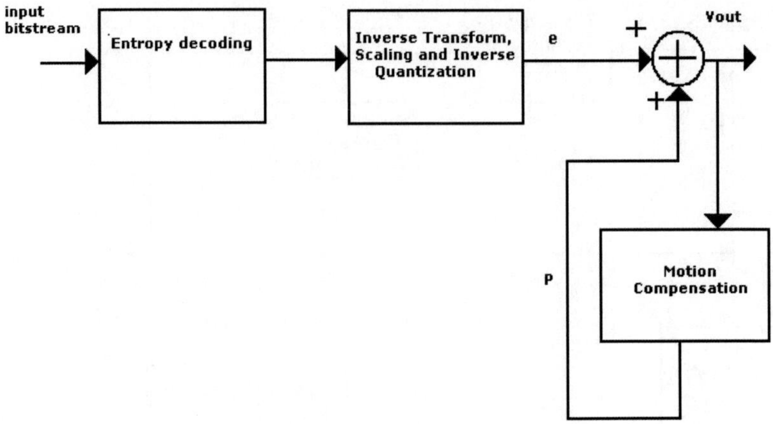

Fig. 2.2 Dirac decoder architecture

2.3 Stages of Encoding and Decoding in Dirac

2.3.1 Wavelet Transform

The 2D discrete wavelet transform provides Dirac with the flexibility to operate at a range of resolutions. This is because wavelets operate on the entire picture at once, rather than focusing on small areas at a time. In Dirac, the discrete wavelet transform plays the same role as the DCT in MPEG-2 in de-correlating data in a roughly frequency-sensitive way, whilst having the advantage of preserving fine details better than block based transforms. Synthesis filters can undo the aliasing introduced by critical sampling and perfectly reconstruct the input. The wavelet transform is constructed by repeated filtering of signals into low- and high-frequency parts. For two-dimensional signals, this filtering occurs both horizontally and vertically. At each stage, the low horizontal / low vertical frequency sub-band is split further, resulting in logarithmic frequency decomposition into sub-bands. [4]

Wavelet transforms have been proven to provide a more efficient technique than block transforms with still images. Within the Dirac wavelet filters, the data is encoded in 3 stages as shown in Fig. 2.3.

Daubechies wavelet filters [29] [30] are used to transform and divide the data in sub-bands which then are quantized with the corresponding RDO (rate distortion optimization) parameters and then variable length encoded. These three stages are then reversed at the decoder. [5]

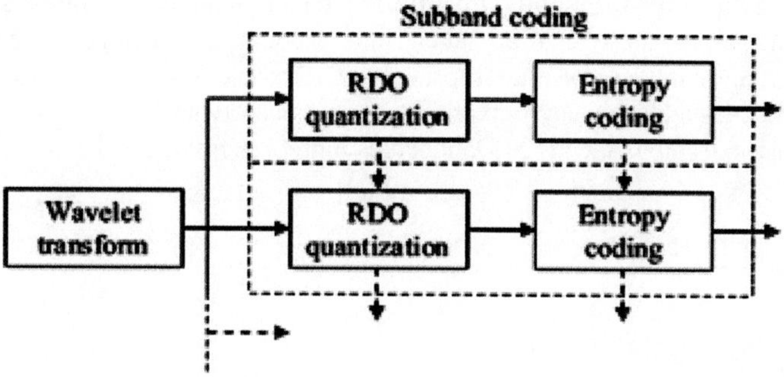

Fig. 2.3 Dirac's wavelet transform architecture [5]

The choice of wavelet filters has an impact on compression performance. Filters are required to have compact impulse response in order to reduce ringing artifacts and other effects so as to represent smooth areas compactly. It also has an impact on encoding and decoding speed in software. There are numerous filters supported by Dirac to allow a tradeoff between complexity and performance. These are configurable in the reference software. [4]

One filter available in Dirac is an approximation of the Daubechies (9, 7) low pass wavelet filter whose lifting stages are defined as follows: [4]

$$s_n^0 = x_{2n}$$

$$d_n^0 = x_{2n+1}$$

$$d_n^1 = d_n^0 - \frac{6497.\left(s_n^0 + s_{n+1}^0\right)}{4096}$$

$$s_n^1 = s_n^0 - \frac{217.\left(d_n^1 + d_{n-1}^1\right)}{4096}$$

$$d_n^2 = d_n^1 + \frac{3616.\left(s_n^1 + s_{n+1}^1\right)}{4096}$$

$$s_n^2 = s_n^1 + \frac{1817.\left(d_n^2 + d_{n-1}^2\right)}{4096} \qquad (2.1)$$

where s denotes sum and d denotes difference.

The numbers are integer approximations of the Daubechies lifting coefficients. This makes the transform fully invertible. The implementation ignores scaling coefficients, since these can be taken into account in quantizer selection by weighting the quantizer noise appropriately. The problem with this filter is that it has four lifting stages, and so it takes longer time in software. [4]
At the other extreme is the (5, 3) Daubechies high pass filter: [4]

$$\mathbf{d}_n^1 = \mathbf{d}_n^0 - \frac{\left(\mathbf{s}_n^0 + \mathbf{s}_{n+1}^0\right)}{2}$$

$$\mathbf{s}_n^1 = \mathbf{s}_n^0 + \frac{\left(\mathbf{d}_n^1 + \mathbf{d}_{n-1}^1\right)}{4} \tag{2.2}$$

The discrete wavelet transform packs most of the information into only a few sub-bands (at low frequency) as shown in Fig. 2.4, which allows compression to be achieved. Most of the energy is concentrated in the LL sub-band. All the other sub-bands can be coarsely quantized.

Fig. 2.4 Stages of wavelet transform [1]

This process can be repeated to achieve higher levels of wavelet transform. In case of two-dimensional images, wavelet filters are normally applied in both vertical and horizontal directions to each image component to produce four so-called sub-bands termed Low-Low (LL), Low-High (LH), High-Low (HL) and High-High (HH). In the case of two dimensions, only the LL band is iteratively decomposed to obtain the decomposition of the two-dimensional spectrum as shown in Fig. 2.5. [4]

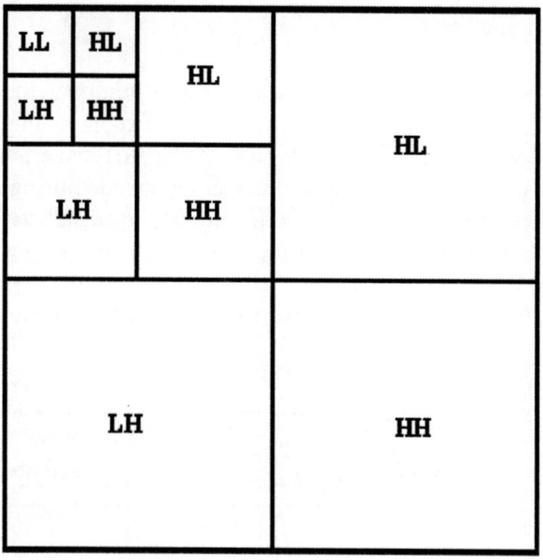

Fig. 2.5 Wavelet transform frequency decomposition [5]

2.3.2 Scaling and Quantization

Scaling involves taking frame data after application of wavelet transform and scaling the coefficients to perform quantization. Quantization employs a rate distortion optimization algorithm to strip information from the frame data that results in as little visual distortion as possible.

Dirac uses dead-zone quantization technique (Fig. 2.6) which differs from uniform quantization by making the first set of quantization steps twice as wide. This method is simple, efficient and allows Dirac to perform coarser quantization on smaller values. [5]

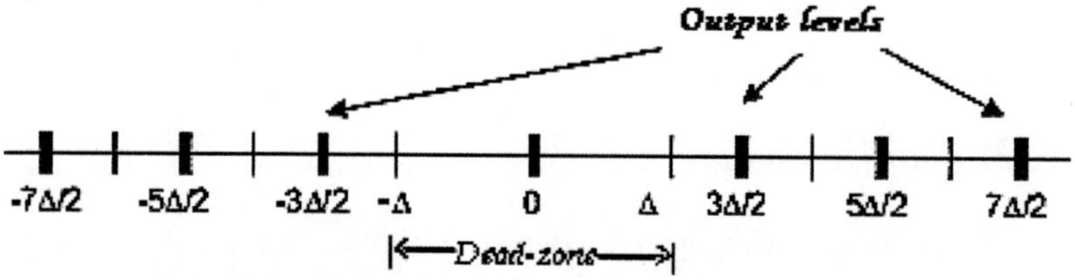

Fig. 2.6 Dead-zone quantizer with quality factor (QF) [5]

2.3.3 Entropy Coding

Entropy coding is applied after wavelet transform to minimize the number of bits used. It consists of three stages: binarization, context modeling and arithmetic coding [5] as shown in Fig. 2.7. The purpose of the first stage is to provide a bit stream with easily analyzable statistics that can be encoded using arithmetic coding, which can adapt to those statistics, reflecting any local statistical features. The context modeling in Dirac is based on the principle that whether a coefficient is small or not is well-predicted by its neighbors and its parents. [3] Arithmetic coding performs lossless compression and is both flexible and efficient.

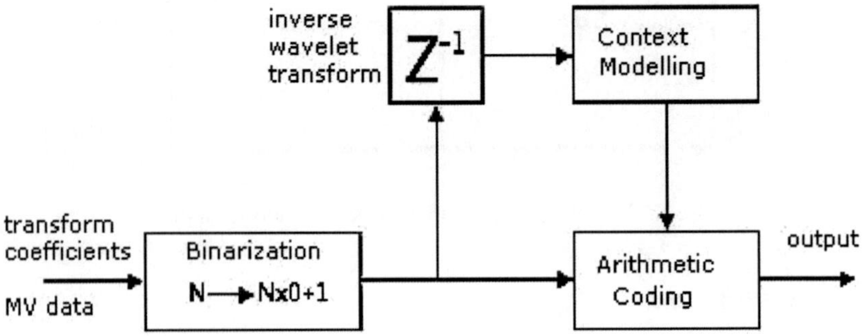

Fig. 2.7 Dirac's entropy coding architecture [6]

The non-zero values in the higher frequency sub-bands of the wavelet transform are often in the same part of the picture as they are in lower frequency sub-bands. Dirac creates statistical models of these correlations and arithmetic coding allows us to exploit these correlations to achieve better compression. The motion information estimated at the encoder also uses statistical modeling and arithmetic coding to compress it into the fewest number of bits. This compressed data is put into the bit stream, to be used by the decoder as part of the compressed video.

2.3.4 Motion Estimation

Motion estimation exploits temporal redundancy in video streams by looking for similarities between adjacent frames. An example of motion estimation technique used in the Dirac reference software is shown in Fig. 2.8.

In the first stage, pixel accurate motion vectors are determined for each block and each reference frame by hierarchical block matching. In the second stage, these pixel-accurate vectors are refined by searching sub-pixel values in the immediate neighborhood. In the final stage, mode decisions are made for each macro-block, determining the macro-block splitting level and the prediction mode

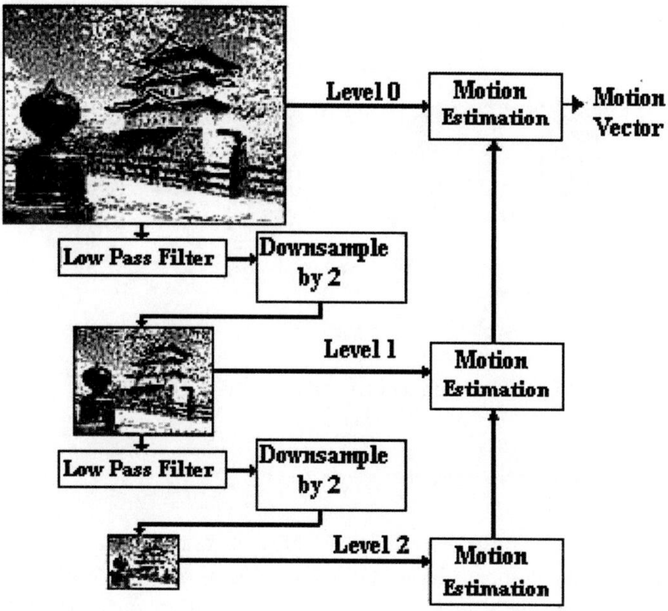

Fig. 2.8 Hierarchical motion estimation [10]

used for each prediction unit. This last stage involves further block matching since block motion vectors are used as candidates for higher-level prediction units. [8]

In its hierarchical motion estimation, Dirac first down converts the size of the current and reference of all types of inter frames (both P and B) using the 12 taps down conversion filter. [9] Down conversion filters are low pass filters that pass only the desired signal and also perform anti-alias filtering prior to decimation. Any suitable low-pass filter can be used including FIR, IIR and CIC filters. [31] The number of down conversion levels depends upon the frame format. [9]

Dirac also defines three types of frames. Intra (I) frames are coded without reference to other frames in the sequence. Level 1 (L1) frames and Level 2 (L2) frames are both inter frames, that is, they are coded with reference to other previously coded frames.

The difference between L1 and L2 frames is that L1 frames are also used as temporal references for other frames, whereas L2 frames are not. [3] A prediction structure for frame coding using a standard group of pictures (GOP) structure [7] is shown in Fig. 2.9. Each frame in Dirac may be predicted from up to two reference frames. Prediction modes can be varied by prediction unit, and there are four possibilities: Intra, Reference 1 only, Reference 2 only, and Reference 1 and 2 (bi-directional prediction). [8]

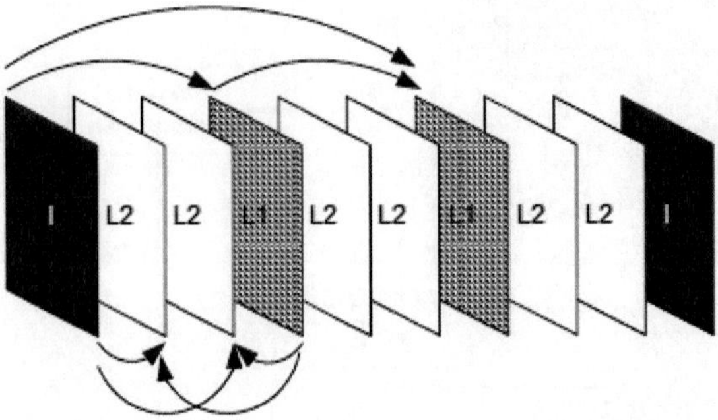

Fig. 2.9 Prediction of L1 and L2 frame in Dirac [7]

2.3.5 Motion Compensation

Motion compensation is used to predict the present frame. Dirac uses overlapped block-based motion compensation (OBMC) to achieve good compression and avoid block-edge artifacts which would be expensive to code using wavelets. OBMC allows interaction of neighboring blocks and is performed with basic blocks arranged into macro-blocks consisting of a 4x4 array of blocks. [8]

There should be an exact number of macro-blocks horizontally and vertically. This is achieved by padding the data. Further padding may also be needed because after motion compensation the wavelet transform is applied, which has its own requirements for divisibility. [4]

Although Dirac is not specifically designed to be scalable, the size of blocks is the only non-scalable feature, and for lower resolution frames, smaller blocks can easily be selected. Dirac's OBMC scheme is based on a separable linear ramp mask. This acts as a weight function on the predicting block. Given a pixel p=p(x,y,t) in frame t, p may fall within only one block or in up to four blocks if it lies at the corner of a block as shown in Fig. 2.10 where the darker-shade areas show overlapping areas. [4]

Each macro-block may be split into prediction units consisting either of 16 individual blocks, or of an array of 4 mid-size blocks, termed sub-macro-bocks, or of a single macro-block-sized block (Fig. 2.11).

OBMC parameters may be changed frame-by-frame, but defaults exist based on frame sizes. The default for both streaming and standard definition resolution is for 12x12 blocks which are overlapped at intervals of 8 pixels vertically and horizontally (the dimensions are scaled appropriately for chroma components of different resolutions). The OBMC overlapping function used is an integer approximation to the raised-cosine function. [8]

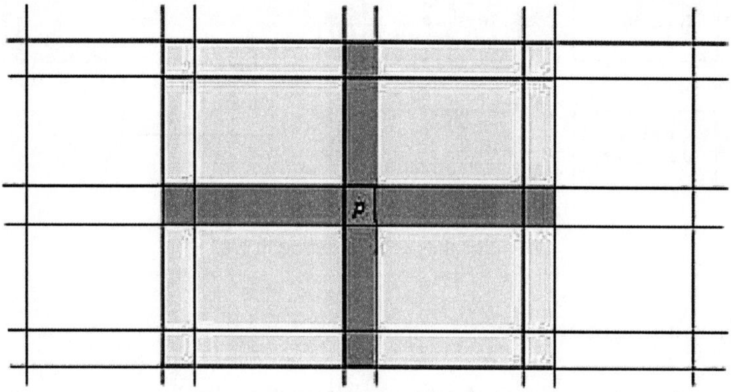

Fig. 2.10 Overlapping blocks in OBMC [4]

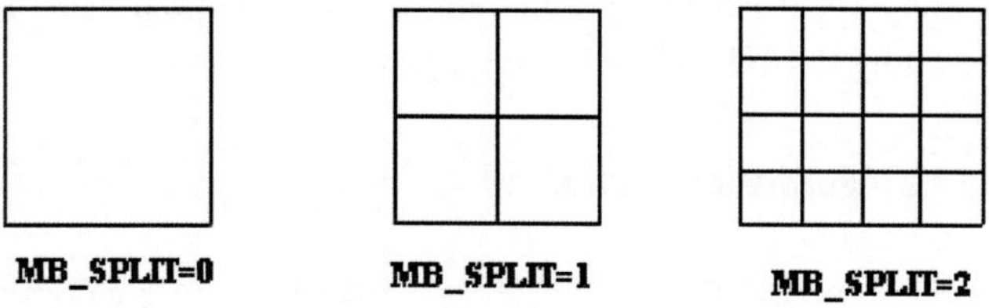

Fig. 2.11 Modes of splitting macro-block into sub-blocks in Dirac [8]

Dirac also provides sub-pixel motion compensation with motion vectors and thereby allows prediction rate up to 1/8th pixel accuracy. However the actual motion vector precision used may be less, depending on the optimum balance, which is largely determined by the bit rate chosen. Techniques such as predicting a frame using only motion information and predicting a frame to be nearly identical to a previous frame at low bit rates are also supported.

2.3.6 Decoder

The decoding process is carried out in three stages as shown in Fig. 2.12. At the first stage, the input encoded bit-stream is decoded by the entropy decoding technique. Next, scaling and inverse quantization is performed. In the final stage, inverse wavelet transform is applied on the data to produce the decoded, uncompressed video output. A trade off is made between video quality and motion vector bit rate. [5]

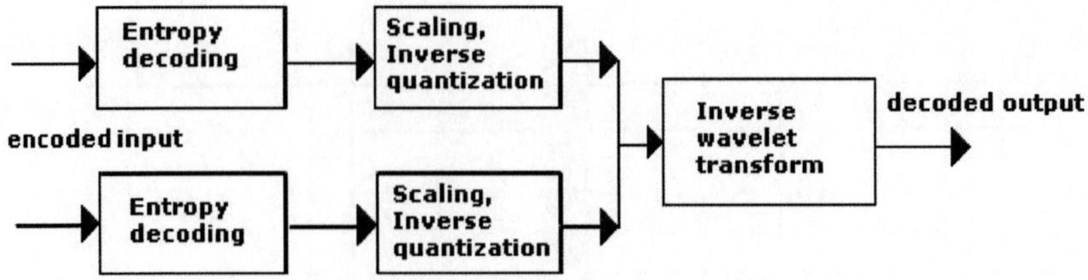

Fig. 2.12 Stages of decoding in Dirac

2.4 Implementation

The Dirac reference software is fully implemented in the C++ programming language which allows object oriented development on all common operating systems. The C++ code compiles to produce libraries for common functions, motion estimation, encoding and decoding, which have an interface that allows them to be called from C. An application programmer's interface can be written in C so that it can be kept simple and integrated with various media players, video processing tools and streaming software. [1]

2.4.1 Code Structure Overview

The Dirac codec has an object-orientated code structure. The encoder consists of objects which take care of the compression of particular 'objects' within a picture sequence. In other words, the compression of a sequence, a frame and a picture component are defined in individual classes.

2.4.2 Simplicity and Relative Speed of Encoding

Due to the relative simplicity of the Dirac reference software, its encoding speed is found to be much faster compared to the H.264 JM 17.1 reference software [11-14]. The decoding speeds of both the codecs are found to be comparable.

There are quite a few research papers [3] [46] [47] suggesting techniques to optimize Dirac's entropy coder. According to one [46], a much faster video codec can be achieved by replacing the original arithmetic coder of the Dirac algorithm with an accurately configured M-coder. The new arithmetic coder is three times faster for high bit rates and even outperforms the original compression performance.

Another paper [47] suggests a rate control algorithm based on optimization of quality factor for Dirac codec. This method exploits the existing constant-quality control, which is governed by a parameter called quality factor (QF) to give a constant bit rate.

In Dirac, the overall trade-off factor is derived from QF, meaning quality or quantization factor. QF is not is a direct measure of quality. Coding with constant QF will ensure constant quality only on homogenous material where the trade-off between distortion and rate is constant. [6]

Picture lambda values are used for rate-distortion control of quantization and motion estimation: They are initially derived from the picture QF, which is either set on the command line and used for all pictures or determined by means of the Rate Control algorithm. However, a number of factors are used to modify the lambda values after motion estimation. [6]

The initial assignation of lambda values is as follows: [6]

$$\begin{aligned} I_{lambda} &= \frac{1}{16} * 10^{\frac{10-QF}{2.5}} \\ L1_{lambda} &= 32 * I_{lambda} \\ L2_{lambda} &= 256 * I_{lambda} \end{aligned} \tag{2.3}$$

These lambda variables are used for quantizer selection in I, L1 and L2 pictures. From these, motion estimation lambdas are derived. The ideal trade-offs may change with different sequences, video resolutions, perceptual weightings, or block sizes. [6]

$$\begin{aligned} L1_{me_{lambda}} &= 2.0 * \sqrt{L1_{lambda}} \\ L2_{me_{lambda}} &= L1_{me_{lambda}} \end{aligned} \tag{2.4}$$

The guiding principles for I, L1 and L2 pictures are as follows: [6]

1. I pictures should be of higher quality than L1 pictures and L1 pictures should be of higher quality than L2 pictures.
2. Motion data and good motion rendition is more significant at lower bit rates (low QFs) than at higher ones (high QFs).

The first principle arises because I pictures are used as references for the L1 and the L2 pictures; L1 pictures are used as references for the L2 pictures. If the quality were to go up from I to L1 or from L1 to L2, then the encoder would need to correct the quantization error introduced in the reference picture and "pushed forward" by motion compensation. This error is noise-like and expensive to code. Also, an error in a single coefficient in the reference picture can spread to several coefficients when that picture is shifted through motion compensation. As a result, the L1 and the L2 lambdas multiply. The aim of the second principle is to stop the quality from falling off a cliff since when QF goes down, lambdas go up. The motion field is not over-smoothed at low bit rates. Even if the quality is lower, there are no poorly corrected areas. L2 pictures have less opportunity to correct motion estimation errors in residual coding. [6]

A mathematical model called the rate–quality factor (R–QF) is derived to generate optimum QF for the current coding frame using the bit rate resulting from the encoding of the previous frame in order to meet the target bit rate.

In another research project [48] different approaches to encoder optimization such as multi-threading, Streaming SIMD (Single Instruction Multiple Data) Extensions (SSE) [49] and compilation with Intel's C/C++ compiler [50] using the Visual Studio add-in [51] have been extensively discussed.

2.5 Results

Objective test methods attempt to quantify the error between a reference and an encoded bit stream. [5] To ensure accuracy of the tests, there is a need to maintain a compatible test bed. This would require both codecs to be tested under the same bit rates. [5] [47]

Since the latest version of Dirac includes a constant bit rate (CBR) mode, the comparison between Dirac and H.264 / MPEG-4 Part 10's [11-14] performance was produced by encoding several test sequences at different bit rates. By utilizing the CBR mode within H.264, we can ensure that H.264 is being encoded at the same bit rate as that of Dirac. [47]

Objective tests are divided into three sections, namely (i) Compression, (ii) Structural similarity index (SSIM) [16], and (iii) Peak to peak signal to noise ratio (PSNR). The test sequences "Miss-America" QCIF (176x144) [23], "Stefan" CIF (352x288) [23] and "Susie" standard-definition (SD) (720x480) [24] are used for evaluation. The two methods are very close and comparable in compression, PSNR and SSIM. Also, a significant improvement in encoding time is achieved by Dirac, compared to H.264 for all the test sequences.

2.5.1 Compression Ratio Test

By evaluating the magnitude of the *.drc and *.264 files, compression ratio results in comparison to the file size of the original sequence are produced from Dirac and H.264 respectively.

Using the CBR mode, it is possible to set a "target rate" for both the codecs and this would prevail over quality i.e. QF in the case of Dirac. This would ensure that both codecs were being used under equal operating environments. In these tests QF has been replaced with the bit rate metric (KBps).

Figures 2.13, 2.14 and 2.15 show a comparison of how Dirac and H.264 perform in compression for QCIF, CIF and SDTV sequences respectively.

Dirac achieves slightly higher compression ratios for lower bit rates than H.264 in the case of QCIF sequences. At higher QCIF bit rates both Dirac and H.264 achieve similar compression.

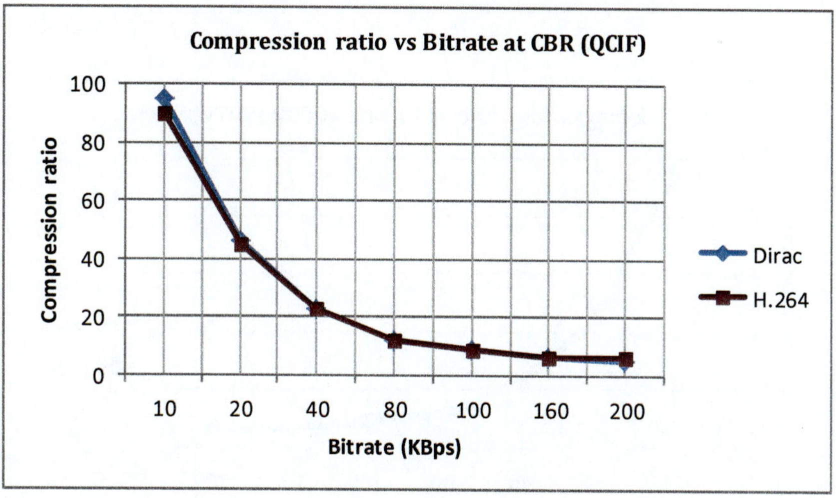

Fig. 2.13 Compression ratio comparison of Dirac and H.264 for "Miss-America" QCIF sequence

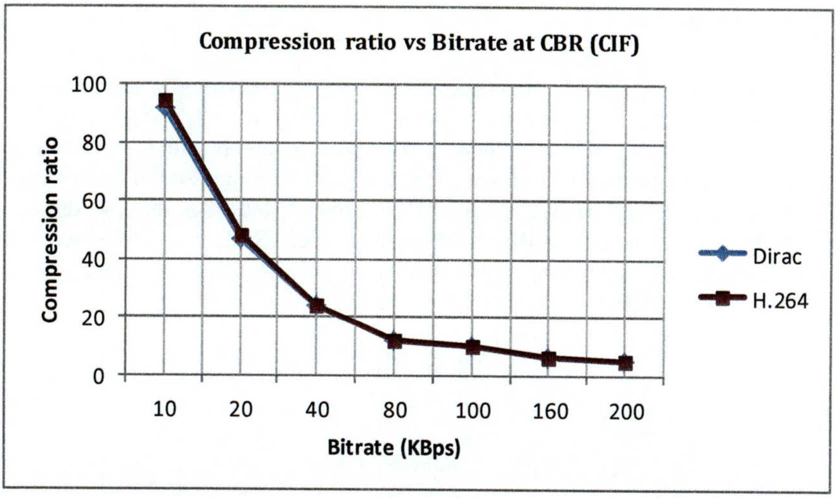

Fig. 2.14 Compression ratio comparison of Dirac and H.264 for "Stefan" CIF sequence

In case of CIF and SD media, H.264 provides slightly better compression at lower bitrates. At higher bit rates, both Dirac and H.264 achieve similar compression.

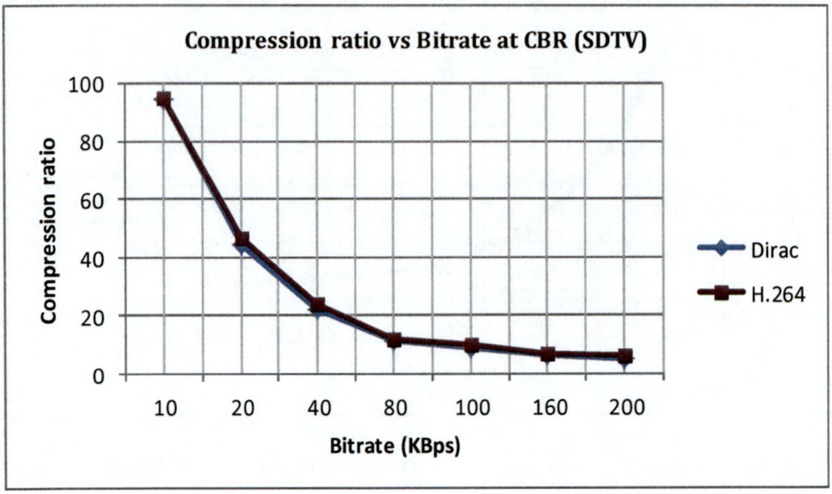

Fig. 2.15 Compression ratio comparison of Dirac and H.264 for "Susie" SDTV sequence

2.5.2 SSIM Test

Structural similarity (SSIM) [16] operates by way of comparing local patterns of pixel intensities that have been normalized for luminance and contrast [16]. This basically means that SSIM is computed based on the combination of luminance similarity, contrast similarity and structural similarity encompassed into one value. The maximum possible value for SSIM is 1, which indicates the encoded sequence is an exact replica of the reference sequence. SSIM is an alternative method of objectively evaluating video quality. [5] H.264 achieves slightly better SSIM than Dirac as seen in Figures 2.15, 2.16 and 2.17.

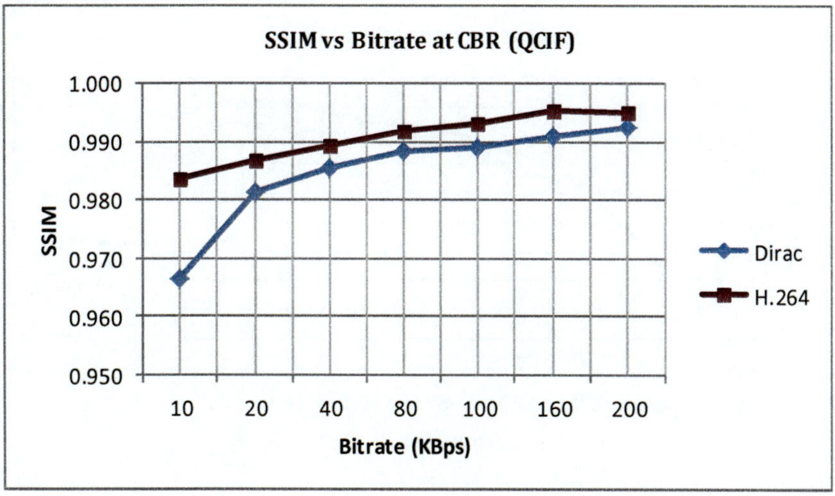

Fig. 2.16 SSIM comparison of Dirac and H.264 for "Miss-America" QCIF sequence

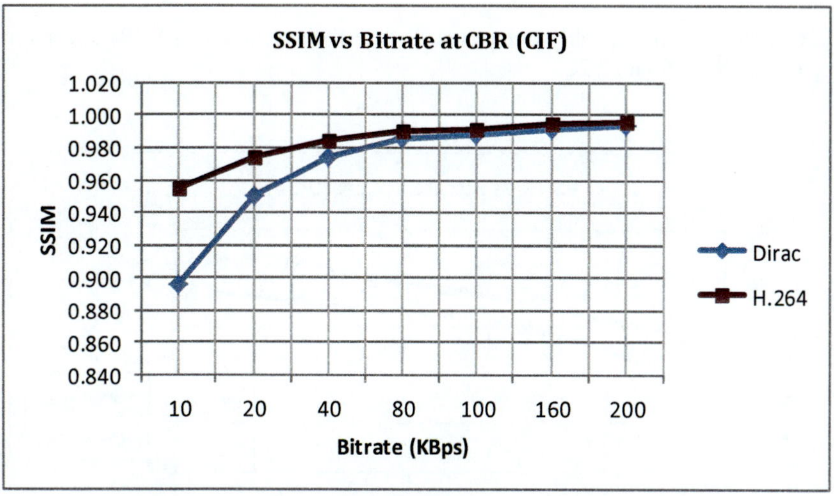

Fig. 2.17 SSIM comparison of Dirac and H.264 for "Stefan" CIF sequence

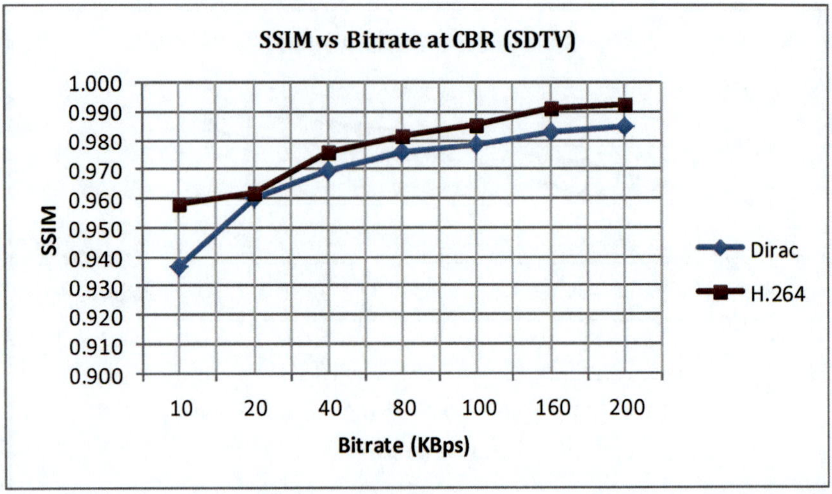

Fig. 2.18 SSIM comparison of Dirac and H.264 for "Susie" SDTV sequence

2.5.3 PSNR Test

H.264 achieves considerably higher PSNR than Dirac (about 3 – 4 dB) as seen in Figures 2.18, 2.19 and 2.20.

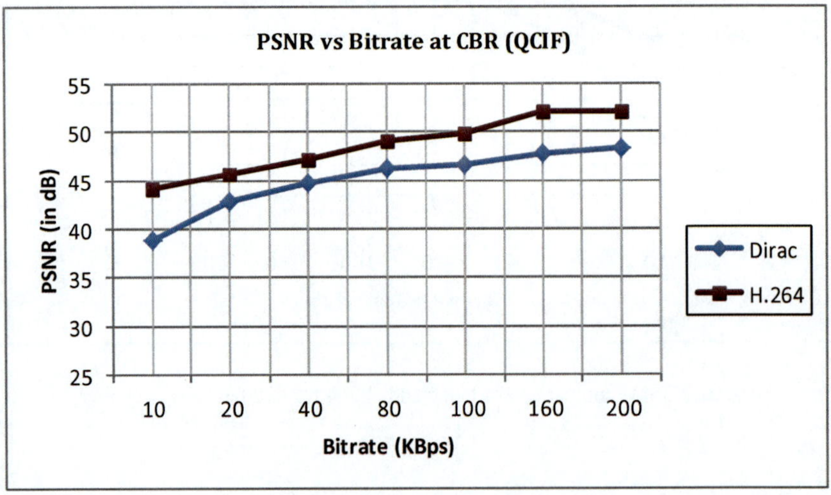

Fig. 2.19 PSNR comparison of Dirac and H.264 for "Miss-America" QCIF sequence

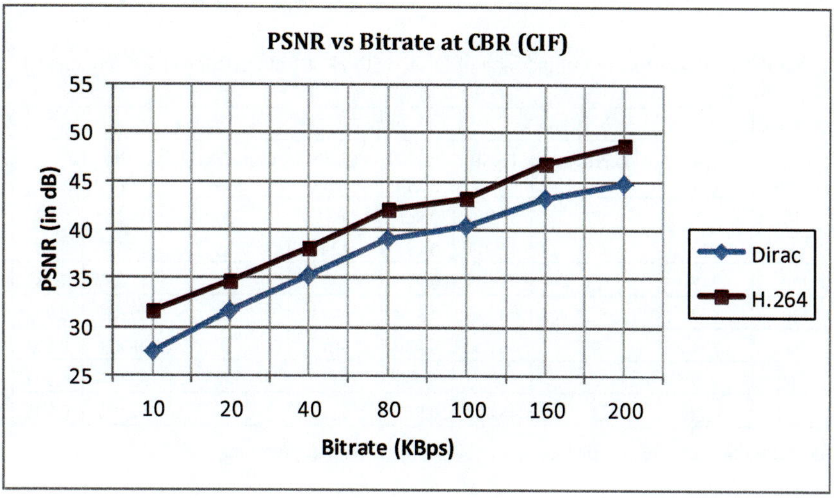

Fig. 2.20 PSNR comparison of Dirac and H.264 for "Stefan" CIF sequence

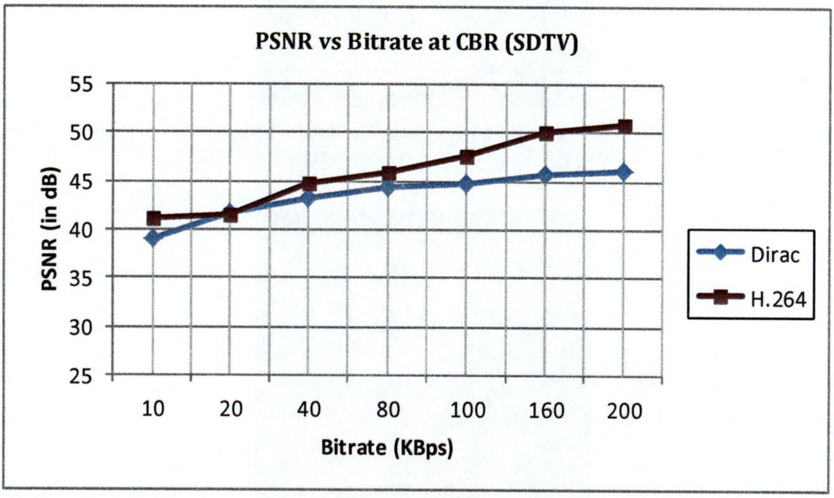

Fig. 2.21 PSNR comparison of Dirac and H.264 for "Susie" SDTV sequence

Tables 2.1, 2.2 and 2.3 and Figures 2.21, 2.22 and 2.23 show the performance comparison of Dirac with H.264 / MPEG-4 Part 10 at constant bit rates (CBR) ranging from 10-200 KBps for QCIF, CIF and SD sequences respectively.

Table 2.1 Performance comparison of Dirac with H.264 at CBR for QCIF sequence

	Dirac				**H.264**			
CBR (KB/s)	Size* (KB)	Compression ratio	PSNR (Y)	SSIM	Size* (KB)	Compression ratio	PSNR (Y)	SSIM
10	59	95	38.913	0.966	63	90	44.162	0.983
20	120	46	42.911	0.981	123	45	45.729	0.987
40	247	23	44.648	0.986	243	23	47.257	0.989
80	477	12	46.180	0.988	481	12	49.054	0.992
100	594	9	46.640	0.989	601	9	49.826	0.993
160	949	6	47.717	0.991	911	6	52.073	0.995
200	1186	5	48.420	0.992	912	6	52.077	0.995

*indicates encoded file size including all 150 frames after compression.

Dirac output sequence, CBR=10KBps
PSNR: 38.913, Bitrate:9.955KBps

H264 output sequence, CBR=10KBps
PSNR: 44.162, Bitrate: 10.603KBps

Fig. 2.22 Comparison of Dirac and H.264 at CBR = 10KBps, QCIF

Table 2.2 Performance comparison of Dirac with H.264 at CBR for CIF sequence

	Dirac				H.264			
CBR (KB/s)	Size* (KB)	Compression ratio	PSNR (Y)	SSIM	Size* (KB)	Compression ratio	PSNR (Y)	SSIM
10	146	92	27.468	0.896	142	94	31.617	0.955
20	285	47	31.613	0.951	282	48	34.650	0.974
40	559	24	35.296	0.975	559	24	38.055	0.984
80	1114	12	39.012	0.986	1112	12	42.103	0.991
100	1386	10	40.343	0.988	1389	10	43.134	0.992
160	2216	6	43.273	0.992	2199	6	46.840	0.995
200	2757	5	44.684	0.994	2731	5	48.729	0.997

*indicates encoded file size including all 90 frames after compression.

Dirac output sequence, CBR=100KBps
PSNR: 40.343, Bitrate: 101.331KBps

H264 output sequence, CBR=100KBps
PSNR: 43.134, Bitrate: 101.546KBps

Fig. 2.23 Comparison of Dirac and H.264 at CBR = 100KBps, CIF

Dirac output sequence, CBR=100KBps
PSNR: 44.676, Bitrate: 103.606KBps

Dirac output sequence, CBR=100KBps
PSNR: 47.491, Bitrate: 96.324KBps

Fig. 2.24 Comparison of Dirac and H.264 at CBR = 100KBps, SDTV

Table 2.3 Performance comparison of Dirac with H.264 at CBR for SD sequence

	Dirac				H.264			
CBR (KB/s)	Size* (KB)	Compression ratio	PSNR (Y)	SSIM	Size* (KB)	Compression ratio	PSNR (Y)	SSIM
10	180	94	39.055	0.937	178	95	41.028	0.958
20	388	44	41.729	0.960	361	47	41.530	0.962
40	751	22	43.220	0.970	701	24	44.814	0.976
80	1470	11	44.276	0.976	1405	12	45.871	0.981
100	1822	9	44.676	0.978	1694	10	47.491	0.986
160	2849	6	45.589	0.983	2562	7	50.016	0.991
200	3539	5	45.988	0.985	2953	6	50.819	0.993

*indicates encoded file size including all 25 frames after compression.

2.6 Conclusions

Overall Dirac codec is very promising. According to BBC R&D [1] [2], Dirac was developed with a view to optimize its performance with compression ratio and perceptual quality at the forefront. Its simplistic nature provides robustness and fast compression which is very beneficial, therefore to a large extent Dirac has succeeded in its aim. [5]

Dirac is a less developed codec and it is creditable that such an early reference code produces good results relative to H.264. SSIM indicates that H.264 has slightly greater improvement in terms of quality. The choice of the codec will depend on the end user's application which will decide if the enormous cost in license fees justifies the additional increase in quality (as in the case of H.264/MPEG-4 Part 10). [5] Both Dirac and H.264 maintain a near constant quality at low bit rates, which is beneficial for applications such as video streaming.

In conclusion, Dirac is an extremely simple yet robust codec and has the potential to achieve compression results very close to H.264, at reduced complexity and without royalty payments. But with these codec implementations, H.264 definitely wins the comparison.

2.7 Future Research

This implementation of the Dirac codec is directed towards high-quality video compression from web video up to ultra HD. However, the standard just defines a video codec and has no mention of any audio compression. It is necessary to associate an audio stream along with the video in order to have meaningful delivery of the video to the end user.

The Dirac video codec can be further improved by integrating it with an audio codec such as MPEG Layer 2 (MP2) [42] or the AAC [25]. MP2 is royalty free, applicable to high quality audio and has performance similar to MP3 [43] at higher bit rates. The Dirac research group at BBC also suggests Vorbis [41] audio codec and FLAC (free lossless audio codec)[40] developed by Xiph.Org Foundation as high quality audio formats available under royalty free terms that can be used with Dirac video codec.

Hence it is possible to transmit by multiplexing the video and audio coded bit streams to create a single bit stream for transmission and de-multiplexing the streams at the receiving end. This can be followed by synchronization of the audio and video during playback so that it can be suitable for various applications.

Acknowledgments. The first author would like to deeply thank Mr. Antriksh Luthra of Ohio State University, USA for his valuable support during the course of writing this chapter.

References

[1] Borer, T., Davies, T.: Dirac video compression using open technology. BBC EBU Technical Review (July 2005)

[2] BBC Research on Dirac, `http://www.bbc.co.uk/rd/projects/dirac/index.shtml`

[3] Eeckhaut, H., et al.: Speeding up Dirac's entropy coder. In: Proc. 5th WSEAS Int. Conf. on Multimedia, Internet and Video Technologies, Greece, pp. 120–125 (August 2005)

[4] The Dirac web page and developer support, `http://diracvideo.org/`

[5] Onthriar, K., Loo, K.K., Xue, Z.: Performance comparison of emerging Dirac video codec with H.264/AVC. In: IEEE International Conference on Digital Telecommunications, ICDT apos, August 29-31, vol. 06, p. 22 (2006)

[6] Davies, T.: The Dirac Algorithm (2008), `http://dirac.sourceforge.net/documentation/algorithm/`

[7] Tun, M., Fernando, W.A.C.: An error-resilient algorithm based on partitioning of the wavelet transform coefficients for a DIRAC video codec. In: Tenth International Conference on Information Visualization, IV 2006, vol. 5-7, pp. 615–620 (July 2006)

[8] Davies, T.: A modified rate-distortion optimization strategy for hybrid wavelet video coding. In: ICASSP Proceedings, IEEE International Conference on Acoustics, Speech and Signal Processing, May 14-19, vol. 2, pp. 14–19 (2006)

[9] Tun, M., Loo, K.K., Cosmas, J.: Semi-hierarchical motion estimation for the Dirac video codec. In: 2008 IEEE International Symposium on Broadband Multimedia Systems and Broadcasting, March 31-April 2, pp. 1–6 (2008)

[10] CMPT 365 Course Slides, School of Computing Science, Simon Fraser University, fig.3, `http://www.cs.sfu.ca/CourseCentral/365/li/material/notes/Chap4/Chap4.3/Chap4.3.html`

[11] Kwon, S.K., Tamhankar, A., Rao, K.R.: Overview of H.264 / MPEG-4 Part 10. J. Visual Communication and Image Representation 17, 186–216 (2006)

[12] Wiegand, T., et al.: Overview of the H.264/AVC video coding standard. IEEE Trans. CSVT 13, 560–576 (2003)

[13] Wiegand, T., Sullivan, G.J.: The H.264 video coding standard. IEEE Signal Processing Magazine 24, 148–153 (2007)

[14] Marpe, D., Wiegand, T., Sullivan, G.J.: The H.264/MPEG-4 AVC standard and its applications. IEEE Communications Magazine 44, 134–143 (2006)

[15] Gargour, C., et al.: A short introduction to wavelets and their applications. IEEE Circuits and Systems Magazine 9, 57–68 (2009)

[16] Wang, Z., et al.: Image quality assessment: From error visibility to structural similarity. IEEE Trans. on Image Processing 13, 600–612 (2004)

[17] Microsoft Windows Media, http://www.microsoft.com/windows/windowsmedia

[18] MPEG-4 Part 2, ISO/IEC 14496-2, International Organization for Standardization, http://www.iso.ch

[19] Dirac software and source code, http://diracvideo.org/download/dirac-research/

[20] http://en.wikipedia.org/wiki/VC-1

[21] Dirac video codec - A programmer's guide, http://dirac.sourceforge.net/documentation/code/programmers_guide/toc.htm

[22] Jia, H., Zhang, L.: Directional diamond search pattern for fast block motion estimation. IEE Electronics Letters 39(22), 1581–1583 (2003)

[23] Video test sequences (YUV 4:2:0), http://trace.eas.asu.edu/yuv/index.html

[24] Video test sequences ITU601, http://www.cipr.rpi.edu/resource/sequences/itu601.html

[25] MPEG–2 advanced audio coding, AAC. International Standard IS 13818–7, ISO/IEC JTC1/SC29 WG11 (1997)

[26] Davidson, G.A., et al.: ATSC video and audio coding. Proceedings of IEEE 94, 60–76 (2006)

[27] Puri, A., Chen, X., Luthra, A.: Video coding using the H.264/MPEG-4 AVC compression standard. Signal Processing: Image Communication 19, 793–849 (2004)

[28] H. 264 AVC JM software, http://iphome.hhi.de/suehring/tml/

[29] Daubechies wavelet, http://en.wikipedia.org/wiki/Daubechies_wavelet

[30] Daubechies wavelet filter design, http://cnx.org/content/m11159/latest/

[31] Digital down converter, http://en.wikipedia.org/wiki/Digital_down_converter

[32] H.264/MPEG-4 AVC, http://en.wikipedia.org/wiki/H.264

[33] Fieldler, M.: Implementation of basic H.264/AVC Decoder. Seminar paper at Chemnitz University of Technology (June 2004)

[34] H.264 encoder and decoder, http://www.adalta.it/Pages/407/266881_266881.jpg

[35] H.264 video compression standard, White paper, Axis communications

[36] MPEG-4: ISO/IEC JTC1/SC29 14496-10: Information technology – Coding of audio-visual objects - Part 10: Advanced Video Coding, ISO/IEC (2005)

[37] Kumar, D., Shastry, P., Basu, A.: Overview of the H.264 / AVC. In: 8th Texas Instruments Developer Conference, India, Bangalore, November 30- December 1 (2005)

[38] Schäfer, R., Wiegand, T., Schwarz, H.: The emerging H.264/AVC standard. EBU Technical Review (January 2003)

[39] Joint Photographic Experts Group, JPEG, http://www.jpeg.org/

[40] FLAC - Free Lossless Audio Codec, http://flac.sourceforge.net/

[41] Vorbis, http://www.vorbis.com/

[42] MPEG Layer II, http://en.wikipedia.org/wiki/MPEG-1_Audio_Layer_II

[43] MP3/MPEG Layer III, http://en.wikipedia.org/wiki/MP3

[44] Borer, T.: Dirac coding: Tutorial and Implementation. In: EBU Networked Media Exchange Seminar (June 2009)

[45] Dirac Pro, http://www.bbc.co.uk/rd/projects/dirac/diracpro.shtml

[46] Eeckhaut, H., et al.: Tuning the M-coder to improve Dirac's Entropy Coding, http://escher.elis.ugent.be/publ/Edocs/DOC/P105_088.pdf

[47] Tun, M., Loo, K.K., Cosmas, J.: Rate control algorithm based on quality factor optimization for Dirac video codec. Signal Processing: Image Communication 23, 649–664 (2008)

[48] Noam, K., Tamir, B.: Dirac video codec: Optimizing software performance using architectural considerations. Technion - Israel Institute of Technology, Electrical Engineering Faculty, Software lab Performance Tuning

[49] Streaming SIMD extensions (SSE), http://msdn.microsoft.com/en-us/library/t467de55%28VS.71%29.aspx

[50] Intel Compilers, http://software.intel.com/en-us/intel-compilers/

[51] Microsoft Visual Studio add-ins, http://en.wikipedia.org/wiki/List_of_Microsoft_Visual_Studio_add-ins

[52] GStreamer, http://www.gstreamer.net/

[53] FFmpeg, http://www.ffmpeg.org/

[54] VLC media player, http://www.videolan.org/vlc/

Abbreviations

AVC - Advanced video coding
CBR – Constant bit rate
CIF - Common intermediate format
DCT – Discrete cosine transform
GOP - Group of picture(s)
HDTV - High definition television
MPEG - Moving Picture Experts Group
MV - Motion vector
OBMC - Overlapped block-based motion compensation
PSNR – Peak signal-to-noise ratio
QCIF - Quarter common intermediate format
QF – Quality factor
RDO - Rate distortion optimization
SAD - Sum of the absolute difference
SD - Standard definition
SIMD - Single Instruction Multiple Data
SSE - Streaming SIMD Extensions
SSIM – Structural Similarity index
VLC – Variable length coding.

Chapter 3
Linear and Non-linear Inverse Pyramidal Image Representation: Algorithms and Applications

Roumen Kountchev[1], Vladimir Todorov[2], and Roumiana Kountcheva[2]

[1] Department of Radio Communications and Video Technologies,
Technical University of Sofia, Sofia 1000, Bulgaria
rkountch@tu-sofia.bg
[2] T&K Engineering, Mladost 3, Pob 12, Sofia 1712, Bulgaria

Abstract. In the chapter is presented one specific approach for image representation, known as Inverse Pyramid Decomposition (IPD), and its main applications. The chapter is arranged as follows: the Introduction reviews the state of the art, comprising the presentation of various pyramidal decompositions and outlining their advantages and demerits. In the next sections are considered in detail the principles of the IPD based on linear (DFT, DCT, WHT, KLT, etc.) and non-linear transforms: deterministic, based on oriented surfaces, and adaptive, based on pyramidal neural networks. Furthermore, the work introduces the non-recursive and recursive implementations of the IPD. Special attention is paid to the main application areas of the IPD: the image compression (lossless, visually lossless and lossy), the multi-view and the multispectral image representation. Significant part of the chapter is devoted to the evaluation and comparison of the new representation with the famous compression standards JPEG and JPEG2000. In the conclusion are outlined the main advantages of the IPD and the trends for future development and investigations.

Keywords: pyramidal image decomposition, reduced inverse spectrum pyramid, pyramidal neural network, multi-view image representation, multispectral images compression.

3.1 Basic Methods for Pyramidal Image Decomposition

The aim of the pyramidal decomposition is to present the image in a compact form through limitation of the decomposition components number based on the permissible value of the error obtained (Ahmed and Rao, Pratt, Rosenfeld). The decomposition could be implemented on the basis of the famous linear transforms of

R. Kountchev et al. (Eds.): Advances in Reasoning-Based Image Processing, ISRL 29, pp. 35–89.
springerlink.com

the kind KLT, PCA, SVD, DFT, DCT, etc. (Ahmed and Rao, Pratt), or on the use of a pyramidal representation (Rabbani and Jones, Topiwala). Decompositions with linear transforms are assigned to the first generation coding methods (Kunt et al.), based on various mathematical models for image representation in the corresponding spectral space. According to Kunt et al. again, pyramidal decompositions are assigned to the second generation of methods for image coding, which better go together with human visual system. In result, these decompositions have higher efficiency and serve as a basis for the development of large number of image compression methods. The Pyramidal Image Decomposition had been an object of multiple investigations and publications, the former of which are related to the initiating works of Tanimoto and Knowlton.

The analysis of the PID methods shows that there are two main approaches, used for the pyramidal image decomposition building. The first approach is based on the multiresolution principle, developed by Burt and Adelson and aimed mainly at the image compression and the Progressive Image Transmission (PIT) presented by Tzou.

The second approach for image decomposition, developed by Vetterli, Smith and Barnwell, Woods and Daubechies, is based on the use of digital filter banks for frequency analysis and synthesis, together with operations for repeated image decimation and interpolation in the frequency bands, corresponding to different areas of the 2D spectrum.

In correspondence with the general approach for pyramidal structures representation given by Topiwala, they could be divided into two main classes:

1. Non-orthogonal pyramids, presented in detail by Rosenfeld, Kunt et al., Wang and Goldberg, Vetterli and Uz, Lu et al., Strintzis and Tzovaras, build on the basis of the difference between the original image and its approximation, and obtained using low-frequency digital filtration, double decimation, interpolation and negative feedback. The last one is a prerequisite to reduce the required approximation accuracy and in result, higher efficiency is obtained. For the implementation of the decimation and interpolation are used two mutually complementing operators: EXPAND – used for the double enlargement of the image size through interpolation, and REDUCE – for the inverse operation (double size reduction through decimation).

The first, basic non-orthogonal pyramid is the Laplacian, LP, presented in detail by Burt and Adelson, Efstratiadis et al., and Chen. It is build on the basis of the Gaussian, which contains the original image in its base. Each of the higher Gaussian pyramid levels is calculated recursively after low-frequency filtration of the preceding one, followed by Decimation operator used to obtain double reduction of the horizontal and vertical image size (Fig. 3.1). On the basis of this structure is build the LP, comprising images, which are the differences between any current and next level of the Gaussian pyramid, doubled in horizontal and vertical direction with the operator Interpolation.

Then for an image of size M×N the total number of pixels in the LP is:

$$MN+MN/4+MN/16+\ldots \approx (4/3)MN$$

This is why such representation is called over-complete (Bovik).

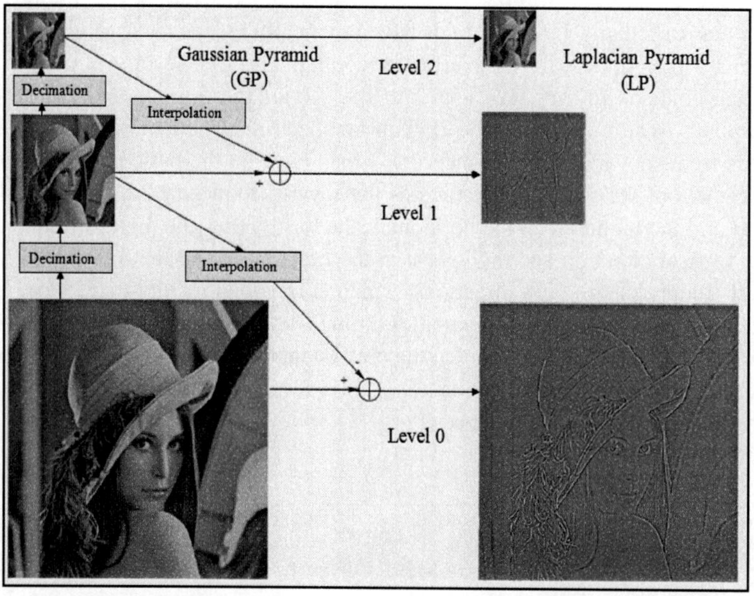

Fig. 3.1 Principle of the Gaussian and Laplacian pyramids building

To the class of the non-orthogonal pyramids are also related the following modifications:

- Mean Pyramid (MP) (Tanimoto);
- Reduced/Enhanced LP (RLP/ELP) (Aiazzi, Muller et al.);
- Content Driven Laplacian Pyramid (CDLP) (Mongatti et al.; Aiazzi et al.)
- Reduced-Sum/Reduced-Difference Pyramid (RSP/RDP) (Wang and Goldberg 1989, 1991);
- S-Transform Pyramid (STP) (Wang and Goldberg 1991);
- Morphological Image Pyramid (MIP) (Kong and Goutsias);
- Hierarchy-Embedded Differential Pyramid (HEDI) (Kim et al.);
- Polynomial Approximation Pyramid (PAP) (Lu et al.);
- Rounding Transform Pyramid (RTP) (Jung et al.);
- Least-Square LP (LSLP) (Unser 1992);
- Polynomial Spline Pyramid (PSP) (Unser 1999);
- Centered Pyramid (CP) (Brigger et al.);
- Stochastic Pyramid (SP) (Meer);
- Pyramid based on Hierarchical Interpolation by Radial Basis Function Networks (HI-RBFN) (Sigitani et al.);
- Contrast Pyramid (Yu).

2. Orthogonal pyramids (Tzou, Mallat 1989, Antonini et al., Joshi et al., Bovik), obtained through sub-band decomposition and discrete wavelet transform (DWT) in particular. These pyramids have practically independent neighbor levels which is a prerequisite for their efficient coding. The first level in a pyramid of this kind contains four images with half the size of the original and each fourth corresponds to definite part of the 2D Fourier spectrum. The first one, LL contains the low-frequency spectrum frequencies, which have on the basic part of the visual information; the second, LH contains the lowest frequencies in the vertical direction and the highest – in the horizontal; the third, HL – the highest frequencies in the vertical direction and the lowest in the horizontal, and the fourth, HH – the highest frequencies in both directions, which correspond to finest structure. Each image is obtained from the original after digital filtration and double decimation in horizontal and vertical direction. On the other side, the LL image is used as a basis for the recursive building of the next pyramid level, or is used as a pyramid top when the decomposition is finished (Fig. 3.2).

a

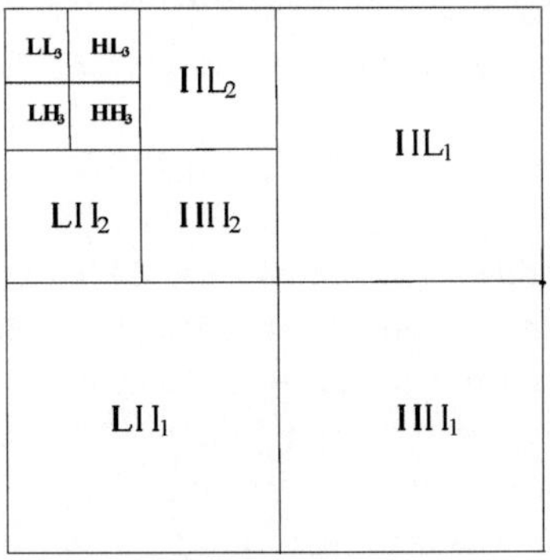

b

Fig. 3.2 Recursive building of a Wavelet pyramid of 3 levels (a); location of the images, corresponding to 10 sub-bands (b).

When the processed image is restored, each of the four images in the highest level is processed with double interpolation and inverse filtration in horizontal and vertical directions and is then summed with the images from the lower levels, restored in similar way. However, the restored image LL from the higher level is used as original for the lower and sets up new four images just as the remaining three images from same decomposition level. The basic advantage of the orthogonal pyramids towards the non-orthogonal (and in particular towards the wavelet pyramids) is that they are complete, (Bovik) i.e. the number of pixels is equal with that of the original image).

To the class of the orthogonal pyramids could be related:

- Multistage Lattice Filter Banks (FB) and Tree Structured FB (Vaidyanathan-1993);
- Quadrature Mirror FB, Quadrature Complex Conjugated FB and M-band perfect reconstruction FB (Smith and Barnwell, Vaidyanathan 1987, Vetterli et al., Unser 1993);
- Octave-band FB, using orthogonal wavelet Haar functions or bi-orthogonal spline functions (Wavelet Decomposition by Tree Structured FB) (Mallat 1990, Rioul and Vetterli, Vetterli and Uz 1992, Froment and Mallat, DeVore et al., Majani, Kim and Li);
- Gauss Pyramid with DWT (Olkkonen and Pesola, Strintzis and Tzovaras);
- Advanced Wavelet Pyramid (WP) (Froment and Mallat, Antonini et al., Egger et al.):
  - Embedded zero-tree wavelet (EZW) (Shapiro);
  - Spatial partitioning of images into hierarchical trees, SPIHT (Topiwala, Efstratiadis et al.);
  - Space-frequency quantization (SFQ) (Nuri);
  - Wavelet packet transforms, WPT (Demaistre and Labit);
  - Trellis coded quantization, TCQ (Topiwala);
  - Compression with Reversible Embedded Wavelets, CREW (Boliek et al.);
  - Embedded block coding with optimized truncation of the embedded bit-streams, EBCOT (Taubman);
  - Embedded Predictive Wavelet Image Coder, EPWIC (Buccigrossi and Simoncelli);
- DCT-H Pyramid (Gibson et al.) and Variable Block Subband DCT Decomposition, VB-SBDCT (Tan and Ghambari);
- Morphological Subband Pyramid, MSP (Toet, Wang et al., Kong and Goutsias) and Morphological Subband Decomposition, MSD (Egger et al.).
- Steerable Pyramid based on bank of steerable filters (Simoncelli and Freeman).
- Shiftable complex directional pyramid decomposition (Nguyen and Oraintara);
- Improved multiresolution image representation, based on the Ridgelet Transform (Do and Vetterli).

The so described two groups of pyramidal structures have high potential for efficient image compression. Together with this however, they have some common disadvantages, related to:

- The principle of their creation, in accordance with which the base of the pyramid is first calculated and then follow the next, higher levels until the pyramid top is reached. Then, in case of "progressive" image transfer (PIT) (Knowlton, Tzou, Wang and Goldberg) the highest pyramid image should be transferred first, in the quality of the coarsest approximation of the processed image. This is why the use of the pyramidal structure of non-inverse kind requires larger delay of the transferred visual information;
- The use of multiple 2D decimation and interpolation together with digital low-frequency or band image filtration result in specific distortions in the restored image. Because of the Gibbs phenomenon, in images, containing high-contrast transitions are generated false concentric circles (Ringing effect), which depend on the structure of the spatial decimation lattice, or Aliasing effect (Velho et al.), and on the approximation accuracy of the phase-frequency and amplitude-frequency characteristics of the used digital filters. As it is known, the implementation of the "ideal" low-frequency or band filtration is practically impossible because of the requirement for strict linearity of the phase-frequency characteristic and for the rectangular shape of the amplitude-frequency characteristic. Furthermore, the reduction of the image size makes their filtration more complicated because of the boarder effect reinforcement (Pratt). On account of this, the levels number in the non-inversed pyramids is usually limited up to 3 or 4. This additionally restricts the abilities of these pyramids for highly efficient compression;
- The quantization of the coefficients' values in the pyramids levels, which ensures higher compression ratio, results in the appearance of specific noises in the restored image and deteriorate its quality (Aiazzi et al. 1997). In this case the use of noise-resistant coding is the transaction, because the compression efficiency is reduced and the coding/decoding is more complicated;

More approaches for efficient image representation are based on:

- Non-linear representation based on normalization of the Wavelet transform coefficients aiming at their better matching the statistical properties of the images and the perceptual sensitivity of the human visual system (Malo et al.), or based on the anisotropic filtration controlled by the visual attention model (Mancas et al.);
- Image representation using hybrid methods based on the Support Vector Machine (SVM) and the Discrete cosine transform (SVM-DCT) or fuzzy logic (SVM-Fuzzy Logic), Artificial Neural Networks and Wavelet Transforms (Kalra);
- Image representation based on Locally Adaptive Resolution (LAR) (Deforges et al.), presented in Chapter 4 of this book.

The analysis of the most renowned contemporary methods for image representation, based on pyramidal structures, shows their big abilities for further investigations and development. In this chapter is presented one new general approach for pyramidal image representation, called by the authors *Inverse Pyramid Decomposition* (Kountchev et al. 2002, 2005). It permits the use of the famous linear orthogonal transforms of deterministic and statistical kind, and of various nonlinear transforms, based on neural networks, morphological and rank filters, etc. In the next sections is described the general principle for building the IDP through deterministic orthogonal transforms and through linear transforms based on neural networks, and the representation of multi-view and multispectral images.

3.2 Basic Principles of the Inverse Pyramid Decomposition

The Inverse Pyramid Decomposition (IPD) is accomplished as follows. The digital image is first processed with some kind of 2D orthogonal transform using limited number of coefficients only. The values of the coefficients, calculated in result of the transform, constitute the first pyramid level. Using these values, the image is restored with inverse orthogonal transform and in result is obtained the coarse approximation of the original image. The approximating image is then subtracted pixel by pixel from the original and the so obtained difference image is divided into 4 sub-images. Each sub-image is processed with the 2D orthogonal transform again (the values of the coefficients constitute the second pyramid level). The processing continues in a similar way until all pyramid levels, consisting of coefficients only, are calculated. The set of coefficients of the orthogonal transform, chosen for every pyramid level, can be different. The image decomposition is stopped when the required image quality is reached - usually earlier than the last possible pyramid level. The coefficients obtained in result of the orthogonal transform from all pyramid levels are sorted in accordance with their frequency, scanned sequentially and losslessly compressed.

3.2.1 Inverse Pyramid Decomposition with Orthogonal Transforms

The matrix, which depicts any digital halftone image, could be represented through IPD based on linear orthogonal or non-linear transforms. In this section, the principle of IPD will be presented, using linear orthogonal transforms. For this, the image matrix is divided into blocks of size $2^n \times 2^n$, as shown on Fig. 3.3.

Each sub-image is then represented with IDP, consisting of r levels ($1 < r \leq n$), obtained as a result of their quad-tree partition. Let $k_p=1,2,..,4^pK$ is the number of a sub-image in the layer p, which contains 4^pK sub-images ($p = 0,1,\dots,r-1$). The matrix of the block $[B(2^n)]$ is represented by the relation:

$$[B(2^n)] = [\hat{B}_0(2^n)] + \sum_{p=1}^{r}[\hat{E}_{p-1}(2^n)] + [E_r(2^n)] \quad \text{for } r \leq n\text{-}1, \tag{3.1}$$

where r is the number of the decomposition components.

All matrices in Eq. 3.1 are of size $2^n \times 2^n$.

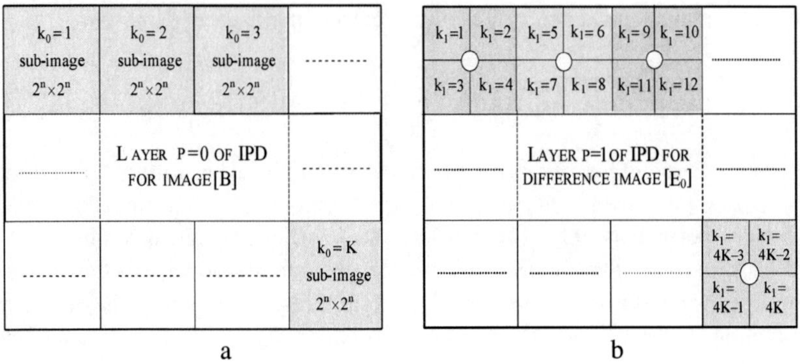

Fig. 3.3 Division of the original image [B] into blocks of size $2^n \times 2^n$ in the decomposition level p=0 (a), and of the difference image $[E_0]$ into sub-blocks of size $2^{n-1} \times 2^{n-1}$ in the decomposition level p=1 (b)

The matrix $[E_r(2^n)]$ contains the errors of a decomposition consisting of (r+1) components. The first component for the lowest level p=0 is the matrix $[\hat{B}_0(2^n)]$, which is the coarse approximation of the block $[B(2^n)]$. It is obtained after 2D inverse orthogonal transform of the block transform $[\tilde{S}'_0(2^n)]$ in correspondence with the relation:

$$[\hat{B}_0(2^n)] = [T_0(2^n)]^{-1}[\hat{S}'_0(2^n)][T_0(2^n)]^{-1} \tag{3.2}$$

where $[T_0(2^n)]^{-1}$ is a matrix of size $2^n \times 2^n$ of the inverse orthogonal transform of $[\hat{S}'_0(2^n)]$.

On the other hand,

$$[\hat{S}'_0(2^n)] = Q_0^{-1}\{[\hat{S}_0(2^n)]\} = Q_0^{-1}\{Q_0\{[\tilde{S}_0(2^n)]\}\} \tag{3.3}$$

Here $Q_0\{\bullet\}$ and $Q_0^{-1}\{\bullet\}$ are operators for the decomposition level p=0, used to perform the quantization/dequantization of the spectrum coefficients $\tilde{s}_0(u,v)$ and $\hat{s}_0(u,v)$, which are the matrix elements of $[\tilde{S}_0(2^n)]$ and $[S_0(2^n)]$ correspondingly.

The first matrix, $[\tilde{S}_0(2^n)] = [m_0(u,v).s_0(u,v)]$ is the transform of the "truncated" orthogonal transform of the block $[B(2^n)]$. Here $m_0(u,v)$ are the elements of the binary matrix-mask $[M_0(2^n)]$, which defines the retained (non-zero) coefficients of $[\tilde{S}_0(2^n)]$, in correspondence with the relation:

$$m_0(u,v)=\begin{cases}1, \text{if } s_0(u,v) - \text{ retained coefficient,}\\ 0 \quad - \quad \text{in all other cases.}\end{cases}$$

The values of $m_0(u,v)$ are set so that the retained coefficients $\tilde{s}_0(u,v) = m_0(u,v).s_0(u,v)$ to be these with maximum mean energy in the transforms $[S_0(2^n)]$ of all blocks. The transform $[S_0(2^n)]$ of the block $[B(2^n)]$ is defined through direct 2D orthogonal transform:

$$[S_0(2^n)] = [T_0(2^n)][B(2^n)][T_0(2^n)] \tag{3.4}$$

where $[T_0(2^n)]$ is a matrix of size $2^n \times 2^n$ corresponding to the decomposition level p=0, used for the implementation of the 2D orthogonal transform (for example, DFT, DCT, WHT, KLT, etc.).

The retained decomposition components in Eq. 3.1 are the approximating matrices $[\hat{E}_{p-1}(2^{n-p})]$ for levels p=1,2,..,r, which contain the sub-matrices $[\hat{E}^{k_p}_{p-1}(2^{n-p})]$ of size $2^{n-p} \times 2^{n-p}$, for $k_p=1,2,\ldots,4^pK$, obtained after quad-tree division of the matrix $[\hat{E}_{p-1}(2^{n-p})]$. Each of its sub-matrices $[\hat{E}^{k_p}_{p-1}(2^{n-p})]$ is defined by the relation:

$$[\hat{E}^{k_p}_{p-1}(2^{n-p})] = [T_p(2^{n-p})]^{-1}[\hat{S}'^{k_p}_p(2^{n-p})][T_p(2^{n-p})]^{-1} \tag{3.5}$$

where 4^p is the number of the quad-tree branches in the decomposition level p. Here $[T_p(2^{n-p})]^{-1}$ is a matrix of size $2^{n-p} \times 2^{n-p}$ corresponding to the level p, which is used to perform the inverse 2D orthogonal transform:

$$[\hat{S}'^{k_p}_p(2^{n-p})] = Q_p^{-1}\{[\hat{S}^{k_p}_p(2^{n-p})]\} = Q_p^{-1}\{Q_p\{[\tilde{S}^{k_p}_p(2^{n-p})]\}\} \tag{3.6}$$

Using the operators $Q_p\{\bullet\}$ and $Q_p^{-1}\{\bullet\}$ in the decomposition level p are quantizated and dequantizated the selected spectrum coefficients $\tilde{s}^{k_p}_p(u,v)$ and $\hat{s}^{k_p}_p(u,v)$, which are elements of the matrices $[\tilde{S}^{k_p}_p(2^{n-p})]$ and $[\hat{S}^{k_p}_p(2^{n-p})]$. The elements $\tilde{s}^{k_p}_p(u,v) = m_p(u,v).s^{k_p}_p(u,v)$ of the first matrix depend on the elements $m_p(u,v)$ of the binary matrix-mask $[M_p(2^{n-p})]$. The retained coefficients of the matrix $[\tilde{S}^{k_p}_p(2^{n-p})]$ are defined in the way already described for the level p=0.

The matrix $[S_p^{k_p}(2^{n-p})]$ is the transform of $[E_{p-1}^{k_p}(2^{n-p})]$ and is defined through direct 2D orthogonal transform:

$$[S_p^{k_p}(2^{n-p})] = [T_p(2^{n-p})][E_{p-1}^{k_p}(2^n)][T_p(2^{n-p})] \tag{3.7}$$

Here $[T_p(2^{n-p})]$ is a matrix of size $2^{n-p}\times 2^{n-p}$ for the decomposition level p used for the 2D orthogonal transform of each block $[E_p^{k_p}(2^{n-p})]$ for $k_p=1,2,\ldots,4^p$ in the difference matrix for the same level, defined by the relation:

$$[E_{p-1}(2^{n-p})] = \begin{cases} [B(2^n)] - [\hat{B}_0(2^n)] & \text{for } p=1; \\ [E_{p-2}(2^{n-p})] - [\hat{E}_{p-2}(2^{n-p})] - & \text{for } p=2,3,..,r. \end{cases} \tag{3.8}$$

In accordance with the decomposition, presented by Eq. 3.1, for each block $[B(2^n)]$ are calculated the following spectrum coefficients:

- all non-zero coefficients of the transform $[\hat{S}'_0(2^n)]$ from the decomposition level p=0 ;
- all non-zero coefficients of the transforms $[\hat{S}'^{k_p}_p(2^{n-p})]$ for $k_p=1,2,..,4^p$ from levels p=1,2,..,r.

The spectrum coefficients, which correspond to same spatial frequency (u,v) from all image sub blocks, are arranged in common data sequences in correspondence with the decomposition level, p. The conversion of the 2D data sequences into one-dimensional data sequence is performed following the recursive Hilbert scan, shown on Fig.3.4. The main advantage of this scan is that it retains very well the correlation between neighboring coefficients in the corresponding data sequences.

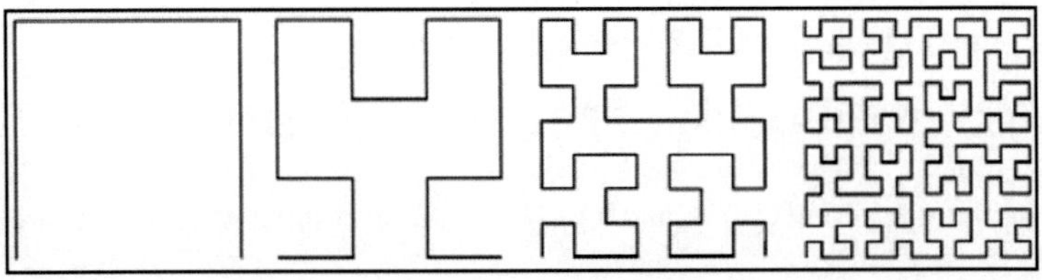

Fig. 3.4 Recursive Hilbert scan of the coefficients in blocks of size $2^n\times 2^n$ for n=1,2,3,4.

The general block diagram of the IPD coder and decoder is shown on Fig. 3.5, where are used the following abbreviations: TOT/IOT – truncated direct and inverse orthogonal transform; Q/Q^{-1} – operators for quantization/dequantization of spectrum coefficients; RLE+HE/RLD+HD - lossless coding of the coefficients' values using run-length encoding/decoding and Huffman encoding/decoding; 2D-1D/1D-2D – rearrangement of coefficients' values from 2D into one-dimensional data sequence and vice versa; Post-Filter – post-filtration in the last decomposition level, used for image quality enhancement. In accordance with this diagram is performed the recursive IPD processing of each image block, using the selected kind of 2D orthogonal transform: statistic (KLT, PCA, SVD) or determined (DFT, DCT, HT, WHT, CHT, etc.).

On Fig. 3.6 are shown the halftone test images "Lena" (256×256 pixels, 8 bpp), corresponding to the IPD decomposition levels, p = 0,1,..,8. In this case was used the 2D-DCT orthogonal transform, with retained 4 DCT coefficients only: (0.0), (1.0), (0.1) and (1.1), without quantization. The quality of the so obtained images, evaluated by their peak signal-to-noise ratio (PSNR) in dB, is higher in every consecutive level (for levels p = 6, 7 and 8 was obtained PSNR = 27.11, 32.10 and 40.85 dB correspondingly). The so described approach for IPD decomposition of halftone images could be generalized for color images as well. In this case, the IPD decomposition is applied on each of the color components (RGB, YCrCb, etc.) individually.

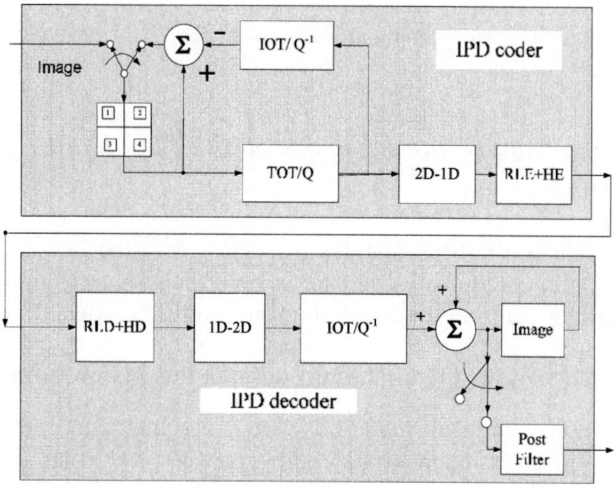

Fig. 3.5 Block diagram for recursive coding/decoding of the IPD levels

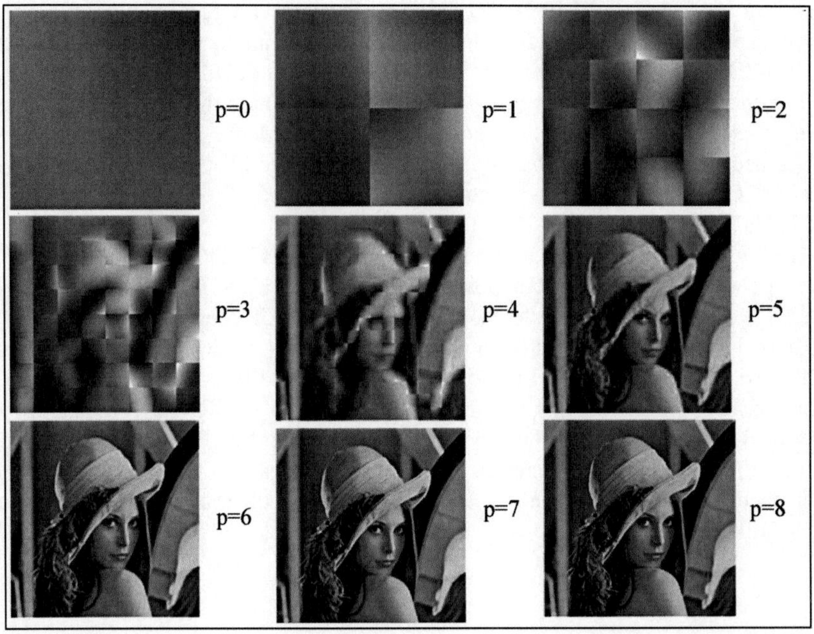

Fig. 3.6 Layered image representation obtained for IPD-DCT with 4 retained coefficients for sub-block

3.2.2 Comparison of the Inverse and the Laplacian Pyramid Decompositions

In order to evaluate the IPD qualities in respect with the famous Laplacian decomposition, in this section are compared two of their main features: the influence of the quantization noise in the restored image and the computational complexity.

3.2.2.1 Influence of the Quantization Noise on the Restored Image Quality

In order to simplify the analysis, in this case is assumed that in the processing of each IPD sub-image is retained one coefficient only – the one, corresponding to spatial frequency (0,0) (i.e., the DC coefficient). Then the full IPD of (n+1) levels is represented by the relation (Kountchev and Kountcheva, 2010):

$$B(i,j)=\overline{B}(i,j)+\sum_{p=1}^{n-1}\overline{E}_{p-1}^{k_p}(i,j)+E_{n-1}(i,j), \qquad k_p=1,2,...,4^p\,;\, i,j=1,2,..,N, \tag{3.9}$$

where $\overline{B}(i,j)$ is defined by the DC coefficient, $s_0(0,0)=\overline{B}$:

$$\overline{B}=M_0[B(i,j)]=(2^n\times 2^n)^{-1}\sum_{i=1}^{2^n}\sum_{j=1}^{2^n} B(i,j) \text{ and } \overline{B}(i,j)=I_0(\overline{B})=\overline{B}$$

for $i,j=1,2,..,2^n$.

Here $M_0(\bullet)$ and $I_0(\bullet)$ are the operators for averaging and zero-level interpolation in the decomposition level p=0, using a window of size $2^n \times 2^n$.

The difference components in the decomposition, represented by Eq. 3.9 are:

$$E_0(i,j) = B(i,j) - \overline{B}(i,j);\ E_{p-1}(i,j) = E_{p-2}(i,j) - \overline{E}_{p-2}^{k_{p-1}}(i,j);$$

$$E(i,j) = E_{n-2}(i,j) - \overline{E}_{n-2}^{k_{n-1}}(i,j),$$

where for $p = 1,2,..,n-1$

$$\overline{E}_{p-1}^{k_p}(i,j) = I_p(\overline{E}_{p-1}^{k_p}),\ \overline{E}_{p-1}^{k_p} = M_p[E_{p-1}^{k_p}(i,j)] = (2^{n-p} \times 2^{n-p})^{-1} \sum\sum_{(i,j)\in W_{k_p}} E_{p-1}(i,j);$$

In this case W_{k_p} is averaging window of size $2^{n-p} \times 2^{n-p}$, and k_p is the serial number of the averaged difference $\overline{E}_{p-1}^{k_p}$, or interpolated $\overline{E}_{p-1}^{k_p}(i,j)$ image in the decomposition level p.

The IPD components from all decomposition levels (p=0 up to p=n) are then quantizated. It is assumed that the influence of the quantization noise could be described using the linear additive model. Then:

$$B'(i,j) = \overline{B}'(i,j) + \sum_{p=1}^{n-1} \overline{E}_{p-1}'^{k_p}(i,j) + E'_{n-1}(i,j), \tag{3.10}$$

where:

$$\overline{B}'(i,j) = I_0\{Q_0^{-1}[Q_0(\overline{B})]\},\ \overline{B} = M_0[B(i,j)],$$

$$\overline{E}_{p-1}'^{k_p}(i,j) = I_p\{Q_p^{-1}[Q_p(\overline{E}_{p-1}^{k_p})]\},\ \overline{E}_{p-1}^{k_p} = M_p[E_{p-1}^{k_p}(i,j)],$$

$$E'_{n-1}(i,j) = Q_n^{-1}\{Q_n[E_{n-1}(i,j)]\}.$$

Here $Q_p(\bullet)$ and $Q_p^{-1}(\bullet)$ are operators for quantization and dequantization in the level p, and $I_p(\bullet)$ and $M_p(\bullet)$ - for interpolation and averaging correspondingly (in the same level). The components $\overline{B}'(i,j), \overline{B}'(i,j),\ \overline{E}_{p-1}'^{k_p}(i,j)$, and $E'(i,j)$ are restored after dequantization, and contain additive noise.

The corresponding dequantizated component for the level p=0 is presented as:

$$\overline{B}'(i,j) = \overline{B}(i,j) + \varepsilon_0(i,j), \tag{3.11}$$

where $\varepsilon_0(i,j)$ is the noise component. By analogy, for the next decomposition levels, p=1,2,...,n is obtained:

$$\overline{E}'^{k_p}_{p-1}(i,j)=\overline{E}^{k_p}_{p-1}(i,j)+M_p\{\varepsilon_{p-1}(i,j)+M_{p-1}[\varepsilon_{p-2}(i,j)+...+M_1[\varepsilon_0(i,j)]]\}+\varepsilon_p(i,j). \quad (3.12)$$

Here $\varepsilon_p(i,j)$ is the quantization noise of the pth IPD component.

The restored image could be represented using the original image and Eqs. (3.10-3.12):

$$B'(i,j)=B(i,j)+e_\Sigma(i,j), \quad (3.13)$$

where the total quantization error, accumulated on the last level p=n, is:

$$\varepsilon_\Sigma(i,j)=\sum_{p=0}^{n}\varepsilon_p(i,j)+M_1[\varepsilon_0(i,j)]+M_2\{\varepsilon_1(i,j)+M_1[\varepsilon_0(i,j)]\}+...$$
$$+2M_{n-1}\{\varepsilon_{n-2}(i,j)+M_{n-2}[\varepsilon_{n-3}(i,j)+....+M_1[\varepsilon_0(i,j)]]\}+\varepsilon_{n-1}(i,j). \quad (3.14)$$

For the LP decomposition (Burt and Adelson) the total quantization error, accumulated on the level p=0 (which corresponds to IPD level p=n), is defined by the relation:

$$\varepsilon_\Sigma^{PL}(i,j)=\sum_{p=0}^{n}\varepsilon_p(i,j)+F[\varepsilon_n(i,j)]+F\{\varepsilon_{n-1}(i,j)+F[\varepsilon_n(i,j)]\}+...$$
$$...+F\{\varepsilon_1(i,j)+F[\varepsilon_2(i,j)+...+F[\varepsilon_n(i,j)]]\}, \quad (3.15)$$

where F(•) is the operator, which represents the filtration of the corresponding LP component.

The comparison for the quantization noise distribution on the corresponding IPD and LP levels shows, that the accumulated noise in the IPD is much lower.

The noise interrelation for levels p=1 of IPD and the corresponding p=(n-1) of LP could be represented as follows:

$$\varepsilon_1(i,j)+M_1[\varepsilon_0(i,j)]<<\varepsilon_{n-1}(i,j)+F[\varepsilon_n(i,j)], \quad (3.16)$$

because

$$\frac{F[\varepsilon_n(i,j)]}{M_1[\varepsilon_0(i,j)]}=(2^{n-2}\times 2^{n-2})\frac{\sum\sum_{(i,j)\in W(2\times 2)}\varepsilon_n(i,j)}{\sum_{(i,j)\in W_{k_1}(2^{n-1}\times 2^{n-1})}\varepsilon_0(i,j)}>1. \quad (3.17)$$

For levels p=2 of IPD and p=(n-2) of LP this is:

$$\varepsilon_2(i,j)+M_2\{\varepsilon_1(i,j)+M_1[\varepsilon_0(i,j)]\}<<\varepsilon_{n-2}(i,j)+F\{\varepsilon_{n-1}(i,j)+F[\varepsilon_n(i,j)]\}, \quad (3.18)$$

because

$$F\{\varepsilon_{n-1}(i,j) + F[\varepsilon_n(i,j)]\} >> M_2\{\varepsilon_1(i,j) + M_1[\varepsilon_0(i,j)]\}, \text{ etc.}$$

Then from the comparison of Eqs. 3.17 and 3.18 follows that $\varepsilon_\Sigma(i,j) \ll \varepsilon_\Sigma^{PL}(i,j)$, i.e. the influence of the quantization noise on the restored image is much lower in IPD.

3.2.2.2 Comparison of the Computational Complexity

The block diagrams of the non-recursive coders/decoders for IPD and LP of 3 levels, for image of size 4×4 pixels are shown on Figs. 3.7 and 3.8. Their comparison shows the higher complexity of the LP coder/decoder. The conclusions drawn this way for image of size 4×4, could be also generalized for images of any size. The evaluation of the computational complexity of the IPD and LP based on the number of basic operations for decomposition block of size 4×4 pixels is given in Table 3.1. The comparison shows the advantages of the IPD regarding the LP.

Table 3.1 Evaluation of the computational complexity of IPD and LP

Pyramid of 3 levels	Summators Σ	Quantizets /dequantizers Q/IQ	Truncated direct /inverse 2D transform TOT/IOT	2D digital filter, F	2D decimation (↓2×2) /interpolation (↑2×2)
IPD	4	3/5	3/5	-	-
LP	5	3/5	-	6	2/4

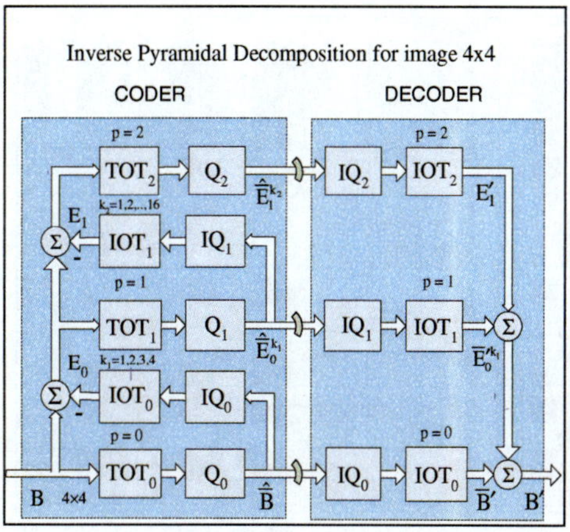

Fig. 3.7. Block diagram of the IPD coder/decoder for image of size 4×4

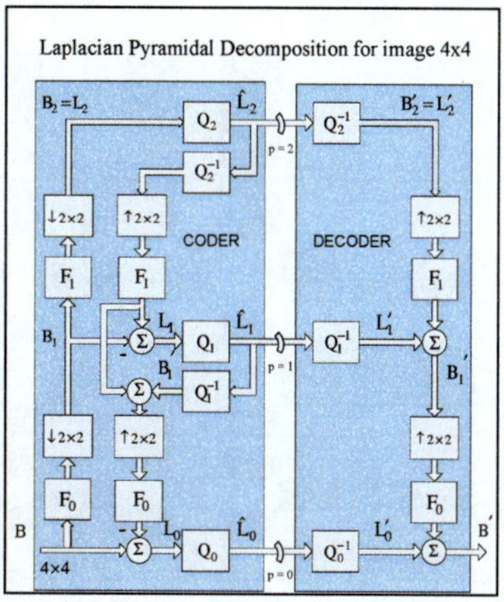

Fig. 3.8. Block diagram of the LP coder/decoder for image of size 4×4.

3.2.3 Reduced Inverse Pyramid Decomposition

As it had already been indicated, the basic disadvantage of the LP is that the needed memory volume is 33 % larger than that of the image size. In order to reduce this volume, various LP modifications had been developed: the reduced LP (RLP), the Reduced-Sum/Reduced-Difference Pyramid (RSP/RDP) and the S-Transform Pyramid (STP) (Aiazzi et al. 1996). For the RSP for example, the needed memory volume is by 8.3% larger than that of the original image (Wang and Goldberg 1989, 1991). Regardless of the so obtained reduction, all these pyramids are overcomplete. In result, the ability of the LP decomposition to offer significant compression ratios together with retained image quality is limited.

The IPD is overcomplete also. To overcome this disadvantage, here is offered the IPD modification, called Reduced IPD.

The Reduced IPD (RIPD) (Kountchev and Kountcheva 2008) permits the needed memory volume to be restricted up to that of the image, and (together with this) to achieve high compression with retained quality of the restored image. For the RIPD building are first defined the IPD components, which will constitute the RIPD levels. These relations could be described on the basis of the analysis of the IPD coefficients for one image block, $[B(2^n)]$.

The values of the coefficients are calculated using the 2D Walsh-Hadamard Transform (WHT) with arranged matrix. The basic functions of the WH functions, corresponding to the spatial frequencies (u,v) of size 4×4 (n=2), are shown on Fig. 3.9.a. For comparison only, on Fig. 3.9.b are shown the basic functions of the 2D discrete cosine transform (DCT) of size 4×4.

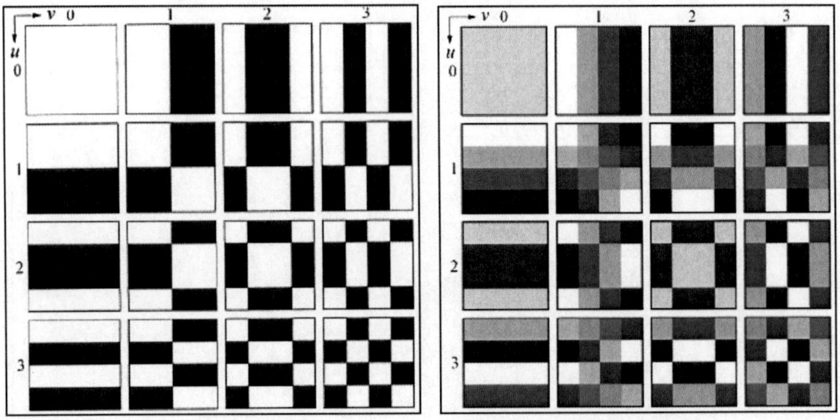

a. 2D-WHT b. 2D-DCT

Fig. 3.9 Basic functions of 2D-WHT and 2D-DCT, of size 4×4

The transform of the basic sub-matrix $[E^{k_p}_{p-1}(2^{n-p})]$ of size $2^{n-p}\times 2^{n-p}$ and with sequential number k_p in the IPD level p=1,2,..,r for the block (or sub-block) $[B(2^n)]$ is defined by the relation:

$$\begin{aligned}
&[S^{k_p}_p(2^{n-p})]=[T_p(2^{n-p})][E^{k_p}_{p-1}(2^{n-p})][T_p(2^{n-p})]=\\
&=[T_p(2^{n-p})][E^{k_p}_{p-2}(2^{n-p})][T_p(2^{n-p})]-[T_p(2^{n-p})][\hat{E}^{k_p}_{p-2}(2^{n-p})][T_p(2^{n-p})]=\\
&=[S^{k_p}_{p-1}(2^{n-p})]-[\hat{S}^{k_p}_{p-1}(2^{n-p})].
\end{aligned} \tag{3.19}$$

In particular, let the retained IPD coefficients for the sub-block k_p in the level p correspond to spatial frequencies (0,0), (1,0), (0,1) and (1,1). Then, the relation between the spectrum coefficients $s^{k_p}_p(u,v)$ of the matrix $[S^{k_p}_p(2^{n-p})]$ of size $2^{n-p}\times 2^{n-p}$ in the decomposition level p and the coefficients $s^{k_{p+1}}_{p+1}(u,v)$ of the matrix $[S^{k_{p+1}}_{p+1}(2^{n-p-1})]$ of size $2^{n-p-1}\times 2^{n-p-1}$ in the next decomposition level (p+1), taking into account the basic functions (0,0), (0,1), (1,0) and (1,1) of the 2D-WHT from Fig. 3.9.a is as follows:

- for u = v = 0

$$s^{k_p}_p(0,0)=s^{k_p}_{p-1}(0,0)-\hat{s}^{k_p}_{p-1}(0,0)=s^{k_{p+1}}_{p+1}(0,0)+s^{k_{p+1}+1}_{p+1}(0,0)+s^{k_{p+1}+2}_{p+1}(0,0)+s^{k_{p+1}+3}_{p+1}(0,0)=0; \tag{3.20}$$

- for u = 0 and v = 1:

$$s^{k_p}_p(0,1)=s^{k_p}_{p-1}(0,1)-\hat{s}^{k_p}_{p-1}(0,1)=s^{k_{p+1}}_{p+1}(0,0)+s^{k_{p+1}+1}_{p+1}(0,0)-s^{k_{p+1}+2}_{p+1}(0,0)-s^{k_{p+1}+3}_{p+1}(0,0)=0, \tag{3.21}$$

- for u = 1 and v = 0:

$$s^{k_p}_p(1,0)=s^{k_p}_{p-1}(1,0)-\hat{s}^{k_p}_{p-1}(1,0)=s^{k_{p+1}}_{p+1}(0,0)-s^{k_{p+1}+1}_{p+1}(0,0)+s^{k_{p+1}+2}_{p+1}(0,0)-s^{k_{p+1}+3}_{p+1}(0,0)=0, \tag{3.22}$$

- for u = v = 1:

$$s^{k_p}_p(1,1)=s^{k_p}_{p-1}(1,1)-\hat{s}^{k_p}_{p-1}(1,1)=s^{k_{p+1}}_{p+1}(0,0)-s^{k_{p+1}+1}_{p+1}(0,0)-s^{k_{p+1}+2}_{p+1}(0,0)+s^{k_{p+1}+3}_{p+1}(0,0)=0. \tag{3.23}$$

As an example, for the relation between the spectrum coefficients from two neighboring levels (p and p+1) and for the case: u=0 and v=1, on Fig. 3.10 are shown the coefficient – origin ("parent") $F_p(0,1)$ and its 4 ancestors: $A_{p+1}(0,0)$, $B_{p+1}(0,0)$, $C_{p+1}(0,0)$ and $D_{p+1}(0,0)$. On the same figure are shown the images of the basic functions: (0,1) in the level p and of (0,0) in the next (p+1) level, for the group of 4 neighbor sub-blocks.

The solution of the system of Eqs. (3.20 – 3.23) in respect to the coefficients $s^{k_{p+1}}_{p+1}(0,0)$, $s^{k_{p+1}+1}_{p+1}(0,0)$, $s^{k_{p+1}+2}_{p+1}(0,0)$ and $s^{k_{p+1}+3}_{p+1}(0,0)$, numbered in correspondence with the "Z" scanning line in the 4 sub- blocks of the decomposition level p+1, is given below:

$$s_{p+1}^{k_{p+1}}(0,0)=s_{p+1}^{k_{p+1}+1}(0,0)=s_{p+1}^{k_{p+1}+2}(0,0)=s_{p+1}^{k_{p+1}+3}(0,0)=0. \quad (3.24)$$

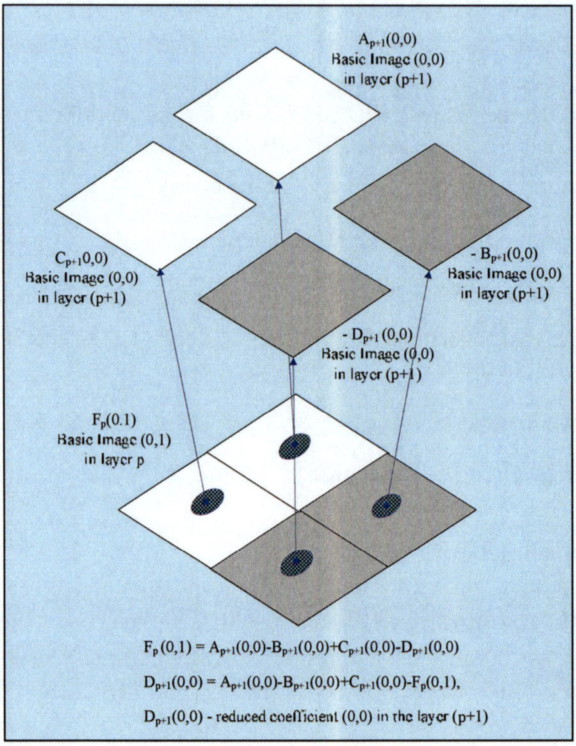

Fig. 3.10 Relations between coefficients in IPD-WHT levels p and p+1.

From this follows the conclusion that the values of the 4 spectrum coefficients with spatial frequency (0,0), calculated for 4 neighboring sub-blocks in each consecutive decomposition level p=1,2,..,n-1 are always equal to zero, regardless of the content of the corresponding block in the preceding level. In result, the number of coefficients from each of the 4 neighboring sub-blocks in every decomposition level p (except p=0) could be reduced by 4.

Then, the total number of spectrum coefficients used to represent each block $[B(2^n)]$ with RIPD-WHT of n levels, on condition that in each transform are retained 4 coefficients with frequencies u,v=0,1, is:

$$N_r = 4+\sum_{p=1}^{n-1}4^{p+1}-\sum_{p=1}^{n-1}4^p = 4+3\sum_{p=1}^{n-1}4^p = 4+3\frac{4}{3}(4^{n-1}-1) = 4^n. \quad (3.25)$$

From this follows that the general number of spectrum coefficients, which build the RIPD, based on the WHT (RIPD-WHT), is equal to the number of pixels in the block $[B(2^n)]$ and the decomposition is full. For same conditions, for IPD-WHT of n levels this number is:

$$N = 4 + \sum_{p=1}^{n-1} 4^{p+1} = (4/3)(4^n - 1) \approx (4/3)4^n,$$

i.e. the reduction of the spectrum coefficients for RIPD-WHT in respect to IPD-WHT is represented by the relation:

$$(N/N_r) \approx (4/3) = 1.33.$$

In Table 3.2 are given the relations between a set of spectrum coefficients in two consecutive pyramid levels p and (p+1) for the case, when their spatial frequencies (u,v) correspond to the basic functions with frequencies u,v=0,1,2,3, shown on Fig. 3.9.a. In correspondence with Table 3.2, the spectrum coefficient (1,2), obtained through 2D-WHT from the sub-block k_p in the level p, is connected with the 4 coefficients (0,1) for i = 0,1,2,3 from the neighboring sub-blocks with number k_{p+1}, $k_{p+1}+1$, $k_{p+2}+2$, $k_{p+3}+3$ in the level p+1 of IDP-WHT. This relation is defined by the equation:

$$s_p^{k_p}(1,2) = s_{p+1}^{k_{p+1}}(0,1) - s_{p+1}^{k_{p+1}+1}(0,1) - s_{p+1}^{k_{p+1}+2}(0,1) + s_{p+1}^{k_{p+1}+3}(0,1) \qquad (3.26)$$

The Eq. 3.26 permits one of the 4 coefficients (0,1) of the sub-blocks $k_{p+1}+i$ for i=0,1,2,3 in the level (p+1) to be reduced, because its value could be calculated if the remaining 3 coefficients from the same level and of the "parent" coefficient from the previous level p, are known.

Table 3.2 Relations between coefficients with frequencies u,v = 0,1,2,3 in IPD-WHT levels p and (p+1).

u → v ↓	IPD-WHT level	0	1	2	3
0	p →	F(0,0)	F(1,0)	F(2,0)	F(3,0)
	p+1 →	A(0,0) B(0,0) C(0,0) D(0,0)	A(0,0) -B(0,0) C(0,0) -D(0,0)	A(1,0)-B(1,0) C(1,0)-D(1,0)	A(1,0) B(1,0) C(1,0) D(1,0)
1	p →	F(0,1)	F(1,1)	F(2,1)	F(3,1)
	p+1 →	A(0,0) B(0,0) -C(0,0) -D(0,0)	A(0,0) -B(0,0) -C(0,0) D(0,0)	A(1,0) -B(1,0) -C(1,0) D(1,0)	A(1,0) B(1,0) -C(1,0) -D(1,0)
2	p →	F(0,2)	F(1,2)	F(2,2)	F(3,2)
	p+1 →	A(0,1) B(0,1) -C(0,1) -D(0,1)	A(0,1) -B(0,1) -C(0,1) D(0,1)	A(1,1)-B(1,1) -C(1,1)D(1,1)	A(1,1) B(1,1) -C(1,1)-D(1,1)
3	p →	F(0,3)	F(1,3)	F(2,3)	F(3,3)
	p+1 →	A(0,1) B(0,1) C(0,1) D(0,1)	A(0,1) -B(0,1) C(0,1) -D(0,1)	A(1,1) -B(1,1) C(1,1) -D(1,1)	A(1,1) B(1,1) C(1,1) D(1,1)

As it was mentioned before, the IPD could be implemented using various orthogonal transforms. There are similar relations between the basic functions and the corresponding coefficients' values. The case of 2D-DCT is presented above.

In Table 3.3 are given the relations between coefficients corresponding to basic functions with spatial frequencies (u,v) in the IPD-DCT levels p and (p+1) for u,v=2,2.

Table 3.3 Relations between coefficients with spatial frequencies (u,v) in IPD-DCT levels p and (p+1)

u→ v ↓	IPD-DCT level	0	2
0	p →	F(0,0)	F(2,0)
	(p+1) →	A(0,0) B(0,0) C(0,0) D(0,0)	A(0,0) B(1,0) - C(1,0) - D(1,0)
2	p →	F(0,2)	F(2,2)
	(p+1) →	A(0,0) - B(0,1) C(0,1) - D(0,1)	A(0,0) -B(1,1) - C(1,1) D(1,1)

On Fig.3.9.b could be noticed that for 2D-DCT of size 4×4 the relations, presented in Table 3.3 refer only to coefficients in the level decomposition p, corresponding to spatial frequencies (0,0), (0,2), (2,0) and (2,2), whose basic functions correspond completely to these of the 2D-WHT. In the next level (p+1) these coefficients are related to the spatial frequencies (0,0), (0,1), (1,0) and (1,1). This conclusion follows from the equation, which represents the kernel of the 2D-DCT for a block of size N×N ($N=2^n$) (Gonzalez and Woods):

$$g(i,j,u,v) = A(u)A(v)\cos\left[\frac{(2i+1)u\pi}{2N}\right]\cos\left[\frac{(2j+1)v\pi}{2N}\right] \qquad (3.27)$$

for i,j,u,v = 0,1,..,N-1

Here i and j are the numbers of the rows and the columns of the block matrix, and coefficients A(u) and A(v) are defined by the relation:

$$A(u/v) = \begin{cases} \sqrt{\frac{1}{N}} & \text{for} \quad u/v = 0; \\ \sqrt{\frac{2}{N}} & \text{for} \quad u/v = 1,2,..,N-1. \end{cases}$$

After dividing the image into 4 sub-images of size $N' \times N'$ pixels, for $N' = N/2 = 2^{n-1}$ in the next pyramid level the Eq. 3.27 for the 2D-DCT kernel is transformed as follows:

$$g(i,j,u,v) = A(u)A(v)\cos\left[\frac{(2i+1)u\pi}{2N'}\right]\cos\left[\frac{(2j+1)v\pi}{2N'}\right] =$$
$$= A(u)A(v)\cos\left[\frac{(2i+1)u\pi}{N}\right]\cos\left[\frac{(2j+1)v\pi}{N}\right], \tag{3.28}$$

for i, j, u, v = 0,.. $N'-1$, and

$$A(u/v) = \begin{cases} \sqrt{\frac{2}{N}} & \text{for} \quad u/v = 0; \\ \sqrt{\frac{4}{N}} & \text{for } u/v = 1,2,..,N/2-1. \end{cases}$$

Therefore, the DCT coefficients with frequencies (0,0), (1,0), (0,1) and (1,1), calculated for 4 neighboring sub-blocks in the IPD-DCT level (p+1) have a precise relation with the coefficients with frequencies (0,0), (2,0), (0,2) and (2,2) with the "parent" block from the lower decomposition level, p.

It is possible to express the DC coefficient of a subimage in level p from the values obtained at the previous level p-1, according to the following relation:

$$s_p^{k_p}(0,0) = \frac{1}{2^{n-p}}\sum_{(i,j)\in W_{k_p}}\sum E_{p-1}^{k_p}(i,j) = \frac{1}{2^{n-p}}\sum_{(i,j)\in W_{k_{p-1}}}\sum [E_{p-2}^{k_{p-1}}(i,j) - \hat{E}_{p-2}^{k_{p-1}}(i,j)] \tag{3.29}$$

where W_{k_p} is the averaging window of size $2^{n-p} \times 2^{n-p}$. The relation between the DC coefficients of 4 neighboring sub-images in level p with these from the preceding (p-1) level is:

$$\sum_{l=0}^{3} s_{k_p}^{l}(0,0) = \frac{1}{2^{n-p}}\sum_{(i,j)\in W_{k_{p-1}}}\sum [E_{p-2}^{k_{p-1}}(i,j) - \hat{E}_{p-2}^{k_{p-1}}(i,j)] = 2[s_{k_{p-1}}(0,0) - s_{k_{p-1}}(0,0)] = 0 \tag{3.30}$$

Here $s_{k_p}^{l}(0,0)$ for l = 0,1,2,3, and p = 1,2,...,r, represents the DC coefficients of the 4 neighboring sub-images in the pyramid level p. From Eq. 3.30 follows:

$$s_{k_p}^{3}(0,0) = -\left[s_{k_p}^{0}(0,0) + s_{k_p}^{1}(0,0) + s_{k_p}^{2}(0,0)\right] \tag{3.31}$$

That is to say, in each group of 4 DC coefficients from the 4 sub-images with same root image, the 4th coefficient could be deduced from the 3 other coefficients belonging to neighboring branches in the quaternary tree. So coefficient $s_{k_p}^{3}(0,0)$ could be omitted in the coding and reconstructed, using Eq. 3.31. In this case, the modeling of the 4 sub-images with same root image requires 3 DC coefficients and 12 AC coefficients instead of 4 DC coefficients and 12 AC coefficients, if Eq. 3.31 is not taken into account. In result is obtained the reduced pyramid IDP-DCT (RIDP-DCT). Its compression ratio is then defined by the expression:

$$C_r(p) = \frac{16}{15}C(p) = \frac{16}{15}\left(3\times 4^{n-p-2}\right) = \frac{4^{n-p}}{5} \tag{3.32}$$

for $p = 1,2,...,r < n\text{-}1$,

where:

$$C(p) = 4^n / 4\sum_{l=0}^{p} 4^l = 4^n /[(4/3)(4^{p+1}-1)] \approx 3\times 4^{n-p-2} \text{ for } p = 0,1,2,...,r$$

is the compression coefficient of IDP-DCT, which depends on the number of decomposition levels, p. The value of $C_r(p)$ for RIDP-DCT is increased by 16/15 compared to the IDP-DCT.

The choice of the orthogonal transform, used in the RIPD, depends on the application and is related to the kind of processed images, the needed compression ratio and the required quality of the restored images.

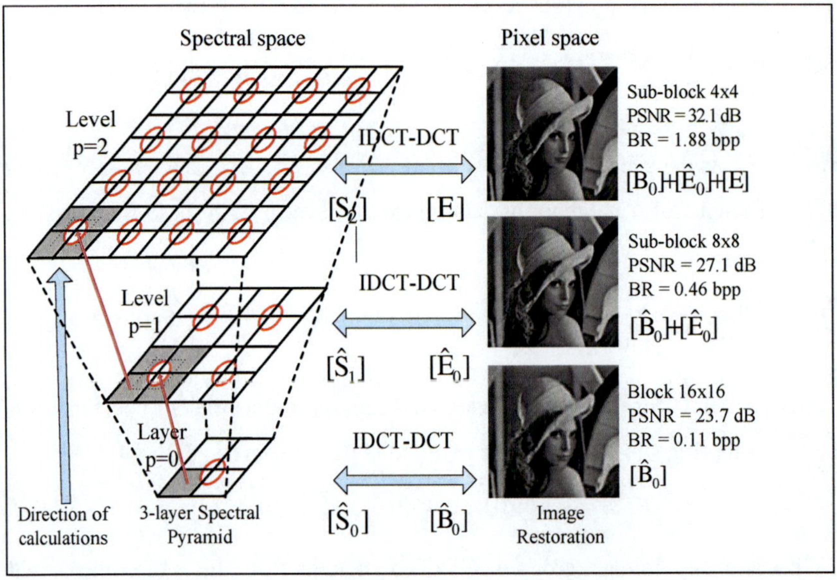

Fig. 3.11. Representation of the test image "Lena (512×512, 8 bpp) based on 3-level RIPD-DCT with initial sub-block of size 16×16 (n=4) and retained 4 low-frequency spectrum coefficients for each sub-block.

On Fig. 3.11 is shown 3-level RIPD-DCT, built on one block of the image "Lena" of size 512×512 pixels and 8 bpp. The image matrix is initially divided into blocks of size 16×16 pixels. The pyramid was built in the spectrum area of each block. It is inverse, because the calculation of the DCT coefficients starts from the top (the level p=0) and continues towards its base (the level p=2). Each restored image corresponds to one of the decomposition levels, built on all blocks.

On the same figure are shown the peak-signal-to-noise ratio (PSNR) in dB, and the number of bits per pixel (Bit Rate, BR), obtained for the set of retained coefficients with frequencies (0,0), (0,1), (1,0) and (1,1) from each image block or sub-block. The values of BR are obtained without quantization or entropy coding of the RIPD-DCT coefficients. After lossless compression of the coefficients' values, based on the Huffman coding, the value of BR is reduced to 1 bpp retaining the PSNR > 32 dB.

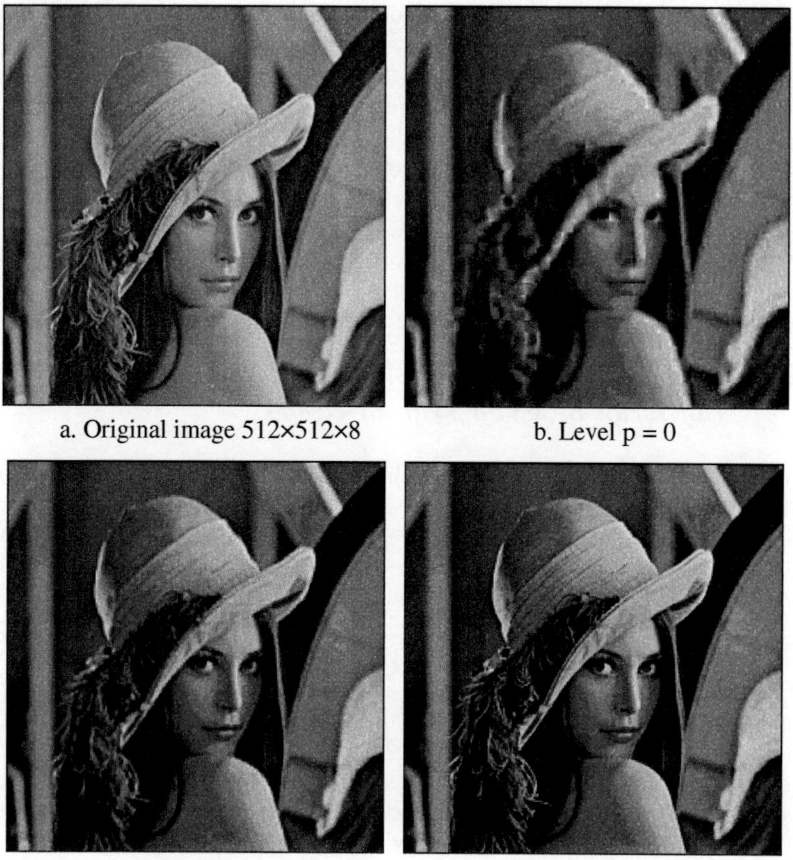

a. Original image 512×512×8 b. Level p = 0

c. Level p = 1 d. Level p = 2

Fig. 3.12 The original test image and the restored images, corresponding to RIPD-DCT levels p = 0,1,2 (in accordance to Fig.3.11).

On Fig.3.12 are shown the original and the restored images from Fig. 3.11 for the RIPD-DCT decomposition levels p = 0,1,2 in all blocks.

Farther growth of the compression retaining the image quality is obtained using algorithms for blocking artifact reduction (Chen and Wu) on the restored images in the last RIPD-DCT level.

The basic advantages of the RIPD are:

1. The compression ratio is approximately 33% higher than that, obtained for same images with IPD-WHT, and 6% higher than that of IPD-DCT. In both cases the quality of the restored image is retained.

2. For the calculation of the RIPD coefficients are used relatively simple relations, what ensures low computational complexity;

3. RIPD could be built on the basis of various orthogonal transforms, which can additionally simplify its implementation. Supplementary enhancement of the compression ratio and lower computation complexity are achieved using a hybrid pyramid RIDP-DCT/WHT. This RIDP modification consists in replacing the DCT transform by the WHT in the low pyramid levels only, where the sub-image size is small (for example, 4×4 and 2×2 pixels). Indeed, for small sub-images, DCT and WHT efficiency is similar in terms of energy concentration of spectral coefficients but WHT is simpler to implement;

4. RIPD offers the ability to build contemporary systems for leveled image transfer with high compression ratio, which is of significant importance for the Internet applications;

5. RIPD could be also used for efficient representation of multi-view and multispectral images and video sequences.

3.2.4 Inverse Pyramid Decomposition with Non-linear Transforms Based on Neural Networks

In the last years a large group of non-linear methods for image representation, based on the artificial neural networks (NN) had been developed (Perry et al., Hu and Hwang, Dony and Haykin, Namphol et al., Kulkarni et al., Jiang, Kouda et al., and Special Issue on Image Compression, IJGVIP). They are easily distinguished from the classic ones, because the NN is trained in the process of coding, which results in higher compression efficiency. The results obtained show that these methods are not able to compete with the famous standards for image compression, JPEG и JPEG2000 (Acharya and Tsai). For example, the adaptive vector quantization (AVQ), based on neural networks of the SOM kind (Hu and Hwang, Kouda et al.), requires the use of code books with large number of vectors, in order to ensure high quality of the restored image, which reduces the compression ratio. In this chapter is presented one new method for Adaptive Inverse Pyramid Decomposition (AIPD) based on NN with error back-propagation (BPNN), in result of which the visual quality of the restored images is higher than that offered by the basic image compression standards.

3.2.4.1 Image Representation with Adaptive Inverse Pyramid Decomposition

The new method for image representation is based on IPD, in which the direct and the inverse transforms in all decomposition levels are performed through 3-layer BPNN (Kountchev, Rubin et al.). The general BPNN structure in the AIPD is chosen to be of 3 layers of the kind $m^2 \times n \times m^2$, and is shown on Fig. 3.13. The input layer is of m^2 elements, which correspond to the input vector components; the hidden layer is of n elements for $n<m^2$, and the output layer is of m^2 elements also, which correspond to the output vector components. The input m^2-dimensional vector is obtained through transformation of the elements m_{ij} in each image block of size m×m into one-dimensional data sequence of length m^2, using the Hilbert scan (Fig.3.4.). In order to compensate the data of the original image, it is represented by a sequence of m^2-dimensional vectors $\vec{X}_1, \vec{X}_2, ..., \vec{X}_K$, which are then transformed into the n-dimensional vectors $\vec{h}_1, \vec{h}_2, ..., \vec{h}_K$. The components of each vector $\vec{h}_k$ for k=1,2,..K represent the neurons in the hidden layer of the trained 3-layer BPNN whose structure is $m^2 \times n \times m^2$.

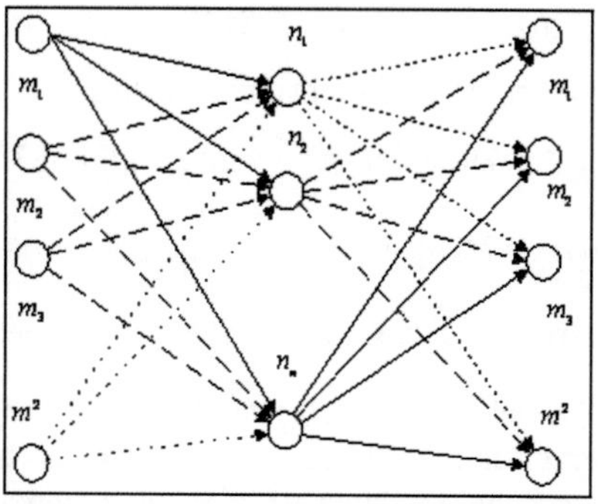

Fig. 3.13 A 3-layer BPNN of $n_n < m^2$ neurons in the hidden layer and m^2 neurons in the input and in the output layer

The vector $\vec{h}_k$ in the output layer is transformed back into the m^2-dimensional output vector $\vec{Y}_k$, which approximates the corresponding input vector $\vec{X}_k$. The approximation error depends on the training algorithm and on the BPNN parameters. The training vectors $\vec{X}_1, \vec{X}_2, ..., \vec{X}_K$ at the BPNN input for the AIPD level p = 0 correspond to all image blocks. For the training was selected the Levenberg-Marquardt (LM) algorithm (Perry et al., Hu and Hwang), which has fast convergence particularly in cases, when high training accuracy is not needed (for this reason it is suitable for the presented case). One more reason for this choice is that the training data have too large volume and significant information surplus. This does not aggravate the training, but makes the process longer. The parameters of the 3-layer BPNN define the configuration of the connections between its inputs and the neurons in the hidden layer, and also between neurons in the hidden and the output layer. These connections are described by corresponding weight matrices and vectors, which contain threshold coefficients, and through functions for nonlinear vector transform. The relation between the input m^2-dimensional vector $\vec{X}_k$ and the corresponding n-dimensional vector $\vec{h}_k$ in the hidden BPNN layer for the AIPD level p=0 is defined by the relation:

$$\vec{h}_k = f([W]_1 \vec{X}_k + \vec{b}_1) \text{ for } k = 1,2,..K, \tag{3.33}$$

where $[W]_1$ is the matrix of the weight coefficients, of size $m^2 \times n$, used for the linear transform of the input vector $\vec{X}_k$; $\vec{b}_1$ is the n-dimensional vector of the threshold coefficients of the hidden layer; f(x) is the nonlinear transfer (activating) function of sigmoid kind, defined in accordance with the relation:

$$f(x) = 1/(1+e^{-x}) \tag{3.34}$$

This function introduces some non-linearity in the network performance, which is revealed stronger for cases, when *x* is out of the interval [-1.5, +1.5].

The relation between the n-dimensional BPNN vector $\vec{h}_k$ in the hidden layer and the corresponding output m^2-dimensional vector $\vec{Y}_k$ for AIPD level p=0, which approximates $\vec{X}_k$, is defined in accordance with Eq. 3.33:

$$\vec{Y}_k = f([W]_2 \vec{h}_k + \vec{b}_2) \text{ for } k = 1,2,..K, \tag{3.35}$$

where $[W]_2$ is the matrix (of size $n \times m^2$) of the weight coefficients of the linear vector transform in the hidden layer $\vec{h}_k$, and $\vec{b}_2$ is the m^2-dimensional vector of the threshold coefficients for the output layer. Unlike the pixels of the halftone images, whose brightness is in the range $[0, 255]$, the components of the input and the output BPNN vectors are normalized in the interval $x_i(k), y_i(k) \in [0,1]$ for $i=1,2..,m^2$. In the same interval are disposed the components of the vectors of the neurons in the hidden layer $h_j(k) \in [0,1]$ for j=1,2..,n, because they are defined by the value of the activating function $f(x) \in [0,1]$.

The so described normalization enhances the BPNN efficiency.

The image representation in compressed form, based on the AIPD-BPNN, needs two consecutive stages: the BPNN training and the coding of the obtained output data. In the stage of the BPNN training for the AIPD level p=0, the vectors $\vec{X}_k$ are used as input and reference vectors, which are compared with the corresponding output vectors. The result obtained is used for the correction of the weight and threshold coefficients towards the MSE minimum. The training is repeated until the MSE value of the output vectors becomes lower than a predefined threshold. For the training of the 3-layer BPNNs for the next AIPD levels (p>0) are used the vectors, obtained after dividing the corresponding difference block

$[E_{k_{p-1}}]$ (or a sub-block) into 4^pK sub-blocks and their transformation into corresponding vectors. In result, the corresponding BPNN is trained for the level p>0 in a way, similar with this for the level p=0. In the second stage the vectors of the hidden BPNN layers for all AIPD levels p=0,1,...,n are losslessly coded. This is performed using two methods: Run-Length Coding (RLC) and Huffman codes of variable length (Acharya and Tsai). The block diagram of the pyramid decomposition for a block of size m×m through 3-layer BPNN for each level p=0,1,2 and entropy coding/decoding is shown on Fig. 3.14.

When the BPNN training for each level p is finished, are defined the corresponding output weight matrix, $[W]_p$ and the threshold vector $[b]_p$. The data is compressed with entropy encoder and then the information, representing the decomposition level p is ready.

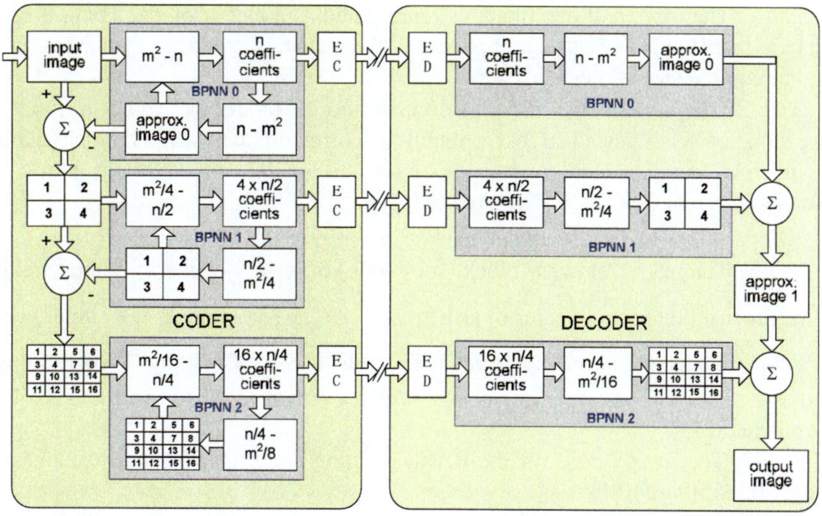

Fig. 3.14 Block diagram of 3-level inverse pyramid image decomposition with 3-layer BPNN; $[b]_p$-vector of the threshold coefficients in the output layer for p=0,1,2; $[W]_p$ – the matrix of the weight coefficients of the hidden to the output BPNN layer.

The coded data comprises:

- The vector of the threshold coefficients of the neurons in the output NN layer (general for all blocks in decomposition level p);
- The matrix of the weight coefficients of the connections of the neurons in the hidden BPNN layer to the output layer (general for all blocks in decomposition level p);
- The vector of the neurons in the hidden BPNN layer, which in the general case is individual for each block in the level p.

In the decoder is first performed the entropy decoding (ED). After that the BPNN is initialized in the level p, appropriating the data for the threshold coefficients of the neurons in the output layer and of the weight coefficients of the neurons used for their connection between the hidden and the output layer. In the end stage of the processing the vector of the neurons in the hidden BPNN layer for each block is transformed into corresponding output vector. On the basis of all output vectors are restored the blocks of the whole image. To simulate the AIPD-BPNN algorithm is necessary to perform the following basic operations: to prepare the input data as a sequence of vectors; to choose the BPNN architecture; to create the BPNN and to initialize its parameters; to train the BPNN using the corresponding input vectors, so that to obtain the needed output vectors; to test the AIPD-BPNN algorithm with images of various kind and to evaluate their quality (objective and subjective evaluation).

The steps of the algorithm, based on the AIPD-BPNN, are:

Coding:

Step 1. The input halftone image is represented as a matrix of size H×V, 8 bpp (in case that H and V are not multiples of 2^n the matrix is expanded with zeros, until the required size is obtained);

Step 2. The input image matrix is divided into K blocks of size m×m ($m=2^n$). The value of *m* is selected so that to retain the correlation between the block pixels as much as possible (for big images of size 1024×1024 or larger the block is usually 16×16 or 32×32, and for smaller images it is 8×8);

Step 3. The AIDP level numbers p are set, starting with p=0;

Step 4. The matrix of every block (sub-block) of $m^2/2^p$ elements in the level p is transformed into input vector of size $(m^2/2^p)\times1$. The so obtained 4^pK input vectors constitute a matrix of size $(m^2/2^p)\times4^pK$, which is used for the BPNN training and as a matrix of the reference vectors, which are then compared with the BPNN output vectors;

Step 5. The matrix used for the BPNN training is normalized transforming its range [0, 255] into [0,1];

Step 6. The BPNN training function and working functions are selected.

Step 7. The criterion for the BPNN training end is defined setting the deviation value (0.01) or the maximum number of training cycles (50000 epochs), after which the training ends;

Step 8. Iterative BPNN tuning is performed, using the function which follows the error gradient. After that the information is saved in a special file, which contains:

- The neurons of the hidden layer, which in general are different for every block (sub-block);
- The threshold coefficients for the output layer;
- The matrix of the weight coefficients between the hidden and the output BPNN layers.

Step 9. The data, described in step 8 is losslessly coded using RLC and Huffman code and is saved in a special file, which contains the compressed data for the level p;

Step 10. The level number p is increased (p=p+1): in case that it is lower than the maximum ($p_{max} \leq n$) is performed step 3, else the processing continues with step 11.

Step 11. One common file is generated, where is stored the data from all levels: $p = 0,1,.., p_{max}$;

Decoding:

Step 1. The decoder receives the sequentially transferred data for the AIPD-BPNN levels $p = 0,1,.., p_{max}$;

Step 2. For every level p are decoded the values of the neurons in the hidden layer for each block (sub-block), the threshold coefficients and the matrix of the weight coefficients for the corresponding output BPNN layer;

Step 3. The decoded data for each AIDP-BPNN level are set for the corresponding BPNN in the decoder;

Step 4. The vector components for each block (sub-block) in the output BPNN layer are restored;

Step 5. The output BPNN vector is transformed into the block (sub-block) matrix;

Step 6. The range [0,1] of the matrix elements is transformed back into [0,255];

For the image representation in accordance with the AIPD-BPNN method was developed new format, which contains the 3 main BPNN components for each pyramid level. The new structure comprises:

- The vector of the neurons values in the hidden layer – individual for each block/sub-block;
- The vector of the threshold coefficients for the output layer – common for all blocks/sub-blocks;
- The matrix of the weight coefficients for the output layer - common for all blocks/sub-blocks.

3.2.4.2 Experimental Results

The experiments with the AIPD-BPNN algorithm were performed with test images of size 512×512, 8 bpp (i.e. 262 144B). In the AIPD level p=0 the image is divided into K blocks of size 8×8 pixels, (K=4096). At the BPNN input for the decomposition level p=0 is passed the training matrix of the input vectors of size 64×4096=262144. The size of each input vector of the hidden BPNN layer is reduced from 64 to 8. The restoration of the output vector in the decoder is performed using these 8 components, together with the vector of the threshold values and the matrix of the weight coefficients in the BPNN output layer. For the level p=0 the size of the data obtained is 266752 B, i.e. - larger than that of the original image (262144 B). As it was already pointed out, the data has high correlation and is efficiently compressed with entropy coding.

For example, the compressed data size for the investigated level (p=0) of the test image "Tracy" is 4374 B (the result is given in Table 3.4). Taking into account the size of the original image, is calculated the compression ratio C_r = 59.93. The quality of the restored test image "Tracy" for p=0 (Table 3.4) is evaluated as PSNR=35,32 dB. In the same Table 3.4 are given the compression ratios obtained with AIPD-BPNN for 8 test images of size (512×512), shown on Fig.3.15. For the mean compression ratio C_r=60 is obtained PSNR>30 dB, i.e. the visual quality of the restored test images is good enough for various applications. On Fig.3.16 are shown the graphic relations PSNR = f(C_r) for each of the 8 test images, compressed in accordance with the AIPD-BPNN decomposition of 3 levels (p=0,1,2).

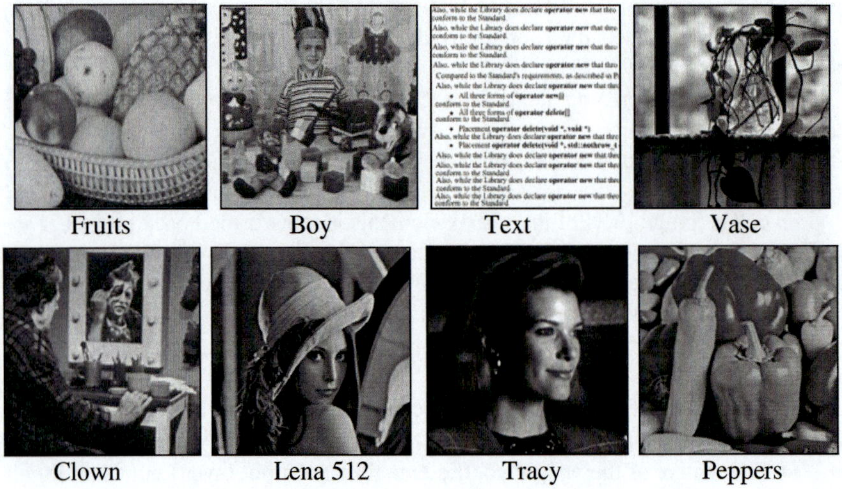

Fig. 3.15 A set of test images

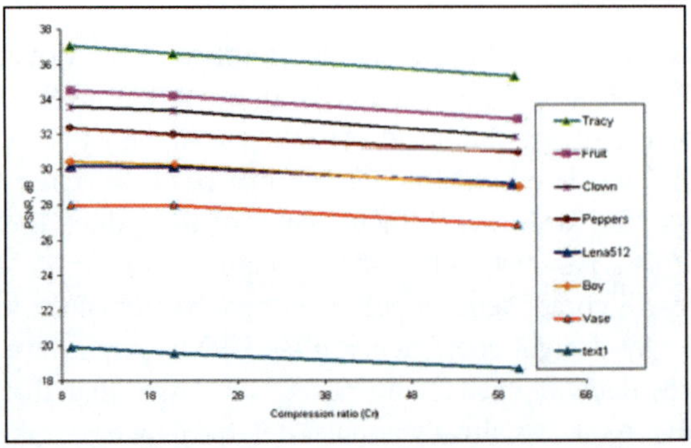

Fig. 3.16 Comparison for 3-level AIPD-BPNN: the right column of points corresponds to p = 0 and the middle and the left - to p = 1 and 2 correspondingly

Table 3.4 Results obtained for the 8 test images after AIDP-BPNN of one layer only (for p=0).

File name	C_r	PSNR [dB]	RMSE	Bits/ pixel (bpp)	Compressed File [B]
Boy	60.40	29.05	9.22	0.1324	4340
Fruit	60.29	32.89	5.79	0.1326	4348
Tracy	59.93	35.32	4.37	0.1334	4374
Vase	60.18	26.83	11.62	0.1329	4356
Clown	60.01	31.81	6.55	0.1333	4368
Peppers	60.23	30.94	7.24	0.1328	4352
Text	60.23	18.69	29.65	0.1328	4352
Lena 512	59.57	29.15	8.89	0.1334	4400

In Table 3.5 the results obtained for AIDP-BPNN are given together with the results for the same set of 8 test images obtained using the software product LuraWave SmartCompress.

Table 3.5 Results obtained for AIDP-BPNN, JPEG and JPEG2000 (LuraWave SmartCompress).

File name	AIPD-BPNN		LuraWave SmartCompress			
			Lura JPEG		Lura JPEG2000	
	C_r	PSNR	C_r	PSNR	C_r	PSNR
Boy	60.40	29.05	28.48	29.33	50.04	29.15
Fruit	60.29	32.89	31.67	32.78	60.00	33.11
Vase	60.18	26.83	35.18	27.00	70.04	27.07
Tracy	59.93	35.32	45.21	35.03	109.3	35.66
Clown	60.01	31.81	31.37	31.88	60.03	31.87
Peppers	60.23	30.94	36.81	31.16	80.02	30.85
Text 1	60.23	18.69	22.37	18.23	30.02	18.21
Lena 512	59.57	29.15	30.75	29.52	60.03	29.31

On Fig. 3.17.a are shown the results for the test image "Boy" after compression with JPEG 2000 (C_r=50) and AIPD-BPNN (C_r= 60).

Fig. 3.17.b presents enlarged parts of the test image, which permits easy visual comparison of the results obtained.

Original | AIPD-BPNN p=0 C_r=60,4; PSNR=29 dB | JPEG (im2jpeg) C_r=29,6; PSNR=28,89 dB

Lura JPEG C_r=28,4; PSNR=29.33 | JPEG2000 C_r=50; PSNR=29,15 dB | AIPD-BPNN C_r=60; PSNR=29 dB

Fig. 3.17.a. The restored test image "Boy" after compression with 5 methods

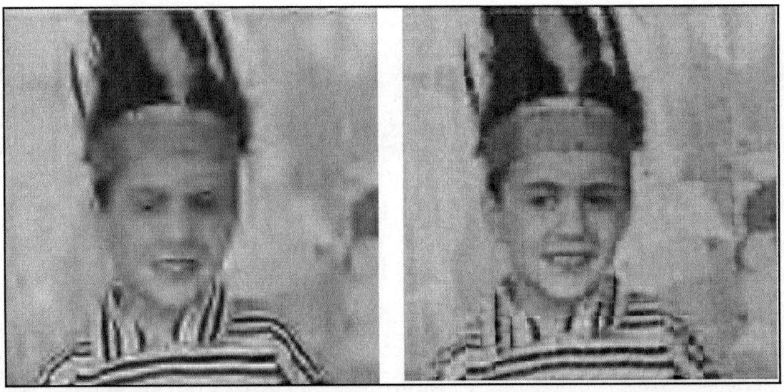

Lura JPEG2000 C_r=50; PSNR=29.15 dB AIPD-BPNN C_r=60 PSNR=29 dB

Fig. 3.17.b. Enlarged part of the restored test image "Boy"

The quality of the restored image in both cases is similar: PSNR = 29 dB. It is easy to notice that for close compression and PSNR the image processed with AIPD-BPNN is not as blurred as that, processed with JPEG 2000, i.e. the quality evaluation with PSNR in this case does not correspond to human perception.

The visual evaluation of the restored images quality shows that AIPD-BPNN ensures better results. The NN architecture used for the experiments comprises 64 neurons in the input layer, 8 neurons in the hidden layer, and 64 neurons in the output layer, used for the zero decomposition level. The chosen proportion for the input vectors was correspondingly: 80% for Training; 10 % for Validation and 10% for Testing. The algorithm modeling was used to compare it with 5 versions of the image compression standards JPEG and JPEG2000. The results obtained show that for same conditions AIPD-BPNN ensures higher visual quality of the restored images. The AIPD-BPNN is asymmetric (the coder is more complicated than the decoder) and this determines it mostly in application areas which do not require real time processing i.e. applications, for which the training time is not crucial. The hardware implementation of the method is beyond the scope of this work. The experiments for the AIPD-BPNN algorithm were performed with sub-blocks of size 8×8 pixels. The computational complexity of the method was compared with that of JPEG and the investigation proved that AIPD-BPNN complexity is comparable with that of JPEG. In general, the computational complexity of the method depends on the training method selected. The new method offers wide opportunities for application areas in the digital image processing, such as the progressive transfer via Internet, saving and searching in large image databases, representation of high definition satellite images (Cherkashyn et al.), etc.

3.3 Multi-view Image Representation Based on the Inverse Pyramidal Decomposition

Scientists and industry increasingly need multi-view representations of objects in the built environment and the demand for such kind of information is ever increasing. Some of the typical application areas are: 3D geographical information systems; hazardous and accident site survey; quality control for production lines; facility or construction management; object data mining, etc. (Kropatsch and Bischof, ISO/IEC). Two different types of image features can be extracted: those that are directly related to the 3-D shape of the part of the object being viewed and features, that result from the 3-D to 2-D down projection (the second one can be ambiguous because part of the 3-D shape information is lost during the projection). The essence of the recognition problem is to relate the structures found in the image with the underlying object models. The pyramidal image representation is one of the frequently used techniques. The object reconstruction at a given pyramid level is based on the feature-based matching approach. The first step required at each level is the extraction of salient features (points and/or lines) together with their topological relations, which is a process controlled by a model of what is expected to be found in the images. Having detected features in two or more images, the correspondence problem has to be solved. The general approach seeks correspondences in object space, because this approach is more flexible with regard to handling occlusions and surface discontinuities. The task-dependent local model of the object surface is then provided, and false correspondences are detected from bad fits to that model in object space (Kim et al. 2006, Mokhtarian and Abbasi). The IPD suits the peculiarities of this basic approach. The IPD-based object representation (and correspondingly - the salient features extraction) is performed in the spectrum domain. The creation of consecutive approximating images with increasing quality suits very well the widely used algorithms for image data mining (Todd, Vazquez et al.). Together with this, the IPD application offers specific advantages when the creation of 3D object model is concerned.

3.3.1 Multi-view 3D Object Representation with Modified IPD

The 3D representation of an object in accordance with the approach, presented below, is done on the basis of its (2N+1) multi-view images. For this are used (2N+1) photo cameras, placed uniformly on a part of an arc at same angle $\alpha=\varphi/(2N+1)$, and the object is on the center of the arc. The angle φ defines the width of the view zone and is usually selected to be in the range 20^0 - 30^0.

For some applications, the view points could be arranged in a line, in a circle or on a part of a sphere. An example multi-view arrangement in a part of a sphere is shown on Fig. 3.18 and typical arrangement in a circle – on Fig. 3.19. The optimum number of view points (correspondingly – the angles between them) depends on the application as well. One of the views is always used as a reference one. For example, if the needed multi-view should represent objects on a theatre scene, the view points should be placed in a plane and their number should correspond to the seats in the simulated hall. In this case, the view points should be placed in a relatively small sector of a sphere. Else, if for example, the application is to represent objects in the way they are seen by an insect, the view points number and positions should be quite larger.

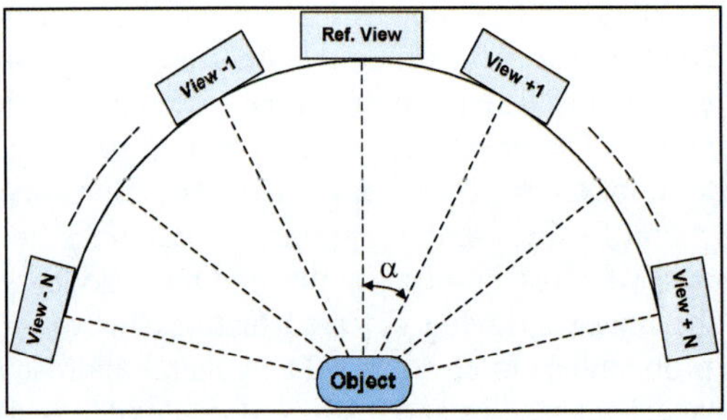

Fig. 3.18. View points arranged on an arc

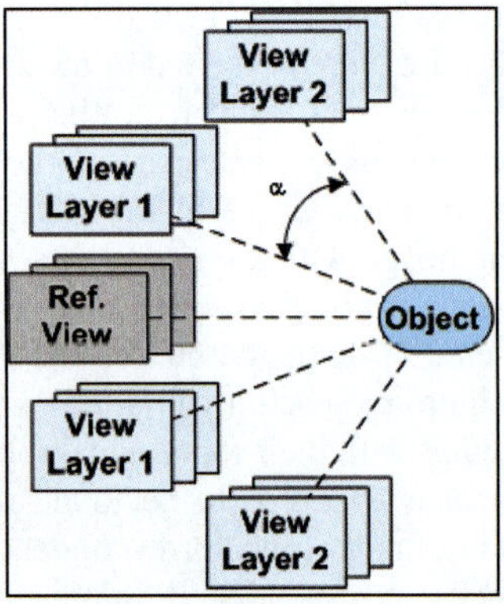

Fig. 3.19. Example view-point arrangement in parallel circles, which build a part of a sphere around the object

Each block of the nth multi-view image of an object in same time moment is represented by the matrix $[B_n]$ of size $2^m \times 2^m$, when n = 0, ±1, ±2, . . , ±N. For all that the matrix $[B_0]$ corresponds to the so-called "reference" image, placed at the middle of the view sequence $[B_n]$, for n=0. In order to decrease the information surplus in the sequence of matrices $[B_n]$ for n = 0, ±1, ±2,.., ±N, is used an IPD modification of 2 levels (Kountchev, Todorov, Kountcheva 2009):

Modified IPD coding:

1. For the IPD level p = 0 is calculated the transform $[S_0^0]$ of the reference image $[B_0]$, using the 2D direct orthogonal transform:

$$[S_0^0] = [T_0][B_0][T_0], \tag{3.36}$$

where $[T_0]$ is the matrix of the direct orthogonal transform of size $2^m \times 2^m$.

2. The matrix of the approximated transform of the reference image is calculated:

$$[\hat{S}_0^0] = [m_0(u,v)s_0^0(u,v)] \tag{3.37}$$

where $m_0(u,v)$ is an element of the matrix-mask $[M_0]$, used to define the retained spectrum coefficients.

3. The matrix of the approximated reference image $[\hat{B}_0]$ is calculated, using the inverse orthogonal transform:

$$[\hat{B}_0] = [T_0]^t[\hat{S}_0^0][T_0]^t \tag{3.38}$$

where $[T_0]^t = [T_0]^{-1}$ is the matrix of the inverse orthogonal transform of size $2^m \times 2^m$.

4. The difference matrix is calculated:

$$[E_0] = [B_0] - [\hat{B}_0] \tag{3.39}$$

and divided into 4 sub-matrices:

$$[E_0] = \begin{bmatrix} [E_0^1] & [E_0^2] \\ [E_0^3] & [E_0^4] \end{bmatrix} \tag{3.40}$$

where $[E_0^i]$ for i = 1,2,3,4 are sub-matrices of size $2^{m-1} \times 2^{m-1}$.

5. For the IPD level p=1 is calculated the transform $[S_0^i]$ of the ith sub-matrix of the difference $[E_0]$, using the direct orthogonal transform:

$$[S_0^i] = [T_1][E_0^i][T_1] \text{ for i=1,2,3,4} \tag{3.41}$$

where $[T_1]$ is the matrix for direct orthogonal transform, of size $2^{m-1} \times 2^{m-1}$.

6. The approximated i^{th} transform is calculated:

$$[\hat{S}_0^i] = [m_1(u,v)\, s_0^i(u,v)] \tag{3.42}$$

where $m_1(u,v)$ is an element of the matrix-mask $[M_1]$ used to set the retained spectrum coefficients.

7. For the level p=1 of the multi-view image $[B_n]$ is calculated the difference:

$$[E_n] = [B_n] - [\hat{B}_0] \text{ for } n = 0,\pm1,\pm2,..,\pm N \tag{3.43}$$

which is divided into 4 sub-matrices:

$$[E_n] = \begin{bmatrix} [E_n^1] & [E_n^2] \\ [E_n^3] & [E_n^4] \end{bmatrix} \tag{3.44}$$

where $[E_n^i]$ for i=1,2,3,4 are sub-matrices of size $2^{m-1}\times 2^{m-1}$.

8. The ith transform $[S_n^i]$ of the difference sub-matrix $[E_n^i]$ is obtained after direct orthogonal transform:

$$[S_n^i] = [T_1][E_n^i][T_1] \text{ for i=1,2,3,4.} \tag{3.45}$$

9. The approximated i^{th} transform (i.e. the spectrum of the difference matrix $[E_n^i]$) is calculated:

$$[\hat{S}_0^i] = [m_1(u,v)\, s_0^i(u,v)] \tag{3.46}$$

where $m_1(u,v)$ is one element of the matrix-mask, used to select the retained spectrum coefficients.

10. The difference matrices of the approximated transforms are calculated:

$$[\Delta\hat{S}_n^i] = [\hat{S}_0^i] - [\hat{S}_n^i] \text{ for n=}\pm1,\pm2,..,\pm\text{N.} \tag{3.47}$$

11. The coefficients of the spectrum matrices $[\hat{S}_0^0]$ and $[\Delta\hat{S}_0^i]$ are losslessly coded for i=1,2,3,4 and n=0,±1,±2,..,±N in the decomposition levels p = 0, 1 of the corresponding (2N+1) pyramids.

Modified IPD decoding:

1. The coefficients of the obtained spectrum matrices $[\hat{S}_0^0]$ and $[\Delta\hat{S}_0^i]$ are losslessly decoded for i=1,2,3,4 and n=±1,±2,..,±N in the decomposition levels p = 0,1 for the corresponding (2N+1) pyramids;

2. The approximated transforms in level p = 1 are restored:

$$[\hat{S}_n^i] = [\hat{S}_0^i] + [\Delta\hat{S}_n^i] \text{ for } n = \pm 1, \pm 2, .., \pm N. \tag{3.48}$$

3. For the reference image (n=0) is calculated each ith approximated sub-matrix $[\hat{E}_0^i]$ of the difference matrix $[\hat{E}_0]$ through inverse orthogonal transform:

$$[\hat{E}_0^i] = [T_1]^t [\hat{S}_0^i][T_1]^t \text{ for } i=1,2,3,4, \tag{3.49}$$

4. For the decomposition level p = 0 is calculated the matrix of the approximated orthogonal transform $[\hat{B}_0]$, through inverse orthogonal transform:

$$[\hat{B}_0] = [T_0]^t [\hat{S}_0^0][T_0]^t \tag{3.50}$$

5. The matrix $[\hat{B}]$ of the restored reference image is calculated (n=0):

$$[\hat{B}] = [\hat{B}_0] + [\hat{E}_0] \tag{3.51}$$

6. The difference matrices $[\hat{E}_n^i]$ of the multi-view images in the decomposition level p=1 are calculated for i=1,2,3,4, through corresponding inverse orthogonal transform:

$$[\hat{E}_n^i] = [T_1]^t [\hat{S}_n^i][T_1]^t \text{ , for } n = \pm 1, \pm 2, .., \pm N. \tag{3.52}$$

7. The matrices $[\hat{B}_n]$ of the restored multi-view images are calculated:

$$[\hat{B}_n] = [\hat{E}_0] + [\hat{B}_0] \text{ for } n = \pm 1, \pm 2, .., \pm N, \tag{3.53}$$

In similar way are decoded the matrices $[\hat{B}_n]$ of all blocks of size $2^m \times 2^m$, which build the multi-view images for n = 0,±1,±2,..,±N. The difference between the basic IPD, and the modification, used for the multi-view processing in the decoding, is represented by Eqs. 3.48 and 3.53. Here, the restoration of the reference view image is performed in the way it is done in the basic IPD, i.e. the two approximations are used directly for the creation of the reference view image, and the remaining views in the same sequence are restored using the coarse approximation for the reference image and the fine approximation, belonging to the corresponding view.

The block diagram of the coder for multi-view object representation based on the 2-level Modified IPD is shown on Fig. 3.20.a, and the block diagram of the decoder - on Fig. 3.20.b. The abbreviations used are: 2D OT – two-dimensional orthogonal transform; 2D IOT - two-dimensional inverse orthogonal transform.

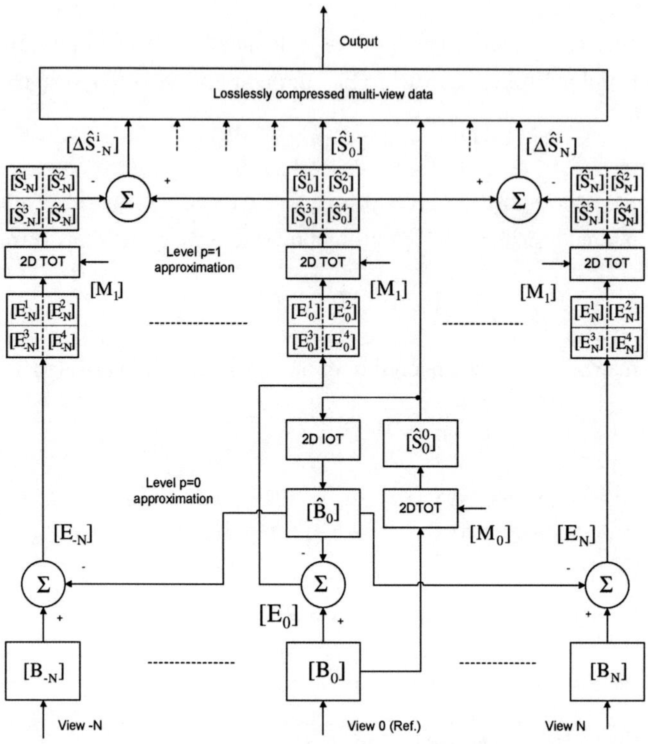

Fig. 3.20.a Block diagram of the coder for multi-view object representation based on the 2-level Modified IPD

The two block diagrams correspond to the methods for coding and decoding of grayscale multi-view images with modified IPD, described above. The two block diagrams represent the processing of one sub-block of the processed image. The coding of color multi-view images is performed in similar way, but it requires the color components to be processed individually. Depending on the color format (RGB, YUV, YC_RC_B, KLT, etc.), and the color sampling format (4:4:4, 4:2:0, 4:1:1, etc.) individual pyramid is built for each component. The approach based on the processing of the reference image and the remaining ones in the group arranged in an arc, is retained. The processing of multi-view images obtained from cameras, arranged in a part of a sphere, is performed in similar way.

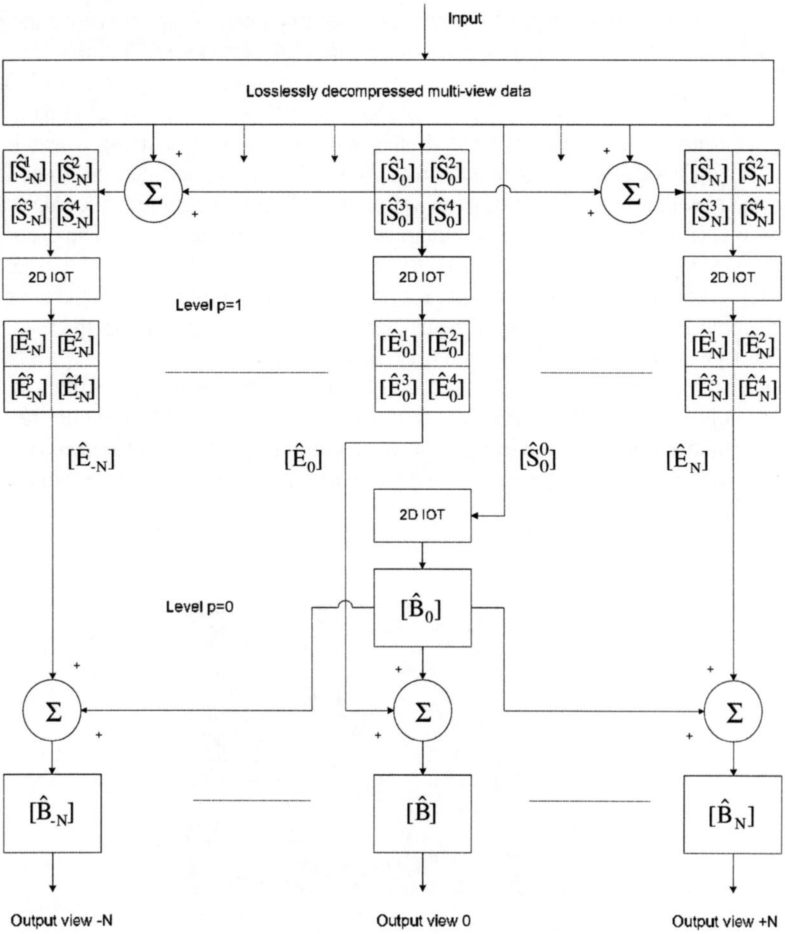

Fig. 3.20.b Block diagram of the decoder for multi-view object representation based on the 2- level Modified IPD

3.3.2 Experimental Results

For higher efficiency the approach presented here is based on the use of a fixed set of transform coefficients (these of lowest spatial frequency). For the experiments was used truncated, 2-level decomposition. In the low decomposition level, a set of 6 coefficients was used, and in the last (higher) was used one coefficient only. In result, more efficient description is achieved. For the experiments was used Modified IDP with the Walsh-Hadamard orthogonal transform (WHT). The views were obtained by moving the photo camera in a line (arc), with an angle of 4^0 between every two adjoining view positions. The total number of views was 11. The reference image was chosen to be the one in the middle of the sequence. Two more view lines

(11 views each) were arranged by moving the photo camera 4^0 up and down in correspondence to the first. The processed images were of size 864×576 pixels, 24 bpp each. The reference image from one of the test groups is shown on Fig. 3.21. The same experiments were performed using DCT instead of the Walsh-Hadamard orthogonal transform. The results were similar: the restored images quality was a little higher (with about 0.2 dB) but the compression (i.e. the representation efficiency) was lower (with about 0,5). Taking into account the lower computational complexity of the WHT, in this paper are given the results obtained for the WHT. The example objects are convex and this permits relatively small number of views to be used for their representation. The experimental results for the first line of test images are given in Table 3.6 below. All experiments were performed transforming the original RGB images into YC_RC_B with sampling format 4:2:0.

On Fig. 3.22 are shown the first (a) and the last (b) image in one of the test sequences (TS). The angle between the first and the last view is 20^0. Despite the apparent similarity between the images, corresponding to the two views placed at the ends of the processed sequence of image views, the difference between them is large (Fig. 3.23).

Fig. 3.21 The reference view for TS 1

Fig. 3.22.a First image in TS 2

Fig. 3.22.b Last image in TS 2

Fig. 3.23 Visualized difference between the first and the last image in TS 2

Similar example is given on Figs. 3.24 and 3.25, which represent view images and the difference between the first and the last one for Test sequence 3 (TS3). For the experiments, the basic sub-image in the low decomposition level was 8×8 pixels and the number of the low-frequency transform coefficients was set to be 4 (the retained coefficients correspond to the low-frequency 2D Walsh-Hadamard functions). The size of the coarse approximation file (level 1) for the reference view was 15418 B and the corresponding PSNR was 37.83 dB. The mean PSNR for the whole group of 11 views for 2-level Modified IDP was 36.32 dB.

The compression ratio was calculated in accordance with the relation:

$$C_r = \frac{(2N+1)4^m b_0}{[L_0 + 4(2N+1)L_1]b_s} \tag{3.54}$$

where b_0 и b_s represent the number of bits for one pixel and one transform coefficient correspondingly; L_0 and L_1 – the number of the retained coefficients for the Modified IDP levels p = 0 and p = 1. The so defined compression ratio does not represent the influence of the lossless coding of the coefficients' values performed for Modified IDP levels p = 0 and p = 1. In the column "L2 file size" is given the size of the corresponding approximations for the higher decomposition level in Bytes. The compression ratio (C_r) was calculated for the whole group of images, i.e. the total data needed for the representation of all 11 views was compared with the uncompressed data for the same images. In the column named "C_r level 2" is given the compression ratio obtained for the corresponding representations of the decomposition level 2 only.

Similar investigation was performed for another 11 views of the same objects, placed in a line positioned at 4^0 higher than the first one. The angles between the adjacent views were 4^0. In this case the reference view was chosen to be at the end of the sequence (next to View No. 10). The results are given in Table 3.7.

The results obtained are close to those given in Table 3.6, but the C_r and the PSNR are a little lower, because the reference view for the second line was set to be this at the end of the sequence and as a result, the correlation between the consecutive views is lower.

Table 3.6 Results for the first line of consecutive views

View No.	C_r Level 2	L2 file size [B]	PSNR L2 [dB]	C_r (group)
Ref.	181.45	6 171	36.83	69.16
1	89.16	12 560	35.55	87.44
2	99.29	11 277	36.25	98.60
3	110.25	10 157	36.54	107.44
4	118.00	9 490	36.58	114.72
5	133.89	8 363	36.53	121.98
6	129.53	8 645	36.51	127.33
7	117.51	9 529	36.45	130.38
8	107.43	10 423	36.24	131.69
9	100.53	11 138	36.37	131.93
10	92.47	12 110	35.81	131.09
		Mean PSNR = 36,32 dB		

Table 3.7 Results obtained for the second line of consecutive views (4^0 up) for Test sequence 1.

View No.	C_r level 2	L2 file size [B]	PSNR L2 [dB]	C_r (group)
Ref.	90.62	16746	35.65	129.98
1	70.80	21088	34.47	126.32
2	75.00	19906	35.56	123.99
3	80.75	18490	34.53	122.74
4	82.23	18157	34.52	121.83
5	86.17	17326	34.53	121.40
6	90.26	16541	34.61	121.34
7	89.36	16708	34.70	121.23
8	88.04	16959	34.81	121.03
9	86.99	17162	35.11	120.78
10	85.16	17532	35.26	120.43
		Mean PSNR = 34.89 dB		

Fig. 3.24.a. First image in TS 3

Fig. 3.24.b. Last image in TS 3

Fig. 3.25. Difference between the first and the last image in TS 3

Additional test was performed for a line of consecutive views positioned at 4^0 down in respect of the first one. The global results are as follows: the PSNR for the whole group (3 lines of views) was 34.8 dB and the compression ratio C_r = 120.1. This means that for the group of 33 color images (one reference image and 32 views arranged in 3 adjoining lines) each of size 864×576 pixels, was achieved a compression ratio $C_r > 120$. The quality of the views was visually lossless, because the errors in images which have a PSNR higher than 32 dB are imperceptible (Fig. 3.26). The tests performed simulated a matrix of 33 views arranged in a rectangle of size 11×3. Best results were obtained for the case, when the reference view was placed in the center of the viewing matrix.

Fig. 3.26. Restored reference image after Mod. IPD compression 100:1.

Fig. 3.27. Restored image after JPEG compression 100:1.

The main advantage of the new approach is that it ensures high compression and very good quality of the restored visual information. In spite of the global approach when multi-view data storage is concerned, each view could be restored individually and used. Compared with the famous JPEG standard for still image compression, the method offers much higher quality of the restored images. For example, the mean PSNR of an image after compression 100:1 is 24.6 dB, but the visual quality of the restored image is very bad (Fig. 3.27). The image from

Fig. 3. 26.a, compressed with JPEG2000-based software gave for same compression a result image with PSNR = 34.4 dB (a little lower than that, obtained for the reference image with the new method), but the computational complexity of JPEG2000 is much higher and the background of the image was visually woollier. For a group of images, comprising all multi-views (the test sequences, used for the investigation), comparison was not done, because JPEG2000 does not offer similar option and the results should be just a sum from all views, i.e. there is no cumulative effect. Additional disadvantage is that JPEG2000 does not offer the ability for retained coefficients reduction, which is possible when the Modified IPD is used, because of the specific relations between the coefficients' values in neighboring decomposition levels.

The described method ensures very efficient description of the multi-view images by using one of them as a reference one. The decomposition has a relatively low computational complexity because it is based on orthogonal transforms (Walsh-Hadamard, DCT etc.). For example, the computational complexity of decompositions, based on wavelet transforms is much higher. The comparison of the computational complexity of the *Modified IPD* and the wavelets-based transforms is given in earlier publications (Kountchev et al. 2005). In the examples was used the WH transform, but DCT or some other transforms are suitable as well. The relations existing between transform coefficients from the consecutive decomposition levels permit significant reduction of the coefficients needed for the high-quality object representation (Kountchev and Kountcheva 2008). The number of necessary views depends on the application. For example, the view area could be restricted to some angle or scale, etc. The experimental results proved the ability to create efficient multi-view object representation based on the Modified IPD. The task is easier when the image of a single object has to be represented. In the examples, presented here, two convex objects were represented and they should be searched together. The significant compression of the data representing the multiple views ensures efficient data storage and together with this - fast access and search in large image databases. The Modified IPD representation is suitable for tasks requiring the analysis of complicated scenes (several objects searched together or context-based search). This is possible, because the lowest level of the pyramidal decomposition consist of sub-images, processed individually. In result, more than one object (described individually) could be searched together. Additional advantage is the similarity of the transform coefficients from any two adjacent decomposition levels, which is a basis for the creation of flexible algorithms for the transformation of the already created object representation into higher or lower scale without using additional views.

3.4 Multispectral Images Representation with Modified IPD

The contemporary research in different application areas sets the task of the efficient archiving of multispectral images. For this, in most cases is necessary to process several images of the same object(s). Multispectral images are characterized by very high spatial, spectral, and radiometric resolution and, hence, by ever-increasing demands of communication and storage resources. Such demands often exceed the system capacity like, for example, in the downlink from satellite to Earth stations, where the channel bandwidth is often much inferior to the intrinsic data rate of the images, some of which must be discarded altogether. In this situation the high-fidelity image compression is a very appealing alternative. As a matter of fact, there has been intense research activity on this topic [105-109], focusing, particularly, on transform-coding techniques, due to their good performance and limited computational complexity. Linear transform coding, however, does not take into account the nonlinear dependences existing among different bands, due to the fact that multiple land covers, each with its own interband statistics, are present in a single image. Based on this observation, a class-based coder was proposed in (Fowler and Fox) that address the problem of interband dependences by segmenting the image into several classes, corresponding as much as possible to the different land covers of the scene. As a consequence, within each class, pixels share the same statistics and exhibit only linear interband dependences, which can be efficiently exploited by conventional transform coding. Satellite-borne sensors have ever higher spatial, spectral and radiometric resolution. With this huge amount of information comes the problem of dealing with large volumes of data. The most critical phase is on-board the satellite, where acquired data easily exceed the capacity of the downlink transmission channel, and often large parts of images must be simply discarded, but similar issues arise in the ground segment, where image archival and dissemination are seriously undermined by the sheer amount of data to be managed. The reasonable approach is to resort to data compression, which allows reducing the data volume by one and even two orders of magnitude without serious effects on the image quality and on their diagnostic value for subsequent automatic processing. To this end, however, is not possible to use the general purpose techniques as they do not exploit the peculiar features of multispectral remote-sensing images, which is why several ad hoc coding schemes have been proposed in recent years. The transform coding is one of the most popular approaches for several reasons. First, transform coding techniques are well established and deeply understood; they provide excellent performances in the compression of images, video and other sources, have a reasonable complexity and besides, are at the core of the famous standards JPEG and JPEG2000, implemented in widely used and easily available coders. The common approach for coding multispectral images in accordance with Markas and Reif is to use some decorrelating transforms along the spectral dimension followed by JPEG2000 on the transform bands with a suitable rate allocation among the bands. Less attention has been devoted to techniques based on vector quantization (VQ) because, despite its theoretical optimality, VQ is too computationally demanding to be of any practical use. Nonetheless, when dealing with multiband images, VQ is a natural candidate, because the elementary semantic unit in such images is the spectral response vector (or spectrum, for short) which collects the image intensities for a given location at all spectral bands. The values of a spectrum at different bands are not simply correlated but strongly dependent, because they are completely determined (but for the noise) by the land covers of the imaged cell. This observation has motivated the search for constrained VQ techniques (Tang et al., Dragotti et al.), which are suboptimal but simpler than full-search VQ, and show promising performances. Multispectral images require large amounts of storage space, and therefore a lot of attention has recently been focused to compress these images. Multispectral images include both spatial and spectral redundancies. Usually we can use vector quantization, prediction and transform coding to reduce redundancies. For example, hybrids transform/vector quantization (VQ) coding scheme (Gersho and Gray, and Aiazzi et al. 2006). Instead, for the spectral redundancies reduction is used Karhunen-Loeve transform (KLT), followed by two-dimensional (2D) discrete cosine transform (DCT) which reduces the spatial redundancies (Dragotti et al., Tang et al.). A quad-tree technique for determining the transform block size and the quantizer for encoding the transform coefficients was applied across KLT-DCT method (Kaarna). In the works of Cagnazzo et al. and Wu was used a wavelet transform (WT) to reduce the spatial redundancies and KLT to reduce the spectral redundancies. After that the data is encoded using the 3-dimensional (3D) SPIHT algorithm. The state-of-the-art analysis shows that despite of the vast investigations and various techniques used for the efficient compression of multispectral images, a recognized general method able to solve the main problems is still not created.

In the work of Kountchev and Nakamatsu is presented one method for representation of multispectral images, based on 2-level Modified IPD The decomposition is similar with that, already presented in Section 3 above, used for the multi-view object representation. The main difference is that the reference image is selected on the basis of the similarity in the processed group of images. For this, should be measured the mutual similarity of each couple of images in the group. The image, which has maximum correlation with the remaining ones, should be the reference. The algorithm used for the selection of the reference image is given below.

3.4.1 Selection of the Reference Image in Multispectral Sequence

The image which will be chosen to be used as a reference one is selected on the basis of the histogram analysis: the image, whose histogram is closest to all the remaining images in the processed set, is selected to be the reference one. The analysis is made using the correlation coefficient. The correlation coefficient ρ_{xy} between vectors $\vec{X} = [x_1, x_2, ..., x_m]^t$ and $\vec{Y} = [y_1, y_2, ..., y_m]^t$, which represent the histograms of the two images in accordance with the work of Bronshtein et al. is:

$$\rho_{x,y} = \sum_{i=1}^{m} (x_i - \overline{x})(y_i - \overline{y}) \Bigg/ \sqrt{\sum_{i=1}^{m} (x_i - \overline{x})^2} \sqrt{\sum_{i=1}^{m} (y_i - \overline{y})^2} \tag{3.55}$$

where $\overline{x} = \frac{1}{m}\sum_{i=1}^{m} x_i$ and $\overline{y} = \frac{1}{m}\sum_{i=1}^{m} y_i$ are the mean values of the two histograms and m is the number of brightness levels for the both spectral images. The decision for the reference image selection is taken after the histograms of the multispectral images had been calculated and then the correlation coefficients for all couples of histograms are calculated and evaluated. For a multispectral image of N components the number of these couples l(p,q) is:

$$L=\sum_{p=1}^{N-1}\sum_{q=p+1}^{N} l(p,q) \tag{3.56}$$

When all L coefficients ρ_{pq} are calculated, is defined the index p_0 in ρ_{p_0q}, for which is satisfied the requirement:

$$\sum_{q=1}^{N}\rho_{p_0q} \geq \sum_{q=1}^{N}\rho_{pq} \text{ for } p,q=1,2,..,N, \tag{3.57}$$

when $p \neq q$ and $p \neq p_0$. The reference image then is $[B_R]=[B_{p_0}]$.

The block diagrams of the coder/decoder for processing of multispectral images based on the 2-level Modified IPD correspond to these from Fig. 3.20, used for the processing of multi-view images, taking into account that the reference image should be selected using Eq. 3.57. The compression ratio Cr for a set of multispectral images is calculated in accordance with Eq. 3.54.

3.4.2 Experimental Results

For the experiments was used the software implementation of the modified IPD method in Visual C++. In accordance with this method was developed a special format for multispectral images compression. For the experiments were used more than 100 sets of multispectral images of size 1000 x 1000 pixels, 24 bpp (each set comprises 3 images, corresponding to the main colors – blue, red and green). One set of 3 test images is shown on Fig. 3.28. On Fig. 3.29 are shown the histograms of the test images from Fig. 3.28. As a reference image for the experiments was used the Test image 1. For the experiments was used 2-levels modified IPD. The size of the sub-image in the lower level was 8x8, and in the next level – 4x4 pixels. The number of coefficients for the lower level was 6, and for the next level - 1.

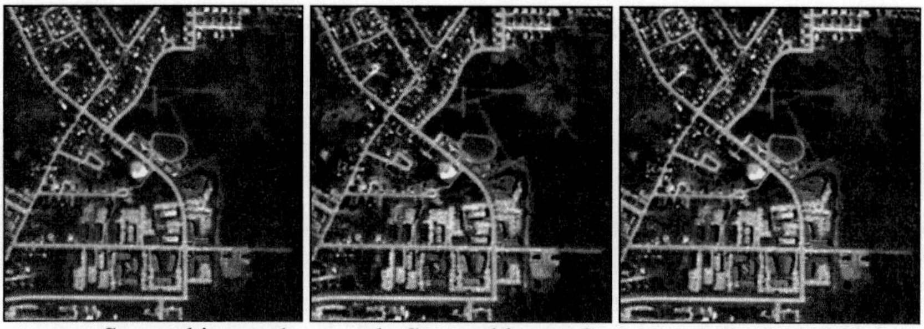

a. Spectral image 1 b. Spectral image 2 c. Spectral image 3

Fig. 3.28 A set of 4 images of size 1000×1000 pixels, 24 bpp each

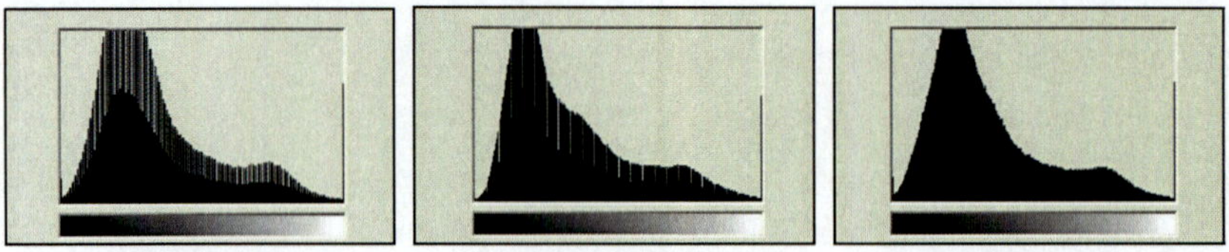

a. Image 1 b. Image 2 c. Image 3

Fig. 3.29 Histograms of the set of test multispectral images from Fig. 3.28

The experimental results for the test sets of spectral images shown on Fig. 3.28, are given in Table 3.8. The size of the compressed first approximation (Lvl1) for the Reference Test image 1 is 45 KB. The size of the next-level approximations (Lvl2) depends on the similarity of the corresponding spectral image and the Reference one. The results for all sets of test images are close to these, given below.

Table 3.8 Results obtained for a set of test spectral images with Modified IPD

Image	Lvl1 [KB]	Lvl2 [KB]	C_r	PSNR [dB]
1	45	26	115	23,5
2		43	69	26,3
3		38	78	25,8

The comparison for the evaluation of the method efficiency was performed with the widely used JPEG and JPEG 2000 standards. The results obtained for the same set of spectral images with JPEG compression are given in Table 3.9, and these with JPEG 2000 – in Table 3.10. The quality of the restored images was selected to be close to that of the images obtained with the IPD compression (exact correspondence is not possible, but the values are close).

Table 3.9 Results obtained for same set of spectral images with JPEG

Image	C_r	PSNR [dB]
1	42	26,4
2	69	27,0
3	78	24,9

Table 3.10 Results obtained for same set of spectral images with JPEG2000

Image	C_r	PSNR [dB]
1	42	28,2
2	69	29,4
3	78	27,1

Despite the higher PSNR calculated for the JPEG2000 images, their visual quality is not as good as that obtained with the IPD coding. For comparison, on Fig. 3.30 are shown enlarged parts of the corresponding restored test images.

The results thus obtained confirmed the expected high efficiency of the new method: for better quality of the restored image the compression ratio was much higher also. The software implementation of the new method for compression of multispectral images based on the modified IPD proved the method efficiency. The

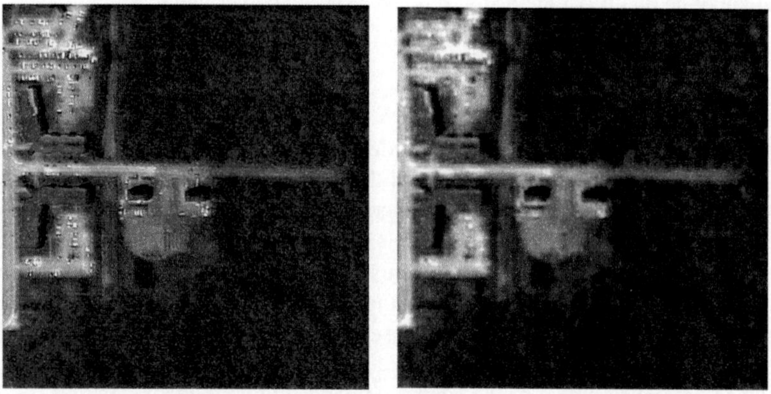

a b

Fig. 3.30 Restored and enlarged parts of Image 3, obtained for the compression ratios in Tables 3.9 and 3.10: (a) original; (b) IPD

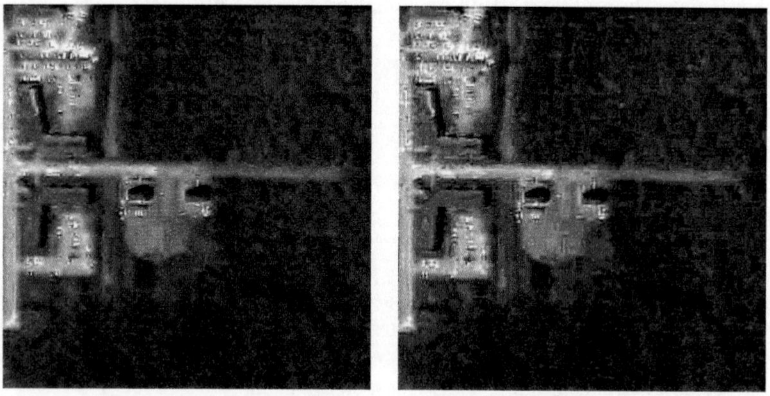

c d

Fig. 3.31 Restored and enlarged parts of Image 3, obtained for the compression ratios in Tables 3.9 and 3.10: (c) JPEG2000; (d) JPEG decomposition flexibility permits the creation of a branched structure, which suits very well the characteristics of multispectral images and makes the most of their similarity. The main advantages of the new method are the high efficiency and the relatively low computational complexity. The high efficiency of the IPD method was proved by the experimental results: for same compression it offers better visual quality than that of the JPEG2000 standard (Acharya and Tsai) and has lower computational complexity. The method is suitable for a wide variety of applications: processing of video sequences, efficient archiving of medical information and satellite images and many others, i.e. in all cases when objects are moving relatively slowly and the quality of the restored images should be very high.

3.5 Conclusions

In this chapter was introduced the main idea of the Inverse pyramid image decomposition. The research and the analysis of the methods for pyramid building and its basic modifications proved its high efficiency and flexibility, which permit it to be successfully used in wide variety of application areas, such as: systems for archiving of visual information, layered transmission and access and processing of multi-view and multispectral images. Other important applications are the systems for search by content in large image databases (Milanova et al.) and the creation of RST-invariant descriptions of the searched objects (Kountchev et al. 2010).

The future development of the adaptive inverse pyramid decomposition (AIPD) will be focused on the abilities for fast tuning and management of the pyramid parameters in accordance with the statistics of the processed images of any kind: grayscale, color, multi-view, multispectral, stereo, etc. The comparison of the presented method for image representation with the famous multiresolution techniques (Bovik) shows that it could be considered as an additional tool for efficient processing of visual information.

References

Acharya, T., Tsai, P.: JPEG 2000 Standard for Image Compression. John Wiley and Sons (2005)

Ahmed, N., Rao, K.: Orthogonal transforms for digital signal processing. Springer, New York (1975)

Aiazzi, B., Alparone, L., Baronti, S.: A reduced Laplacian pyramid for lossless and progressive image communication. IEEE Trans. on Communication 44(1), 18–22 (1996)

Aiazzi, B., Alparone, L., Baronti, S.: A reduced Laplacian pyramid for lossless and progressive image communication. IEEE Trans. on Commun. 44(1), 18–22 (1996)

Aiazzi, B., Alparone, L., Baronti, B., Lotti, F.: Lossless image compression by quantization feedback in Content-Driven enhanced Laplacian pyramid. IEEE Trans. Image Processing 6, 831–844 (1997)

Aiazzi, B., Baronti, S., Lastri, C.: Remote sensing image coding. In: Barni, M. (ed.) Document and Image Compression, ch. 15, pp. 389–412. CRC Taylor&Francis (2006)

Antonini, M., Barlaud, M., Mathieu, P., Daubechies, I.: Image coding using wavelet transform. IEEE Trans. Image Processing 1, 205–220 (1992)

Boliek, M., Gormish, M., Schwartz, E., Keith, A.: A next generation image compression and manipulation using CREW. In: Proc. IEEE ICIP (1997)

Bovik, A.: Multiscale image decomposition and wavelets. In: The Essential Guide to Image Processing, pp. 123–142. Academic Press, NY (2009)

Brigger, P., Muller, F., Illgner, K., Unser, M.: Centered pyramids. IEEE Trans. on Image Processing 8(9), 1254–1264 (1999)

Bronshtein, I., Semendyayev, K., Musiol, G., Muehlig, H.: Handbook of mathematics, 5th edn. Springer, Heidelberg (2007)

Buccigrossi, R., Simoncelli, E.: Image compression via joint statistical characterization in the wavelet domain. GRASP Laboratory Technical Report No 414, pp. 1–23. University of Pennsylvania (1997)

Burt, P., Adelson, E.: The Laplacian pyramid as a compact image code. IEEE Trans. on Comm., COM 31(4), 532–540 (1983)

Cagnazzo, M., Parrilli, S., Poggi, G., Verdoliva, L.: Improved class-based coding of multispectral images with shape-adaptive wavelet transform. IEEE Geoscience and Remote Sensing Letters 4(4), 565–570 (2007)

Chen, C.: Laplacian pyramid image data compression. In: IEEE IC on ASSP, vol. 2, pp. 737–739 (1987)

Chen, T., Wu, H.: Artifact reduction by post-processing in image compression. In: Wu, H., Rao, K. (eds.) Digital Video Image Quality and Perceptual Coding, ch. 15. CRC Press, Taylor and Francis Group, LLC, Boca Raton (2006)

Cherkashyn, V., He, D., Kountchev, R.: A novel adaptive representation method AIPR/BPNN of satellite visible very high definition images. Journal of Communication and Computer 7(9), 55–66 (2010)

Daubechies, I.: Ten lectures on wavelets. SIAM, Philadelphia (1992)

Deforges, O., Babel, M., Bedat, L., Ronsin, J.: Color LAR codec: a color image representation and compression scheme based on local resolution adjustment and self-extracting region representation. IEEE Trans. on Circuits and Systems for Video Technology 17(8), 974–987 (2007)

Demaistre, N., Labit, C.: Progressive image transmission using wavelet packets. In: Proc. ICIP 1996, pp. 953–956 (1996)

DeVore, R., Jarwerth, B., Lucier, B.: Image compression through wavelet transform coding. IEEE Trans. Information Theory 38, 719–746 (1992)

Do, M., Vetterli, M.: Contourlets. In: Welland, G. (ed.) Beyond wavelets. Academic Press, NY (2003)

Dony, R., Haykin, S.: Neural network approaches to image compression. Proc. of the IEEE 23(2), 289–303 (1995)

Dragotti, P., Poggi, G., Ragozini, A.: Compression of multispectral images by three-dimensional SPIHT algorithm. IEEE Trans. Geosci. Remote Sens. 38(1), 416–428 (2000)

Efstratiadis, S., Tzovaras, D., Strintzis, M.: Hierarchical image compression using partition priority and multiple distribution entropy coding. IEEE Trans. Image Processing 5, 1111–1124 (1996)

Egger, O., Fleury, P., Ebrahimi, T.: High-performance compression of visual information-A tutorial review-Part I: Still Pictures. Processing of the IEEE 87(6), 976–1011 (1999)

Fowler, J., Fox, D.: Embedded wavelet-based coding of 3D oceanographic images with land masses. IEEE Trans. Geosci. Remote Sens. 39(2), 284–290 (2001)

Froment, J., Mallat, S.: Second generation image coding with wavelets. In: Chui, C. (ed.) Wavelets: A Tutorial in Theory and Applications, vol. 2. Acad. Press, NY (1992)

Gelli, G., Poggi, G.: Compression of multispectral images by spectral classification and transform coding. IEEE Trans. Image Processing 8(4), 476–489 (1999)

Gersho, A., Gray, R.: Vector quantization and signal compression. Kluwer AP (1992)

Gonzalez, R., Woods, R.: Digital image processing. Prentice-Hall (2001)

Gibson, J., Berger, T., Lookabaugh, T., Lindberg, D., Baker, R.: Digital compression for multimedia. Morgan Kaufmann (1998)

Hu, Y., Hwang, J.: Handbook of neural network signal processing. CRC Press, LLC (2002)

ISO/IEC JTC1/SC29/Wg11 m12542: Multi-view video coding based on lattice-like pyramid GOP structure (2005)

Jiang, J.: Image compressing with neural networks - A survey. In: Signal Processing: Image Communication, vol. 14(9), pp. 737–760. Elsevier (1999)

Joshi, R., Ficher, T., Bamberger, R.: Comparison of different methods of classification in subband coding of images. In: Proc. SPIE Still Image Compression, vol. 2418, pp. 154–163 (1995)

Jung, H., Choi, T., Prost, R.: Rounding transform for lossless image coding. In: Proc. IC for Image Processing 1996, pp. 65–68 (1996)

Kaarna, A.: Integer PCA and wavelet transform for lossless compression of multispectral images. In: Proc. of IGARSS 2001, pp. 1853–1855 (2001)

Kalra, K.: Image Compression Graphical User Interface, Karmaa Lab, Indian Institute of Technology, Kanpur, `http://www.iitk.ac.in/karmaa`

Kim, W., Balsara, P., Harper, D., Park, J.: Hierarchy embedded differential image for progressive transmission using lossless compression. IEEE Trans. on Circuits and Systems for Video Techn. 5(1), 2–13 (1995)

Kim, H., Li, C.: Lossless and lossy image compression using biorthogonal wavelet transforms with multiplierless operations. IEEE Trans. on CAS-II. Analog and Digital Signal Processing 45(8), 1113–1118 (1998)

Kim, S., Lee, S., Ho, Y.: Three-dimensional natural video system based on layered representation of depth maps. IEEE Trans. on Consumer Electronics 52(3), 1035–1042 (2006)

Knowlton, K.: Progressive transmission of gray scale and binary pictures by simple, efficient and lossless encoding scheme. Proc. IEEE 68, 885–896 (1980)

Kong, X., Goutsias, J.: A study of pyramidal techniques for image representation and compression. Journal of Visual Communication and Image Representation 5(2), 190–203 (1994)

Kouda, N., et al.: Image compression by layered quantum neural networks. Neural Processing Lett. 16, 67–80 (2002)

Kountchev, R., Haese-Coat, V., Ronsin, J.: Inverse pyramidal decomposition with multiple DCT. In: Signal Processing: Image Communication, vol. 17(2), pp. 201–218. Elsevier (2002)

Kountchev, R., Milanova, M., Ford, C., Kountcheva, R.: Multi-layer image transmission with inverse pyramidal decomposition. In: Halgamuge, S., Wang, L. (eds.) Computational Intelligence for Modeling and Predictions, vol. 2(13). Springer, Heidelberg (2005)

Kountchev, R., Kountcheva, R.: Image representation with reduced spectrum pyramid. In: Tsihrintzis, G., Virvou, M., Howlett, R., Jain, L. (eds.) New Directions in Intelligent Interactive Multimedia, pp. 275–284. Springer, Heidelberg (2008)

Kountchev, R., Kountcheva, R.: Comparison of the structures of the inverse difference and Laplacian pyramids for image decomposition. In: XLV Intern. Scientific Conf. on Information, Communication and Energy Systems and Technologies, pp. 33–36. SPI, Macedonia (2010)

Kountchev, R., Nakamatsu, K.: Compression of multispectral images with inverse pyramid decomposition. In: Setchi, R., Jordanov, I., Howlett, R.J., Jain, L.C. (eds.) KES 2010. LNCS, vol. 6278, pp. 215–224. Springer, Heidelberg (2010)

Kountchev, R., Rubin, S., Milanova, M., Todorov, V.l., Kountcheva, R.: Non-linear Image representation based on IDP with NN. WSEAS Trans. on Signal Processing 9(5), 315–325 (2009)

Kountchev, R., Todorov, V.l., Kountcheva, R.: Multi-view Object Representation with inverse difference pyramid decomposition. WSEAS Trans. on Signal Processing 9(5), 315–325 (2009)

Kountchev, R., Todorov, V.l., Kountcheva, R.: RSCT-invariant object representation with modified Mellin-Fourier transform. WSEAS Trans. on Signal Processing 4(6), 196–207 (2010)

Kropatsch, W., Bischof, H. (eds.): Digital image analysis: selected techniques and applications. Springer, Heidelberg (2001)

Kulkarni, S., Verma, B., Blumenstein, M.: Image compression using a direct solution method based on neural network. In: The 10th Australian Joint Conference on Artificial Intelligence, Perth, Australia, pp. 114–119 (1997)

Kunt, M., Ikonomopoulos, A., Kocher, M.: Second-generation image-coding technique. Proc. of IEEE 73(4), 549–574 (1985)

Lu, C., Chen, A., Wen, K.: Polynomial approximation coding for progressive image transmission. Journal of Visual Communication and Image Representation 8, 317–324 (1997)

Malo, J., Epifanio, I., Navarro, R., Simoncelli, E.: Nonlinear image representation for efficient perceptual coding. IEEE Trans. on Image Processing 15(1), 68–80 (2006)

Majani, E.: Biorthogonal wavelets for image compression. In: Proc. SPIE Visual Commun. Image Process. Conf., Chicago, IL, pp. 478–488 (1994)

Mallat, S.: A theory for multiresolution signal decomposition: the Wavelet representation. IEEE Trans. on Pattern Analysis and Machine Intelligence PAMI-II, 7, 674–693 (1989)

Mallat, S.: Multifrequency channel decompositions of images and wavelet models. IEEE Trans. ASSP 37, 2091–2110 (1990)

Mancas, M., Gosselin, B., Macq, B.: Perceptual image representation. EURASIP Journal on Image and Video Processing, 1–9 (2007)

Markas, T., Reif, J.: Multispectral image compression algorithms. In: Storer, J., Cohn, M. (eds.), pp. 391–400. IEEE Computer Society Press (1993)

Meer, P.: Stochastic image pyramids. In: Computer Vision, Graphics and Image Processing, vol. 45, pp. 269–294 (1989)

Milanova, M., Kountchev, R., Rubin, S., Todorov, V., Kountcheva, R.: Content Based Image Retrieval Using Adaptive Inverse Pyramid Representation. In: Salvendy, G., Smith, M.J. (eds.) HCI International 2009. LNCS, vol. 5618, pp. 304–314. Springer, Heidelberg (2009)

Mokhtarian, F., Abbasi, S.: Automatic selection of optimal views in multi-view object recognition. In: British Machine Vision Conf., pp. 272–281 (2000)

Mongatti, G., Alparone, L., Benelli, G., Baronti, S., Lotti, F., Casini, A.: Progressive image transmission by content driven Laplacian pyramid encoding. IEE Processings-1 139(5), 495–500 (1992)

Muller, F., Illgner, K., Praefcke, W.: Embedded Laplacian pyramid still image coding using zerotrees. In: Proc. SPIE 2669, Still Image Processing II, San Jose, pp. 158–168 (1996)

Namphol, A., et al.: Image compression with a hierarchical neural network. IEEE Transactions on Aerospace and Electronic Systems 32(1), 327–337 (1996)

Nguyen, T., Oraintara, S.: A shift-invariant multiscale multidirection image decomposition. In: Proc. IEEE International Conf. on Acoustics, Speech, and Signal Processing, France, pp. 153–156 (2006)

Nuri, V.: Space-frequency adaptive subband image coding. IEEE Trans. on CAS -II: Analog and Digital Signal Processing 45(8), 1168–1173 (1998)

Olkkonen, H., Pesola, P.: Gaussian pyramid wavelet transform for multiresolution analysis of images. Graphical Models and Image Processing 58(4), 394–398 (1996)

Perry, S., Wong, H., Guan, L.: Adaptive image processing: a computational intelligence perspective. CRC Press, LLC (2002)

Pratt, W.: Digital image processing. Wiley Interscience, New York (2007)

Rabbani, M., Jones, P.: Digital image compression techniques. Books, SPIE Tutorial Texts Series, vol. TT7. SPIE Opt. Eng. Press (1991)

Rioul, O., Vetterli, M.: Wavelets and signal processing. IEEE Signal Processing Magazin 6, 14–38 (1991)

Rosenfeld, A.: Multiresolution image processing and analysis. Springer, NY (1984)

Shapiro, J.: Embedded image coding using zerotrees of wavelet coefficients. IEEE Trans. on SP 41(12), 3445–3462 (1993)

Sigitani, T., Iiguni, Y., Maeda, H.: Image interpolation for progressive transmission by using radial basis function networks. IEEE Trans. on Neural Networks 10(2), 381–390 (1999)

Simoncelli, E., Freeman, W.: The steerable pyramid: A flexible architecture for multi-scale derivative computation 3, 444–447 (1995)

Smith, M., Barnwell, T.: Exact reconstruction techniques for tree structured subband coders. IEEE Trans. on ASSP, ASSP-34, 434–441 (1986)

Strintzis, M., Tzovaras, D.: Optimal pyramidal decomposition for progressive multiresolutional signal coding using optimal quantizers. IEEE Trans. on Signal Processing 46(4), 1054–1068 (1998)

Special Issue on Image Compression, International Journal on Graphics, Vision and Image Processing (2007), http://www.icgst.com

Tan, K., Ghambari, M.: Layered image coding using the DCT pyramid. IEEE Trans. on Image Processing 4(4), 512–516 (1995)

Tang, X., Pearlman, W., Modestino, J.: Hyperspectral image compression using three dimensional wavelet coding. In: Proc. SPIE, vol. 5022, pp. 1037–1047 (2003)

Taubman, D.: High performance scalable image compression with EBCOT. IEEE Trans. Image Processing 9, 1158–1170 (2000)

Todd, J.: The visual perception of 3D shape. Trends in Cognitive Science 8(3), 115–121 (2004)

Toet, A.: A morphological pyramidal image decomposition. Pattern Recognition Lett. 9, 255–261 (1989)

Tzou, K.: Progressive image transmission: A review and comparison of techniques. Optical Eng. 26(7), 581–589 (1987)

Topiwala, P.: Wavelet image and video compression. Kluwer Acad. Publ., NY (1998)

Tanimoto, S.: Image transmission with gross information first. In: Computer,Graphics and Image Processing, vol. 9, pp. 72–76 (1979)

Unser, M.: An improved least squares Laplacian pyramid for image compression. Signal Processing 27, 187–203 (1992)

Unser, M.: On the optimality of ideal filters for pyramid and wavelet signal approxi-mation. IEEE Trans. on SP 41 (1993)

Unser, M.: Splines: A perfect fit for signal and image processing. IEEE Signal Processing Magazine 11, 22–38 (1999)

Vaidyanathan, P.: Quadrature mirror filter banks, M-band extensions and perfect reconstruction technique. IEEE Trans. on ASSP 4, 4–20 (1987)

Vaidyanathan, P.: Multirare systems and filter banks. Prentice-Hall, NJ (1993)

Vazquez, P., Feixas, M., Sbert, M., Heidrich, W.: Automatic view selection using viewpoint entropy and its applications to image-based modeling. Computer Graphics Forum 22(4), 689–700 (2003)

Velho, L., Frery, A., Gomes, J.: Image processing for computer graphics and vision, 2nd edn. Springer, Heidelberg (2008)

Vetterli, M.: Multi-dimensional sub-band coding: some theory and applications. Signal Processing 6, 97–112 (1984)

Vetterli, M., Uz, K.: Multiresolution coding techniques for digital television: A Review, Multidimensional systems and signal processing, vol. 3, pp. 161–187. Kluwer Acad. Publ. (1992)

Vetterli, M., Kovačevic, J., LeGall, D.: Perfect reconstruction filter banks for HDTV representation and coding. Image Communication 2, 349–364 (1990)

Wang, L., Goldberg, M.: Progressive image transmission by transform coefficient residual error quantization. IEEE Trans. on Communications 36, 75–87 (1988)

Wang, L., Goldberg, M.: Reduced-difference pyramid: A data structure for progressive image transmission. Opt. Eng. 28, 708–716 (1989)

Wang, L., Goldberg, M.: Comparative performance of pyramid data structures for progressive image transmission. IEEE Trans. Commun. 39(4), 540–548 (1991)

Wang, D., Haese-Coat, V., Bruno, A., Ronsin, J.: Texture classification and segmentation based on iterative morphological decomposition. Journal of Visual Communication and Image Representation 4(3), 197–214 (1993)

Woods, J. (ed.): Subband image coding. Kluwer Acad. Publ., NY (1991)

Wu, J., Wu, C.: Multispectral image compression using 3-dimensional transform zerob-lock coding. Chinese Optic Letters 2(6), 1–4 (2004)

Yu, T.: Novel contrast pyramid coding of images. In: Proc. of the 1995 IEEE International Conference on Image Processing, pp. 592–595 (1995)

Chapter 4
Preserving Data Integrity of Encoded Medical Images: The LAR Compression Framework

Marie Babel*, François Pasteau, Clément Strauss, Maxime Pelcat, Laurent Bédat, Médéric Blestel, and Olivier Déforges

Abstract. Through the development of medical imaging systems and their integration into a complete information system, the need for advanced joint coding and network services becomes predominant. PACS (Picture Archiving and Communication System) aims to acquire, store and compress, retrieve, present and distribute medical images. These systems have to be accessible via the Internet or wireless channels. Thus protection processes against transmission errors have to be added to get a powerful joint source-channel coding tool. Moreover, these sensitive data require confidentiality and privacy for archiving and transmission purposes, leading to use cryptography and data embedding solutions.

This chapter introduces data integrity protection and developed dedicated tools of content protection and secure bitstream transmission for medical encoded image purposes. In particular, the LAR image coding method is defined together with advanced securization services.

4.1 Introduction

Nowadays, easy-used communication systems have emphasized the development of various innovative technologies including digital image handling, such as digital cameras, PDAs or mobile phones. This naturally leads to implement image compression systems used for general purposes like digital storage, broadcasting and display. JPEG, JPEG 2000 and now JPEG XR have become international standards for image compression needs, providing efficient solutions at different complexity levels. Nevertheless, if JPEG 2000 is proved to be the most efficient coding scheme, its intrinsic complexity prevents its implementation on embedded systems that are limited in terms of computational capacity and/or memory. In addition, usages associated with image compression systems are evolving, and tend to require more and more advanced functionalities and services that are not always well addressed by current norms. As a consequence, designing an image compression framework still remains a relevant issue.

---

* European University of Brittany (UEB), France - INSA, IRISA, UMR 6074, F-35708 RENNES, `marie.babel@insa-rennes.fr`

R. Kountchev et al. (Eds.): Advances in Reasoning-Based Image Processing, ISRL 29, pp. 91–125.
springerlink.com

The JPEG committee has started to work on new technologies to define the next generation of image compression systems. This future standard, named JPEG AIC (Advanced Image Coding), aims at defining a complete coding scheme able to provide advanced functionalities such as lossy to lossless compression, scalability, robustness, error resilience, embed-ability, content description for image handling at object level.

However, the JPEG committee has decided to first support solutions adapted to particular applications. A call of proposal has been then issued, within the framework of JPEG AIC, and restricted to medical image coders.

Indeed, the introduction of medical imaging management systems to hospitals (PACS : Picture Communication and Information System) is leading to the design of dedicated information systems to facilitate the access to images and provide additional information to help to exploit and understand them. Implementing PACS requires an ad hoc protocol describing the way images are acquired, transferred, stored and displayed. DICOM (Digital Image Communications Management) provides a standard that specifies the way in which to manage these images [31].

The need for efficient image compression quickly becomes apparent. In particular, dataset size is exploding, because of the evolution of medical image acquisition technology together with changes in medical usage [20,19]. From the compression point of view, the challenge lies in finding coding solutions dedicated to the storage or communication of images and associated information that will be compliant with the memory and computation capacities of the final workstations.

The design of a new medical image compression scheme requires many dedicated services. Medical images usually go with private metadata that have to remain confidential. In particular, to insure reliable transfers, flexible and generic scheduling and identification processes have to be integrated for database distribution purposes to take account of secure remote network access together with future developments in network technologies. Fast browsing tools, including the segmentation process and scalability, are therefore needed.

In this context, we propose the Locally Adaptive Resolution (LAR) codec as a contribution to the relative calls for technologies. The LAR method relies on a dedicated quadtree content-based representation that is exploited for compression purposes. Multiresolution extensions have been developed and have shown their efficiency, from low bit rates up to lossless image compression. In particular, the scalable LAR coder outperforms state-of-the-art solutions as a lossless encoder system for medical images.

An original hierarchical self-extracting region representation has also been elaborated: a segmentation process is automatically run at both coder and decoder using the quadtree knowledge as segmentation cues. This leads to a free segmentation representation well adapted for image handling and encoding at region level. Moreover, the inherent structure of the LAR codec can be used for advanced functionalities such as content securization purposes. In particular, hierarchical selective encryption techniques have been adapted to our coding scheme and data hiding system based on the LAR multiresolution description allows efficient content protection. In this study, we show the specific framework of our coding scheme for data integrity preservation purposes, both in terms of metadata embedding and secure transmission.

This chapter does not aim at providing an exhaustive state-of-the-art study, but tends to present a content-based coding solution as a response to the medical needs in terms of data integrity preservation. Theses needs are progressively introduced and illustrated all along this chapter. To first understand the different ways to protect content in an image, section 2 introduces first cryptography and data embedding processes. Section 3 looks into securization processes of coded image transmission, where the Long Term Evolution (LTE) use case is presented. Then section 4 shows the LAR medical framework together with its dedicated functionalities.

4.2 How to Protect Content in an Image?

Huge amount of medical data are stored on different media and are exchanged over various networks. Often, these visual data contain private, confidential or proprietary informations. As a consequence, techniques especially designed for these data are required so that to provide security functionalities such as privacy, integrity, or authentication. Multimedia security is aimed towards these technologies and applications [15].

Despite of the spectacular increase in the Internet bandwidth and the low cost of high storage capacity, compression rates of image codec are still of interest. In this way, an image codec must provide both compression efficiency and additional services. These services are content protection and data embedding. In one hand, content protection consists in preserving data integrity and masking data content. The commonly used methods to obtain these protections are respectively hashing and ciphering. In the other hand, the embedding of hired data aims to protect copyrights or add metadata to a document.

Besides watermarking, steganography, and techniques for assessing data integrity and authenticity, providing confidentiality and privacy for visual data is among the most important topics in the area of multimedia security. Applications range from digital rights management to secured personal communications, such as medical materials.

In this section, basic concepts of image encryption and steganography are given.

4.2.1 Cryptography

In many situations, there is a strong need for security against unauthorized interpretation of coded data. This secrecy requirement is in fact an imperative functionality that has to be found within medical field when communicating medical information over any untrusted medium.

One of the techniques for ensuring privacy of sensitive data is Cryptography. Cryptography aims at protecting data from theft or alteration and can be also used for user authentication. Three types of cryptographic schemes are typically developed: secret key (or symmetric) cryptography, public-key (or asymmetric) cryptography, and hash functions. A complete survey of cryptography principles and techniques has been realized in [7].

This section is dedicated to joint cryptography and image coding frameworks.

4.2.1.1 Cryptography and Images

Contrary to classical encryption [59], security may not be the most important aim for an encryption system devoted to images. Depending on the type of applications, other properties (such as speed or bitstream compliance after encryption) might be equally important. In that context, naive or hard encryption consists of putting in the whole image data bitstream into a standard encryption system, without taking care of its nature. However, considering the typical size of a digital image compared to a text message, the naive algorithm usually cannot meet the speed requirements for real-time digital image processing or transmission applications. In contrast, soft or selective encryption trades off security for computational complexity. They are designed to protect multimedia content and fulfil the security requirements for a particular multimedia application.

Research is focused on fast encryption procedures specifically designed to the targeted environment. There are two levels of security for digital image encryption: low level and high-level security encryption. In low-level security encryption, the encrypted image shows degraded visual quality compared to the original one, but the content of the image remains still visible and understandable to the viewers. In the high-level security case, the content is completely scrambled and the image just looks like random noise. In this case, the image is not understandable to the viewers at all.

In order to make databases of high resolution images, such as medical or art pictures, accessible over the Internet, advanced functionalities combining scalability and security have to be both integrated. Indeed, scalability is a way to make database browsing easier and allow interactivity, thanks to a specific hierarchical organization of data. As for the confidentiality and privacy of visual data, both are obtained from a dedicated encryption process [70, 27]. The joint use of the two concepts aims at providing hierarchical access to the data, through a protection policy dependant on the levels of hierarchy.

On the other hand, selective encryption techniques process only parts of the compressed data, enabling low computational solutions [43, 71]. In spite of the low amount of encrypted data, without the knowledge of the encryption key, the decoding stage only reconstructs noisy images.

4.2.1.2 Selective Cryptography

Because of the large medical image sizes, dedicated scalable encoders have been defined, typically JPEG2000 and SPIHT. Accordingly, encryption processes should also be scalable in terms of level security. Equivalent issue has been described to address IPTV secure transmission problem [37].

Selective encryption [71] thus aims at avoiding the encryption of all bits of a digital image and yet ensuring a secure encryption. The key point is to encrypt only a small part of the bitstream. Consequently, the amount of encrypted data, especially when images are losslessly coded, still remains low in comparison to the global bitstream. The complexity associated with this technique is then naturally low.

The canonical framework for selective encryption has been modeled by Vandroogenbroeck et al [71] and is shown on figure 4.1.a. The image is first compressed. Afterwards, the algorithm only encrypts part of the bitstream with a well-proven ciphering technique: incidentally, a message (a watermark) can be added at this step. To ensure full compliance with any decoder, the bitstream should only be altered at carefully chosen places. With the decryption key, the receiver decrypts the bitstream and decompresses the image. When the decryption key is unknown, the receiver will still be able to decompress the image, but this image will significantly differs from the original, as depicted in figure 4.1.b.

Methods for selective encryption [74] proposed recently include DCT-based methods, Fourier-based methods, SCAN-based methods, chaos-based methods and quadtree-based methods. These methods have to be fast to meet the applications requirements and try to keep the compression ratio as good as without encryption.

A complete overview on this topic can be found in [43].

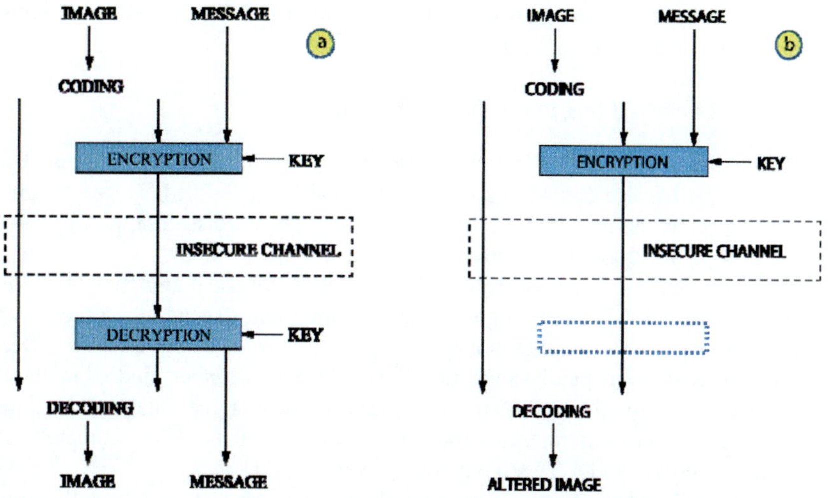

Fig. 4.1 Selective encryption / decryption mechanism with (a) or without (b) encryption key.

4.2.2 *Data Hiding and Image Coding*

Data hiding aims at hiding covert information into a given content. From this matter of fact, two main solutions can be used: steganography and watermarking. Steganography is the process of hiding a secret message, in such a way that an eavesdropper cannot detect the presence hiden data. As for watermarking, it achieves embedding information process into an image so that the message remains difficult to remove [30].

Steganography methods usually do not need to provide strong security against removing or modification of the hidden message. Watermarking methods need to to be very robust to attempts to remove or modify a hidden message.

The security associated to data hiding remains a key issue. Within the watermark only attack (WOA) framework, robust solutions have been discussed, in particular when using secure spread-spectrum watermarking [44]. Security aspects will not be developed in this section.

Data embedding hides data (*i.e.* the payload) in a digital picture so that to be as unnoticeable as possible. For that purpose, image quality should be high after data embedding. The measurement of data embedding algorithm performances is done using three criteria [66]: first, the payload capacity limit, i.e. the maximal amount of data that can be embedded, then visual quality, to measure the distortions introduced by the algorithm, and, at least, complexity, or the computational cost of the algorithm.

In order to fulfil these requirements, techniques have been developed both in the direct domain or in a transformed domain. These latter are both described in following sections.

In terms of pedagogical supports, readers can refer to [17], where lecture notes and associated software are available.

4.2.2.1 Data Embedding in the Direct Domain

Pixel-based methods rely on pixel modification following specific patterns. The first technique of data embedding consists of modifying the LSB (Least Significant Bit) of pixels in the picture [14]. It has good capacity-distortion performance but with the major drawback of being fragile.

Another solution consists of using patchwork as a statistical approach [14]. Selected pixels are divided in two groups and are modified depending on the group they belong to, in order to respect a specific pattern. The detection uses the difference between mean of pixel values in this two groups. Another kind of methods uses fractal code modification by adding similarities in the image [13]. This method is adapted to watermarking (detection), and is robust to JPEG compression (better robustness with DCT use) but not to geometrical attacks.

Recent developments take into account block structures so that to be fully compliant with standard image and video codecs [40].

4.2.2.2 Data Embedding in the Transform Domain

Most of data hiding techniques use the transformed domain, and especially the frequency one. As a matter of fact, the Fourier Transform has very interesting properties of invariance under geometrical transformations [58]. The spread spectrum technique applies successively a FFT (Fast Fourier Transform) and then a FMT (Fourier-Mellin Transform) on the image to reveal invariant areas. Then the payload is spread over these areas, which can be either in the amplitude [58] or in the phase [57] of the image.

The frequency domain can be obtained by the DCT (Discrete Cosine Transform). A blind method using DCT coefficients inversion produces quite good invisibility but bad robustness. Ciphered data can also be inserted in DCT coefficients by addition [75]. The problem is that block-based DCT has inner sensitivity to geometrical modifications. Nevertheless, the spread spectrum technique combined to DCT (instead of FFT) shows efficiency and robustness against geometrical attacks [21].

Joint compression-insertion remains a key issue. Corresponding methods are classically frequency-based methods using the transformation performed by the still image coder. As an example, JPEG2000 is based on the DWT (Discrete Wavelet Transform), and dedicated watermarking frameworks appeared. One consists in inserting pseudo-random watermark [73, 36] by addition to the largest coefficients in the subbands. This DWT watermark approach is robust to many distortions like compression, large variance additive noise and resolution reduction, whereas DCT is not [45]. Recent studies based on Human Visual System present solution relied on a tradeoff between invisibility and robustness [61, 5].

Many other methods exist, like watermarking using the DLT [32] (Discrete Laguerre Transform) instead of DCT, with almost the same results, the Fresnel transform [42] (like FFT, but with a multichannel approach), and also data hiding using the IHWT [67] (Integer Haar Wavelet Transform, also called S-Transform) that insert one bit using two integers. Papers based on more complex approaches such as the quaternion Fourier transforms [69] have also demonstrated their efficiency at the expense of algorithm complexity.

4.3 Secure Transmission of Encoded Bitstreams

Through the development of PACS (Picture Archiving and Communication Systems), health care systems have come to rely on digital information. Furthermore, future medical applications will have to integrate access to generalized databases that contain the personal medical information of each patient. Efficient image management consequently becomes a key issue in designing such a system. Given this situation, two main elements must be considered at the same time: compression and security strategies specific to image handling [51].

Naturally, teleradiology systems integrate the notion of security [63]. In particular, they must guarantee the integrity (to prevent the alteration of data), authentication (to check the sender) and the confidentiality (to prevent unauthorized access) of the medical data at all times. The availability of the information can be ensured by the Internet Protocol (IP). Moreover, wireless transmissions play an increasingly important part in the achievement of this goal, [55, 54, 72]] especially in the context of emergency medicine [35, 18]. However, this access to information must be accompanied by security primitives. Typically, for privacy and authentication purposes, the traditional security solutions integrated in the DICOM standard cover encryption processes and digital signatures [53].

In this section, both robust wireless transmission and IP networks are tackled. Classic error resilience tools used as channel coding features are first described. IP packets loss compensation issue is then addressed. Finally, the Long Term Evolution (LTE) telecommunication standard is described as an application case of the latter content protection solution.

4.3.1 Error Resilience and Channel Coding

Commonly, robust wireless transmission can be achieved through the use of error resilience processes at both source and channel coding. At the source coding side, the entropy coder is often the less robust part of the coder. When using arithmetic entropy coder such as MQ coder used in JPEG2000 format, a single bitshift in the bitstream is enough to create important visual artefacts at the decoding side. Therefore to ensure a proper decoding of the bitstream, different kinds of markers need to be added. First, to prevent the decoder from desynchronizing and therefore error from propagating, synchronisation markers need to be added. Moreover, specific error detection markers can be used to detect errors during decoding and discard bitstreams affected by this error. Such synchronization and error detection markers have already been implemented as SEGMARK, ERTERM and RESTART markers in the JPEG2000 codec [64] as well as in the LAR codec.

At the channel coding side, error robustness is achieved by using error correcting codes, such as Reed Solomon [56] and convolutive codes [26]. Such error correcting codes add redundant data in the bitstream in order to detect and possibly correct transmission errors. Depending on the channel characteristics, error correcting codes have to be tuned to achieve good performance in error correction while keeping a small amount of redundant data. These error correcting codes are usually computationally expensive and fast codes like LDPC and turbo codes can often be used instead.

As described above, at both source and channel coding, error resilience is performed by adding extra data to the bitstream. Such overhead has to be taken in consideration while performing image compression and has to remain as low as possible to maintain an acceptable bit rate.

VCDemo software [62] can be used to illustrate how transmission errors impact the visual quality of decoded images and videos. It contains JPEG and JPEG2000 image codecs as well as MPEG2 and H264 video codecs. Some error correcting mechanisms can be used and their impact on image quality can be observed.

4.3.2 IP Packets Securization Processes

Very few works cover the loss of entire IP packets in medical data transmissions [48]. In a more general framework such as image transmission, most studies relate to the implementation of error control coding e.g. Reed-Solomon codes to compensate for packet loss by avoiding retransmissions [48, 24].

By adjusting the correction capacities and, thus, the rates of redundancy, it is possible to adapt to both a scalable source and an unreliable transmission channel. This is the purpose of Unequal Error Protection (UEP) codes which are now mature and proposed in standardization processes [25]. The specific problem of medical image integrity is very often the volume of the data being transmitted (cf lossless coding, 3D-4D acquisition etc.). Within this framework, UEP must meet algorithmic complexity requirements to satisfy real time constraints.

Roughly speaking, the most important part of the image is more protected by redundant information than non significant data. Figure 4.2 illustrates the associated framework. From image coding process, both bitstream data and codec properties are available for an advanced analysis stage. Then, a hierarchy can be extracted from the bitstream, so that the UEP strategy stage can add adequate redundancy. As a consequence, fine granularity can be obtained for good adaptation both to the hierarchy of the image and to the channel properties as joint source channel coding.

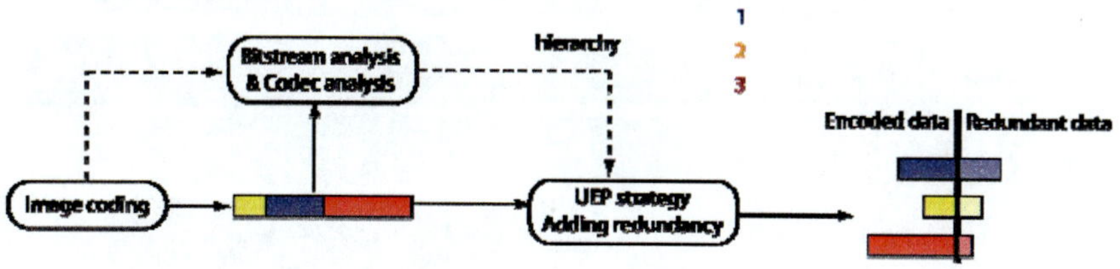

Fig. 4.2 UEP principles: hierarchy and redundancy

A great deal of research work has been done in this area over the past decade. In particular, the working draft of JPEG2000 WireLess (JPWL) [25] proposes concentrated unequal protection on the main header and the tile header with the characteristic that any error on these parts of the stream is fatal for decoding. In this solution, conventional Reed-Solomon error correction codes are applied to a symbol level to provide protection [24]. A very strong protection obviously improves the chances of success in decoding when binary losses occur but it also guarantees the integrity of the headers whether the properties of the channel are good or very bad. Furthermore, performance evaluation and protection on a symbol level are far removed from the real channels like wireless channels as can be seen for example through the variations in the protocol IEEE802.xx (WLAN or

WiMax). Typically, these standards are divided into 7 layers according to the Open Systems Interconnection (OSI) model description, as depicted on figure 4.3. More precisely, the approach never considers the effectiveness of the mechanisms operated on the level of Media Access Control (MAC) layer and physical (PHY) layer such as the Hybrid ARQ (Automatic Query Request - H-ARQ) combining efficient channel coding (turbo-code) and retransmission. Likewise, the working draft does not consider the exploratory research carried out over the past ten years on unequal error protection [6] or the new representations based on a multiple description of information [39]. Classically, when designing joint source-channel coding UEP schemes, we consider the PHY and MAC layers as effective to deliver true symbols so as to focus all our attention of unequal protection at the transmission unit level *i.e.* the packet level.

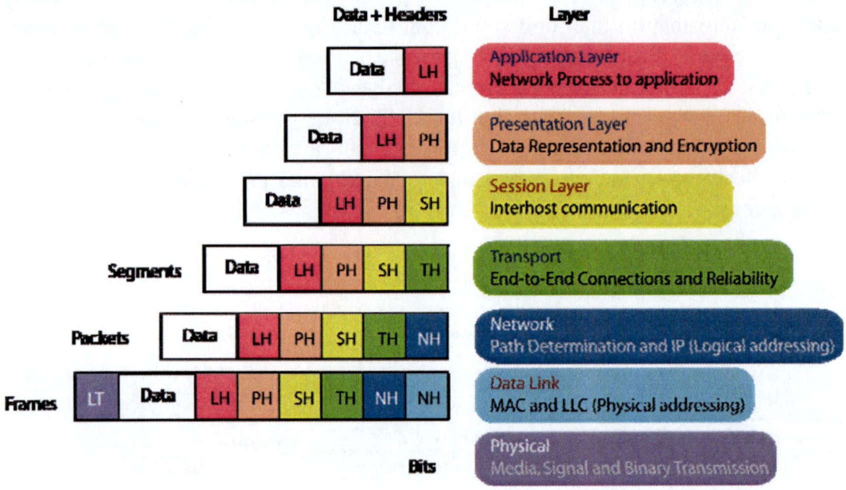

Fig. 4.3 Open Systems Interconnection (OSI) model description: 7 layers

4.3.3 LTE Standard Application Case: Securization Process for Advanced Functionalities

Nowadays, wireless communications and their applications are undergoing major expansion and they have captured media attention as well as the imagination of the public. However, wireless channels are known to generate a high number of errors which perturb complex multimedia applications such as image or video transmission. For these reasons, designing a suitable system for image transmission over wireless channel remains a major issue. In particular, if the new telecommunication standard, the LTE (Long Term Evolution) one, proposes advanced functionalities, it requires accurate securization processes so that to ensure sufficient end-to-end Quality Of Service whatever the transmission conditions are.

4.3.3.1 Evolution of Telecommunication Standards

Terrestrial mobile telecommunications started in the early 1980s using various analog systems developed in Japan and Europe. The Global System for Mobile communications (GSM) digital standard was subsequently developed by the European Telecommunications Standards Institute (ETSI) in the early 1990s. Available in 219 countries, GSM belongs to the second generation mobile phone system. It can provide an international mobility to its users by using inter-operator roaming. The success of GSM promoted the creation of the Third Generation Partnership Project (3GPP), a standard-developing organization dedicated to supporting GSM evolution and creating new telecommunication standards, in particular a Third Generation Telecommunication System (3G) [52].

The existence of multiple vendors and operators and the necessity of interoperability when roaming and limited frequency resources, justify the use of unified telecommunication standards such as GSM and 3G. Each decade, a new generation of standards multiplies the data rate available to its user by ten (Figure 4.4). The driving force behind the creation of new standards is the radio spectrum which is an expensive resource shared by many interfering technologies. Spectrum use is coordinated by ITU-R (International Telecommunication Union, Radio Communication Sector), an international organization which defines technology families and assigns their spectral bands to frequencies that fit the International Mobile Telecommunications (IMT) requirements.

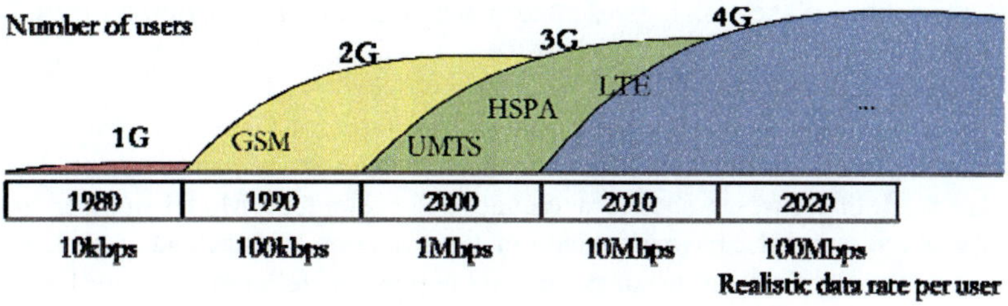

Fig. 4.4 3GPP Standard Generation

Radio access networks must constantly improve to accommodate the tremendous evolution of mobile electronic devices and internet services. Thus, 3GPP unceasingly updates its technologies and adds new standards. Universal Mobile Telecommunications System (UMTS) is the first release of the 3G standard. Evolutions of UMTS such as High Speed Packet Access (HSPA), High Speed Packet Access Plus (HSPA+) or 3.5G have been released as standards. The 3GPP Long Term Evolution (LTE) is the 3GPP standard released subsequent to HSPA+. It is designed to support the forecasted ten-fold growth of traffic per mobile between 2008 and 2015 [52] and the new dominance of internet data over voice in mobile systems. The LTE standardization process started in 2004 and a new enhancement of LTE named LTE-Advanced is currently being standardized.

A LTE terrestrial base station computational center is known as an evolved NodeB or **eNodeB**, where a NodeB is the name of a UMTS base station. An eNodeB can handle the communication of a few base stations, with each base station covering a geographic zone called a cell. The user mobile terminals (commonly mobile phones) are called User Equipment (**UE**). At any given time, a UE is located in one or more overlapping cells and communicates with a preferred cell; the one with the best air transmission properties. LTE is a duplex system, as communication flows in both directions between UEs and eNodeBs. The radio link between the eNodeB and the UE is called the downlink and the opposite link between UE and its eNodeB is called uplink. These links are asymmetric in data rates because most internet services necessitate a higher data rate for the downlink than for the uplink.

LTE also supports data broadcast (television for example) with a spectral efficiency over 1 bit/s/Hz. The broadcasted data cannot be handled like the user data because it is sent in real-time and must work in worst channel conditions without packet retransmission.

4.3.3.2 LTE Radio Link Protocol Layers

The information sent over a LTE radio link is divided in two categories:

- the user-plane which provides data and control information irrespective of LTE technology,
- the control-plane which gives control and signaling information for the LTE radio link.

The protocol layers of LTE are displayed in Figures 4.5 and 4.6. User plane and control plane significantly differ but the lower layers remain common for both planes. Both figures associate a unique OSI Reference Model number to each layer. Layers 1 and 2 have identical functions in control-plane and user-plane even if parameters differ (for instance, the modulation constellation). Layers 1 and 2 are subdivided into different layers that require adapted securization processes, both in terms of content protection and error resilience tools.

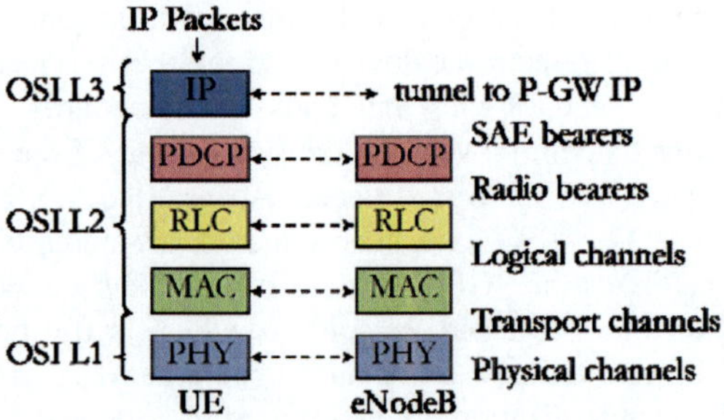

Fig. 4.5 User plane: Protocol Layers of LTE Radio Link

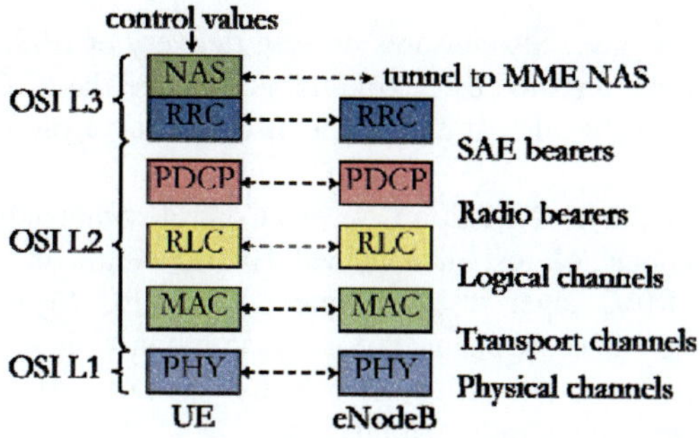

Fig. 4.6 Control plane: Protocol Layers of LTE Radio Link

In particular, the physical layer organization of the LTE standard is illustrated on Figure 4.7. It corresponds to the Release 9 LTE physical layer in the eNodeB, i.e. the signal processing part of the LTE standard that 3GPP finalized in December 2009. The physical layer, OSI layer 1, uplink and downlink baseband processing must share the eNodeB digital signal processing resources. The downlink baseband process is itself divided into channel coding that prepares the bit stream for transmission and symbol processing that adapts the signal to the transmission technology. The uplink baseband process performs the corresponding decoding. The OSI layer 2 controls the physical layer parameters.

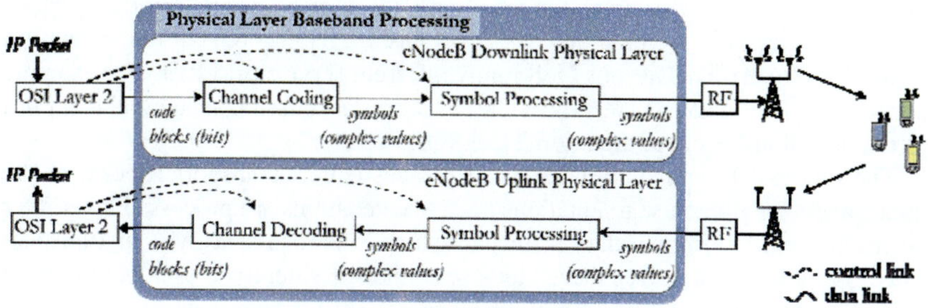

Fig. 4.7 LTE PHY layer overview (OSI L1)

The role of each layer is defined as follows:

- PDCP layer [4] or layer 2 Packet Data Convergence Protocol is responsible for data ciphering and IP header compression to reduce the IP header overhead.
- RLC layer [3] or layer 2 Radio Link Control performs the data concatenation and then generates the segmentation of packets from IP-Packets of random sizes which comprise a Transport Block (TB) of size adapted to the radio

transfer. The RLC layer also ensures ordered delivery of IP-Packets; Transport Block order can be modified by the radio link. Finally, the RLC layer handles a retransmission scheme of lost data through a first level of Automatic Repeat reQuests (ARQ).

- MAC layer [2] or layer 2 Medium Access Control commands a low level retransmission scheme of lost data named Hybrid Automatic Repeat reQuest (HARQ). The MAC layer also multiplexes the RLC logical channels into HARQ protected transport channels for transmission to lower layers. Finally, the MAC layer contains the scheduler, which is the primary decision maker for both downlink and uplink radio parameters.
- Physical layer (PHY) [1] or layer 1 comprises all the radio technology required to transmit bits over the LTE radio link. This layer creates physical channels to carry information between eNodeBs and UEs and maps the MAC transport channels to these physical channels.

Layer 3 differs between control and user planes. Control plane handles all information specific to the radio technology while the User plane carries IP data from system end to system end. More information can be found in [22] and [60].

The LTE system exhibits a high reliability while limiting the error correction overhead. Indeed it uses two level of error concealment; HARQ and ARQ. HARQ is employed for frequent and localized transmission errors while ARQ is used for rare but lengthy transmission errors.The retransmission in LTE is determined by the target service: LTE ensures different **Qualities of Service (QoS)** depending on the target service. For instance, the maximal LTE-allowed packet error loss rate is 10^{-2} for conversational voice and 10^{-6} for transfers based on TCP (Transmission Control Protocol). The various QoS imply different service priorities. For example during a TCP/IP data transfer, the TCP packet retransmission system adds a third error correction system to the two LTE ARQs.

The physical layer manipulates bit sequences called Transport Blocks. In the user plane, many block segmentations and concatenations are processed layer after layer between the original data in IP packets and the data sent over air transmission. Figure 4.8 summarizes these block operations. Evidently, these operations do not reflect the entire bit transformation process including ciphering, retransmitting, ordering, and so on.

A very interesting implementation of a LTE simulator has been developed by TU Wien's Institute of Communications and Radio Frequency Engineering [46] and can be download on the laboratory's web site. The simulators are released under the terms of an academic, non-commercial use license.

The LTE provides a transmission framework with efficient error resilience mechanisms. However, recent researches have been more focused on complete joint source-channel coding scheme so that to ensure an even higher QoS. Typically, since error can remain from the transmission, the source codec must also contain error concealment mechanisms.

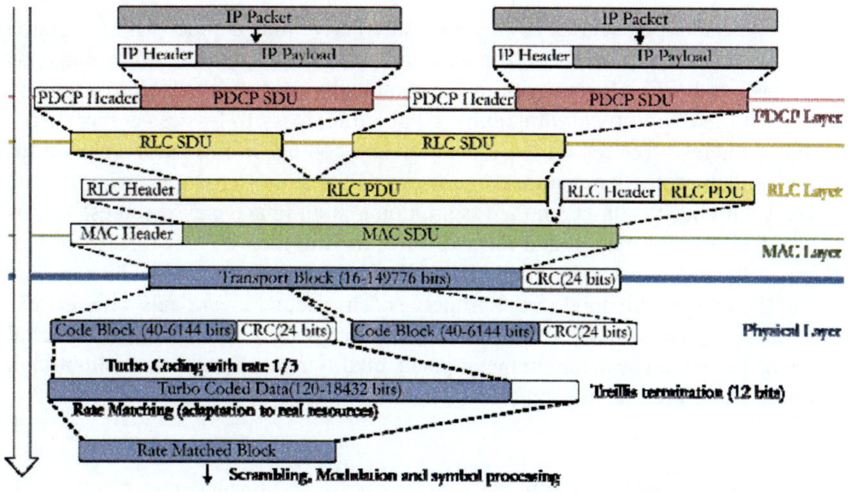

Fig. 4.8 Data Blocks Segmentation and Concatenations

4.4 Application Example: LAR Medical Framework

PACS-based systems tend to manage secure and heterogeneous networks such as wire and/or wireless ones, together with innovative image compression schemes. The design of a new medical image compression scheme requires then many dedicated services. Medical images usually go with private metadata that have to remain confidential. In particular, to insure reliable transfers, flexible and generic scheduling and identification processes have to be integrated for database distribution purposes to take account of secure remote network access together with future developments in network technologies. Fast browsing tools, including the segmentation process and scalability, are therefore needed.

This is the background against which IETR Laboratory proposes the content-based Locally Adaptive Resolution (LAR) codec. The LAR method has already been proposed as a response to the call for contributions of technologies [9, 10] within the JPEG committee. In this section, we focus on LAR medical image processing, and give some specific uses allowed by our compression systems. The LAR coding scheme can be seen as a package of coding tools aiming at different levels of user services. In this context, we focus on a specific scheme called Interleaved S+P and its associated data protection tools.

4.4.1 LAR Codec Overview

The LAR method was initially introduced for lossy image coding [23]. The philosophy behind this coder is not to outperform JPEG2000 in compression; the goal is to propose an open source, royalty free, alternative image coder with integrated services. While keeping the compression performances in the same range as JPEG2000 or JPEG XR, but with lower complexity, our coder also provides services such as scalability, cryptography, data hiding, lossy to lossless compression, region of interest, free region representation and coding. In this paragraph, we focus on content protection features.

The LAR codec is based on the assumption that an image can be represented as layers of basic information and local texture, relying then on a two-layer system (Figure 4.9). The first layer, called Flat coder, leads to construct a low bit-rate version of the image with good visual properties. The second layer deals with the texture that is encoded through a texture coder through DCT-based system (spectral coder) or pyramidal system, aiming at visual quality enhancement at medium/high bit-rates. Therefore, the method offers a natural basic SNR scalability.

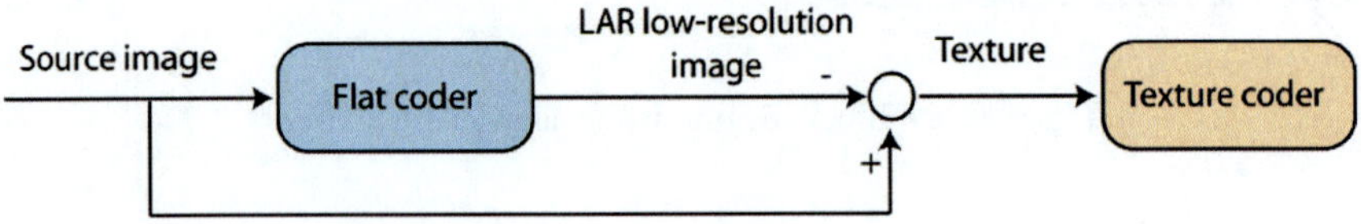

Fig. 4.9 General scheme of two-layer LAR coder

The LAR codec tries to combine both efficient compression in a lossy or lossless context and advanced functionalities and services. For this purpose, we defined three different profiles for user-friendly usage (Figure 4.10).

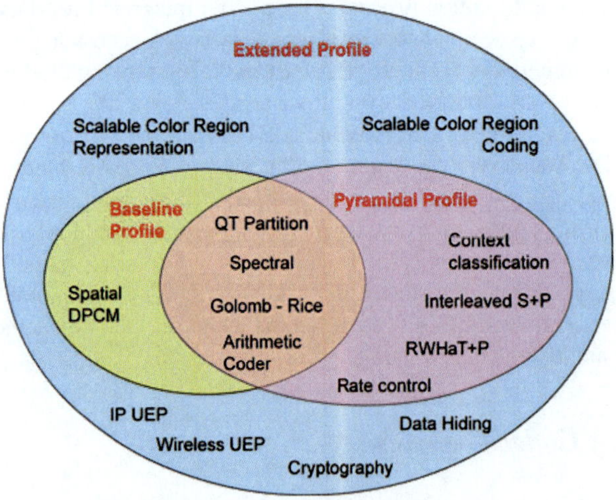

Fig. 4.10 Specific coding parts for LAR profiles

The baseline profile is dedicated to low bit-rate encoding. In the context of medical image compression, this profile is clearly not appropriate. As medical image compression requires lossless solutions, we then focus the discussion on functionalities and technical features provided by the pyramidal and extended profiles dedicated to content protection: cryptography, steganography, error resilience, hierarchical securized processes. In this context, the Interleaved S+P coding tool, based on a two interlaced pyramidal representation, is used for coding purposes [11].

4.4.2 Principles and Properties

The LAR codec relies on a two-layer system. The first layer, called FLAT coder, constructs a low bit-rate version of the image. The second layer deals with the texture that is encoded through a texture coder, aimed at visual quality enhancement at medium/high bit-rates. Therefore, the method offers a natural basic SNR scalability.

The basic idea is that local resolution, in other words pixel size, can depend on local activity, estimated through a local morphological gradient. This image decomposition into two sets of data is thus performed conditionally to a specific quadtree data structure, encoded in the Flat coding stage. Thanks to this type of block decomposition, their size implicitly gives the nature of the given block: smallest blocks are located upon edges whereas large blocks map homogeneous areas (Figure 4.11). Then, the main feature of the FLAT coder consists of preserving contours while smoothing homogeneous parts of the image.

This quadtree partition is the key system of the LAR codec. Consequently, this coding part is required whatever the chosen profile.

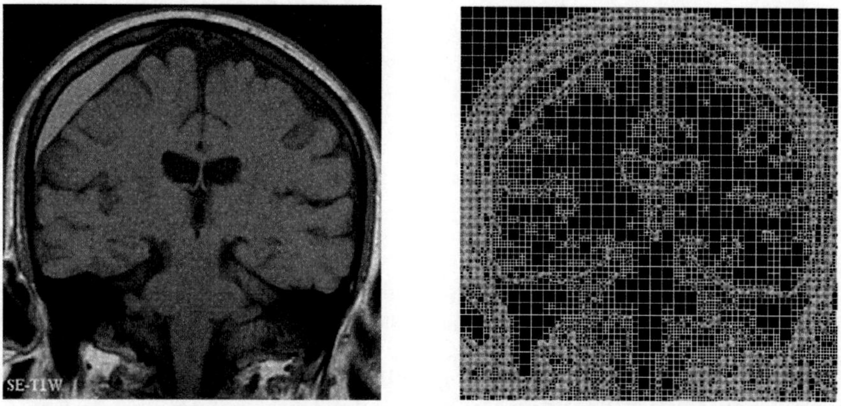

Fig. 4.11 Original image and associated Quadtree partitions obtained with a given value of activity detection parameter

4.4.2.1 Lossy to Lossless Scalable Solution

Scalable image decompression is an important feature in the medical field, which sometimes uses very large images. Scalability enables progressive image reconstruction by integrating successive compressed sub-streams in the decoding process.

Scalability is generally first characterized by its nature: resolution (multi-size representation) and/or SNR (progressive quality enhancement). Just like JPEG2000, the LAR codec supports both of them. The main difference is that the LAR provides multiresolution "edge oriented" quality enhancement. The lossy or lossless coding process involves two-pass dyadic pyramidal decomposition. The first pass, leading to a low bit-rate image, encodes the overall information in the image, preserving main contours, while smoothing homogeneous areas. The second pass adds the local texture in these areas as shown on Figure 4.12.

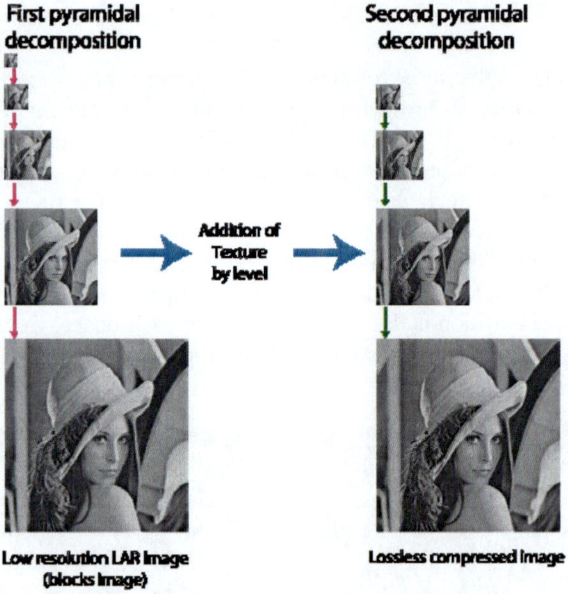

Fig. 4.12 Pyramidal representation of an image

The second important feature for scalability concerns granularity. Scalability granularity defines which elementary amount of data can be independently decoded. Among existing standards, JPEG2000 offers the finest grain scalability. On the other hand, JPEG provides no scalability at all (except in its progressive mode), while JPEG-XR enables up to 4 scalability levels. In LAR, the number of dyadic resolution levels N is adjustable, with two quality levels per resolution. Therefore, the number of elementary scalable sub-streams is 2N. The first pyramid pass provides an image with variable-sized blocks. LAR also contains some interpolation / post-processing steps that can smooth homogeneous areas while retaining sharp edges.

4.4.2.2 Hierarchical Colour Region Representation and Coding

For colour images, we have designed an original hierarchical region-based representation technique adapted to the LAR coding method. An initial solution was proposed in [23]. To avoid the prohibitive cost of region shape descriptions, the most suitable solution consists of performing the segmentation directly, in both the coder and decoder, using only a low bit-rate compressed image resulting from the FLAT coder (or first partial pyramidal decomposition). Natural extensions of this particular process have also made it possible to address medium and high quality encoding and the region-level encoding of chromatic images. Another direct application for self-extracting region representation is found in a coding scheme with local enhancement in Regions Of Interest (ROI). Actual works aim at providing a fully multiresolution version of our segmentation process: indeed this region representation can be connected to the pyramidal decomposition in order to build a highly scalable compression solution.

The extended profile also proposes the use of dedicated steganography and cryptography processes, which will be presented in next sections. To sum up, the interoperability of coding and representation operations leads to an interactive coding tool. The main features of the LAR coding parts are depicted in Figure 4.13.

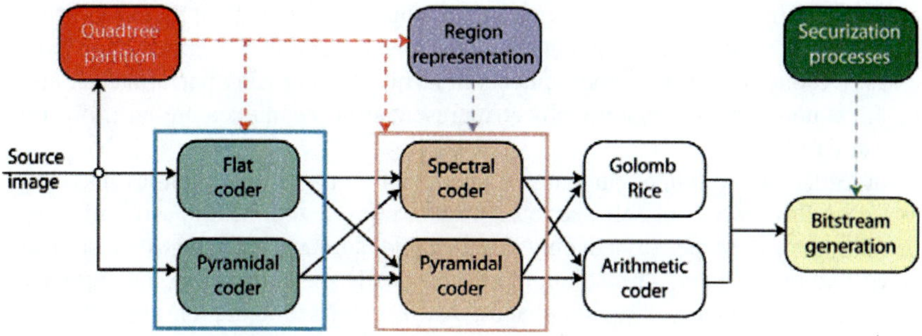

Fig. 4.13 Block diagram of extended LAR coder profile

4.4.2.3 Region and Object Representation

Current image and video compression standards rely only on information theory. They are based on prediction and decorrelation optimization techniques without any consideration of source content. To get higher semantic representation, Kunt first introduced the concept of second generation image and video coding [41]. It refers to content-based representation and compression at region/object level. To obtain a flexible view with various levels of accuracy, a hierarchical representation is generally used, going from a fine level comprising many regions, to a coarse level comprising only a few objects.

Regions are defined as convex parts of an image sharing a common feature (motion, textures, etc). Objects are defined as entities with a semantic meaning inside an image. For region representation, two kinds of information are necessary: shape (contours) and content (texture). For video purposes, motion constitutes a third dimension.

The region-based approach tends to link digital systems and human vision as regards image processing and perception. This type of approach provides advanced functionalities such as interaction between objects and regions, or scene composition. Another important advantage is the ability, for a given coding scheme, of both increasing compression quality on highly visually sensitive areas of images (ROI) and decreasing the compression quality on less significant parts (background). The actual limited bandwidth of channels compared to the data volume required for image transmission leads to a compromise between bit-rate and quality. Once the ROIs are defined and identified, this rate/quality bias can be not only generally but also locally adjusted for each ROI: compression algorithms then introduce only low visual distortions in each ROI, while the image background can be represented with high visual distortions.

Despite the benefits of region-based approaches in terms of high level semantic description, some limitations to common techniques have restricted their use.

The first one is the generally limited compression performances achieved, due to the region description cost: most of the existing methods suggest sending a segmentation map from the coder to the decoder. As the number of regions increases, the overhead costs become significant. The second limitation concerns complexity: most of the existing methods rely on complex segmentation processes. Despite increasing improvements in terms of processing performance, most of the state-of-the-art region / object representation techniques are still too time consuming.

Indeed, LAR provides an unusual method for low cost region-level coding, based on the concept of self-extracting region representation. It consists of a segmentation process performed only from highly compressed images in both the coder and the decoder. This solution prevents costly transmission of the segmentation map to provide the region shapes. An original segmentation algorithm has been designed, leading to an efficient hierarchical region-based description of the image. The process ensures full compliance between the shape of regions and their content encoding. One direct issue is ROI coding: an ROI is rapidly and easily defined as a set of regions in either the coder or the decoder. Local image quality enhancement is then achieved by allowing the second pyramidal decomposition pass only for blocks inside the ROI. Another application of an ROI description is a full encryption process (see below), which can be applied only to the ROI.

The segmentation process is optional. It can be performed on-line or off-line. From a complexity point of view, the segmentation process is of low complexity compared with common segmentation techniques. The main reason is that the LAR segmentation process starts from the block-level representation, given by the quadtree, instead of elementary pixels.

As the LAR solution relies on a compromise between coding and representation, coding key issues are partially solved. In particular, the complexity of the segmentation process has been evaluated and restricted, so that it has been pipelined and prototyped onto embedded multicore system platforms [28].

To avoid the segmentation process at the decoder side, another solution consists of transmitting the binary ROI map. The corresponding cost is limited, as ROIs are described at block-level: a full region-map composed of 120 regions is encoded at around 0.07 bpp, whereas the cost of a binary ROI image, whatever the ROI shape, is less than 0.01 bpp.

JPEG2000 also proposes ROI functionalities, but its technical solution significantly differs from the LAR one. To sum up, ROI in LAR has improved features, for example:

- ROI can represent any shape,
- ROI enhancement accurately matches the shape,
- the encoding cost of the shape is insignificant (a few bytes),
- several ROIs can be defined in the same image,
- any quality ratio between ROI and background can be defined.

4.4.3 Content Protection Features

Whatever the storage or channel transmission used, medical applications require secure transmission of patient data. Embedding them in an invisible way within the image itself remains an interesting solution.

We also deal with security concerns by encrypting the inserted data. Whereas the embedding scheme can be made public, the use of a decryption key will be mandatory to decipher the inserted data.

4.4.3.1 Steganography

Data embedding is one of the new services expected within the framework of medical image compression. It consists of hiding data (payload) in a cover image. Applications of data embedding range from steganography to metadata insertion. They differ in the amount of data to be inserted and the degree of robustness to hacking.

From a signal processing point of view, it uses the image as a communication channel to transmit data. The capacity of the channel for a specific embedding scheme gives the size of the payload that can be inserted. A fine balance has to be achieved between this payload and the artefacts introduced in the image. This being so, different embedding schemes are compared on a payload vs. PSNR basis. Of course, the overall visual quality can be assessed.

The target application is the storage of data related to a given medical image. That data can consist of patient ID, time stamps, or the medical report, transcribed or in audio form. The idea is to avoid having to store several files about specific images by having all the necessary information directly stored within the image data.

We therefore propose a data embedding service that aims to insert a high payload in an image seen either as a cover or a carrier, such as a medical report in audio form. For this purpose, audio data, after coding and ciphering, is inserted in a corresponding medical image. The embedded image is then transmitted using usual channels. Of course, this scheme is compliant with any error protection framework that might be used. When retrieval of audio data is requested, the data embedding scheme is reversed, and both the original image and the audio data are losslessly recovered. To avoid significant perceptually distortions, the data hiding mapping is powered by the quadtree: distortions are less perceptible in homogeneous areas than upon edges as shown in figure 4.14.

In this context, we studied the Difference Expansion (DE) method, introduced by Tian [68] that embeds one bit per pixel pair based on S Transform. As the LAR Interleaved S+P algorithm and DE both use S-Transform during their computation, we have combined both techniques to perform the data insertion without degrading coding performance. In order to adjust the DE algorithm to LAR Interleaved S+P, some minor modifications are introduced compared with the original DE method. In particular, we power the insertion process by the quadtree partition, which means that the insertion is dependent on the image content. Another important improvement is that in the initial DE method, positions of possible "extensible" difference have to be encoded, adding a significant overhead. In our coding scheme, these positions can be directly deduced from the quadtree, and are then not transmitted [49].

We show preliminary results on an angiography 512-squared medical image (Figure 4.15). A payload of 63598 bits is inserted, with a PSNR of 40 dB. Considering a 1 MP image, the payload can be up to 300 kbits. It corresponds roughly to an audio message of 200 s when using a 1.5 kbits voice compression rate. Of course, as many images are taken during the same medical examination, the length of the corresponding audio files is extended. Our embedding scheme is an efficient adaptation of a useful technique to our image coder. It performs well, allowing high payload and minimum distortion, as shown on zoomed parts of the images from the figure 4.15.

From a compression point of view, the data hiding process does not affect the coding efficiency: the total coding cost is about equal to the initial lossless encoding cost of the source image plus the inserted payload.

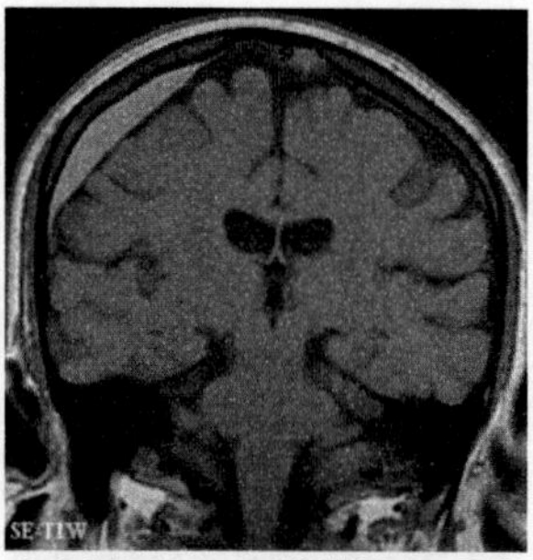

(a) 2x2 blocks, $\wp$=19528 bits, PSNR=35 dB

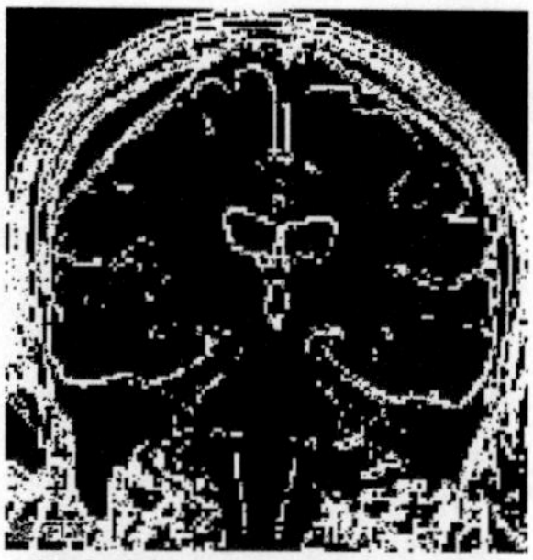

(b) 2x2 blocks

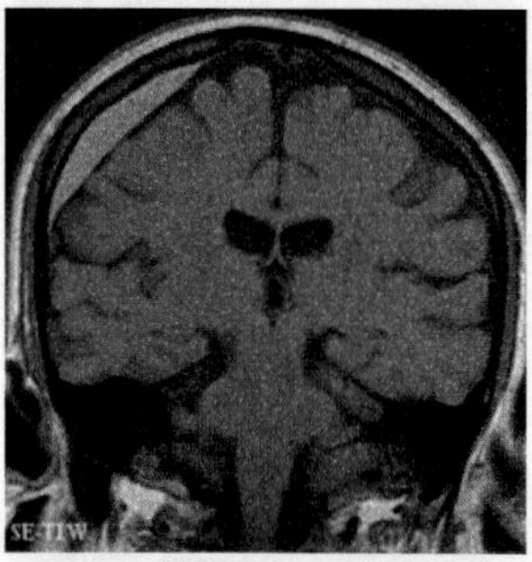

(c) 4x4 blocks, $\wp$=29087bits, PSNR=44 dB.

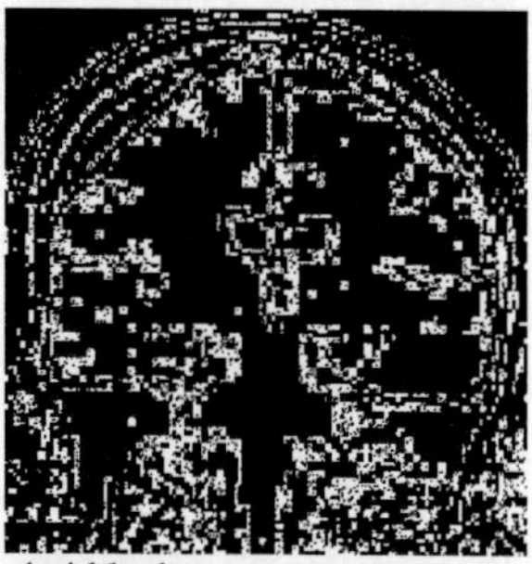

(d) 4x4 blocks

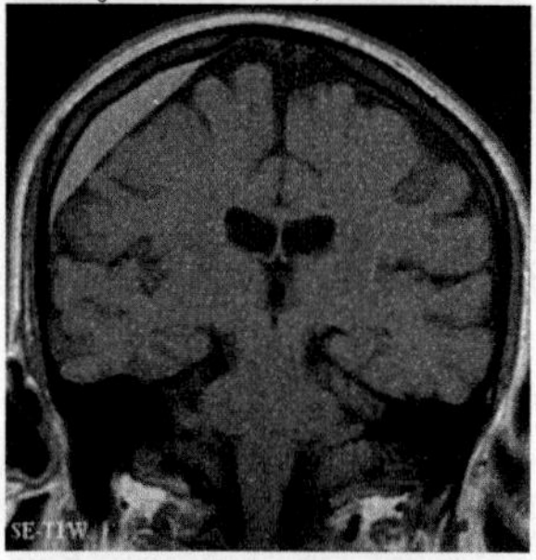

(e) 8x8 blocks, $\wp$=27971 bits, PSNR=48 dB

(f) 8x8 blocks

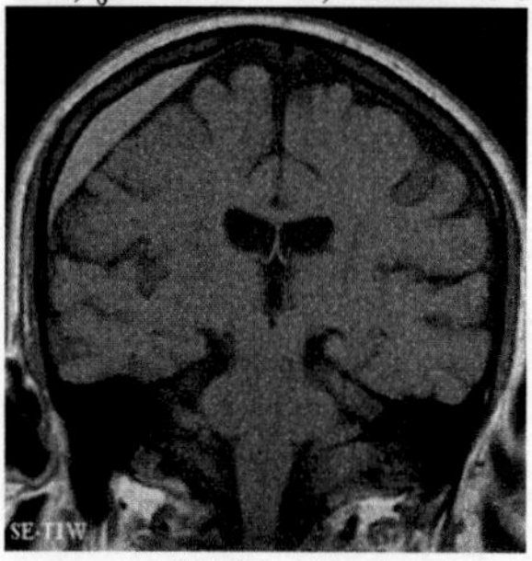

(g) 4x4 up to 16x16 blocks, $\wp$=90126bits, PSNR=42dB

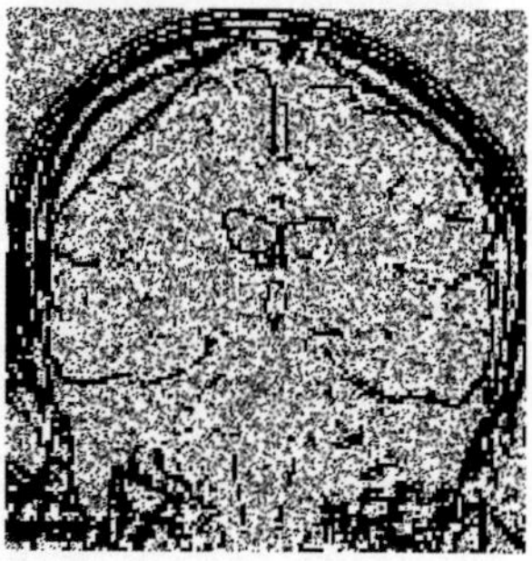

(h) 4x4 up to 16x16 blocks

Fig. 4.14 Visual quality versus watermarked block sizes. For each image, position of modified pixels has been extracted (in white onto black background).

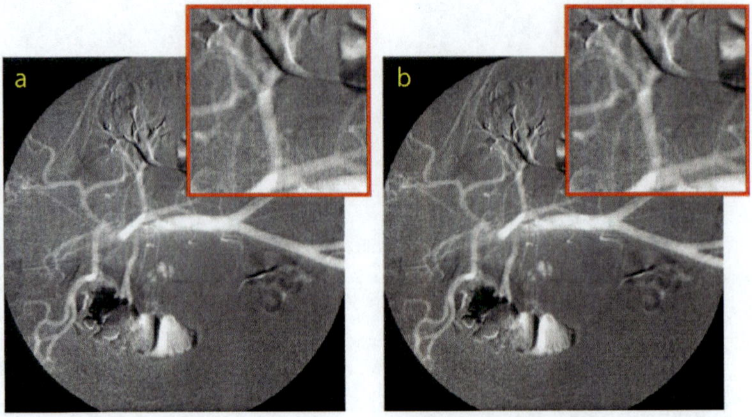

Fig. 4.15 a) Source image - b) Image with inserted payload

4.4.3.2 Cryptography

Besides watermarking, steganography, and techniques for assessing data integrity and authenticity, the provision of confidentiality and privacy for visual data is one of the most important topics in the area of multimedia security in the medical field. Image encryption lies somewhere between data encryption and image coding. Specifically, as the amount of data to be considered is several orders of magnitude greater than the amount for ordinary data, more challenges are to be dealt with. The main challenge is the encryption speed, which can be a bottleneck for some applications in terms of computation time or in terms of computer resources required. A secondary challenge is to maintain the compliance of the encrypted bitstream with the chosen image standard used to compress it.

Partial encryption addresses the first aforementioned challenge. Our partial encryption scheme is based mainly on the following idea: the quadtree used to partition the image is necessary to rebuild the image. This has been backed up by theoretical and experimental work. As a result, the quadtree partition can be considered to be the key itself, and there is no need to encrypt the remaining bitstream. The key obtained is thus as long as usual encryption key and its security has been shown to be good. If further security is requested, the quadtree partition can be ciphered using a public encryption scheme, to avoid the transmission of an encryption key, as depicted in Figure 4.16 [50]. This system has the following properties: it is embedded in the original bit-stream at no cost, and allows for multilevel access authorization combined with a state-of-the-art still picture codec. Multilevel quadtree decomposition provides a natural way to select the quality of the decoded picture.

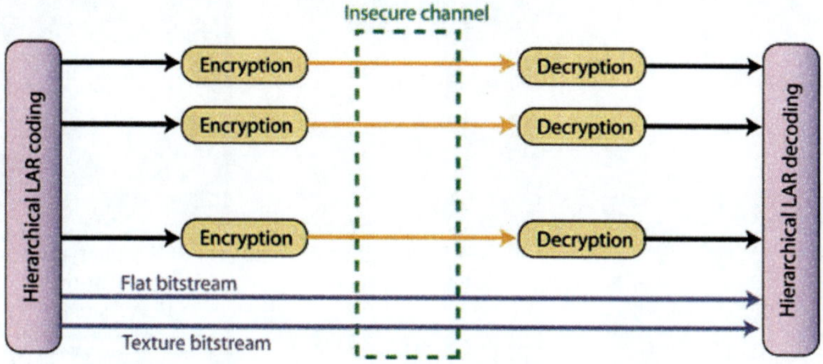

Fig. 4.16 LAR hierarchical selective encryption principle

Selective encryption goes a bit further than partial encryption. The idea is to cipher only a small fraction of the bitstream, the main component, which gives the added advantage of obtaining a valid compliant bitstream. This property allows the user to see a picture even without the key. Of course, the picture must be as different to the original one as possible.

Our selective encryption scheme uses also the quadtree partition as a basis [29]. The data required in the compression framework to build the flat picture are also used. The general idea is to encrypt several levels of the hierarchical pyramid. The process begins at the bottom of the pyramid. Depending on the depth of the encryption, the quality of the image rebuilt without the encryption key varies. The encryption itself is performed by a well-known secure data encryption scheme. One main property of our selective encryption scheme is that the level of encryption (i.e. the level of the details remaining visible to the viewer) can be fully customized. Hierarchical image encryption is obtained by deciding which level will be decrypted by supplying only the keys corresponding to those levels. This refines the quality of the image given to different categories of viewers. The encryption of a given level of the partition prevents the recovery of any additional visually-significant data (Figure 4.17). From a distortion point of view, it appears that encrypting higher levels (smaller blocks) increases the PSNR, and at the same time, the encrypting cost. From a security point of view, as the level increases, the search space for a brute force attack increases drastically.

As our research is focused on fast encryption procedures specifically tailored to the target environment, we use the pyramidal profile with Interleaved S+P configuration. Our encryption tools allow a fine selection of tradeoffs between encryption computing cost, hierarchical aspects, compliance and the quality of encrypted pictures.

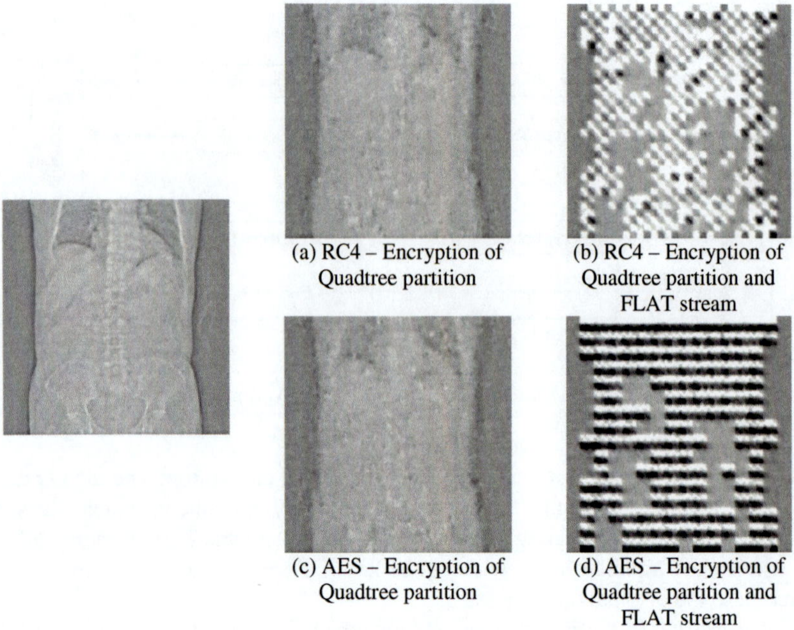

Fig. 4.17 Visual comparison between original image and image obtained from partially encrypted LAR encoded streams without encryption key.

4.4.3.3 Scalable ROI Protection and Encoding for Medical Use

Designing semantic models becomes a key feature in medical image management [65]. Different scenarios can be investigated. We present only one scenario suitable for image storage and off-line decoding. This scenario involves the following processing steps.

- At the coder side, the specialist defines the ROI in the image and chooses the option "lossless mode with encrypted ROI". The resultant stream is given in Figure 4.18.
- At the decoder side, the image can be partially decoded until the lossless ROI has been reconstructed or fully decoded. Figure 4.19 shows the overall process.

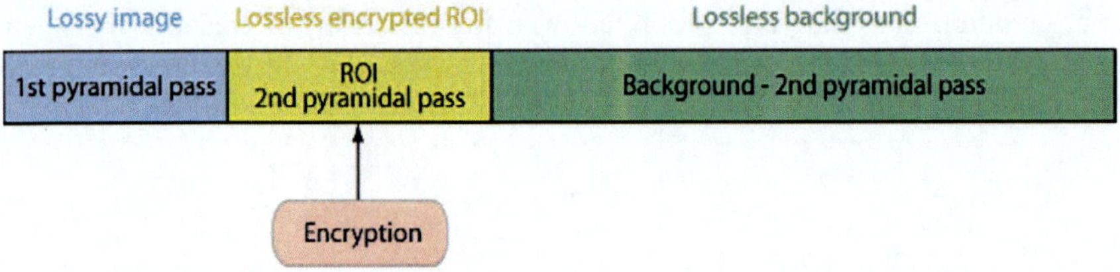

Fig. 4.18 Substream composition for lossless compression with an encrypted ROI

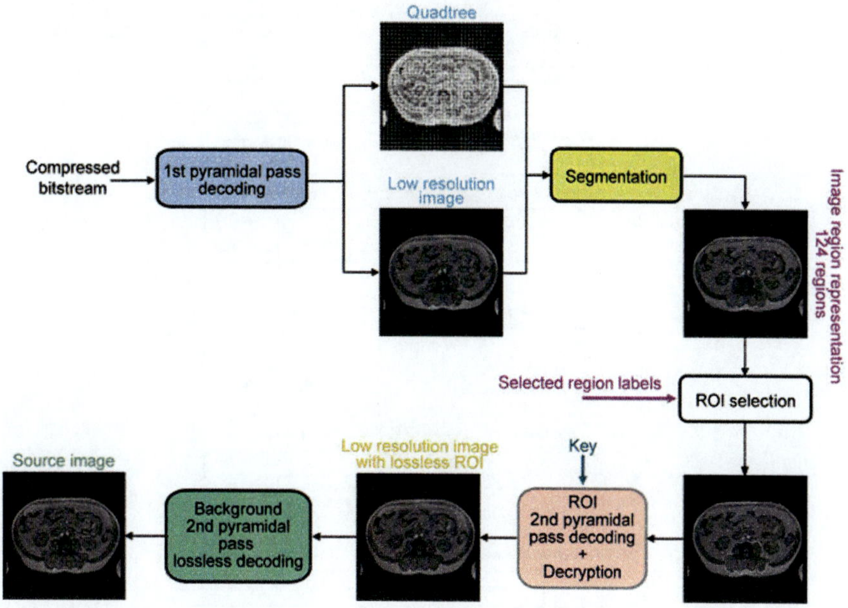

Fig. 4.19 Overall decoding scheme for lossless compression with an encrypted ROI}

4.4.3.4 Client-Server Application and Hierarchical Access Policy

For medical use, together with PACS systems, images and videos databases are a powerful collaborative tool. However, the main concern when considering these applications lies in the secure accessing of images. The objective is therefore to design a medical image database accessible through a client-server process that includes and combines a hierarchical description of images and a hierarchical secured access.

A corresponding client-server application [8] has been then designed. Every client will be authorized to browse the low-resolution image database and the server application will verify the user access level for each image and ROI request. ROIs can be encrypted or not, depending on the security level required.

If a client application sends a request that does not match the user access level, the server application will reduce the image resolution according to access policy. The exchange protocol is depicted in Figure 4.20.

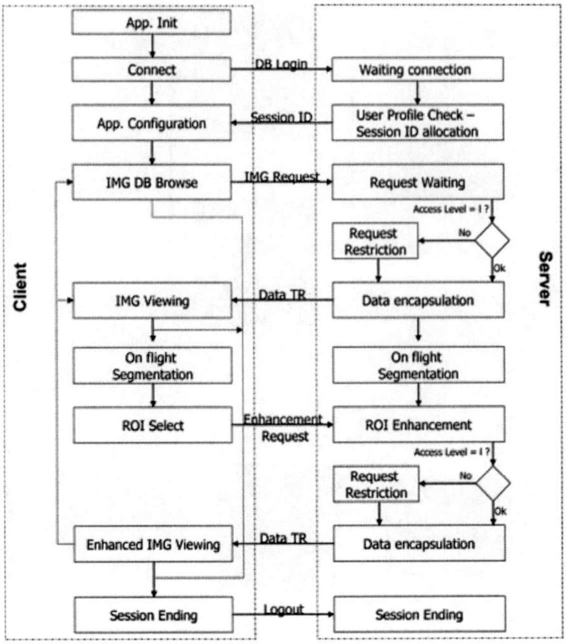

Fig. 4.20 Exchange protocol for client-server application

4.4.4 Transmission Error Protection - Error Resilience

Interest in remote medical applications has been rapidly increasing. Telemedicine aims to speed up the diagnosis process, reduce risks of infection or failure, enhance mobility and reduce patient discomfort.

Although wire networks are traditionally used for teleradiology or telesurgery purposes, the rapid growth of wireless technologies provides new potential for remote applications. To ensure optimal visualization of transmitted images, there are two possible ways of protecting the bitstreams.

Firstly, protecting the encoded bit-stream against error transmission is required when using networks with no guaranteed quality of service (QoS). In particular, the availability of the information can be ensured by the Internet protocol (IP). We focused our studies on two topics, namely the loss of entire IP packets and the transmission over wireless channel.

Secondly, we develop error resilience strategies adapted to our compression scheme. UEP solutions used together with proper resynchronization processes and robust encoding naturally leads to optimal conditions for the transmission of sensitive data.

4.4.4.1 UEP strategies

Limited bandwidth and SNR are the main features of a wireless channel. Therefore, both compression and secure transmission of sensitive data are simultaneously required. The pyramidal version of the LAR method and an Unequal Error Protection strategy are applied respectively to compress and protect the original image. The UEP strategy takes account of the sensitivity of the substreams requiring protection and then optimizes the redundancy rate. In our application, we used the Reed Solomon Error Correcting Code RS-ECC, mixed with symbol block interleaving for simulated transmission over the COST27 TU channel [34] (Figure 4.21). When compared to the JPWL system, we show that the proposed layout is better than the JPWL system, especially transmission conditions are bad (SNR < 21 dB).

Other simulation tests have been designed for MIMO systems using a similar framework, and have shown the ability of our codec to be easily adapted to bad transmission conditions, while keeping reasonable additional redundancy. At this point, comparisons with other methods remain difficult. Both SISO and MIMO transmissions simulation tools were provided by the French X-LIM Laboratory [16].

Current developments are focused on the study of the LTE transmission system [47], and its combination with LAR coded bitstreams.

These preliminary tests have been carried out without implementing basic error resilience features, such as resynchronization process, that should greatly improve our results. Some related solutions are presented below.

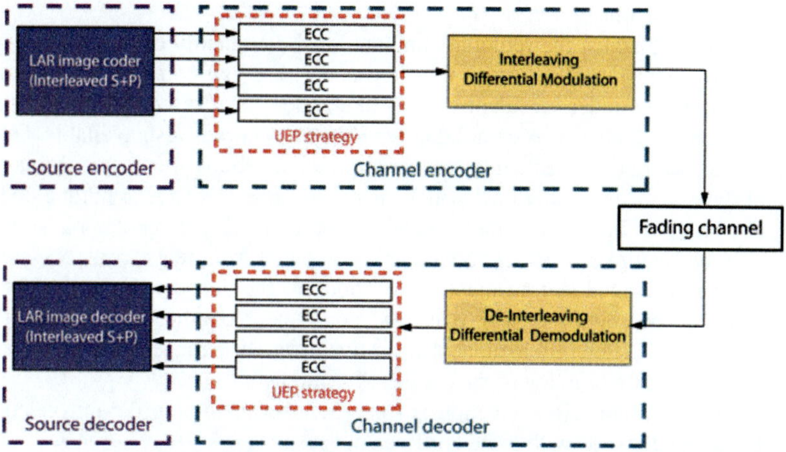

Fig. 4.21 Overall layout of the multi-layer transmission/compression system

In other words, compensating IP packet loss also requires a UEP process, which uses an exact and discrete Radon transform, called the Mojette transform [12]. The frame-like definition of this transform allows redundancies that can be further used for image description and image communication (Figure 4.22), for QoS purposes.

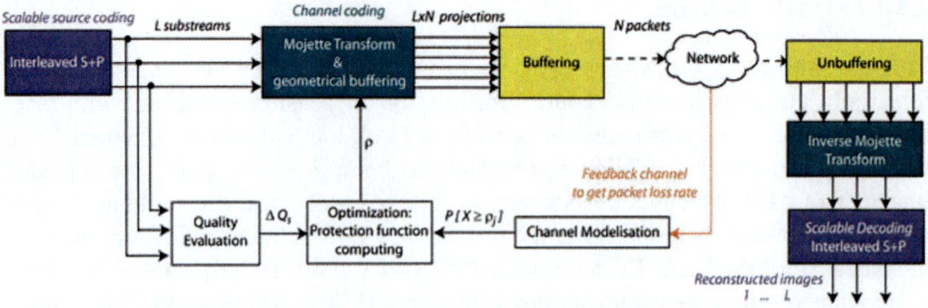

Fig. 4.22 General joint LAR-Mojette coding scheme

4.4.4.2 Error Resilience

When only considering source coding, some simple adapted solutions of error resilience can be implemented. Introducing resynchronization markers remains the easiest way of adding error resilience to an encoding process. In this respect, the idea is to adapt marker definition to the used entropy coder. Although generic markers that fit any entropy encoder can be implemented, we have designed specific markers adapted to our particular arithmetic Q15-like coder [38] together with the LAR bistream organization. Hence, different intra and inter substream markers have been defined. These distinct markers can also be used as stream identifiers to design an intelligent resynchronization process: if we consider entire IP packet loss, the system is automatically able to identify the lost packet and ask for its retransmission.

In addition, to adjust the required computational complexity of our system, we then simply adapt the classic Golomb-Rice coder, for low complex application, and the arithmetic coder, or adaptive MQ like coder, or adaptive Q15 coder, for better compression results. A semi-adaptive Huffman coder is also available. One-pass solution can be implemented with an a priori codebook: for medical images of same type (e.g. mammograms) which share the same statistics, a unique codebook can be used. Two-passes methods lead to build an adapted codebook, so that to reduce the final rate. If exact codebook is computed from real errors, two solutions can be envisaged to transmit this codebook to the decoder. First, the entire codebook can be sent, implying a natural consequent overhead. Secondly, the codebook can be efficiently estimated from five different parameters, which characterized the distribution law of the codebook symbols.

Moreover, internal error detection is freely realized thanks to the introduction of forbidden codewords within the Huffman coder.

Online entropy decoding can also take advantage of the properties of the coded residual errors. These errors are naturally bounded by the adaptive quadtree decomposition. As soon as this bound is not respected, an error can be detected. Thus an intrinsic MQF like decoding process [33] is also available for free.

In terms of complexity, the Q15-LAR coder is 2.5 times faster than the arithmetic coder, and the semi-adaptive Huffman coder is 2 times faster than the Q15-LAR coder.

Finally, as previously mentioned, these error resilience techniques can be coupled with UEP strategies, for optimal protection features.

4.5 Conclusion

This chapter was dedicated to joint medical image coding and securization framework. General principles and notions have been described and the joint source-channel coding context has been emphasized. Cryptography and data hiding were shown to be efficient solutions for content securization. In terms of error resilience, source-based together with channel-based coding have been developed. As an example of standard implementation of transmission process, the Long Term Evolution has been studied.

In the medical context, the LAR coding scheme has been developed to face the secure transmission issues. Embedded functionalities such as adapted selective cryptography, human vision-based steganography coupled with Unequal Error Protection and error resilience tools have been designed. The idea is to maintain good coding properties together with embedded Quality Of Service oriented system. This framework has been evaluated by the JPEG committee and has shown its global efficiency.

However, the exchange of medical data remains a key research topic. As for the moment, PACS oriented frameworks have limitations in terms of securization process durability. If classical medical frameworks use image coding schemes such as JPEG, JPEG2000, JPEGXR, securization processes act as only additional features. A complete joint system should be built in such a manner that both coding and secure properties would benefit from each other. This remains an open research area!

References

[1] 36.211, G.T.: Evolved Universal Terrestrial Radio Access (E-UTRA); physical channels and modulation, Release 9 (2009)

[2] 36.321, G.T.: Evolved Universal Terrestrial Radio Access (E-UTRA); medium access control (MAC) protocol specification, Release 9 (2009)

[3] 36.322, G.T.: Evolved Universal Terrestrial Radio Access (E-UTRA); radio link control (RLC) protocol specification, Release 9 (2009)

[4] 36.323, G.T.: Evolved Universal Terrestrial Radio Access (E-UTRA); packet data convergence protocol (PDCP) specification, Release 9 (2009)

[5] Abdulfetah, A.A., Sun, X., A.N. Mohammad, H.Y.: Robust Adaptive Image Watermarking using Visual Models in DWT and DCT Domain. Information Technology Journal 9(3), 460–466 (2010)

[6] Albanese, A., Blmer, J., Edmonds, J., Luby, M., Sudan, M.: Priority Encoding Transmission. IEEE Transaction on Information Theory 42(6), 1737–1744 (1996)

[7] Anderson, R.: Security Engineering - A Guide to Building Dependable Distributed Systems. Wiley (2008)

[8] Babel, M., Bédat, L., Déforges, O., Motsch, J.: Context-Based Scalable Coding and Representation of High Resolution Art Pictures for Remote Data Access. In: Proc. of the IEEE International Conference on Multimedia and Expo, ICME 2007, pp. 460–463 (2007)

[9] Babel, M., Déforges, O.: WG1N4870 - Response to call for AIC techniques and evaluation methods. Tech. rep., ISO/ITU JPEG commitee, San Francisco (2009)

[10] Babel, M., Déforges, O., Bédat, L., Strauss, C., Pasteau, F., Motsch, J.: WG1N5315 – Response to Call for AIC evaluation methodologies and compression technologies for medical images: LAR Codec. Tech. rep., ISO/ITU JPEG commitee, Boston, USA (2010)

[11] Babel, M., Déforges, O., Ronsin, J.: Interleaved S+P Pyramidal Decomposition with Refined Prediction Model. In: IEEE International Conference on Image Processing, ICIP 2005, Genova, Italy, vol. 2, pp. 750–753 (2005)

[12] Babel, M., Parrein, B., Déforges, O., Normand, N., Guédon, J.P., Coat, V.: Joint source channel coding: secured and progressive transmission of compressed medical images on the Internet. Computerized Medical Imaging and Graphics 32(4), 258–269 (2008)

[13] Bas, P., Marc Chassery, J., Davoine, F.: Using the fractal code to watermark images. In: Proc. Int. Conf. Image Processing ICIP, pp. 469–473 (1998)

[14] Bender, W., Butera, W., Gruhl, D., Hwang, R., Paiz, F.J., Pogreb, S.: Applications for data hiding. IBM Systems Journal 39, 547–568 (2000)

[15] Bender, W., Gruhl, D., Morimoto, N., Lu, A.: Techniques for data hiding. IBM Systems Journal 35(3/4), 313–336 (1996)

[16] Boeglen, H.: IT++ library for numerical communications simulations (2007), http://herve.boeglen.free.fr/itppwindows/

[17] Cayre, F., Chappelier, V., Jegou, H.: Signal processing and information theory library (2010), http://www.balistic-lab.org/

[18] Chu, Y., Ganz, A.: Wista: a wireless telemedicine system for disaster patient care. Mobile Networks and Applications 12, 201–214 (2007)

[19] Clunie, D.: DICOM Research Applications, Life at the Fringe of Reality. In: SPIE Medical Imaging, USA (2009)

[20] Clunie, D.: DICOM support for compression schemes - more than JPEG. In: 5th Annual Medical Imaging Informatics and Teleradiology Conference, USA (2009)

[21] Cox, I.J., Member, S., Kilian, J., Leighton, F.T., Shamoon, T.: Secure spread spectrum watermarking for multimedia. IEEE Transactions on Image Processing 6, 1673–1687 (1997)

[22] Dahlman, E., Parkvall, S., Skold, J., Beming, P.: 3G Evolution: HSPA and LTE for Mobile Broadband. Academic Press Inc. (2007)

[23] Déforges, O., Babel, M., Bédat, L., Ronsin, J.: Color LAR Codec: A Color Image Representation and Compression Scheme Based on Local Resolution Adjustment and Self-Extracting Region Representation. IEEE Trans. on Circuits and Systems for Video Technology 17(8), 974–987 (2007)

[24] Dufaux, F., Nicholson, D.: JWL: JPEG 2000 for wireless applications. In: SPIE Proc. Applications of Digital Image Processing XXVII, vol. 5558, pp. 309–318 (2004)

[25] Editors, J.: JPEG 2000 image coding system - Part 11: Wireless JPEG2000 Committee Draft. ISO/IEC CD 15444-11 / ITU-T SG8 (2004)

[26] Elias, P.: Coding for Noisy Channels. Convention Record 4, 37–49 (1955)
[27] Ferguson, N., Schneier, B.: Practical Cryptography. Wiley (2003)
[28] Flécher, E., Raulet, M., Roquier, G., Babel, M., Déforges, O.: Framework For Efficient Cosimulation And Fast Prototyping on Multi-Components With AAA Methodology: LAR Codec Study Case. In: Proc. of the 15th European Signal Processing Conference (Eusipco 2007), Poznan, Poland, pp. 1667–1671 (2007)
[29] Fonteneau, C., Motsch, J., Babel, M., Déforges, O.: A Hierarchical Selective Encryption Technique in a Scalable Image Codec. In: Proc. of International Conference in Communications (2008)
[30] Furon, T., Cayre, F., Fontaine, C.: Watermarking Security in Digital Audio Watermarking Techniques and Technologies: Applications and Benchmarks. In: Cvejic, Seppanen (eds.). Idea Group Publishing (2007)
[31] Gibaud, B.: The DICOM standard: a brief overview. In: Molecular Imaging: Computer Reconstruction and Practice. NATO Science for Peace and Security Series, pp. 229–238. Springer, Heidelberg (2008)
[32] Gilani, M., Skodras, A.N.: DLT-Based Digital Image Watermarking. In: Proc. First IEEE Balkan Conference on Signal Processing, Communications, Circuits and Systems, Istanbul, Turkey (2000)
[33] Grangetto, M., Magli, E., Olmo, G.: A syntax-preserving error resilience tool for JPEG 2000 based on error correcting arithmetic coding. IEEE Trans. on Image Processing 15(4), 807–818 (2006)
[34] Hamidouche, W., Olivier, C., Babel, M., Déforges, O., Boeglen, H., Lorenz, P.: LAR Image transmission over fading channels: a hierarchical protection solution. In: Proc. of The Second International Conference on Communication Theory, Reliability, and Quality of Service, Colmar France, pp. 1–4 (2009)
[35] Hashmi, N., Myung, D., Gaynor, M., Moulton, S.: A sensorbased, web service-enabled, emergency medical response system. In: Workshop on End-to-End, Sense-and-Respond Systems, Applications and Services, pp. 25–29 (2005)
[36] Hsu, C.T., Wu, J.L.: Multiresolution watermarking for digital images. IEEE Transactions on Circuits and Systems II: Analog and Digital Signal Processing 45(8), 1097–1101 (1998)
[37] Hwang, S.O.: Content and service protection for iptv. IEEE Transactions on Broadcasting 55(2), 425–436 (2009)
[38] I.-T.I.-T.T.: 81 (JPEG-1)-based still-image coding using an alternative arithmetic coder. Tech. rep., ISO/ITU JPEG commitee (2005)
[39] Wolf, J.K., Wyner, A.D., Ziv, J.: Source coding for multiple description. Bell System Technical Journal 59(8), 1417–1426 (1980)
[40] Kang, J.S., You, Y., Sung, M.Y.: Steganography using block-based adaptive threshold. In: 22nd International Symposium on Computer and Information Sciences, ISCIS 2007, pp. 1–7 (2007)
[41] Kunt, M., Ikonomopoulos, A., Kocher, M.: Second Generation Image Coding Techniques. Proceedings of the IEEE 73(4), 549–575 (1985)
[42] Li, J., Zhang, X., Liu, S., Ren, X.: An adaptive secure watermarking scheme for images in spatial domain using fresnel transform. In: 1st International Conference on Information Science and Engineering (ICISE), pp. 1630–1633 (2009)
[43] Liu, X., Eskicioglu, A.M.: Selective encryption of multimedia content in distribution networks: challenges and new directions. In: Conf. Communications, Internet, and Information Technology, pp. 527–533 (2003)

[44] Mathon, B., Bas, P., Cayre, F., Macq, B.: Comparison of secure spread-spectrum modulations applied to still image watermarking. Annals of Telecommunication 11-12, 810–813 (2009)
[45] Meerwald, P., Uhl, A.: A survey of wavelet-domain watermarking algorithms. In: Proceedings of SPIE, Electronic Imaging, Security and Watermarking of Multimedia Contents III, pp. 505–516. SPIE (2001)
[46] Mehlführer, C., Wrulich, M., Ikuno, J.C., Bosanska, D., Rupp, M.: Simulating the long term evolution physical layer. In: Proc. of the 17th European Signal Processing Conference (EUSIPCO 2009), Glasgow, Scotland (2009)
[47] Mehlführer, C., Wrulich, M., Ikuno, J.C., Bosanska, D., Rupp, M.: Simulating the Long Term Evolution Physical Layer. In: Proc. of the 17th European Signal Processing Conference (2009)
[48] Mohr, A., Riskin, E.A., Ladner, R.E.: Unequal Loss Protection: Graceful degradation of image quality over packet erasure channels through forward error correction. Journal on Selected Areas in Communications 18(6), 819–828 (2000)
[49] Motsch, J., Babel, M., Déforges, O.: Joint Lossless Coding and Reversible Data Embedding in a Multiresolution Still Image Coder. In: Proc. of European Signal Processing Conference, EUSIPCO, Glasgow, UK, pp. 1–4 (2009)
[50] Motsch, J., Déforges, O., Babel, M.: Embedding Multilevel Image Encryption in the LAR Codec. In: IEEE Communications International Conference 2006, Bucharest, Romania (2006)
[51] Norcen, R., Podesser, M., Pommer, A., Schmidt, H.P., Uhl, A.: Confidential storage and transmission of medical image data. Computers in Biology and Medicine 33(3), 277–297 (2003)
[52] Norman, T.: The road to LTE for GSM and UMTS operators. Tech. rep., Analysys Mason (2009)
[53] Oosterwijk, H.: The DICOM standard, overview and characteristics. Tech. rep., Ringholm Whitepapers (2004)
[54] Pattichis, C., Kyriacou, E., Voskarides, S., Pattichis, M., Istepanian, R., Schizas, C.: Wireless telemedicine systems: An overview. IEEE Antennas and Propagation Magazine 44(2), 143–153 (2002)
[55] Pedersen, P.C., Sebastian, D.: Wireless Technology Applications in a Rural Hospital. In: 2004 American Telemedicine Association Annual Meeting (2004)
[56] Reed, I., Solomon, G.: Polynomial Codes Over Certain Finite Fields. Journal of the Society of Industrial and Applied Mathematics (SIAM) 2, 300–304 (1960)
[57] Ruanaidh, J., Dowling, W., Boland, F.: Phase watermarking of digital images. In: International Conference on Image Processing, vol. 3, pp. 239–242 (1996)
[58] Ruanaidh, J.J.K., Pun, O.: D'informatique Rotation, scale and translation invariant digital image watermarking. IEEE International Conference on Image Processing ICIP1997, 536–539 (1997)
[59] Schneier, B.: Applied Cryptography, 2nd edn. John Wiley & Sons (1996)
[60] Sesia, S., Toufik, I., Baker, M.: LTE, The UMTS Long Term Evolution: From Theory to Practice. Wiley (2009)
[61] Shinohara, M., Motoyoshi, F., Uchida, O., Nakanishi, S.: Wavelet-based robust digital watermarking considering human visual system. In: Proceedings of the 2007 Annual Conference on International Conference on Computer Engineering and Applications, pp. 177–180 (2007)

[62] Signal & Information Processing Lab, D.U.o.T.: Image and Video Compression Learning Tool VcDemo (2004), http://siplab.tudelft.nl/content/image-and-video-compression-learning-tool-vcdemo

[63] Sneha, S., Dulipovici, A.: Strategies forWorking with Digital Medical Images. In: HICSS 2006: Proceedings of the 39th Annual Hawaii International Conference on System Sciences, vol. 5, p. 100 (2006)

[64] Taubman, D.S., Marcellin, M.W.: JPEG2000: Image Compression Fundamentals, Standards, and Practice. Kluwer Academic Publishers (2001)

[65] Temal, L., Dojat, M., Kassel, G., Gibaud, B.: Towards an ontology for sharing medical images and regions of interest in neuroimaging. Journal of Biomedical Informatics 41(5), 766–778 (2008)

[66] Tian, J.: Reversible data embedding using a difference expansion. IEEE Transactions on Circuits and Systems for Video Technology 13, 890–896 (2003)

[67] Tian, J., Wells Jr., R.O.: Reversible data-embedding with a hierarchical structure. In: 2004 International Conference on Image Processing, ICIP, vol. 5, pp. 3419–3422 (2004)

[68] Tian, J., Wells, R.O.: Reversible data-embedding with a hierarchical structure. In: ICIP, vol. 5, pp. 3419–3422 (2004)

[69] Tsui, T.K., Zhang, X.P., Androutsos, D.: Color image watermarking using multidimensional Fourier transforms. IEEE Transactions on Information Forensics and Security 3(1), 16–28 (2008)

[70] Uhl, A., Pommer, A.: Image and Video Encryption - From Digital Rights Management to Secured Personal Communication. In: Advances in Information Security, vol. 15. Springer, Heidelberg (2005)

[71] Van Droogenbroeck, M., Benedett, R.: Techniques for a selective encryption of uncompressed and compressed images. In: ACIVS Advanced Concepts for Intelligent Vision Systems, Ghent, Belgium, pp. 90–97 (2002)

[72] Vucetic, J.: Telemedicine: The Future of Wireless Internet Applications. In: Southeast Wireless 2003 (2003)

[73] Xia, X.G., Boncelet, C.G., Arce, G.R.: A multiresolution watermark for digital images. In: IEEE International Conference on Image Processing (ICIP), pp. 548–551 (1997)

[74] Yang, M., Bourbakis, N., Li, S.: Data, image and video encryption. IEEE Potentials, 28–34 (2004)

[75] Zhao, J., Koch, E.: Embedding robust labels into images for copyright protection. In: Proceedings of the International Congress on Intellectual Property Rights for Specialized Information, Knowledge and New Technologies, pp. 242–251 (1995)

Chapter 5
Image Processing in Medicine

Baigalmaa Tsagaan and Hiromasa Nakatani

Shizuoka University, Japan

5.1 Introduction

The development of medical imaging, such as x-ray computed tomographic (CT), magnetic resonance imaging (MRI) or ultrasound (US) imaging etc., has undergone revolutionary changes over the past three decades. Recently developed CT and MRI scanners are more powerful than previous machines providing the sharpest images with high resolution ever seen, without absorbing much radiation during procedures. Medical imaging is an important part of routine care nowadays[1]. It allows physicians to know what is going on inside a patient's ever-complex body.

These have not only been driven by advanced acquisition technologies of imaging, but also by significant challenges in computational image processing and analysis techniques[2]. For example, in cancer examine using CT images, shape analysis and texture classification techniques are used to assist the physician's diagnosis and to analyze cancer's risk. In surgery planning, three dimensional (3D) volumetric visualization of CT and MRI has become the standard for diagnostic care, and there is an increasing demand for surgical navigation systems.

These widespread aspects in medical imaging require the knowledge and application of image processing, pattern recognition and visualization methods[3]. Image segmentation, enhancement and quantification analysis are essential for an automated recognition and diagnosis. 3D visualization techniques are used broadly through many applications, ranging from simple camera calibration to the virtual endoscopic view of medical images. The use of various information of one patient requires advanced reason-based algorithms for image registration and visualization. Thus, in this chapter we focus on the image processing, pattern analysis and computer vision methods in medicine.

This chapter is organized as follows: We start with brief overview of medical image acquisition systems and move to general approaches of image processing and vision applications in medicine. First part reviews conventional issues of medical imaging: image modalities, image reconstruction, use of medical imaging in diagnostic practice. Second part emphasizes those methods that are appropriate when medical images are the subjects of image processing and analysis. Overview of segmentation and registration algorithms is presented briefly.

R. Kountchev et al. (Eds.): Advances in Reasoning-Based Image Processing, ISRL 29, pp. 127–146.
springerlink.com

The final section of the chapter presents a more detailed view on the recent practices incorporating interdisciplinary fields of computer aided diagnosis (CAD), computer-assisted surgery (CAS) systems and virtual endoscopy which encompass knowledge from medicine, image processing, pattern recognition and computer vision. This part gives an example of navigation system for paranasal sinus surgery. The recent issues of developments for medical imaging systems are summarized in the end of chapter.

5.2 Overview of Medical Imaging

In recent medical treatment, the imaging devices which visualize internal organs of human body are dispensable for the early diagnosis of the disease. The medical imaging is used to define normal or abnormal structures in the body and assist in procedures by helping to accurately guide the placement of instruments. The past 30 years have been remarkable developments in medical imaging technology. Academic side and industry have made huge investment in developing the technology needed to acquire images from multiple imaging modalities. There is very wide range of medical imaging modalities nowadays, such as X-ray, CT, MRI, PET, Ultrasonic imaging and so on. These modalities have greatly increased anatomical knowledge of human body for medical research and are a critical component in diagnosis and treatment planning.

5.2.1 Imaging Modality

In 1895, the invention of X-ray by was a remarkable discovery, one of the most important medical advancements in human history. X-ray technology allows doctors to see straight through human tissue to examine broken bones or cavities. X-rays were already being utilized clinically for visualizing bone fractures in the US in early 1896 and was a mainstream of medical imaging till 1970's. In 1971, CT was invented and followed by PET scanning in 1977. A CT scan is a large donut-shaped X-ray machine that takes X-ray images at many different angles around the body[4]. These images are processed by a computer to generate cross-sectional slice views of the body.

Been invented in 1980's, the use of MRI scanner has grown tremendously in just few decades[5,6]. MRI scanner uses powerful magnets to polarize and excite hydrogen nuclei in water molecules in human tissue, producing a detectable signal which is spatially encoded, that result images of the body. Unlike CT, which uses only X-ray attenuation to generate image contrast, MRI contrast related to photon density, relaxation times, flow, and other parameters. By variation of scanning parameters, tissue contrast can be altered and enhanced in various ways to detect different features. MRI can generate cross-sectional images in any plane. In the past, CT was limited to acquiring images in the axial plane. However, the development of multi-detector CT scanners with near-isotropic resolution allows the CT scanner to produce data that can be retrospectively reconstructed in any plane with minimal loss of image quality. The same tomographic reconstruction technique is used to generate slice images in CT and MRI.

Because CT and MRI are sensitive to different tissue properties, the appearance of the images obtained with the two techniques differ markedly. In CT, X-ray images are sensitive to tissue density and atomic composition, the image quality is poor at soft tissues. In MRI, while any nucleus with a net nuclear spin can be used, the proton of the hydrogen atom returns a large signal. This nucleus, present in water molecules, allows the excellent soft-tissue contrast achievable with MRI.

CT can be enhanced by use of contrast agents containing elements of a higher atomic number such as iodine or barium. Contrast agents for MRI have paramagnetic properties, e.g., gadolinium and manganese.

For purposes of tumor detection and identification in the brain, MRI is generally superior[7]. However, in the case of solid tumors of the abdomen and chest, CT is often preferred due to less motion artifact. Furthermore, CT usually is more widely available, faster, and less expensive. MRI is also best suited for cases when a patient is to undergo the exam several times successively in the short term. Unlike CT, MRI does not involve the use of ionizing radiation and is therefore less-invasive imaging technique.

Imaging modalities of nuclear medicine, such as Positron Emission Tomography (PET) and Single Photon Emission Computed Tomography (SPECT) encompasses both diagnostic imaging and treatment of disease[8]. Nuclear medicine uses certain properties of isotopes and the energetic particles emitted from radiopharmaceutical material to diagnose or treat various pathology. Different from the typical concept of anatomic radiology, nuclear medicine enables assessment of physiology. This function-based approach to medical evaluation has useful applications in oncology, neurology, and cardiology. Gamma rays are used in SPECT and PET to detect regions of biologic activity that may be associated with disease[9]. Relatively short-lived positron isotope is administered to the patient in SPECT imaging. In PET, emitting isotope is incorporated with an organic substance such as glucose, which can be used as a marker of metabolic utilization. Isotopes are often preferentially absorbed by biologically active tissue in the body, and can be used to identify tumors, fracture points in bone, metastasis, or infection. The radioactive gamma rays are emitted through the body as the natural decaying process of these isotopes takes place. The emissions of the gamma rays are captured by detectors that surround the body. A dual detector head gamma camera combined with a CT scanner, which provides localization of functional SPECT data, is termed a SPECT/CT, and has shown utility in advancing the field of molecular imaging. Modern scanners that combine SPECT or PET with CT, or MRI, can easily optimize the image reconstruction involved with positron imaging. This is performed on the same equipment without physically moving the patient off of the gantry. The resultant hybrid of functional and anatomic imaging information is a useful tool in non-invasive diagnosis and patient management.

Ultrasound detects subtle changes in acoustic impedance at tissue boundaries and diffraction patterns in different tissues, providing discrimination of different tissue types[10]. Doppler ultrasound provides images of flowing blood.

5.2.2 Image Reconstruction

In medical imaging, images could be acquired in the continuous domain such as on X-ray film, or in discrete space as in MRI. The location of each measurement is called a pixel in 2D, a voxel in 3D discrete images, respectively. In medical imaging, a variety of practical reconstruction algorithms have been developed to implement the process of reconstruction of a 3D object from its projections[11].

The mathematical basis for the tomographic imaging was laid down by Johann Radon. It is applied in CT scan to obtain cross-sectional images of the patient. In X-ray CT, the projection of an object at given angle θ is equal to a set of line integrals as it is shown in Figure 5.1. Here, the line integral represents the total attenuation of the beam of X-rays that pass through the object in a line path and result an image is 2D model of the attenuation coefficient $\mu(\theta)$. This integral data is collected as a series of parallel rays. Attenuation occurs exponentially in every pixel of tissue:

$$I = I_0 \, exp(- \int \mu(t,\theta) ds) \tag{5.1}$$

where I is detected radiation, I_0 is an initial radiation intensity, ds is thickness of reconstruction, $\mu(t,\theta)$ indicates the attenuation coefficient of the tissue at position t, across projection at angle θ. As Eq. (1) presents that the detected X-ray contains information relating to every pixel in the path of beam.

The reconstruction problem is to decode these mixes of the information into spatial information throughout the subject, by combining information from every

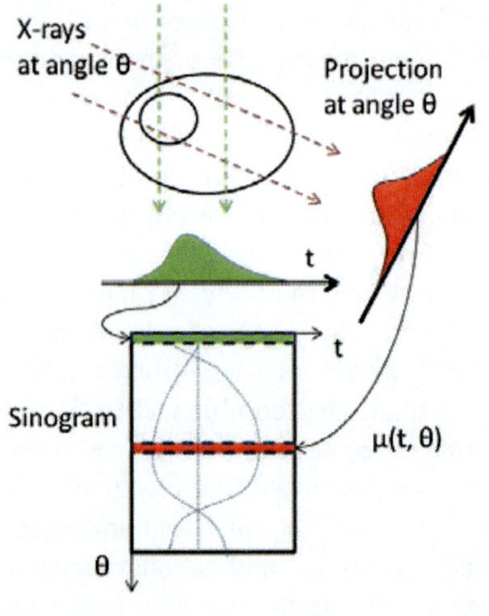

Fig. 5.1 A set of line integrals through the object is obtained in each projection.

path through the object. Thus, the artifact may occur due to the underlying physics of the energy-tissue interaction. It is called a partial volume effect in medical imaging, where multiple tissues contribute to a single pixel resulting in a blurring of intensity across boundaries. A higher resolution decreases this effect, as it better resolves the tissue. The method to correct for the partial volume effect is referred to as partial volume correction[12]. Further misrepresentations of tissue structures, so called artifacts, could occur in any modality of medical images, those that caused by errors in data acquisition such as patient motion. In clinical practice, physicians naturally learn to recognize these artifacts to avoid mistaking them for actual pathology.

5.2.3 Image Format

Medical imaging techniques produce very large amount of data, especially from CT, MRI modalities. In medical imaging, electronic Picture Archiving and Communication Systems (PACS) have been developed in an attempt to provide economical storage, rapid retrieval of images, access to images acquired with multiple modalities, and simultaneous access at multiple sites[13]. A PACS consists of four major components: the imaging modalities such as CT and MRI, a secured network for the transmission of patient information, workstations for interpreting and reviewing images, and archives for the storage and retrieval of images and reports. The universal format for PACS is Digital Imaging and Communications in Medicine (DICOM)[14]. DICOM includes a file format definition and a network communications protocol, thus patient data in DICOM format can be exchanged between two entities. DICOM has been widely adopted by hospitals and it is nowadays a common format in medical imaging applications.

5.2.4 Diagnostic Practice Using Medical Images

Medical imaging technologies now provide rich sources of data on the physical properties and biological function of tissues at wide range of spatial resolutions. Each successive generation of medical imaging system has acquired images faster, with higher resolution and improved image quality. At present, physician's diagnosis by using medical images depends greatly on his or her subjective skill which has achieved through long-term clinical experience. Therefore, there is constant request for image processing device that automatically detects the tumor or region of interest (ROI) from the large amount of images and help physician's decision quantitatively by excavating useful information from original images. Table 5.1 roughly categorizes the clinical application of medical imaging modalities.

Table 5.1 Clinical application of medical imaging modalities

Imaging Modality	Dimension; Resolution	Safety	Common Examination
X-ray, CT	2D/3D; high	invasive	chest X-ray, mammography, brain CT, CT mammography, abdominal CT, chest CT, CT colonography, cardiac CT angiogram, etc.
MRI	3D; high	less-invasive	brain MRI, MRI angiography, MR spectroscopy, functional MRI, real-time MRI,
PET/SPECT	3D; low	less-invasive, but involves exposure to ionizing radiation	mostly tumor imaging, brain imaging, myocardial perfusion scan, thyroid imaging, bone imaging, functional brain imaging
Ultrasound	2D; medium	less-invasive	obstetric ultrasound, echocardiography, abdominal sonography, intracerebal ultrasonography, intravascular ultrasound

5.3 Conventional Approaches of Image Processing

In a typical imaging application of medicine, image processing may include four stages: data acquisition, image enhancement, feature extraction, and decision making in the application or visualization. The goal of image acquisition is to capture the suitable signal from target. At this stage, the main concern is to avoid losing information, while reducing the artifacts such as partial volume effect. The goal of image enhancement is to eliminate or reduce extraneous components such as noise from the image. Feature extraction means identifying and measuring a number of parameters or features that best the ROI in an image such as tumor region. Finally, the extracted information must be used in the application e.g. in the visualization systems, or in the detection of computer-aided diagnosis.

This section will describe an overview of image segmentation and registration problems which are most important procedures in any medical imaging system. Methods and applications are described briefly. A full description of competing methods is beyond the scope of this section and the readers are referred to references for additional details. Section focuses on providing the reader an introduction to the different applications of segmentation and registration in medical imaging and the various issues that must be confronted. Also, section refers only to the most commonly used radiological modalities for imaging anatomy: X-ray CT, MRI.

5.3.1 Image Segmentation

Image segmentation plays crucial role in many medical imaging applications involving measurement of tissue volumes, treatment planning[15], registration[16], computer-aided diagnosis[17], computer-assisted surgery[18] systems and 3D visualization[60]. For example, it is a key component in following applications: In study of brain development or brain functional mapping; in detection of micro-classifications on mammograms; in detection of coronary borders in angiograms; in surgery simulations or planning *etc.*

However, there is no gold-standard that yields satisfactory segmentation results for all medical imaging application. General imaging artifacts such as noise, partial volume effects, and patient's motion can also have significant consequences on the performance of segmentation methods. Methods that are specialized to particular applications can often achieve better performance by taking into account prior knowledge on a gray level appearance or shape characteristics. Since there is no general solution to the image segmentation problem, segmentation techniques often have to be combined with domain knowledge in order to effectively solve the existing problem.

Segmentation techniques can be divided into several categories, depending on the classification scheme, imaging modality, and specific application:

- Manual, semiautomatic or interactive, and automatic
- Pixel or region-based (thresholding, region growing, edge-based, watershed, morphological), knowledge model -based (expectation/maximization algorithm, Bayesian prior model[19], probability functional, 3D atlas mapped[20]), deformable model-based (snakes, deformable surfaces, level-set method[21],)
- Classical (thresholding, edge-based, region growing), fuzzy clustering, statistical atlas mapped[22], hierarchical, neural network techniques.

The simplest way to obtain good segmentation results is segmentation by man. However, manual procedure can be laborious and time consuming for large population studies. The type of interaction required by segmentation methods can range from the selection of a seed point for a region growing algorithm, to a manual delineation of entire structure. Even automated segmentation methods typically require some interaction for specifying initial parameters that can significantly affect performance. An automated segmentation method needs to reconcile gray level appearance of tissue, characteristics of imaging modality, geometry of anatomy.

Pixel or region-based segmentation is performed by partitioning the image into clusters of pixels that have strong similarity in a feature space. Basic operation is to examine each pixel and assign it to the cluster that best represents the value of its characteristic feature vector of interest. The clustering algorithm shows high performance, when it incorporates statistical priori-knowledge. For example, expectation/maximization (EM) algorithm[31] applies the clustering principles with the underlying assumption that the data follows a Gaussian mixture model. It iterates between computing the posterior probabilities and computing maximum likelihood estimation of the means, co variances, and mixing coefficients of the mixture model.

A number of deformable model- based segmentation techniques can be found in the literatures, such as active contours, snakes[23], deformable models[24], Fourier surface models[25], coupled surface propagation using level set methods. A lot of work has presented deformable model that combined with domain of prior-knowledge.

To illustrate the advantages of use of priori-knowledge, the results of kidney segmentation method that combines deformable model[26] with prior-knowledge, is shown in Figure 5.2. In this work, the segmentation uses a prior-knowledge of shape curvature of kidney in order to deform initial shape onto a desired boundary of kidney. Figure 5.2A present the original CT images of abdomen in the coronal plane. Figure 5.2B shows a result of kidney segmentation obtained by deformable model, using only intensity information derived from the images. Considerable improvement is evident from the results of segmentation incorporating priori-knowledge, as it is shown in Figure 5,2C. The main advantages of deformable models are their ability to directly generate closed parametric curves or surfaces from images and their incorporation of a smoothness constraint that provides robustness to noise and spurious edges. A disadvantage is that they require manual interaction to place an initial model and choose appropriate parameters.

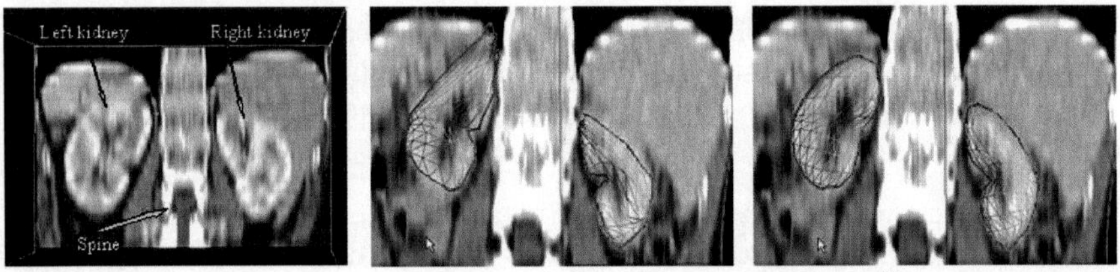

A B C

Fig. 5.2 Segmentation of kidney region using deformable model with priori-knowledge of shape. Figure 5.2A presents the original CT image of abdomen in the coronal plane. Figure 5.2B shows segmented surfaces of kidney using only image intensity information. Figure 5.2C shows the resultant segmentation that incorporates priori-knowledge of shape of kidney.

Recently, there has been proposed a 3D volume-based segmentation that uses atlas information to guide segmentation, mostly for the MRI segmentation of the human brain[27] that does labeling images according to tissue type of brain such as white matter, gray matter, and cerebrospinal fliud[20]. Basics of atlas-guided approaches are similar to classifiers except they are implemented in the spatial domain of the image rather than in a feature space and these deal segmentation as a registration problem[28]. An advantage of atlas-guided approaches is that gives robust accurate segmentations of even in complex structures which is in difficult to apply other techniques due to anatomical variability.

Traditionally, most segmentation techniques use one modality of images. Performance of these techniques can be improved by combining images from multimodality sources or integrating images over time. Especially in the brain segmentation issue, many algorithms have been presented using in multi-modality images: k-means[29], neural networks algorithms[30], EM algorithms[31]. Kapur *et al.* presented a segmentation method of brain tissues for evaluation of cortical thinning' aging[32] that successfully combines the strengths of three techniques: EM algorithm, binary mathematical morphology, and active contour models. These multi-modality techniques require images to be properly registered, in order to reduce noise and increase performance of the segmentation.

Segmentation is an important step in many medical imaging applications. The selection of the suitable technique for a given application is difficult task. It depends on careful analysis of image quality in terms of its modality and definition of a segmentation goal. Usually, the combination of a several techniques is necessary and integrating of images from different modalities and priori-knowledge helps to improve its performance.

5.3.2 Image Registration

Medical imaging is about establishing shape, structure, size and spatial relationships of anatomical structures within the patient[33]. Establishing the correspondence of spatial information in medical images is a fundamental to image interpretation and analysis. Registration methods compute spatial transformations between coordinate systems that establish correspondence between points or regions within image, or between physical space and images. It is common for patient to be imaged multiple times or imaged with different modalities.

Registration techniques in medical imaging can be divided into classes in many ways, depending on the spatial dimension, imaging modality, data presentation, optimization scheme and specific application:

- 2D/2D, 2D/3D, 3D/3D
- Inter-modality(registration between images of different modality[34]), intra-modality (comparison within same modality images)
- Rigid (landmark-based, surface-based[35]), non-rigid[36] (elastic or fluid registration, finite element methods using mechanical model, intensity-based method[37])
- In-vivo(registration of intra-operative images or surgical instrument into pre-operative images), out-vivo(registration of pre-operative images, registration of scanning devices in the operating room before intervention)

Although a great survey of the registration approaches in medical imaging can be found in Hanjal[16], this section try to cover above categories briefly.

Registration issue depends on the number of spatial dimensions involved. Most current works focus on the 3D/3D registration of two images. 3D/3D registration normally applies to the registration of two tomographic images. Careful calibration of each scanning device is required to determine image scaling and size of the voxels in each modality. An alignment of single tomographic slice to spatial data would be 2D/3D registration. 2D/2D registration may apply to separate slices from one tomographic data. Compared to 3D/3D registration, 2D/2D registration is less complex and faster. Most 2D/3D applications concern intra-operative procedures within the operating room, so speed issues need to be addressed as constrained.

Inter-modality registration enables the combination of complementary information from different modalities, and intra-modality registration enables accurate comparison between images from same modalities. An example of the use of registering different modalities can be found in radiotherapy treatment planning. For example, the use of MRI and PET combined would be beneficial[38], as the former is better suited for delineation of tumor tissue, while the latter is needed for accurate computation of the radiation dose. Registration of the images from any combination will benefit the physician. Time series of images are acquired for various reasons, such as monitoring of bone growth in children, monitoring of tumor growth, post-operative monitoring of healing. If two images need to be compared, registration will be necessary except in instances of ultra-short time series, where the patient does not leave the scanner between scanning procedures.

Registration algorithms compute spatial transformation between coordinate systems of image or physical space of patient. When only translations and rotations are allowed by registration then it is called rigid registration. The goal of rigid registration is to find the six degrees of freedom (three rotations and three translations) of transformation, which maps any point in source image into the corresponding point in the target image. Conventional rigid registration algorithm is a landmark-based registration, in which coordinate system of landmark points is translated into those of corresponding points in the other image. Such algorithms transformation that optimizes the average distance between each landmark. Iterative Closest Point algorithm[39] is the well-known method for point-based registration.

In many applications a rigid transformation is sufficient to achieve the spatial relationship between two images. For example, brain images of the same subject can be related by a rigid transformation since the motion of the brain is constrained by the skull[40]. However there are many other applications where non-rigid transformations are required to describe the spatial relationship between images adequately. In intra-modality registration non-rigid transformations are required to accommodate any tissue deformation due to interventions or changes over time.

Any non-rigid registration can be described by three components: a transformation which relates the target and source images, a similarity measures between target and source images, and optimization which determines the optimal transformation parameters as a function of the similarity measure. In non-rigid registration, more degree of freedom is required than that of rigid registration. Registration of images to an atlas or images from another individual, or registration of tissue that deforms over time is an example of non-rigid registrations. By adding additional degree of freedom such a linear transformation model can be extended to nonlinear transformation model. Several non-rigid registration techniques in the area of medical imaging are presented in the past: B-splines[41], elastic registration, finite element methods using mechanical models. There are a large number of applications for non-rigid registration: Correction of image acquisition errors, for example in MRI; Intra-modality registration of breast mask region in mammography over time; Motion analysis of brain region during intervention in neurosurgery[42] etc.

Establishing this correspondence allows the image to be used to guide, direct, and monitor therapy. In the last few years, image registration techniques have entered routine clinical use in image-guided neurosurgery systems and computer-assisted orthopedic surgery.

5.3.3 Visualization

3D medical image is a stack of 2D slice images that have a regular number of image pixels in regular size within slice, but not regular between slices. Interpolation of slice images into a volume with isotopic pixel resolution and visualization of volume in an optimal chosen plane makes the medical imaging system convenient to exam the target structure. In this way, a quantitative measurement of length and examination of cross sectional area of target can be made for diagnosis, treatment or for intervention.

Typical screen layout for diagnostic software shows one 3D volume and three views (axial, coronal, sagittal views) of multiplanar reconstruction[43]. In which, 3D volume visualization can be done in two ways: surface rendering, direct volume rendering[44]. In surface rendering, target structure or ROI is segmented from volume data. Then the segmented region is constructed as a 3D model volume and displayed on the screen. Multiple models can be constructed from various regions, allowing different colors to represent each anatomical component such as bone, muscle, and cartilage. The marching cubes algorithm[45] is a common technique for surface rendering.

Direct volume rendering is a computationally intensive task that may be performed in several ways. In volume rendering, transparency and colors are used to allow a better representation of the volume to be displayed in single image. Perspective projection methods, such as maximum-intensity projection (MIP) or minimum-intensity projection (mIP) address to volume rendering method.

5.4 Application

As it has been described previously, advances of medical imaging and computer-based technology have greatly increased interpretation of medical images. Nowadays, a number of interdisciplinary complex applications are onstage that aim to provide a computer output as a second opinion to assist physician's diagnosis or to contribute surgeon's intervention or just to simulate throughout the human body. These applications include various categories of image and pattern processing algorithms, such as segmentation, registration, classification, modeling, rendering and so on.

This section will briefly introduce basics of recent advanced computer-based systems in medicine, such as computer-aided diagnosis (CAD), computer assisted surgery (CAS) systems and virtual endoscopy in conjunction with one example.

5.4.1 CAD, CAS and Virtual Endoscopy

In a brief, CAD is a combination of computerized algorithms that is developed and optimized to assist radiologists in the diagnosis of possible abnormality, mostly tumor lesion. CAD system inputs medical images of the target structure and highlights conspicuous regions within the input image in terms of tumor diagnosis. Computerized procedure of typical CAD system includes image pre-processing, segmentation or detection of ROI, feature extraction and classification. Each step requires intelligent image processing algorithms. Specially, a few thousand images are required to optimize the classification stage of the system. Basically, after the detection of suspicious region, every region is evaluated for the probability of true positive. There are so many scoring procedures proposed so far: nearest-neighbor rule, minimum distance classifier, Bayesian classifier, Support Vector Machine, radial basis function network etc. If the detected structures have reached a certain threshold level, they are highlighted in the resultant image for radiologist.

Today, CAD systems are available routinely: In the mammography screening, CAD uses breast images and highlights microcalcification clusters and hyperdense structures in the breast soft tissue[51]; In colonography, CAD uses abdominal CT images and detects the polyps by identifying bump-like shapes on an inner lining of the colon, rejecting haustral fold shapes of a normal colon wall[46, 47]; In the diagnosis of lung cancer, CAD uses CT images and detects small round lesions[48, 49], In the coronary CT angiography, CAD automatically detects coronary artery disease of deformity. The routine application of CAD systems helps physician to realize suspicious small changes in an image at the early stage of cancer development. Early detection of tumor lesion extends the survival-rate of patients, by making early therapy possible[50]. Today's CAD systems cannot detect 100% of pathological abnormalities. However its sensitivity depends on application[51]. Achievement of high sensitivity decreases the specificity of the CAD. Therefore the benefits of using CAD remain uncertain and most CAD systems play supporting role. The physician is always responsible for the final interpretation of a medical image.

Computer assisted surgery (CAS), also known as, image guided surgical navigation, represents a surgical concept and set of computer-based procedures that include image processing and real-time sensing technologies for pre-surgical planning, and for guiding surgery.

An accurate model of the surgical target should be acquired in the CAS. Therefore, the medical image of the target has to be scanned before intervention and uploaded into the computer system for the further image processing. In a case of using several inter-modality or intra-modality image datasets, they have to be combined with appropriate image registration techniques. During the intervention, the gathered dataset will be rendered as a virtual 3D model of the patient, and this model is manipulated by a surgeon to provide views from any point within the volume of target. Thus the surgeon can better assess the case and establish a more accurate diagnostic. The surgical intervention will be planned and simulated virtually, before actual surgery takes place.

Particularly, CAS fits most of the surgeon's needs in areas with limited surgical access and requiring high-precision actions, such as middle-ear surgery[52], in minimally invasive brain microsurgery. An application of CAS is widespread in routine interventions of hip replacement[53] or bone segment navigation[54] in orthopedics, where CAS is useful for pre-planning and guiding the correct anatomical position of displaced bone fragments in fractures. Basically CAS improves surgeon performance and decreases the risk of surgical errors and reduces the operating time.

Virtual endoscopy (VE) provides endoscopic simulation of patient specific organs similar to those produced by conventional endoscopic procedures. Typical routine of endoscopic procedures are invasive and often uncomfortable for patients. Use of VE avoids the risks that associated with real endoscopy when used prior to performing an actual endoscopic exam. Moreover, non-reachable body regions through real endoscopy can be explored with VE[55].

Overall process of development of VE systems may consist of following steps. Acquired 3D images are input into computer. Some image pre-processing is performed in order to prepare initial images for the modeling. This pre-processing step includes interpolation of the dataset into isotropic volume, multimodality spatial registration, and segmentation of target structure. The segmented region is then converted to a polygonal surface representation. The endoscopic display procedure is then simulated in two ways: A pre-determined fly-through-path views that are rendered in an animation; Real-time display using an interactive simulator.

A number of investigators have been working in this field: virtual colonoscopy[56], 3D fly-through of carotid arteries[57], patient specific 3D organ visualizations and interactive organ fly-through[58], simulated endoscopy to a variety of intra-parenchymal visualizations[59]. Recent work characterizes a rapidly maturing development and evaluation of VE in various applications. Studies on VE have grown tremendously after a release of visible human datasets (VHD) from the National Library of Medicine, NIH, USA[60]. VHD are multimodality (CRYO, CT, MRI) whole body images of male and female that has isotropic high resolution and available for free of charge. VHD is well-suited dataset not only to develop VE simulation, but also to evaluate effectiveness of image processing methods for applications in clinical diagnosis and therapy.

5.4.2 Image-Guided Navigation for Paranasal Sinus Surgery

A nasal area has very complex structure covered by face bones, hence it is very difficult to operate surgery in a small visual field and it demands high surgical skills in the endoscopic paranasal sinus surgery[61]. Furthermore, very important organs, such as brain and optic nerves, exist in the neighborhood of operation target. Tracking of surgical instruments on the preoperative CT image is highly required to prevent surgical accidents as well as to obtain accurate image guidance[62, 63], Generally, navigation proceeds by tracking a pointer whose position can be estimated based on the head band fitted rigidly to the patient in conventional systems[64, 65]. The head set must be worn by the patient during the CT scan and during the intervention. Head set allows precise registration; however, its usage implicates invasiveness, lots of restriction on the equipments. Furthermore, time-consuming set-up procedures of head set are main drawbacks in this navigation. Recently, some clinical experiences of the use of electromagnetic or optoelectric navigation systems which are developed particularly for the surgical navigation in paranasal sinus surgery, have been reported not only the advancements but almost necessities of navigation systems, at least for difficult surgical procedures[66, 67].

Lately, Tsagaan et.al. have presented marker-less navigation system for par nasal sinus surgery[68, 69]. Main solution of the system relates to establishing marker-less registration between preoperative images and a patient in the surgical room. Thus, a frameless tracking of a surgical tool is realized is realized during intervention. Before the intervention, patient's facial surface is acquired by optical-based 3D range scanner and registered to the facial surface extracted from the preoperative images. The use of optical 3D range device and facial surface allows the system to achieve easy-to-use and semi-automatic registration with less invasiveness. Once registration of 3D scanning device is done in the operation room, the tracking of surgical tool is done intra-operatively by using 3D range scanner, thus, it does not need to any marker on the patient.

A schema of data processing of the CAS system is divided in two parts: preoperative procedures and intra-operative procedures, as it is shown in Figure 5.3. Procedures of (1) and (2) are done preoperatively, whereas, procedures of (3), (4), and (5) are intra-operative steps, respectively.

(1) Regular clinical CT images of patient that are taken without any markers are sufficient for the navigation propose.
(2) As for preoperative image processing, the facial skin surface is extracted from the 3D CT images. In particular, zero crossing edge detection[70] and threshold techniques have adapted to extract of facial skin surface from CT images. After extraction of whole facial surface, an appropriate region in the nose area of facial surface is set as ROI for following registration procedure.
(3) Registration of 3D range device in the operation room. The scanning of facial surface is done by 3D range scanner in the operation room before intervention. Then the scanned facial surface is registered into the facial ROI which has extracted in step (2). This registration establishes the relation between preoperative CT images and range scanning device (patient's physical space) in the operating room. Iterative Closest Point algorithm[39] is employed to match above mentioned two face surfaces.

(4) During intervention, 3D range image of the surgical instrument is measured. After an each measurement, the positions of spherical markers which are attached to the surgical instrument are calculated from the obtained range image. An instrument position is estimated based on the spatial relationship of these attached markers[71].
(5) As a consequence of two transformations in (3) and (4), the position and orientation of a surgical tool is determined in the preoperative image space. Visualization of a derived position of surgical tool is done in the preoperative images as an image-guidance for surgeon.

Figure 5.4 shows navigation results of the presented system. In Figure 5.4, two upper and lower left figures present preoperative CT images that are visualized in tri-planar (axial, coronal, sagittal) display view during the intervention. Crosshairs on each plane indicate the position of the tip of surgical tool in the preoperative CT image space. The surgical tool (red) together with above mentioned tri-planar images is shown in perspective view in the lower right figure. Face surface data taken before the surgery, and then used for the registration, is shown also in a green.

As a conclusion, we may say that, the main advantages of the presented navigation system are (a) marker-less on the patient's body, (b) an easy semiautomatic registration, (c) frameless during surgery, thus, it is feasible to update a registration and to restart the tracking when the patient moves.

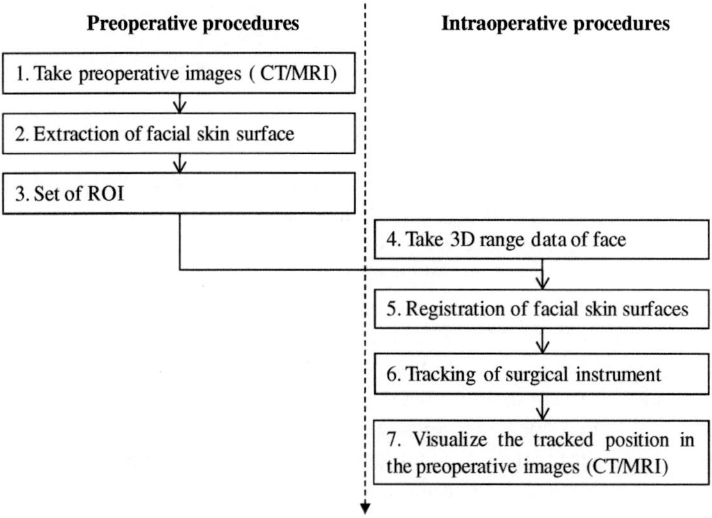

Fig. 5.3 Schematic flow of data processing in image-guided surgery

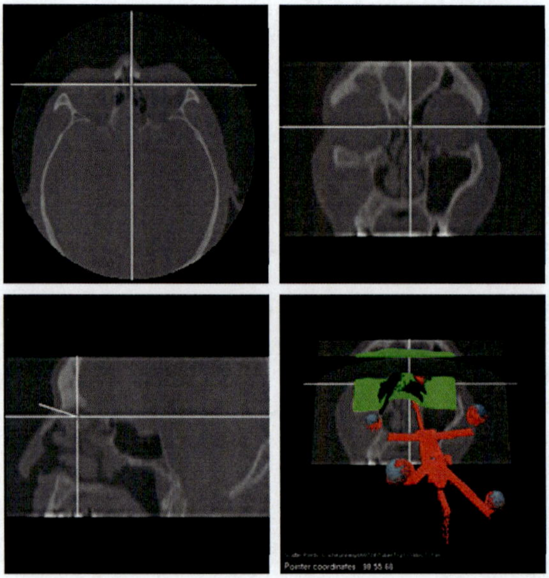

Fig. 5.4 Results of an image-guided intervention for paranasal sinus surgery. Upper and lower-left figures present preoperative CT images that are visualized in tri-planar view (axial, coronal, sagittal), respectively. Crosshairs on each picture indicate the position of surgical tool in the preoperative CT that derived after proper registration. Lower right figure presents a surgical tool (in red) and intra-operative 3D surface of face (in green) in conjunction with preoperative CT images in a perspective view.

5.5 Summary

Nowadays, the amount of data obtained in medical imaging is very extensive. With increasing size and number of medical images, the use of computers for facilitating clinical work has become necessary. Within the current clinical setting, medical imaging is a vital component of a large number of applications: throughout the clinical track of events; in the area of diagnosis, treatment planning, evaluation of surgical or radiotherapy procedures. This chapter described several aspects of a fast-moving medical imaging field and its current state of the arts.

Recent innovations in image processing techniques do not just enable better use of images, and it also opens up new application or new possibilities for physician: Segmentation of serial images enables to monitor subtle changes due to disease progression or treatment; Registration enables a surgeon to use the pre-operative images to guide intervention, that significantly improves surgeon performance and decreases the risk of surgical errors. The performance of CAD and CAS proves that they can provide accurate and reproducible measurements for clinical use as a second opinion.

Although a rapid progress of development has been made toward successful solution of technical problems of medical imaging and toward realization of the presented applications, a clinical acceptance of developed techniques depends on its computational cost, a sufficient validity and ease of use; nevertheless new technologies might produce unexpected risks for the patient. At the same time, ethical issues involved for each newly developed device or technology have to be discussed by all means.

References

1. Udupa, J.K., Herman, G.T.: 3D imaging in medicine. CRC Press (2000)
2. Dhawan, P.A.: Medical imaging analysis. Wiley-IEEE (2003)
3. Bankman, I.: Handbook of medical imaging: Processing and analysis. Academic Press (2000)
4. Napel, S.A.: Basic principles of spiral CT. In: Fishman, E.K., Jeffrey, R.B. (eds.) Principles and techniques of 3D spiral CT angiography, pp. 167–182. Raven Press (1995)
5. Lauterbur, P.C.: Image formation by induced local interactions: Examples of employing nuclear magnetic resonance. Nature 242, 190–191 (1973)
6. Filler, A.G.: The history, development, and impact of computed imaging in neurological diagnosis and neurosurgery: CT, MRI, DTI. Int. J. Neurosurgery 7(1) (2010)
7. Deck, M.D., Henschke, C., Lee, B.C., DZimmerman, R., et al.: Computed tomography versus magnetic resonance imaging of the brain. A collaborative interinstitutional study. Clin. Imaging 13(1), 2–15 (1989)
8. http://www.snm.org/
9. Bailey, D.L., Townsend, D.W.: Positron emission tomography: basic sciences. Springer, Heidelberg (2005)
10. Wells, P.N.T.: Ultrasound imaging: review. Phys. Med. Biol. 51, R83–R98 (2006)
11. Herman, G.T.: Fundamentals of computerized tomography: Image reconstruction from projection. Springer, Heidelberg (2009)
12. Rousset, O.G., Ma, Y., Evans, A.C.: Correction for partial volume effects in PET: Principle and validation. J. of Nuclear Medicine 39(5), 904–911 (1998)
13. Choplin, R.: Picture archiving and communication systems: an overview. Radiographics 12, 127–129 (1992)
14. http://medical.nema.org/
15. Khoo, V.S., Dearnaley, D.P., Finnigan, D.J., Padhani, A., et al.: Magnetic resonance imaging: Considerations and applications in radiotheraphy treatment planning. Radiother. Oncology 42, 1–15 (1997)
16. Hajnal, J.V., Hill, D.L.G., Hawkes, D.J.: Medical image registration. CRC Press (2001)
17. Taylor, P.: Invited review: computer aids for decision-making in diagnostic radiology. Brit. J. Radiol. 68, 945–957 (1995)
18. Ayache, N., Cinquin, P., Cohen, I., Cohen, L., et al.: Segmentation of complex three-dimensional medical objects: a challenge and a requirement for computer-assisted surgery planning and performance. In: Taylor, R.H., Lavallee, S., Burdea, G.C., Mosges, R. (eds.) Computer integrated surgery: technology and clinical applications, pp. 59–74. MIT Press (1996)

19. Yan, M.X.H., Karp, J.S.: An adaptive Bayesian approach to three-dimensional MR brain segmentation. In: XIVth Int. Conf. Infor. Proc. in Med. Imag., pp. 201–213 (1995)
20. Andreasen, N.C., Rajarethinam, R., Cizadlo, T., et al.: Automatic atlas-based vol-ume estimation of human brain regions from MR images. J. Comp. Assist. Tom. 20, 98–106 (1996)
21. Osher, S., Fedkiw, P.R.: Level set methods and dynamic implicit surfaces. Springer, Heidelberg (2002)
22. Rajapakse, J.C., Giedd, J.N., Rapoport, J.L.: Statistical approach to segmentation of single-channel cerebral MR images. IEEE T. Med. Imag. 16, 176–186 (1997)
23. Kass, M., Witkin, A., Terzopoulos, D.: Snakes: Active contour models. Int. J. Comp. Vision 1, 321–331 (1988)
24. Davatzikos, C., Bryan, R.N.: Using a deformable surface model to obtain a shape representation of the cortex. IEEE T. Med. Imag. 15, 785–795 (1996)
25. Staib, L.H., Duncan, J.S.: Boundary finding with parametrically deformable contour models. IEEE T. Pattern Anal. Mach. Intell. 14, 1061–1075 (1992)
26. Tsagaan, B., Shimizu, A., Kobatake, H., Miyakawa, K.: Development of extraction method of kidneys from abdominal CT images using a three-dimensional de-formable model. Systems and Computers in Japan, 37–46 (2003)
27. Atkins, M.S., Mackiewich, B.T.: Fully automatic segmentation of the brain in MRI. IEEE T. Med. Imag. 17, 98–109 (1998)
28. Kikinis, R., Shenton, M.E., Losifescu, D.V., McCarley, R.W., et al.: A Digital brain atlas for surgical planning, model-driven segmentation, and teaching. IEEE T. Vis. and Comp. Graph. 2(3), 232–241 (1996)
29. Pham, D.L., Prince, J.L.: An adaptive fuzzy c-means algorithm for image segmentation in the presence of intensity in homogeneities. Patt. Rec. Let., 57–68 (1999)
30. Wismüller, A., Vietze, F., Dersch, D.R.: 'Segmentation with Neural Networks. In: Bankman, I.N., Frank, J., Brody, W., Zerhouni, E. (eds.) Handbook of medical imaging. Academic Press (2000)
31. Kay, J.: The EM algorithm in medical imaging. Stat. Methods Med. Res. 6(1), 55–75 (1997)
32. Kapur, T., Grimson, E., Wells, W., Kikinis, R.: Segmentation of brain tissue from magnetic resonance images. Med. Im. Anal. 1, 109–127 (1996)
33. Maintz, J.B.A., Viergever, M.A.: A survey of medical image registration. Med. Im. Anal. 2, 1–36 (1998)
34. Wells, W.M., et al.: Multi-modal volume registration by maximization of mutual Information. Med. Im. Anal. 1, 35–51 (1996)
35. Fischl, B., et al.: High-resolution inter-subject averaging and a coordinate system for the cortical surface. Human Brain Mapping 8, 272–284 (1999)
36. Risholm, P., Pieper, S., Samset, E., Wells, W.M.: Summarizing and Visualizing Uncertainty in Non-Rigid Registration. Med. Imag. Comp. Comp. Assist. Interv. 13(Pt 2), 554–561 (2010)
37. Thévenaz, P., Ruttimann, U.E., Unser, M.: A pyramid approach to subpixel registration based on intensity. IEEE T. Imag. Process. 7, 27–41 (1998)
38. Studholme, C., Hill, D.L.G., Hawkes, D.J.: 'Automated 3D MR and PET brain image registration. Comp. Assist. Radiology, 248–253 (1995)
39. Besl, P.J., McKay, N.D.: A method for registration of 3-D shapes. IEEE T. Pattern Anal. Mach. Intelli. 14(2), 239–256 (1992)
40. Davatzikos, C.: Nonlinear registration of brain images using deformable models. In: Mathematical methods in biomedical image analysis, pp. 94–103. IEEE Computer Society Press (1996)

41. Oguro, S., Tokuda, J., Elhawary, H., Haker, S., Kikinis, R., et al.: MRI signal intensity based B-spline nonrigid registration for pre- and intraoperative imaging during prostate brachytherapy. J. Magn. Reson. Imag. 30(5), 1052–1058 (2009)
42. Grimson, W.E.L., et al.: An automatic registration method for frameless stereotaxy, image guided surgery, and enhanced reality visualization. IEEE T. Med. Imag. 15(2), 129–140 (1996)
43. `http://www.vtk.org/`
44. Rusinek, H., Mourino, M.R., Firooznia, H., Weinreb, J.C., Chase, N.E.: Volumetric rendering of MR images. Radiology 171, 269–272 (1989)
45. Lorensen, W.E., Cline, H.E.: Marching cubes: A high resolution 3D surface construction algorithm. Computer Graphics (SIGGRAPH 1987 Proc.) 21, 163–169 (1987)
46. Yoshida, H., Näppi, J., Nagata, K., Choi, J.R., Rockey, D.C.: Comparison of fully automated CAD with unaided human reading in CT colonography. In: Proc. Eight Int. Symp. Virtual Colonoscopy, pp. 96–97 (2007)
47. Petrick, N., Haider, M., Summers, R.M., Yeshwant, S.C., et al.: CT colonography with computer-aided detection as a second reader: observer performance study. Radiology 246(1), 148–56 (2008)
48. Murao, K., Ozawa, A., Yamanaka, T., et al.: Integrated CAD tools for CT lung cancer screening: automatic detection and real-time comparison with the past images on PACS. Radiology 221, 726 (2001)
49. Nakazawa, T., Goto, Y., Nakagawa, T., et al.: New CAD (computer-aided detection) system for lung cancer screening using CT image. Radiology 221, 727 (2001)
50. `http://www.cancer.org/`
51. Gilbert, F.J., Astley, S.M., Gillan, M.G.C., Agbaje, O.F., et al.: Single reading with computer-aided detection for screening mammography. The New England J. of Medicine 359, 1675–1684 (2008)
52. Berlinger, N.: Robotic surgery-squeezing into tight places. New England J. of Medicine 354, 2099–2101 (2006)
53. Haaker, R.G., Stockheim, M., Kamp, M., et al.: Computer-assisted navigation increases precision of component placement in total knee arthroplasty. Clin Orthop Relat. Res. 433, 152–9 (2005)
54. Marmulla, R., Niederdellmann, H.: Computer-assisted bone segment navigation. J. Cranio-Maxillofac. Surg. 26, 347–359 (1998)
55. Geiger, B., Kikinis, R.: Simulation of endoscopy, AAAI Spring Symposium Series: Applications of Comp. Vis. Med. Imag. Proc., pp. 138–140 (1994)
56. Vining, V.C., Shifrin, R.Y., Grishaw, E.K., et al.: Virtual colonoscopy. Radiology 193, 446 (1994)
57. Lorensen, W.E., Jolesz, F.A., Kikinis, R.: The exploration of cross-sectional data with a virtual endoscope. In: Satava, R., Morgan, K. (eds.) Interactive Technology and the New Paradigm for Healthcare, pp. 221–230. IOS Press, Ohmsha (1995)
58. Robb, R.A., Hanson, D.P.: The ANALYZE software system for visualization and analysis in surgery simulation. In: Lavalle, S., Taylor, R., Burdea, G., Mosges, R. (eds.) Computer Integrated Surgery. MIT Press (1993)
59. Rubin, G.D., Beaulieu, C.F., Argiro, V., Ringl, H., et al.: Perspective volume rendering of CT and MR images: Applications for endoscopic imaging. Radiology 199, 321–330 (1996)
60. `http://www.nlm.nih.gov/`
61. Rice, D.H., Schaefer, S.D.: Endoscopic paranasal sinus surgery, pp. 159–235. Raven Press (1993)

62. Tomoda, K., Murata, H., Ishimasa, H., Yamashita, J.: The evaluation of navigation surgery in nose and paranasal sinuses. Int. J. Comp. Assist. Radiology and Surgery 1, 311–312 (2006)
63. Caversaccio, M., Bachler, R., Ladrach, K., Schroth, G., et al.: Frameless computer-aided surgery system for revision endoscopic sinus surgery. Otolaryngol. Head Neck. Surg. 122(6), 808–813 (2000)
64. Grevers, G., Menauer, F., Leunig, A., Caversaccio, M., Kastenbauer, E.: Navigation surgery in diseases of the paranasal sinuses. Laryngorhinootologie 78(1), 41–46 (1999)
65. Kherani, S., Javer, A.R., Woodham, J.D., Stevens, H.E.: Choosing a computer-assisted surgical system for sinus surgery. J. Otolaryngol. 32(3), 190–197 (2003)
66. Kherani, S., Stammberver, H., Lackner, A., Reittner, P.: Image guided surgery of paranasal sinuses and anterior skull base-five years experience with the Insta-Trak-System. Rhinolgy 40, 1–9 (2002)
67. Yamashita, J., Yamauchi, Y., Mochimaru, M., Fukui, Y., Yokoyama, K.: Real-time 3-D model-based navigation system for endoscopic paranasal sinus surgery. IEEE T. Biomed. Eng. 46(1), 107–116 (1999)
68. Tsagaan, B., Iwami, K., Abe, K., Nakatani, H., et al.: Development of navigation system for paranasal sinus surgery. In: Int. Symp. Comp. Methods on Biomechanics and Biomedical Engineering, vol. 1, pp. 1–8 (2006)
69. Tsagaan, B., Abe, K., Iwami, K., Nakatani, H., et al.: Newly developed navigation system for paranasal sinus surgery. J. Comp. Assist. Radiology and Surgery 1(1), 502–503 (2006)
70. Horn, B.: Robot Vision. ch.8. MIT Press (1986)
71. Ohta, N., Kanatani, K.: Optimal estimation of three-dimensional rotation and reliability evaluation. In: Proc. Computer Vision, vol. 1, pp. 175–187 (1998)

List of Abbreviations

CT	Computed Tomography
MRI	Magnetic Resonance Imaging
PET	Positron Emission Tomography
SPECT	Single Photon Emission Computed Tomography
US	Ultrasound
2D	Two Dimensional
3D	Three Dimensional
PACS	Picture Archiving and Communication Systems
DICOM	Digital Imaging and Communications in Medicine
ROI	Region of Interest
EM	Expectation and Maximization
CAD	Computer-Aided Diagnosis
CAS	Computer-Assisted Surgery
VE	Virtual Endoscopy
VHD	Virtual Human Datasets

Chapter 6
Attention in Image Sequences: Biology, Computational Models, and Applications

Mariofanna Milanova and Engin Mendi

University of Arkansas at Little Rock
Department of Computer Science, University of Arkansas at Little Rock, AR, 72204, USA
{mgmilanova,esmendi}@ualr.edu

6.1 Introduction

The ability to automatically detect visually interesting regions in images and video has many practical applications, especially in the design of active machine vision and automatic visual surveillance systems.

The human system is exposed to a variety of visual data, from which it actively selects and analyzes relevant visual information in an efficient and effortless manner. Humans employ attention to try to limit the amount of information that needs to be processed in order to speed up search and recognition. Elazary and Itti correctly point out that we rarely look at the sky when searching for our car. [9].

The term saliency was used by Tsotsos et al. [10] and Olshausen et al. [11] in their work on visual attention and by Itti et al. [1] in their work on rapid scene analysis. Saliency has also been refereed to as visual attention [10], unpredictability, rarity, or surprise. Many of the saliency models use results from psychology and neurobiology to construct plausible mechanisms for guiding attention. These models are biologically based models. More recently, a number of models attempt to explain attention based on more mathematically motivated principles that address the goal of computation. The development of affordable and efficient eye-tracing system led to a number of computational models attempting to account for the data and to address the question of what attracts attention.

It is well known that search and recognition behavior in humans can be explained through the combination of bottom-up information from the incoming visual scene [1] and top-down information from visual knowledge of the target and the scene (Hayhoe and Ballard, 2005) [12]. The exact interaction between the two processes still remains elusive. A saliency map is a topologically organized map that indicates interesting regions in an image based on the spatial organization of the features and an agent's current goal. These maps can be entirely stimulus driven, or bottom-up, if the model lacks a specific goal. There are numerous areas of the primate brain that contain saliency maps, such as the frontal eye fields, superior colliculus, and lateral intraparietal sulcus [13].

R. Kountchev et al. (Eds.): Advances in Reasoning-Based Image Processing, ISRL 29, pp. 147–170.
springerlink.com

Most models can be grouped in bottom up, top-bottom and hybrid approaches.

Bottom- up. Methods falling in this category are stimulus driven. The idea is to seek for the so-called "visual pop out" saliency. To model this behavior, various approaches were proposed, such as center-surround operation [1] or graph based activation maps.

Frintrop et al. [14] present a method inspired by Itti's method, but they compute center surround differences with square filters and use integral images to speed up the calculations. Hou and Zhang in [5] proposed a method based on residual of images in the frequency domain. In [5] many saliency detectors from a frequency domain perspective are presented. Tingjun et al. proposed the attention model using infrared images [23].

Top-bottom. Top-down visual attention processes are consider driven by the observer's goal when analyzing a scene [15].Object detection can be seen as a case of top-down saliency detection where the predefined task is given by the object class to be detected as a target.

Hybrid. Most of the saliency methods are hybrid models combining bottom –up and top-down approaches [16], [17].

Hybrid models are structured in two levels, top-down layer filters out noisy regions in saliency maps created by the bottom-up layer. Chen et al. [4] combined a face and text detector with multiple visual attention measurements. The proposed image attention model is based on three attributes (region of interest, attention value and minimal perceptible size). The authors adopted the Itti's et al. model presented in [1] and generate the three channel saliency maps: color contrasts, intensity contracts, and orientation contrast.

Wang and Li [17] combine spectral residual for bottom –up analysis with features capturing similarity and continuity based on Gestalt principles.

Recent approaches suggest that saliency can be learned from manually labeled examples Liu et al in [18] formulate salient object detection as an image segmentation problem where they separate the salient object from the image background.

The presented supervised approach in [19] is for learning to detect a salient object in an image or sequential images. First the authors model the salient object detection problem by a condition random field (CRF) where a group of salient features is compiled through CRF learning. Second, a new set of local, regional, and global salient is proposed to define a genetic salient object. The authors also constructed very large image database with 20,000 well –labeled images for training and evaluation. The data base is called MSRA Salient Object Database and is presented in [24]. The authors have developed a voting strategy by labeling a "ground truth" salient object in the image by multiple users.

The figure–ground separation task is similar to salient object detection in that it also has a goal to find the objects. The main difference is that salient object detection algorithms detect objects automatically without any prior needed knowledge of the category, shape, or size. The figure –ground segregation algorithms require the information of the category of objects or user interactions. In addition, the visual features adopted for the detection greatly differ.

In [19] Bruce and Tsotsos present an attention framework for stereo vision.

6.2 Computational Models of Visual Attention

6.2.1 A Taxonomy of Computational Model of Bottom-Up Visual Attention

There has been a growing interest in the subject since 1998 when the first computational and biologically plausible model of bottom-up visual attention was published by L. Itti and C. Koch.

The main idea in bottom-up visual attention approach is that the attention is in general unconsciously driven by a low level stimulus in the scene such as intensity, contrast and motion. This approach consists of the following three steps. The first step is feature extraction, in which multiple low-level features, such as intensity, color, orientation, texture and motion are extracted from the image at multiple scales. The second step is saliency competition. After normalization and linear/non linear combination a master map [21] or a salient map [1] is computed to represent the saliency of each image pixel. Last a few key locations on the saliency map are identified by winner-take-all, or inhibition of return, or other nonlinear operations.

These models can be classified into three different categories. Examples and the main properties of models of each category can be seen in Table 6.1.

- Hierarchical models (HM) characterized by a hierarchical decomposition, whether it involves a Gaussian, a Fourier based or wavelet decomposition. The difference of Gaussian is applied on the computed subbands to estimate the salience decomposition level. Different techniques are then used to aggregate this information across levels in order to build a unique saliency map.
- Statistical models (SM) are based on a probabilistic framework deduced from the content of the current image. The saliency is then defined as a measure of the deviation between the features of a current location and features present in its neighborhood.
- Bayesian models (BM) are based on combination of bottom-up saliency with prior knowledge. This prior knowledge concerns the statistic of visual features in natural scene, its layout or its spectral signature. This is probably one of the most important factors that affect our perception. Prior knowledge coming from our perceptual learning would help the visual system to understand the visual scene and it could be compared to a visual priming effect that would facilitate the scene perception.

Table 6.1 Main features of computational models of bottom-up visual attention

HM	Visual dimension	Operations	Prior knowledge
Itti at al [1]	Intensity two chromatic channels, orientations, flicker	Dyadic Gaussian and Gabor pyramid, center/surround filters, peak to peak normalization, pooling.	None
Le Meur at al. [26]	Luminance, two chromatic channels, motion	Oriented subband decomposition in the Fourier domain, contrast sensitivity functions, masking, center/surround filters, long-term normalization, pooling.	None
Bur et al. [27]	Intensity, two chromatic channels, orientations, contrast	Dyadic Gaussian and Gabor pyramids, center/ filters, long-term normalization, pooling.	None
SM	**Visual dimension**	**Operations**	**Prior knowledge**
Oliva et al. [28]	R,G,B	Saliency of location is inversely proportional of its occurrence probability in the image, The probability distribution in only based on the statistic of the current image.	Past search experience in similar scene
Bruce et al. [29]	R,G.B	Saliency is based on the self-information computation. Joint probability of the feature, deduced from a given neighborhood.	None
Gao et al. [30]	Intensity, two chromatic channels, orientation, and motion	Gaussian and Gabor pyramids. Center/surround filters. Saliency is assessed by using the Kulback-Leiber divergence between the local position and its neighborhood.	None

Table 6.1 *(continued)*

BM	Visual dimension	Operations	Prior knowledge
Zang et al. [8]	Luminance, two chromatic channels	Saliency is based on the self-information computation.	Probability distribution estimation
Kanan et al. [6]	LMS color space	Saliency is based on fixation –based approach.	

All the computational models of visual attention described in Table 6.1 are still a very basic description of the human vision. However, a promising trend seems to emerge with models based on a Bayesain framework. In this category we can add the work of Itti and Baidi concerning the theory of surprise [31]. They proposed a Bayesian definition of surprise in order to measure the distance between posterior and prior beliefs of the observers. It was proved that a measure by surprise has the capability to attract human attention. Fig. 6.1 shows saliency maps when we implement different attention models. The models provide a saliency map, i.e. a localized representation of saliency.

From biological viewpoint numerous evidence suggests that there is no locus in the brain where a unique saliency map would be located. The concept of saliency map is more of an abstract representation updated at each computational level of the brain. The update would take into account information coming from the low-level visual features but also from our knowledge, our memory and our expectation. Fecteau and Munoz [32] introduced the concept of priory map such map is a combined representation of bottom-up and top-down salience. This approach is related to idea that visual and cognitive processes are strongly tied. Short review of hybrid model of visual attention is presented in Section 6.2.2.

(a)

(b)

Fig. 6.1 a) original images, b) Itti-Koch Model [1], c) frequency-tuned saliency model [2], d) global rarity based attention model [3], e) local contrast based attention model [3], f) graph-based [4], g) spectral residual approach [5], h) salient region detection and segmentation [6], i) natural statistics [7], j) Bayesian model [8]

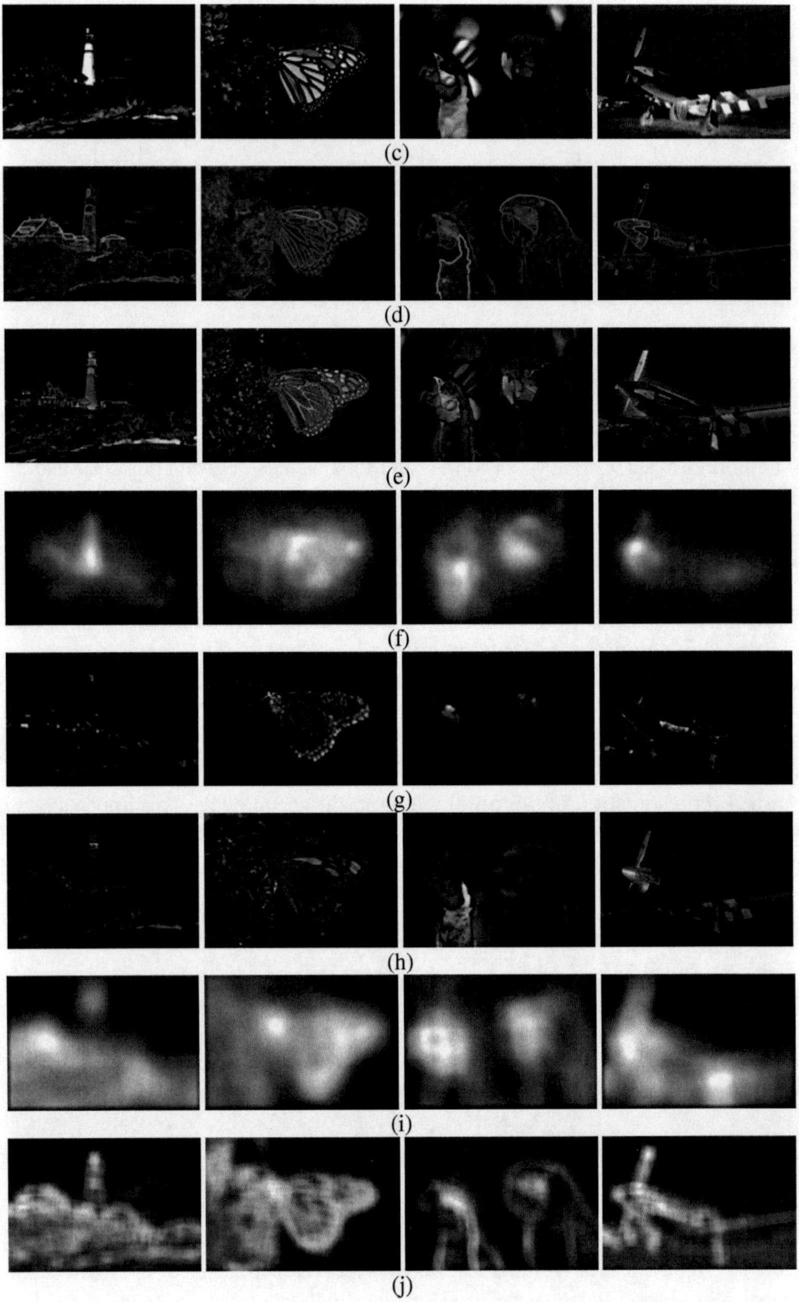

Fig. 6.1 *(continued)*

6.2.1.1 Itti-Koch Model

The Itti-Koch saliency model includes twelve feature channels sensitive to color contrast (red/green and blue/yellow, temporal luminance flicker, luminance contrast, four orientations (0°, 45°, 90°, 135°), and four oriented motion energies (up, down, left, right) [1]. These features detect spatial outliers in the image space, using a center – surround architecture inspired from biological receptive fields (RF). The RF of a neuron is classically defined as the area of visual space within which stimuli such as bars or edges can elicit responses from the neuron.

All feature maps contribute to a unique saliency map representing the conspicuity of each location in the visual field. Itti-Koch's model relay only on local measurements.

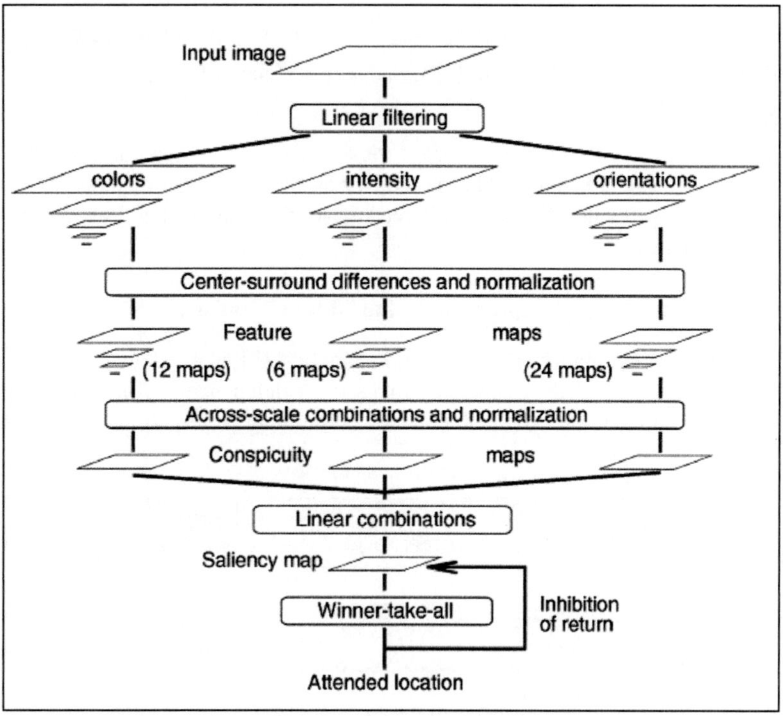

Fig. 6.2 Itti-Koch saliency model

An important feature of the Itti-Koch model is its incorporation of inhibition of return (IOR) once a point has been attended to its saliency will be reduced so that it is not looked at again (see Fig. 6.2).

6.2.1.2 Frequency-Tuned Salient Region Detection

Frequency-tuned saliency model [2], [36] finds low-level bottom–up saliency. It is inspired by the biological concept of center-surround contracts sensitivity of human visual system. The proposed approach offers three advantages over existing methods: uniformly highlighted salient regions with well defined boundaries, full resolution and computational efficiency.

Saliency maps are produced from the color and luminance features of the image. Saliency map S is formulated for the image I as follows:

$$S(x, y) = \left\| I_{\mu} - I_{w}(x, y) \right\| \tag{6.1}$$

I_{μ} is the mean pixel value of the image, $I_{w}\ (x, y)$ is the corresponding pixel vector value of the Gaussian blurred image from the original image and $\|.\|$ is the Euclidean distance. Each pixel location is the Lab color space vector, i.e. $[L, a, b]^{T}$.

Blurred image is a Gaussian blurred version (using 5x5 separable binominal kernel) of the original image. The method finds the Euclidean distance between the Lab pixel vector in a Gaussian filtered image with the average Lab vector for the input image.

6.2.1.3 Saliency Map Computation Using Natural Statistics

Saliency map computation using natural statistics [7] is used in classification problem. The model produces saliency maps from Independent Component Analysis (ICA) features of the LMS color spaces of the images. (Fig. 6.3)

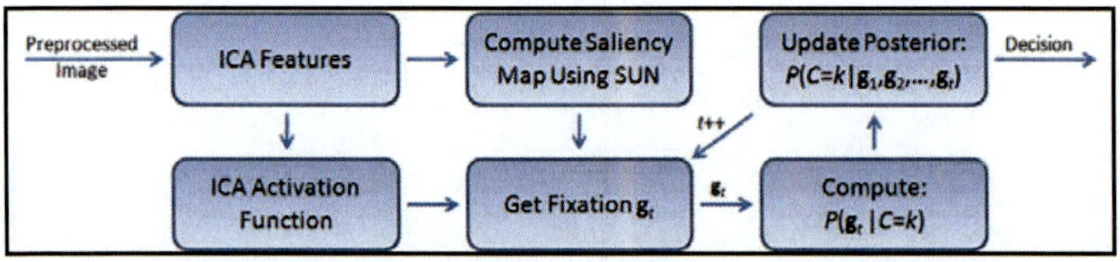

Fig. 6.3 An overview of the model during classification

The model is based upon sparse visual features capturing the statistical regularities in natural scenes and sequential fixation-based visual attention. First, images are converted from the default RGB color space to LMS color space. Sparse ICA features are then extracted from the images using FastICA [34]. These features are used to compute a saliency map which is treated as a probability distribution and locations are randomly sampled from the map.

6.2.1.4 Additional Bottom-Up Attention Models

Harel et al. [4] present method called Graph-Based Visual Saliency (GB). In method GB, the initial steps for creating feature maps are similar to Itti-Koch Model, with the difference that fewer levels of the pyramid are used to find center surround differences. The special frequencies retained are within the range $[\pi/128, \pi/8]$. Approximately 98% of the high frequencies are discarded for a 2D image. As illustrated in Fig. 6.1f, there is slightly more high frequency content that in result image using Koch model.

In a method presented by Hou and Zhang, the input image is resized to 64x64 pixels (via low-pass filtering and downsampling) based on the dispute that the spatial resolution of pre-attentive vision is very limited [5]. The resulting frequency content of the resized image therefore varies according to the original size of the image. For example, with input image of size 320 x 320 pixels, the retained frequencies are limited to the range $[0, \pi/5]$. As seen in Fig. 6.1.g, higher frequencies are smoothed out.

In a method presented by Achanta et al. [6], a difference of means filter is used to estimate center surround contrast. The lowest frequencies retained depend on the size of the largest surround filter (which is half of the image's smaller dimension) and the highest frequencies depend on the size of the smallest center filter (which is one pixel). This method effectively retains the entire range of frequencies $[0, \pi]$ with a notch at DC. All the high frequencies from the original image are retained in the saliency map but not all low frequencies (see Fig. 6.1h).

6.2.2 Hybrid Computational Models of Visual Attention

Once salient regions are determined by the pre-attentive bottom-up features, top-down factors guide a user to select one region to visually focus on. Unlike the bottom –up features, this top-down guidance is task-dependent. Feature such as object distance from a viewer, image coverage and novelty have also been treated as top-down factors. Lee et al. [33] proposed real-time framework using combination of bottom-up (stimulus driven) features and top-down (goal-directed) context. The framework first build features maps using features such as luminance, hue, depth, size and motion. The feature maps are then integrated into a single saliency map using the center – surround difference. Finally, the top-down contexts are inferred from the user's spatial and temporal behaviors during interactive navigation and used to select the most attended object among candidates produced in the object saliency map.

Peters and Itti present computation gaze-prediction attention model that includes bottom –up and top-down component [25].The novel top-down component is based on the idea of capturing eye-positions.

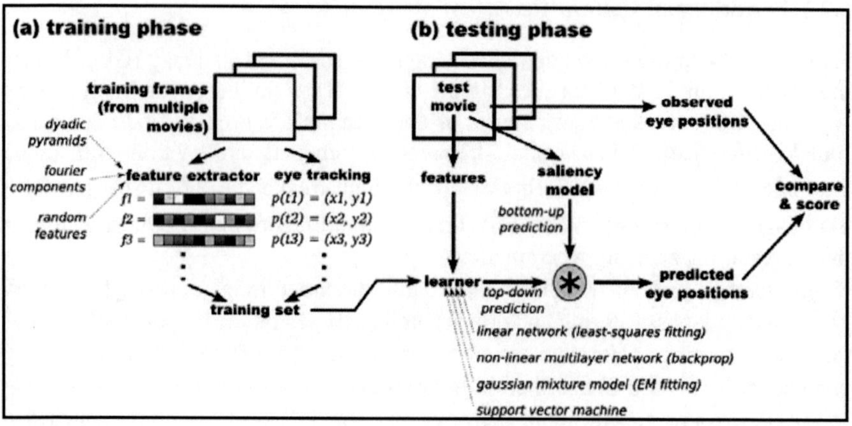

Fig. 6.4 Gaze – prediction attention model

Fig. 6.4 shows attention model proposed Peters and Itti [25]. First the authors implement training phase using set of feature vectors and eye positions corresponding to individual frames from several video game clips which are recorded while observers interactively played the games (Fig. 6.4a). The training set is used to learn a mapping between feature vectors and eye positions. Then in (b) the testing phase the authors use a different video game clip to test the model.

6.2.2.1 Hybrid Model Based on Sparse Learning

Barlow's hypothesis is that the purpose of early visual processing is to transform the highly redundant sensory input into more efficient factorial code [41]. Milanova et al. [42] proposed hybrid model of visual attention. The approach presented in this paper extends the Itti-Kock attention model and Olshausen's algorithm [43] to incorporate conjunction search and temporal aspects of sequences of natural images. The proposed model integrate model of conjunction search. Conjunction search (a search for a unique combination of two features – e.g, orientation and spatial frequency – among distractions that share only one of these features) examines how the system combines features into perceptual wholes. Attentional guidance does not depend solely on local visual features, but must also include the effects of interactions among features. The idea is to group filters (basis components) which become responsible for extracting similar features.

In natural time-varying images, temporal correlations are highly significant. Let suppose that the input consists of different sequences of k images each, a given sequence being denoted by the vectors $I(t)$ for $t=1,\ldots,k$. A set of k basis matrices $M(t)$ for $t=1,\ldots k$ will be used. For each t, $M(t)$ will be used to capture the statistical structure of the time step in the training sequences.

$$I(x,y,t)=\sum_i\sum_{t'}a_i(t')M_i(x,y,t-t')+\nu(x,y,t)=$$
$$=\sum_i a_i(t)*M_i(x,y,t)+\nu(x,y,t) \tag{6.2}$$

where * denotes convolution over time. The time-varying coefficient, $a_i(t)$, represents the amount by which the basis function M_i is multiplied to model the structure around time t in the moving image sequence. The noise is used to model additional uncertainty and is not captures by this model. The goal of contextual features extraction is to find a matrix M and to infer for each image the proper coefficients a_i. Rather than making prior assumption about the shape or form of the basis functions; the bases are adapted to the data using an algorithm that maximizes the log-probability of the data under the model.

Maximizing the posterior distribution over the coefficients is equivalent to minimizing: coefficients $a_i(t)$.This is accomplished by gradient descent:

$$a_i(t+1)=a_i(t)+\lambda_N\sum_{x,y}M_i(x,y,t)*e(x,y,t)-$$
$$-\beta/\sigma S'(a_i(t)/\sigma) \tag{6.3}$$

where $S(a)=\beta\log(1+(a/\sigma^2)$

$$e(x,y,t)=I(x,y,t)-\sum_i a_i(t)M_i(x,y,t) \tag{6.4}$$

This is the residual error $e(x,y,t)$ between the input at time t and its reconstruction. The Eq. (6.3) can be presented as:

$$a_i(\tau+1)=a_i(\tau)+\lambda_N\sum_t M_i(x,y,t)^T I(\tau-\kappa+t)-$$
$$-Wa(\tau)-\beta/\sigma S'(a_i(t)/\sigma) \tag{6.5}$$

where $W=\sum_{t=1}^{k}M(x,y,t)^T M(x,y,t)$ and τ represents the current time instant.

In summary, the current spatiotemporal response is determined by three factors: the previous response $a_i(\tau)$, the past k inputs $I(\tau)$ ...... $I(\tau-k+1)$, and lateral inhibition due to recurrent term W and some non-linear self inhibition term. The matrix W represents the topological lateral connection between the coefficients. The decomposition coefficients a_i and the corresponding basis functions are used as new context features.

Proposed algorithm

Step 1: Using the Independent Component Analysis (ICA) algorithm [42] we received an initial set of basis functions for the above learning rule (Eq. 6.5)
Step 2: For the implementation of the learning rule (Eq. 6.5), we interpret it in terms of a local network implementation and suggested the model of Cellular Neural Network (CNN), introduced in 1988 by Chua and Yang. [48]
Step 3: For the initial step, the basis functions received in step 1 were used as lateral connections. The learning rule for the active neurons, including the center neuron becomes:

$$\mathbf{w}_i(t+1) = \mathbf{w}_i(t) + \gamma h(i,c,t)(\mathbf{M}_i(t) - \mathbf{w}_i(t)) \tag{6.6}$$

where the neighborhood function *h(i,c,t)* implements a family of "Mexican hat" functions. The basis functions are similar to those obtained by sparse learning, but in our model they have a particular order.

The proposed algorithm is depicted in Fig. 6.5.

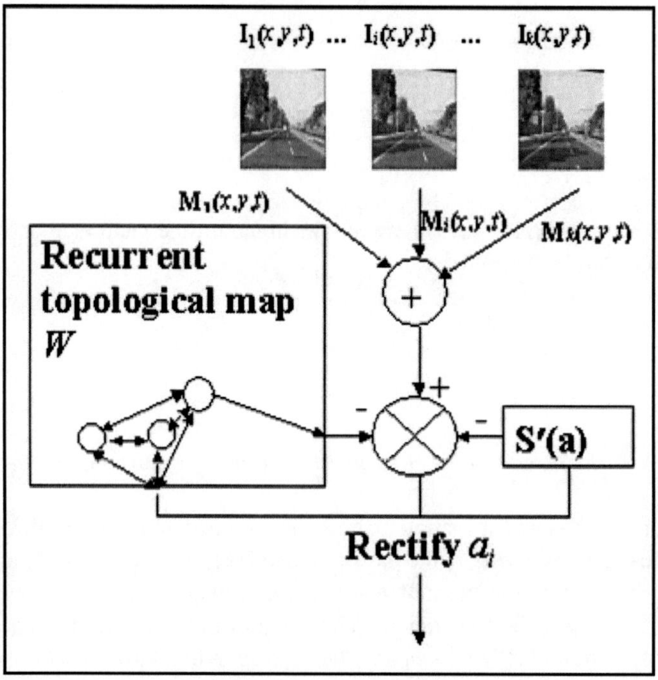

Fig. 6.5 Diagram of the Perceptual Learning Model

Step 4: Top-down task relevance model

The new top down component is based on the hypothesis that image resolution exponentially decreases from the fovea to the retina periphery [42]. This hypothesis can be represented computationally with different resolutions. The visual attention points may be considered as the most highlighted areas of the Visual Attention model. These points are the most salient regions in the image. When going further from these points of attention, the resolution of the other areas dramatically decrease. Different authors work with different filters and different kernel size to mimic this perceptual behavior [49]. These models ignore contextual information representation.

In our top-bottom component, we define the higher attention level areas by using the eye-tracking system. When the set of regions of interest is selected, these regions need to be represented with the highest quality while the remaining parts of the processed image could be represented with a lower quality. In result, higher compression is obtained. The adaptive compression technique proposed is based on new image decomposition called Inverse Difference Pyramid (IDP) [50].

The main idea is that the decomposition is performed starting with low resolution and calculating the coarse approximation of the processed image with some kind of 2D orthogonal transform such as for example, Walsh-Hadamard (WH), Discrete Cosine (DC) or Wavelet Transforms. The calculation of the coefficients in the lowest decomposition layer is performed dividing the image in sub-images of size 64 x 64 (or 32 x 32) pixels and performing the transform with restricted number of 2D coefficients only. Then, using the values of the calculated coefficients, the image is restored performing the inverse transform. The obtained approximation is subtracted from the original and then the difference image is divided in sub-images of smaller size: 32 x 32 (or 16 x 16) pixels correspondingly. The processing follows the same pattern. The decomposition ends when the quality of the restored image is high enough for the application performed.

The IDP decomposition presented in brief above permits the creation of regions of interest because some of the initial sub-images will be represented by the total number of decomposition layers available while the remaining parts will be represented by one or two decomposition layers only (i.e. the corresponding pyramid decomposition is truncated).

The eye movements were recorded with a head mounted ASL model 6000 Eye-tracking Interface system. Four subjects participated in this experiment. They were seated 80 cm in front of the screen using a chin rest to assure minimal head movements. Fixation locations were obtained with the built-in fixation detection mechanism. The full quality of the processed image is preserved only for the selected objects (the Road Sign and the Car, shown in Fig. 6.6). These objects were compressed with lossless IDP, while the remaining parts of the image were compressed with lossy compression (lossy IDP). Due to the Pyramidal layered structure of the IDP decomposition, we can create images with different quality within one picture frame. In Table 6.2, the compression ratios are given obtained for the selected objects, in this case the road sign (triangle) and the car and for the whole picture.

Fig. 6.6 A small patch of the image around each fixation was extracted

Table 6.2 Multiresolution image representation

Image	Picture size	Compression ratio	PSNR [dB]
Street (Fig. 5)	432 x 323	102,22	22,63
Car	104 x 81	4,14	40,72
Road sign	82 x 89	3,31	32,01

6.3 Selected Datasets

For the purpose of obtaining the right selection of datasets, it is necessary to research what kind of information each dataset provides.

6.3.1 LABELME

Labelme is an open source annotational tool [50], [51]. There are three ways to download the images and annotations: (1) via the LabelMe Matlab toolbox, allowing the user to customize the portion of the database that is to be downloaded, (2) by clicking on links pointing to a set of large tar files, (3) via the LabelMe Matlab toolbox, without directly downloading the images.

6.3.2 Amsterdam Library of Object Images (ALOI)

ALOI dataset contains more than 48,000 images of 1,000 objects, under various illumination conditions [52] [53]. It is possible to embed information within the file name, e.g naming images. This is sufficient for classification. This is done in a file naming pattern known as Caltech-256 [54].

6.3.3 Spatially Independent, Variable Area, and Lighting (SIVAL)

The SIVAL dataset contents 1,500 images equally divided in 25 object categories, such as WD40 can, shoe, apple, tea box, etc. There is only one salient object per image, with variations on the scale, position in the image, illumination condition and background. The Groud truth representation is available in the form of object masks [55].

6.3.4 MSRA

Liu et al. [19] [24] present MSRA Salient Object Database that consists of one data set of 20,000 images labeled from three users and second dataset of 5,000 images labeled from nine users. In these datasets each image contains an unambiguous salient object. These salient objects differ in category, color, shape, size, etc. This image database is different from the UIUC Cars dataset [56] or from the PASCAL VOC 2006 dataset, where images containing a specific category of objects are collected together.

6.3.5 Caltech

The Caltech – 101 dataset contains 101 diverse classes (e.g faces, beacvers, anchors, ets) with a large amount of intra-class appearance and shape variability [54]. Outlines of the objects in the pictures: are presented in the following dataset: Caltech-256 [57].

6.3.6 PASCAL VOC

PASCAL VOC datasets are couple of datasets originally provided for Visual Object Classes Challenge Competition [58] [59]. PASCAL has the following features: 1) images are annotated using metadata and multiple bounding boxes for the selected objects are available as well as labels for 20 object classes (for example: person, bird, cat, boat) 2) The images are much more challenging with respect to visual complexity, they contain multiple, ambiguous, often small objects and very cluttered backgrounds. All images have appropriate metadata annotations, where bounding boxes for the objects are available.

6.4 Software Implementations of Attention Modeling

There is a variety of software tools implementation of visual attention models. This section presents short description of the existing systems.

6.4.1 Itti-Koch Model

The *iLAb Neuromorphc Vision C++ Toolkit* was developed at the University of California and at Caltech [47]. It is based on the original idea first advanced by Koch and Ullman [60].

Neuromorphic models are computational neuroscience algorithms whose architecture and function is closely inspired from biological brains. The iLab Neuromorphic Vision C++ Toolkit comprises not only base classes for images, neurons, and brain areas, but also fully-developed models, such as the model of bottom-up visual attention and model based of Bayesian surprise [35]. This Toolkit includes a set of C++ classes implementing a range of vision algorithms for use in attention models.

6.4.2 Matlab Implementations

There is a number of visual attention models implemented in Matlab, a commercial multi-purpose numerical computing environment developed by MathWorks. Image and visual representation group (IVRG) of École Polytechnique Fédérale de Lausanne provided a Matlab implementation of a visual attention model based on frequency-tuned salient region detection [36].

Kanan and Cottrell [22] developed a saliency map computation using natural statistics and presented Matlab code for computing the features and for generating saliency maps.

Itti and Baldi [31] proposed attention model based on bayesian surprise in which surprise represents an approximation to human attentional allocation. A Matlabtoolkit for the proposed model is available in [61].

Bruce [62] [63] developed a Matlab code for attention model motivated by information maximization. Localized saliency computation of the model serves to maximize information sampled from one's environment.

Bruce and Tsotsos [64] [65] also extended model of attention based on information maximization in the spatiotemporal domain by proposing a distributed representation for visual saliency comprised of localized hierarchical saliency computation.

6.4.3 TarzaNN

Laboratory for Active and Attentive Vision (LAAV) developed TarzaNN [66]. TarzaNN is a general purpose neural network simulator for visual attention modeling.TarzaNN is a neural network simulator that abstracts from single neurons to layers of neurons and was designed specifically to implement visual attention models.

6.4.4 Model Proposed by Matei Mancas

Matei Mancas presents The Saliency Toolbox for still images and a very simple video attention top-down model [67].

Mancas provide codes of computational attention models and video attention top-down model is for video sequences.

6.4.5 JAMF

JAMF offers a unique combination of highly – optimized algorithms [68]. JAMF was developed at the Neurobiopsychology Labs of the Institute of Cognitive Science at the University of Osnabrueck. JAMF is open-source software downloadable from the following website: http://jamf.eu/jamf/.

6.4.6 LabVIEW

LabVIEW is commercial software developed by National Instrument. It is based on G, a graphical dataflow programming language. This tool has additional Machine Vision Module with wide range of functionalities. The additional module is useful for standard machine vision tasks, but does not include any of the recently developed attention models. In addition to the licensing cost and the nature of the G language, understanding its non- standard language paradigm also places an additional load on new users. LabVIEW software includes Machine Vision Module [69].

6.4.7 Attention Models Evaluation and Top-Down Models

A free mouse-tracking utility was set-up at the TCTS Lab of FPMs. You can upload your images and get the mouse-tracking results on these images. You may also upload entire sets of specific images and than ask for the top-down model to the website administrator.

This tool is called Validattention and it is available at `http://tcts.fpms.ac.be/~mousetrack`.

6.5 Applications

The capability to predict the location onto which an observer will focus his attention presents a strong interest. There are many applications for visual attention for example automatic image cropping [37], adaptive image display on small devices [38], image/video compression, advertising design [39] and content based image browsing.

Mancas' work demonstrates a wide applicability of the attention models [3]. He groups the applications in 6 groups including, Medical Imaging, machine vision, Image Coding and Enhancement, Image ergonomics and High level attention applications, such as object tracing and recognition.

Recently, new applications have been considered:

- Quality assessment : the idea relies on the fact that an artifact appearing on a salient region is more annoying then an artifact appearing in the background [40]
- Robust classification [22]
- Content base image retrieval [44]

Kanan and Cottrell develop SUN used the LabelMe dataset to train a classifier using features inspired by the properties of neurons in primary visual cortex. Torralba et al. (2006) gathered eye movement data from people who were told to look for particular objects (mugs, paintings, and pedestrians) in natural images. The authors used their data to evaluate how well SUN predicts the subject's eye movements when it is given the very same images, which SUN has never seen before. We compared our model to Torralba et al.'s Contextual Guidance Model, which is one of the few models with a comparable ability to predict human eye movements in natural images.

6.6 Example

The SaliencyToolbox [45] is a collection of Matlab functions and scripts for computing the saliency map for an image, for determining the extent of a proto-object, and for serially scanning the image with the focus of attention. It can be downloaded at http://www.saliencytoolbox.net/. To access to the toolbox, SaliencyToolbox directory including its subdirectories must be added to the Matlab path:

addpath(genpath('< SaliencyToolbox path>'));

To start graphical user interface (GUI) of the toolbox, following command must be typed in Matlab:

guiSaliency;

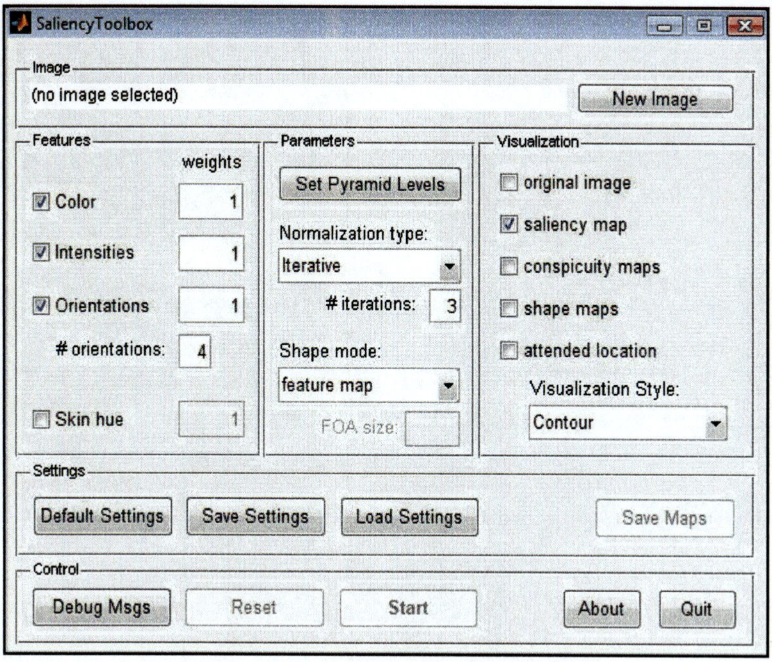

Fig. 6.7. GUI of the SaliencyToolbox

"New Image" button allows the user to select an input image. Once the image is selected, saliency computation starts by "Start" button. The toolbox generates saliency, conspicuity and shape maps as well as attended locations of the input image. Fig. 6.8 shows an example [46] output of the toolbox.

There is also the command-line version of the program:

runSaliency('inpit image');

Since binaries for the most common architectures are available with the toolbox, most of the time there is no need to compile the mex files. If the binaries for operating system of the user and CPU combinations are not included in the SaliencyToolbox/bin directory, compilation may occur. Compilation details under different operating systems can be found at the documentation of the toolbox.

(http://www.saliencytoolbox.net/doc/index.html)

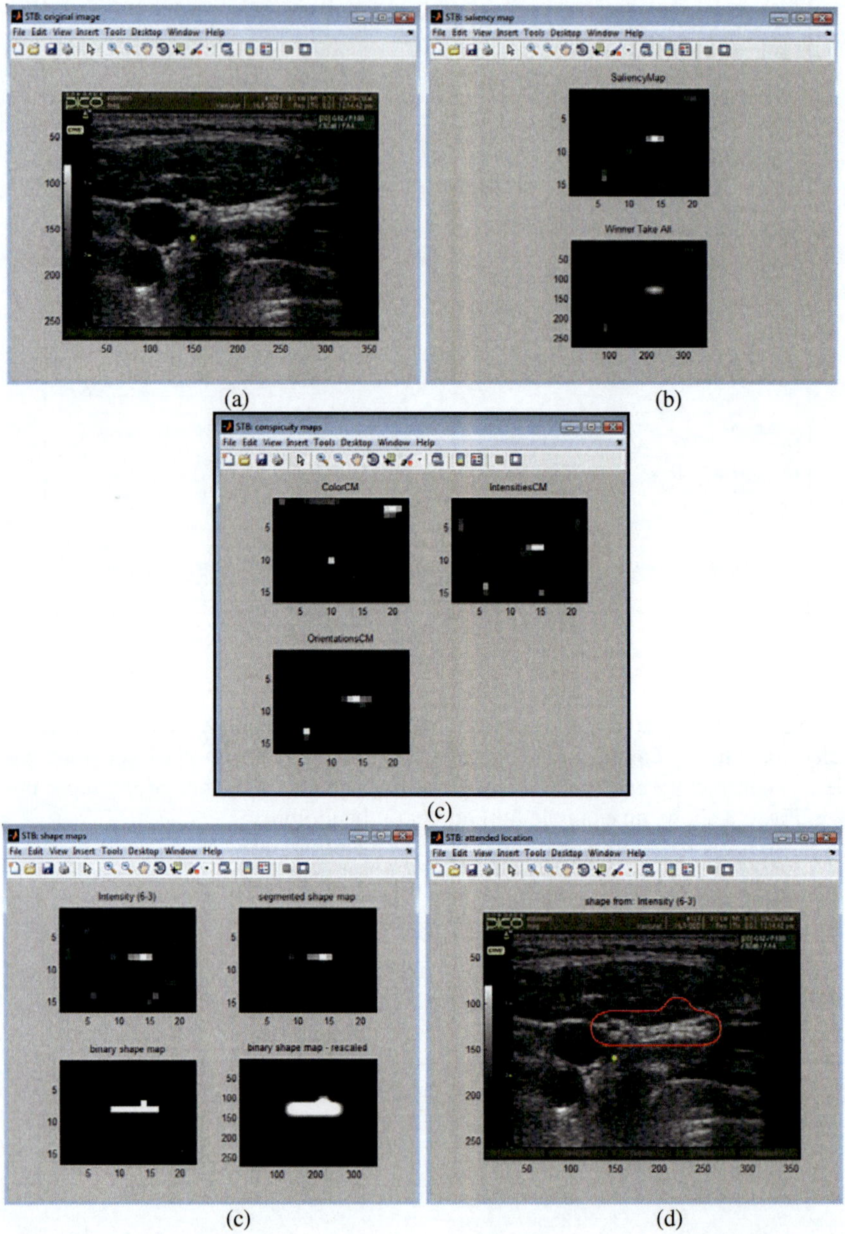

Fig. 6.8 a) Input image, outputs of the toolbox: b) saliency maps, c) conspicuity maps, d) shape maps, e) attended location.

References

1. Itti, L., Koch, C., Niebur, E.: A model of saliency-based visual attention for rapid scene analysis. IEEE Transactions on Pattern Analysis and Machine Intelligence 20(11), 1254–1259 (1998)
2. Achanta, R., Hemami, S., Estrada, F., Süsstrunk, S.: Frequency-tuned Salient Region Detection. In: IEEE International Conference on Computer Vision and Pattern Recognition (CVPR), Miami Beach, Florida (June 2009)
3. Mancas, M.: Computational Attention: Modelisation&Application to Audio and Image Processing, PhD. Thesis, University of Mons (2007)
4. Harel, J., Koch, C., Perona, P.: Graph-Based Visual Saliency. In: Proceedings of Neural Information Processing Systems, NIPS (2006)
5. Hou, X., Zhang, L.: Saliency detection: A spectral residual approach. In: IEEE Conference on Computer Vision and Pattern Recognition (2007)
6. Achanta, R., Estrada, F.J., Wils, P., Süsstrunk, S.: Salient Region Detection and Segmentation. In: Gasteratos, A., Vincze, M., Tsotsos, J.K. (eds.) ICVS 2008. LNCS, vol. 5008, pp. 66–75. Springer, Heidelberg (2008)
7. Kanan, C., Cottrell, G.W.: Robust Classification of Objects, Faces, and Flowers Using Natural Image Statistics. In: Proceedings of the IEEE Conference on Computer Vision and Pattern Recognition, CVPR (2010)
8. Zhang, L., Tong, M.H., Marks, T.K., Shan, H., Cottrell, G.W.: SUN: A Bayesian framework for saliency using natural statistics. Journal of Vision 8(7), 32, 1–20 (2008)
9. Elazary, L., Itti, L.: A Bayesian model for efficient visual search and recognition. Visual Research 50, 1138–1352 (2010)
10. Tsotsos, J.K., Gulhane, S.M., et al.: Modeling visul attention via selective tuning. Artificial Intelligence 78(1-2), 507–545 (1995)
11. Olshausen, B., Anderson, C., Van Essen, D.: A neurobiological model of visual attention and invariant pattern recognition based on dynamic routing of information. Journal of Neuroscience 13, 470–4719 (1993)
12. Hayhoe, M., Ballard, D.H.: Eye Movements in Natural Behavior. Trends in Cognitive Sciences 9(4), 188–193 (2005)
13. Glimcher, P.: Making choices: the neurophysiology of visual-saccadic dcission making. Trends in Neuro- sciences 24, 654–659 (2001)
14. Frintrop, S., Klodt, M., Rome, E.: A real- time visual attention system using integral images. In: International Conference on Computer Vision Systems (2007)
15. Marchesottu, L., Cifarelli, C., Csurka, C.: A framework for visual saliency detection with applications to image thumbnalling. In: IEEE ICCV, pp. 2232–2239 (2009)
16. Chen, L.Q., Xie, X., Fan, X., Ma, W.Y., Zhang, H.J., Zhou, H.Q.: A visual attention model for adapting images on small displays. ACM Multimedia Systems Journal 9(4) (2003)
17. Wang, Z., Li, B.: A two –stage approach to saliency detection in images. In: ICASSP, pp. 964–968 (2008)
18. Liu, T., Sun, J., Zheng, N., Tang, X., Shum, H.: Video Attention: Learning to detect a salient object. In: CVPR (2007)

19. Liu, T., Yuan, Z., Sun, J., Wang, J., Zheng, N., Tang, X., Shum, H.Y.: Learning to detect salient object. IEEE Transactions on Pattern Analysis and Machine Intelligence 33(2), 353–367 (2011)
20. Bruce, Tsotsos: An Attention Framework for Stereo Vision. Computer and Robot Vision, 88–95 (2005)
21. Treiosman, A., Gelade, G.: A feature-integration theory of attention. Cognitive Psychology 12(1), 97–136 (1980)
22. Kanan, C., Cottrell, G.W.: Robust Classification of Objects, Faces, and Flowers Using Natural Image Statistics. In: Proceedings of the IEEE Conference on Computer Vision and Pattern Recognition, CVPR (2010), Kanan's Web page `http://cseweb.ucsd.edu/~ckanan/NIMBLE.html`
23. Tingjun, L., Zhang, F., Cai, X., Huang, Q., Guo, Q.: The Model of Visual Attention Infrared Target Detection Algorithm. In: International Conference on Communications and Mobile Computing, pp. 87–91 (2010)
24. MSRA Salient Object Database, `http://research.microsoft.com/en-us/um/people/jiansun/SalientObject/salient_object.htm`
25. Peters, R., Itti, L.: Beyond bottom-up: Incorporating task-dependent influences into a computational model of spatial attention. In: CVPR (2007)
26. Le Meur, O., Le, C.P., Barba, D., Thoreau, D.: Predicting visual fixation on video based on low-level visual features. Visual Research 47(19), 2483–2498 (2007)
27. Bur, A., Hügli, H.: Optimal Cue Combination for Saliency Computation: A Comparison with Human Vision. In: Mira, J., Álvarez, J.R. (eds.) IWINAC 2007. LNCS, vol. 4528, pp. 109–118. Springer, Heidelberg (2007)
28. Oliva, A., Torralba, A., Castelhano, M.S., Henderson, J.: Top-down control of visual attention in object detection. In: IEEE ICIP, vol. 1, pp. 253–256 (2003)
29. Bruce, N.D.B., Tsotsos, J.K.: Saliency, Attention, and Visual Search: An Information Theoretic Approach. Journal of Vision 9(3), 1–24 (2009), `http://journalofvision.org/9/3/5/`, doi:10.1167/9.3.5
30. Gao, D., Mahadevan, V., Vasconcelos, N.: On the plausibility of the discriminant center-surround hypothesis for visual saliency. Journal of Vision 8(7), 1–18 (2008), `http://www.svcl.ucsd.edu/projects/discsalbu/`
31. Itti, L., Baldi, P.F.: Bayesian surprise attracts human attention. In: Advance in Neural Information Processing Systems, Cambridge, MA, pp. 547–554 (2006)
32. Fecteau, J.H., Munoz, D.P.: Salience, relevance, and firing: a priority map for target selection. Trends in cognitive science 10, 382–390 (2006)
33. Lee, S., Kim, G., Choi, S.: Real -time tracking of visually attended objects in virtual environments. IEEE Transaction on Visualization and Computer Graphics 15(1), 6–19 (2009)
34. Koldovský, Z., Tichavský, P., Oja, E.: Efficient Variant Of Algorithm FastICA For Independent Component Analysis Attaining The Cramér-Rao Lower Bound. IEEE Trans. on Neural Networks 17, 1090–1095 (2006)
35. Itii, I., Baldi, P.: A principal approach to detecting surprising events in video. In: Proc. IEEE Conference on Computer Vision and Pattern Recognition (CVPR), San Diego, CA, pp. 631–637 (2005)
36. Achanta, R., Hemami, S., Estrada, F., Süsstrunk, S.: Frequency-tuned Salient Region Detection. In: IEEE International Conference on Computer Vision and Pattern Recognition (CVPR) (2009)

37. Santella, A., Agravala, D., et al.: Gaze-based interaction for semi-automatioc photo cropping. In: CHI, pp. 771–780 (2006)
38. Chen, L., Xie, X., Fan, X., Ma, W., Shang, H., Zhou, H.: A visual attention mode for adapting images on small displays, Technical report, Microsoft Research, Redmond, WA (2002)
39. Itti, L.: Models of Bottom-Up and Top-Dawn Visual Attention, Ph.D thesis, California Institute of technology, Pasadena (2000)
40. Larson, E.C., Cuong, V., Chandler, M.: Can visual fixation patterns improve image fidelity assessment? In: IEEE International Conference on Image Processing (2008)
41. Barlow, H.B.: What is the computational goal of the neocortex? In: Koch, C., Davis, J.L. (eds.) Large-Scale Neuronal Theories of the Brain, pp. 1–22. MIT Press, Cambridge (1994)
42. Milanova, M., Rubin, S., Kountchev, R., Todorov, V., Kountcheva, R.: Combined visual attention model for video sequences. In: ICPR, pp. 1–4 (2008)
43. Olshausen, B.: Sparse Codes and Spikes. In: Rao, R.P.N.B., Olshausen, A., Lewicki, M. (eds.) Probabilistic Models of Perception and Brain Function. MIT Press
44. Bamidele, A., Stentiford, F.W., Morphett, J.: An attention based approach to content based image retrieval. British Telecommunications Advanced Research Technology Journal on Intelligent 22(3) (2004)
45. Walther, D., Koch, C.: Modeling attention to salient proto-objects. Neural Networks 19, 1395–1407 (2006)
46. Mendi, E., Milanova, M.: Image Segmentation with Active Contours based on Selective Visual Attention. In: 8th WSEAS International Conference on Signal Processing (SIP 2009) including 3rd WSEAS International Symposium on Wavelets Theory and Applications in Applied Mathematics, Signal Processing & Modern Science (WAV 2009), May 30-June 1, pp. 79–84 (2009)
47. Bottom-Up Visual Attention Home Page, `http://ilab.usc.edu/bu/`
48. Chua, L.O., Yang, L.: Cellular Neural Networks:Theory and Applications. IEEE Trans. On Circuits and Systems 35, 99–120 (1988)
49. Mancas, B.G., Macq, B.: Perceptual Image Representation. EURASIP. Journal on Image and Visual Proccesing (2007)
50. LabelMe in, `http://labelme.csail.mit.edu/instructions.html`
51. Russell, B.C., Torralba, A., Murphy, K.P., Freeman, W.T.: LabelMe: a database and web-based tool for image annotation. International Journal of Computer Vision 77, 1–3 (2008)
52. Amsterdam Library of Objects Images (ALOI), `http://staff.science.uva.nl/~aloi/`
53. Geusebroek, J.M., Burghouts, G.J., Smeulders, A.W.M.: The Amsterdam library of object images. Int. Journal of Computer Vision 61(1), 103–112 (2005)
54. Caltech dataset, `http://www.vision.caltech.edu/html-files/archive.html`
55. Sival Image Repositorty, `http://www.cs.wustl.edu/~sg/accio/SIVAL.html`
56. UIUC Image Database for Car Detection, `http://cogcomp.cs.illinois.edu/Data/Car/`

57. Caltech256,
http://www.vision.caltech.edu/Image_Datasets/Caltech256/
58. Everingham, M., Van Gool, L., Williams, C.K.I., Winn, J., Zisserman, A.: International Journal of Computer Vision 88(2), 303–338 (2010)
59. PASCAL, http://pascallin.ecs.soton.ac.uk/challenges/VOC/
60. Koch, C., Ullman, S.: Shifts in selective visual attention: towards the underlying neural circuitry. Human Neurobiology 4, 219–227 (1985)
61. Bayesian Surprise Toolkit for Matlab,
http://sourceforge.net/projects/surprise-mltk
62. Bruce, N.D.B., Tsotsos, J.K.: Saliency, Attention, and Visual Search: An Information Theoretic Approach. Journal of Vision 9(3), 1–24 (2009)
63. Neil Bruce's web page,
http://www-sop.inria.fr/members/Neil.Bruce/
64. Selective Tuning and Saliency,
http://web.me.com/john.tsotsos/Visual_Attention/ST_and_Saliency.html
65. Bruce, N., Tsotsos, J.K.: Spatiotemporal Saliency: Towards a Hierarchical Representation of Visual Saliency. In: 5th Int. Workshop on Attention in Cognitive Systems, Santorini Greece, May 12 (2008)
66. Centre for Vision Research (CVR) at York University,
http://www.cvr.yorku.ca/home/
67. Attention Models Comparison and Validation,
http://www.tcts.fpms.ac.be/attention/index_old.php#validation
68. Steger, J., Wilming, N., Wolfsteller, F., Höning, N., König, P.: The JAMF Attention Modelling Framework. In: Paletta, L., Tsotsos, J.K. (eds.) WAPCV 2008. LNCS, vol. 5395, pp. 153–165. Springer, Heidelberg (2009)
69. LabVIEW for Machine Vision,
http://sine.ni.com/nips/cds/view/p/lang/en/nid/10419

Part II
Pattern Recognition, Image Data Mining and Intelligent Systems

Chapter 7
Visual Mobile Robots Perception for Motion Control

Alexander Bekiarski

Department of Radio Communications and Video Technologies,
Technical University of Sofia, Sofia 1000, Bulgaria
aabbv@tu-sofia.bg

Abstract. Visual perception methods are developed first mainly for human perception description and understanding. The results of these researches are now very popular for robots visual perception modeling. In this chapter is present first a brief review of the basic visual perception methods suitable for intelligent mobile robots applications. The analysis of these methods is directed to the mobile robot motion control, where the visual perception is used for objects or human body localization like: Bayesian visual perception methods for localization; log-polar visual perception; area of robot observation mapping using visual perception; landmark-based finding and localization with visual perception etc. The development of an algorithm for mobile robot visual perception is proposed based on the features of log-polar transformation to represent some of the objects and scene fragments in area of mobile robot area of observation in a more simple form for the image processing. The features and advantages of the proposed algorithm are demonstrated with the popular for the mobile robots visual perception situation of motion control in a road or corridor with outdoor road edges, painted lane separation lines or indoor two side existing room or corridor lines. The proposed algorithm is tested with suitable simulations and the experiments with real mobile robots like Pioneer 3-DX (Mobil Robots INC), WiFiBot and Lego Robot Mindstorms NXT. The results are summarized and presented as graphics, test images and comparative tables in the conclusion.

Keywords: Visual perception, intelligent robots, visual mobile robots motion control, visual tracking, visual navigation.

7.1 The Principles and Basic Model of Mobile Robot Visual Perception

Mobile robots visual perception can be considered as physical or mathematical models of human visual perception [1, 2, 3]. By means of such models some

R. Kountchev et al. (Eds.): Advances in Reasoning-Based Image Processing, ISRL 29, pp. 173–209.
springerlink.com

mobile robot vision related concepts, such as visual attention [4], scene description by perception [5], visual mobile robot motion control [6] etc. are successfully defined and applied in many robotics application.

The fundamental concepts of visual perception are defined as two basic types of visual perceptions: mind-dependent and mind-independent visual perception, depending of involving not only the human eye but also the human brain [7]. The mind-dependent visual perception modeling of the human perception can be determined as "ill defined" or "not completely defined" [8], in comparison of mind-independent modelling of the human perception [9], mainly because it involves the eye and also the brain. This is because when a human say that perceive something, the meaning is that he can recall relevant properties of it. If the human cannot remember, he cannot claim that he is perceived, although he may suspect that corresponding image information was on his retina. In this sense the mobile robot visual perception modeling must involve together the "robot eye" or visual sensors and also the "robot brain" or memory. Using this concept in visual mobile robot perception is more precise but required more computational resources and efforts. The most popular and simple way in mobile robot visual perception modeling is to consider the mobile robot visual perception system only as "robot eye" or visual sensors, which capture and processing the incoming visual information in area of mobile robot visual observation. This assumption reduces the consideration of mobile robot visual perception only in the range of visual information processing.

One of the important and useful tasks in mobile robot visual perception is visual attention [10], which is based on the probability theory and the assumption of existing in the human visual perception the early stage of visual process [11], where an observer builds up an unbiased understanding of the environment, without involvement of tasks specific bias. This early stage may serve as basis for modeling later stages of mobile robot visual perception, which may involve task specific bias. The probabilistic theory can be seen as a unifying theory describing visual processes of human or mobile robot. A feature of this approach is that it is not necessary to define strictly the visual attention and perception, since this will be established naturally as result of the vision model.

The geometry of the basic probabilistic vision model is presented in Fig.7.1 [12]. In Fig. 7.1 the mobile robot visual perception sensor or observer is viewing a vertical plane from the point denoted by P. By viewing, the observer or mobile robot visual sensor has chance to receive visual data from all directions with equal probability in the first instance. That is, the observer visually take a survey of all locations on the plane with uniform probability density as to the angle θ that defines the viewing direction, without any preference for one direction over another. This consideration ensures that there is no visual bias at the beginning of perception as to the direction, from which information in the environment may be obtained. In particular, the unbiasedness above is with respect to the direction, and therefore the model concerns the very starting instance of a vision mobile robot process, before any directional preference for certain information in the environment can be executed, such as in visual search, object recognition or object tracking, for example. The starting point of the theory is based on basic human vision experience and it is not an ad hoc approach trying to explain a certain vision perception cases.

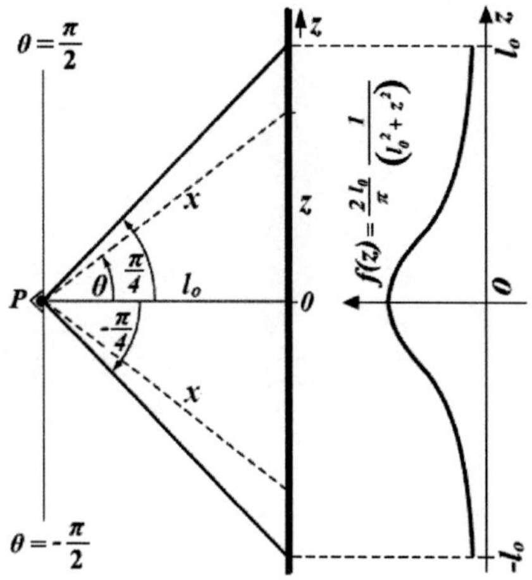

Fig. 7.1 The geometry of visual perception model from top view, where P represents the position of eye or mobile robot visual sensor, perceiving a vertical plane with a distance l_0 to the eye and $f(z)$ is the probability density function of the visual perception.

From the model on Fig. 7.1 is evident that the probability is the same for each single differential visual resolution angle $d\theta$ of the mobile robot visual perception system. This means that the probability density function (pdf), which belongs to the angle θ is uniformly distributed. In Fig. 7.1 is presented a concrete case of mobile robot visual perception for an angle $\theta = \pm\pi/4$. The angle θ is a random variable in the terminology of the probability theory. Since θ is trigonometrically related with each point on the plane, the distances x or z, which are indicated in Fig. 7.1, are also random variables.

The briefly presented mobile robot visual perception probabilistic model can be used as simple not connected to a chosen direction visual robot observation and mainly as a means of comparison with other more complex visual mobile robot perceptual models. For example, other direction dependent situations in mobile robot visual perception can be modeled using suitable, usually not uniformly distributed angle θ [13] shown in Fig. 7.2.

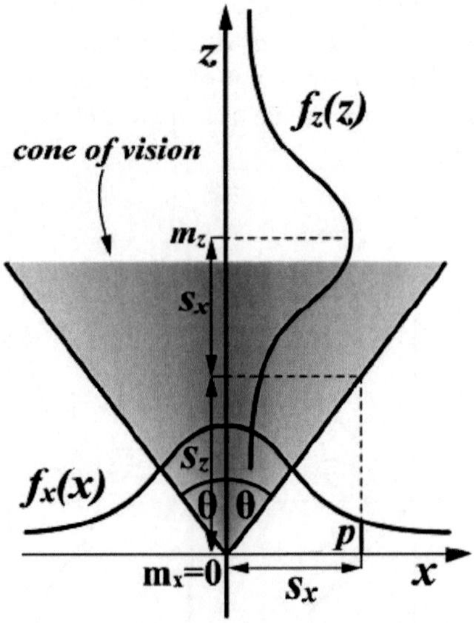

Fig. 7.2 The Gaussian mobile robot visual perception model in forward direction in the shape of a cone with the angle 2θ and forward direction z .

The mobile robot navigation or motion control is one of the major fields of study in autonomous mobile robots [14, 15, 16] in which is possible to apply the mentioned probabilistic model and represent the mobile robot visual perception like the human-like vision process in applications of mobile robot motion control [17, 18, 19, 20]. This approach of an autonomously moving robot with human-like navigation belongs to an emerging robotics technology, which is known as perceptual robotics [21, 22]. From the human-like behaviour viewpoint, perceptual robotics is fellow counterpart of emotional robotics, which is found in a number of applications in practice [23]. Due to its merits, the perceptual robotics can also have various applications in practice.

It is possible to use the described visual perception model in the case of circular geometry from the central point of the circle as is shown in Fig. 7.3. In this case the probability density function (pdf) of the visual perception becomes uniform.

The basic form and the modifications of the presented mobile robot visual perception model lead to the following conclusions:

- the mobile robot visual perception in area of observation depend from the angle θ that defines the viewing direction;
- the algorithms for processing of visual information can be performed only in area defined from the angle θ ;
- depending on the choice of probability density function (pdf), which belongs to the angle θ is possible to select in area of mobile robot visual perception a direction which is important in mobile robot motion control tasks for objects detection and tracking;

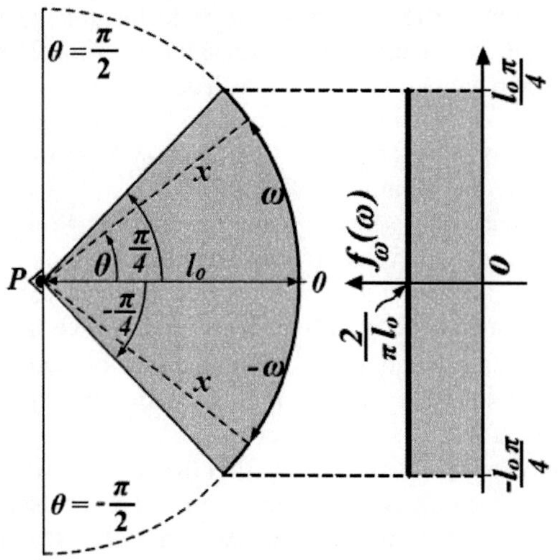

Fig. 7.3 Mobile robot visual perception model for circular geometry with uniform probability density function (pdf) of the visual perception

- choosing the visual perception model with circular geometry allow to use uniform probability density function (pdf) in mobile robots visual perception systems and combined this advantage with the performances of circular coordinate systems like polar or log-polar systems for image presentation [24, 25, 26, 27].

The mentioned in conclusion advantages of the described visual perception model are used in the development of an algorithm for visual mobile robot perception for motion control representing the perceived from the mobile robot visual sensors images in circular polar or log-polar coordinate system.

7.2 Log-Polar Visual Mobile Robot Perception Principles and Properties

7.2.1 Definition of Log-Polar Transformation for Mobile Robot Visual Perception

The log-polar visual perception is a class of methods that represent and process visual information with a space-variant resolution inspired by the visual system of mammals [28], [29], [30]. It can be applied also in mobile robots visual perception systems as an alternative to the conventional approaches in robotics, mainly in the ones where real-time constraints make it necessary to utilize resource-economic image representations and processing methodologies. The suitable applications of log-polar visual perception in robotic vision are: visual attention, target tracking, motion estimation, and 3D perception.

The visual perception robot systems have to deal with large amounts of information coming from the surrounding environment in area of mobile robot observation.

When real-time operation is required, as happens with mobile robots in dynamic and unstructured environments, image acquisition and processing must be performed in a very short time (a few milliseconds) in order to provide a sufficiently fast response to external stimulus.

Appropriate visual robot sensor geometries and image representations are essential for the efficiency of visual robot perception. In biological systems, for instance, the visual perception system of many animals exhibits a non-uniform structure, where some of the receptive fields represent certain parts of the visual field more densely and acutely [31]ïï[32]ïïIn the case of mammals, whose eyes are able to move, retinas present a unique high resolution area in the center of the visual field, called the fovea.

The distribution of receptive fields within the retina is fixed and the fovea can be redirected to other targets by ocular movements. The same structure can be commonly used also in robot visual perception systems with moving cameras, applying pan-tilt devices.

The log-polar image geometry was first motivated by its resemblance with the structure of the retina of some biological vision systems and by its data compression qualities. It can be adapted successful also for the mobile robot visual sensors given visual information in form of Cartesian plane (x, y), represented in complex space z by the variables or coordinates x and y:

$$z = x + jy \tag{7.1}$$

Then the log-polar transformation is a conformal mapping from the points on the visual robot sensor plane (x, y) to points in the log-polar plane (u,v), represented also in complex space w by the variables or coordinates u and v:

$$w = u + jv, \tag{7.2}$$

where

u and v are the log-polar coordinates of eccentricity or radius and angle, respectively.

The complex log-polar transformation or mapping is defined as:

$$w = \log(z) \tag{7.3}$$

7.2.2 Log-Polar Transformation of Image Points in Visual Mobile Robot Perception System

The log-polar representation of image points in visual mobile robot perception system is the transformation first from Cartesian coordinates (x,y) of initial images of mobile robot visual sensor to focal polar plane with polar coordinates (r, θ) and then to the cortical Cartesian plane with Cartesian coordinates (u,v):

$$(x, y) \rightarrow (r, \theta) \rightarrow (u, v), \tag{7.4}$$

where

r is the radius in polar coordinate system (r, θ):

$$r = \sqrt{x^2 + y^2} \quad ; \tag{7.5}$$

θ - angle in polar coordinate system (r, θ):

$$\theta = \arctan\frac{y}{x} \quad ; \tag{7.6}$$

u, v - coordinates in log-polar system (u, v):

$$u = \log(r); \qquad v = \theta \tag{7.7}$$

If the coordinates (x_p, y_p) of the each image point P, given from the mobile robot visual sensors, are in Cartesian plane (x, y), it is necessary first to determine the polar coordinates (r_p, θ_p) in polar plane (r, θ) of the image point P using the equations (7.5) and (7.6) for the Cartesian to polar coordinates transformation (7.5):

$$r_p = \sqrt{x_p^2 + y_p^2} \tag{7.8}$$

$$\theta_p = \arctan\frac{y_p}{x_p}, \tag{7.9}$$

Due to infinity density of pixels in the image center the log-polar transforma-tion cannot be physically and practically implemented for mobile robot visual perception purposes. For this reason, the visual perception mobile robot system in log-polar plane is divided in two areas:

- central part (equivalent to the fovea in animals and human visual sys-tems or so-called central blind spot), which is described in polar coordinates, to hold up the finite number of pixels in the center;
- peripheral part (equivalent to the retina in animals and human visual systems), which is described in log-polar coordinates, to significantly reduction of the amount of data required to be processed in visual mobile robot perception sys-tem, since log-polar transformation collapses the original Cartesian video frames into log-polar images with much smaller dimensions.

This representation of visual perception mobile robot system in log-polar plane divided in two areas, briefly named as fovea and retina, is shown in Fig. 7.4 and is described with the following equations:

$$u = u_{FB}\frac{r}{k} \qquad \text{if} \quad r < k \quad \text{(fovea)} \tag{7.10}$$

$$u = u_{FB} + \frac{1}{l}\log\frac{r}{k} \qquad \text{if} \quad r \geq k \qquad \text{(retina)} \tag{7.11}$$

$$v = \theta\,, \tag{7.12}$$

where

u_{FB} is the fovea to retina boundary;

k and l - scaling constants between Cartesian and log-polar coordinates: k is radius of fovea in Cartesian pixel dimensions and l is exponential function to ensure the log-polar view field matching to the Cartesian image size.

It is obviously from Fig. 7.4 that central part of log-polar plane (equivalent to the fovea in animals and human visual systems or so-called central blind spot) is not suitable for mobile robots visual perception applications.

Therefore, only peripheral part of log-polar plane (equivalent to the retina in animals and human visual systems) is considered next in cases of visual mobile robots perception and general equations (7.10), (7.11) and (7.12) can be simplified and modified in the following more practical equations applicable in the mobile robot visual perception systems:

$$u = \log\frac{r}{r_0} = \log\frac{\sqrt{x^2 + y^2}}{r_0} \tag{7.13}$$

$$v = \theta = \arctan\frac{y}{x} \tag{7.14}$$

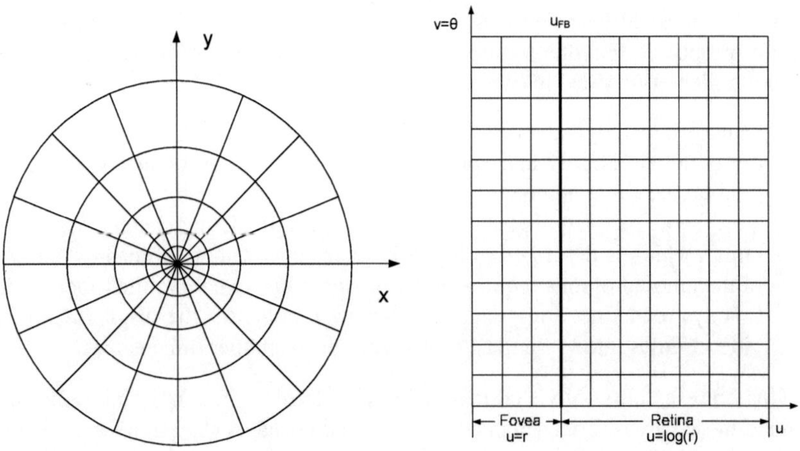

Fig. 7.4 The visual mobile robot perception system representation of in log-polar plane divided in two areas, named as fovea and retina

In case of sampled and quantized discrete images representation the mapping of image points in visual mobile robot perception system between Cartesian and log-polar coordinates is shown more precise in Fig. 7.5. Each discrete receptive element (RE), shown with bold lines in Fig. 7.5, from mobile robot visual input image (left in Fig. 7.5.) is mapped as a corresponding rectangle in log-polar plane (right in Fig. 7.5.).

It can be seen from Fig. 7.5, that in peripheral part (left in Fig. 7.5.) of mobile robot visual input image (equivalent to the retina in animals and human visual systems) many Cartesian pixels, forming the receptive elements (RE), are transformed or collapsed into one pixel of the output log-polar image, which lead to image data and time image processing reduction, when the mobile robot visual perception algorithms are performed in log-polar coordinates.

In the central part (left in Fig. 7.5.) of mobile robot visual input image (equivalent to the fovea in animals and human visual systems or so-called central blind spot) the opposite effect takes place. Since the uniform structure of the mobile robot visual sensor has a finite resolution, the receptive fields near the center or fixation point of mobile robot visual observation become smaller than the Cartesian pixels in input image from the visual mobile robot sensor. Therefore, the information near the center of the Cartesian image results in a highly redundant area in the log-polar image. It is called oversampling, because the Cartesian images are oversampled in central area. In order to allow only a reasonable amount of oversampling in the output log-polar image in the mobile robot applications, the mapping is limited by some inner radius $r_{\min} = r_0$ (Fig. 7.5.) that forms the "blind spot".

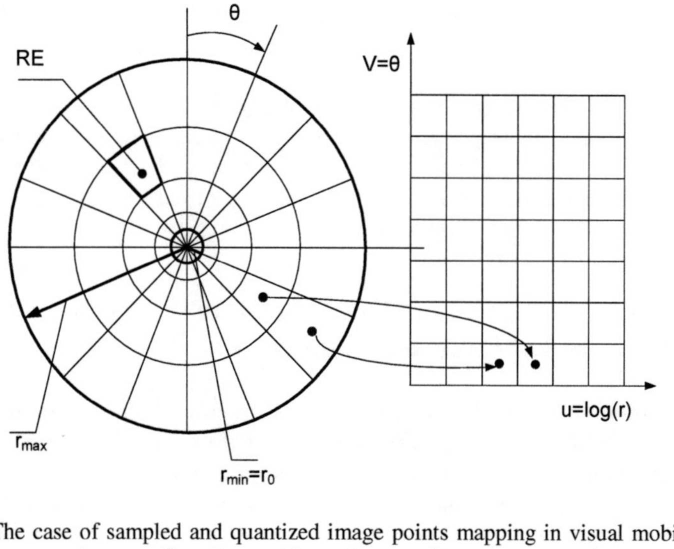

Fig. 7.5 The case of sampled and quantized image points mapping in visual mobile robot perception system between Cartesian and log-polar coordinates representation

7.2.3 The Properties of Log-Polar Transformation Suitable for Visual Mobiles Robot Perception System

The log-polar visual space transformation shows properties important and suitable for mobile robots visual perception systems [29, 30]. The main advantages of log-polar presentation of visual information are:

- reduction of the size of log-polar images, representing visual information with a space-variant resolution inspired by the biological visual systems;
- remarkable mathematical property of log-polar images includes simplification of image rotation and scaling along the optical axis;
- conformal space mapping using for log-polar image transformation preserves oriented angles between curves and neighborhood relationships, almost everywhere, with respect to the original image, allowing with this property to predicts that image processing operations developed for Cartesian images can be applied directly to log-polar images;
- reducing the amount of data to be processed and simplifies several vision algorithms, making possible real time execution of image processing algorithms and their hardware implementation on a single chip.

When compared to the usual Cartesian images, the log-polar images allow faster sampling rates on artificial vision systems without reducing the size of the field of view and the resolution on the central part of the retina (fovea). It has been found that the log-polar geometry also provides important algorithmic benefits. For instance in mobile robot visual perception, it is shown that the use of log-polar images increases the size range of objects that can be tracked using a simple translation model.

The above mentioned properties of log-polar image representation are applicable also in mobile robots visual perception systems along with the conventional approaches in robotics, estimation and comparing their efficiency mainly in real-time mobile robots application, where it is necessary to utilize resource-economic image representations and processing algorithms.

7.2.4 Visual Perception of Objects Rotation in Log-Polar Mobile Robot Visual Perception Systems

Some operations, that in a Cartesian plane of mobile robot perception systems present complications of calculation, are converted in simple expressions in log-polar plane. One of these operations is objects rotation accomplished in log-polar mobile robot visual perception systems easily as a mere translation of dealing in Cartesian coordinates.

The mobile robot visual perception of objects rotation as translation in log-polar plane can be asserted and presented with some simple examples of visual mobile robot observation of objects rotation in area of mobile robot visual sensors. An example of an object in Cartesian plane (x, y) without and with rotation of an angle α concerning a center of rotation located in (r_0, θ_0) is presented in Fig 7.6.

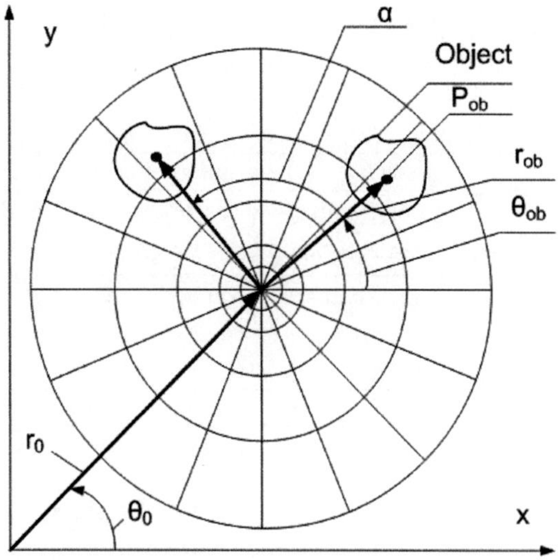

Fig. 7.6 An example of an object in Cartesian plane (x, y) without and with rotation of an angle α concerning a center of rotation located in (r_0, θ_0)

In a simplest example the object on Fig. 7.6, perceived with the mobile robot visual system, can be substituted or described only as a point P_{ob} , usually with its centre of gravity in Cartesian $P_{ob}(x_{ob}, y_{ob})$, polar $P_{ob}(r_{ob}, \theta_{ob})$ and log-polar coordinates $P_{ob}(u_{ob}, v_{ob})$, respectively:

$$P_{ob} = r_{ob} \exp(j\theta_{ob}) \tag{7.15}$$

$$P_{ob} = \log(r) + j\theta_{ob} \tag{7.16}$$

$$u_{ob} = \log(r_{ob}) \tag{7.17}$$

$$v_{ob} = \theta_{ob} \tag{7.18}$$

If α is the angle of object rotation concerning a center of rotation located in (r_0, θ_0) as is shown in Fig 7.6, then from Fig. 7.6 and using the equations (7.15), (7.16), (7.17) and (7.18) is possible to determine the visual perception of mobile robot as a new position of the centre of gravity $P_{ob}(r_{ob},\theta_{ob})$ of the rotated object in polar and log-polar coordinates, respectively:

$$P_{ob}=\frac{r_{ob}\sin(\theta_{ob}+\alpha)-r_0\sin(\theta_{ob}+\alpha)+r_0\sin\theta_0}{\sin\arctan\dfrac{r_{ob}\sin(\theta_{ob}+\alpha)-r_0\sin(\theta_{ob}+\alpha)+r_0\sin\theta_0}{r_{ob}\cos(\theta_{ob}+\alpha)-r_0\cos(\theta_{ob}+\alpha)+r_0\cos\theta_0}}\times$$
$$\times\exp j\arctan\frac{r_{ob}\sin(\theta_{ob}+\alpha)-r_0\sin(\theta_{ob}+\alpha)+r_0\sin\theta_0}{r_{ob}\cos(\theta_{ob}+\alpha)-r_0\cos(\theta_{ob}+\alpha)+r_0\cos\theta_0} \tag{7.19}$$

$$P_{ob}=\log\frac{r_{ob}\sin(\theta_{ob}+\alpha)-r_0\sin(\theta_{ob}+\alpha)+r_0\sin\theta_0}{\sin\arctan\dfrac{r_{ob}\sin(\theta_{ob}+\alpha)-r_0\sin(\theta_{ob}+\alpha)+r_0\sin\theta_0}{r_{ob}\cos(\theta_{ob}+\alpha)-r_0\cos(\theta_{ob}+\alpha)+r_0\cos\theta_0}}+$$
$$+\,j\arctan\frac{r_{ob}\sin(\theta_{ob}+\alpha)-r_0\sin(\theta_{ob}+\alpha)+r_0\sin\theta_0}{r_{ob}\cos(\theta_{ob}+\alpha)-r_0\cos(\theta_{ob}+\alpha)+r_0\cos\theta_0} \tag{7.20}$$

$$u_{ob}=\log\frac{r_{ob}\sin(\theta_{ob}+\alpha)-r_0\sin(\theta_{ob}+\alpha)+r_0\sin\theta_0}{\sin\arctan\dfrac{r_{ob}\sin(\theta_{ob}+\alpha)-r_0\sin(\theta_{ob}+\alpha)+r_0\sin\theta_0}{r_{ob}\cos(\theta_{ob}+\alpha)-r_0\cos(\theta_{ob}+\alpha)+r_0\cos\theta_0}} \tag{7.21}$$

$$v_{ob}=\arctan\frac{r_{ob}\sin(\theta_{ob}+\alpha)-r_0\sin(\theta_{ob}+\alpha)+r_0\sin\theta_0}{r_{ob}\cos(\theta_{ob}+\alpha)-r_0\cos(\theta_{ob}+\alpha)+r_0\cos\theta_0} \tag{7.22}$$

It is seen from the equations (7.19), (7.20), (7.21) and (7.22) that the general form of rotation in polar and log-polar coordinates is also quite complex and difficult to perform in a mobile robot visual perception system as it is in Cartesian coordinates. However, there are particular cases in mobile robot perception where the rotation of an object is transformed as a translation, using the log-polar transformation and in these cases is possible to expound and demonstrate the advantages of the transformation of the rotation to translation like a positive characteristic of the log-polar representation in the applications of mobile robot visual perception.

If the rotation of an object, perceived with mobile robot visual sensor, is considered as the case of rotations concerning the optic axis of visual perception system as is seen from Fig.7.7, then it is possible to suppose that the center of rotation is strictly or approximately very close to zero:

$$r_0=0 \text{ or } r_0\approx 0 \tag{7.23}$$

The assumption (7.23) gives the right to rewrite the equations (7.21) and (7.22) in a more and very simple way like:

$$u_{ob}=\log(r_{ob}) \tag{7.24}$$

$$v_{ob}=\theta_{ob}+\alpha \tag{7.25}$$

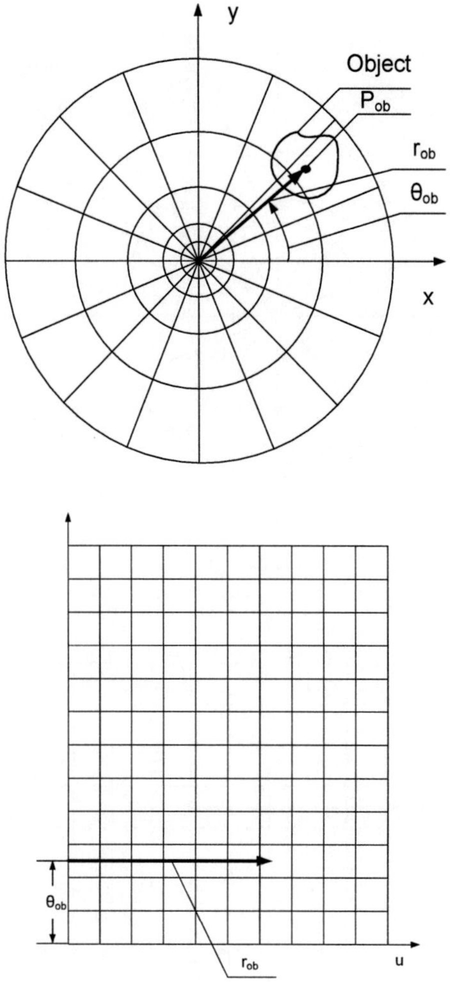

Fig. 7.7 The rotation of an object, perceived with mobile robot visual sensor, concerning the optic axis of visual perception system, has been transformed in the log-polar visual robot perception as a translation concerning the axis v and of a value depending of angle α of the object rotation.

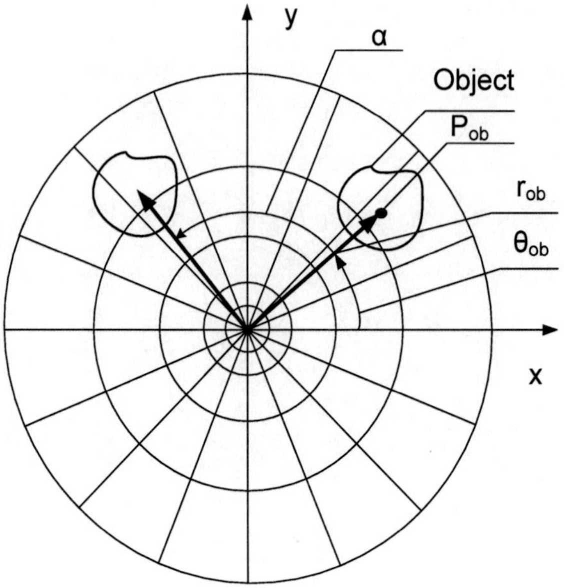

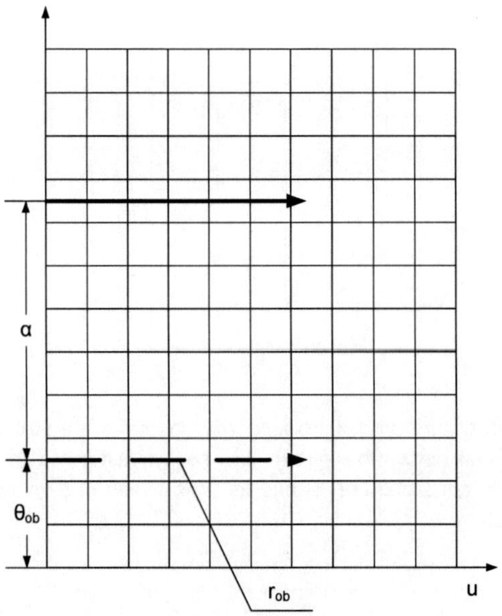

Fig. 7.7 *(continued)*

It is seen from the comparison of equations (7.17), (7.18) and (7.24), (7.25) and from Fig.7.7, that the visual mobile robot perception of an object rotation in Cartesian visual mobile robot sensor plane has been transformed in the log-polar plane, in visual robot perception as a mere translation concerning the axis v and of a value depending of angle α of the object rotation.

The similar results are achieved also with performing the log-polar transformation to real images without and with rotation captured from mobile robot visual perception sensors shown in Fig.7.8, where Fig. 7.8 (a) is input image with rotation angle $\alpha = 0$; Fig. 7.8 (b) – log-polar image with rotation angle $\alpha = 0$; Fig. 7.8 (c) – restored image with rotation angle $\alpha = 0$; Fig. 7.8 (d) – input image with rotation angle $\alpha \neq 0$; Fig. 7.8 (e) – log-polar image with rotation angle $\alpha \neq 0$; Fig. 7.8 (f) – restored image with rotation angle $\alpha \neq 0$.

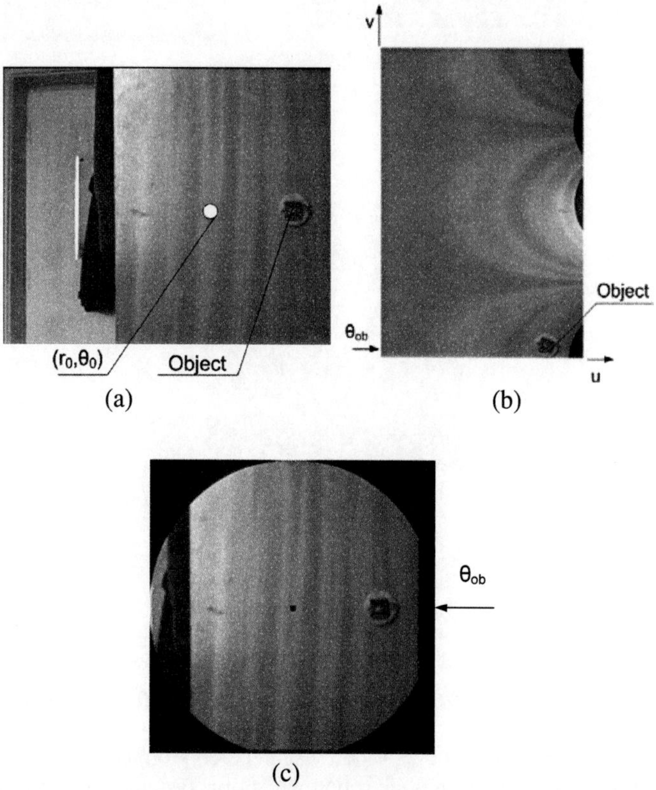

Fig. 7.8 The result of rotation of an object concerning the optic axis of visual perception system in real images captured from mobile robot visual perception sensors: (a) – input image with rotation angle $\alpha = 0$; (b) – log-polar image with rotation angle $\alpha = 0$; (c) – restored image with rotation angle $\alpha = 0$; (d) – input image with rotation angle $\alpha \neq 0$; (e) – log-polar image with rotation angle $\alpha \neq 0$; (f) – restored image with rotation angle $\alpha \neq 0$

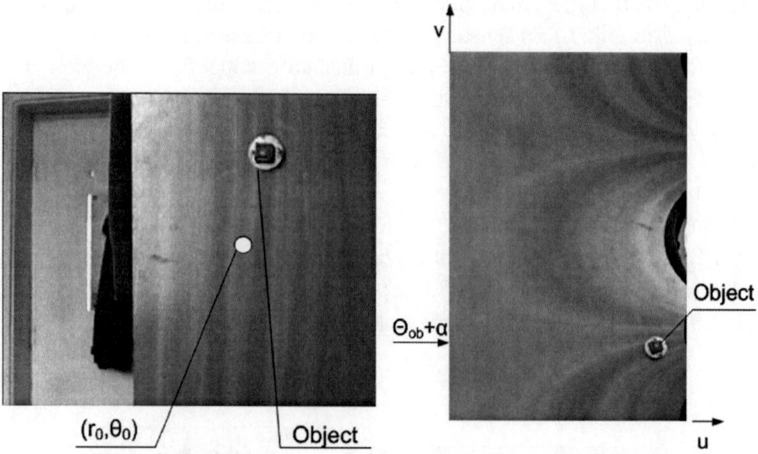

(d) (e)

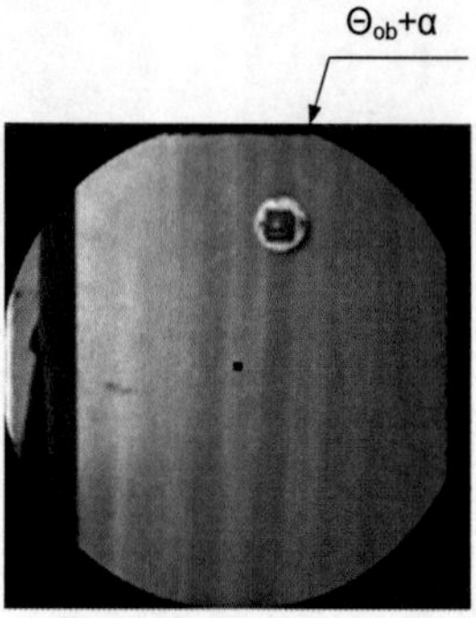

(f)

Fig. 7.8 *(Continued)*

7.2.5 Visual Perception of Objects Translation and Scaling in Log-Polar Mobile Robot Visual Perception Systems

Other operation useful in mobile robot visual perception is objects translation, which in a Cartesian plane of mobile robot perception systems present like rotation the complications of calculations. In log-polar plane this operation is performed from mobile robot visual perception systems with very simple expressions in comparison with Cartesian coordinates. The similar advantages and relations between translation and scaling using log-polar representation in mobile robot visual perception can be found in applications of objects scaling if this operation is necessary and is a part of the in processing algorithm of visual information in mobile robot visual perception systems.

In the case of mobile robot visual perception of object translation can be represented also as in the case of rotation as simple translation only of a point P_{ob} of the object representing centre of gravity in Cartesian $P_{ob}(x_{ob}, y_{ob})$, polar $P_{ob}(r_{ob}, \theta_{ob})$ and log-polar coordinates $P_{ob}(u_{ob}, v_{ob})$ and described with the equations (7.15), (7.16), (7.17) and (7.18). In general case translation or scaling of the centre of gravity $P_{ob}(r_{ob}, \theta_{ob})$ of object with a factor K_{sc} is considered concerning a center of rotation located in polar coordinates (r_0, θ_0) and is presented with the following equations:

$$P_{ob} = \frac{K_{sc} r_{ob} \sin\theta_{ob} + (1-K_{sc}) r_0 \sin\theta_0}{\sin\arctan\dfrac{K_{sc} r_{ob} \sin\theta_{ob} + (1-K_{sc}) r_0 \sin\theta_0}{K_{sc} r_{ob} \cos\theta_{ob} + (1-K_{sc}) r_0 \cos\theta_0}} \times \\ \times \exp j \arctan\frac{K_{sc} r_{ob} \sin\theta_{ob} + (1-K_{sc}) r_0 \sin\theta_0}{K_{sc} r_{ob} \cos\theta_{ob} + (1-K_{sc}) r_0 \cos\theta_0} \tag{7.26}$$

$$P_{ob} = \log \frac{K_{sc} r_{ob} \sin\theta_{ob} + (1-K_{sc}) r_0 \sin\theta_0}{\sin\arctan\dfrac{K_{sc} r_{ob} \sin\theta_{ob} + (1-K_{sc}) r_0 \sin\theta_0}{K_{sc} r_{ob} \cos\theta_{ob} + (1-K_{sc}) r_0 \cos\theta_0}} + \\ + j \arctan\frac{K_{sc} r_{ob} \sin\theta_{ob} + (1-K_{sc}) r_0 \sin\theta_0}{K_{sc} r_{ob} \cos\theta_{ob} + (1-K_{sc}) r_0 \cos\theta_0} \tag{7.27}$$

$$u_{ob} = \log \frac{K_{sc} r_{ob} \sin\theta_{ob} + (1-K_{sc}) r_0 \sin\theta_0}{\sin\arctan\dfrac{K_{sc} r_{ob} \sin\theta_{ob} + (1-K_{sc}) r_0 \sin\theta_0}{K_{sc} r_{ob} \cos\theta_{ob} + (1-K_{sc}) r_0 \cos\theta_0}} \tag{7.28}$$

$$v_{ob} = \arctan\frac{K_{sc} r_{ob} \sin\theta_{ob} + (1-K_{sc}) r_0 \sin\theta_0}{K_{sc} r_{ob} \cos\theta_{ob} + (1-K_{sc}) r_0 \cos\theta_0} \tag{7.29}$$

It is seen from the equations (7.26), (7.27), (7.28) and (7.29) that the general form of translation and scaling with a chosen scaling factor K_{sc} in polar and log-polar coordinates is also quite complex and difficult to perform in a mobile robot visual perception system as it is in Cartesian coordinates. However, there are particular cases in mobile robot perception where the translation and scaling of an object is transformed as a translation, using the log-polar transformation and in these cases is possible to expound and demonstrate the advantages of the transformation of the translation and scaling to translation like a positive characteristic of the log-polar representation in the applications of mobile robot visual perception.

If the translation and scaling of an object, perceived with mobile robot visual sensor, is considered as the case of translation and scaling concerning the optic axis of visual perception system as is seen from Fig. 7.9, then it is possible to suppose that the center of rotation is strictly or approximately very close to zero:

$$r_0 = 0 \text{ or } r_0 \approx 0 \tag{7.30}$$

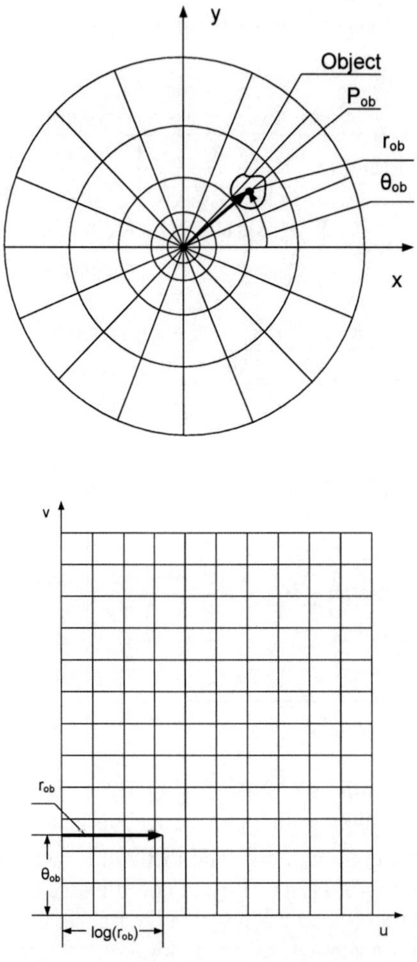

Fig. 7.9 The translation and scaling of an object concerning the optic axis of visual perception system in Cartesian visual mobile robot sensor plane is transformed as a translation concerning the axis u and of a value depending of the scaling factor K_{sc} of the object translation in the log-polar plane

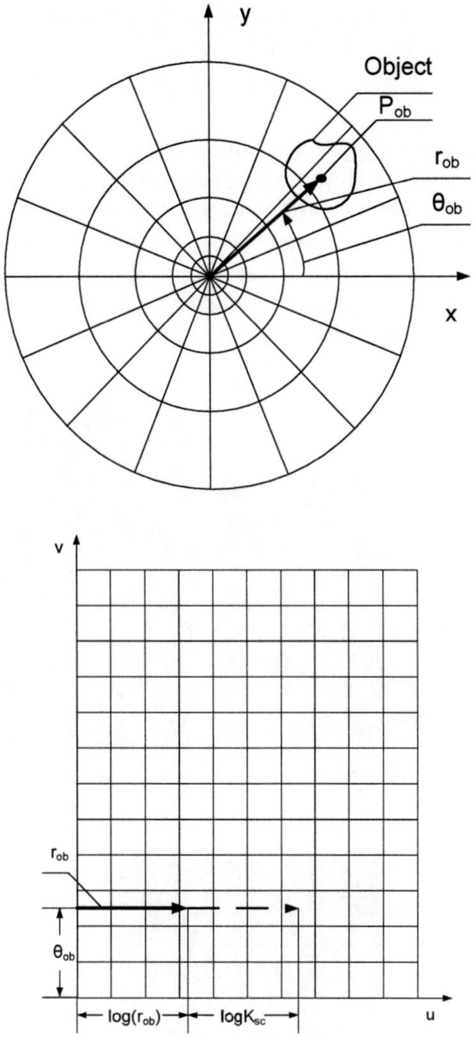

Fig. 7.9 *(continued)*

The assumption (7.30) gives the right to rewrite the equations (7.28) and (7.29) in a more and very simple way like:

$$u_{ob} = \log(r_{ob}) + \log K_{sc} \tag{7.31}$$

$$v_{ob} = \theta_{ob} \tag{7.32}$$

It is seen from the comparison of equations (7.17), (7.18) and (7.31), (7.32) and from Fig. 7.9, that the visual mobile robot perception of an object translation and scaling in Cartesian visual mobile robot sensor plane has been transformed in the log-polar plane, in visual robot perception as a mere translation concerning the axis u and of a value depending of the scaling factor K_{sc} of the object translation and scaling.

The similar results are achieved also with performing the log-polar transformation to real images without and with translation and scaling captured from mobile robot visual perception sensors shown in Fig.7.10, where: Fig. 7.10 (a) is input image without translation and scaling; Fig. 7.10 (b) – log-polar image without translation and scaling; Fig. 7.10 (c) – restored image without translation and scaling; Fig. 7.10 (d) – input image with translation and scaling; Fig. 7.10 (e) – log-polar image with translation and scaling; Fig. 7.10 (f) – restored image with translation and scaling.

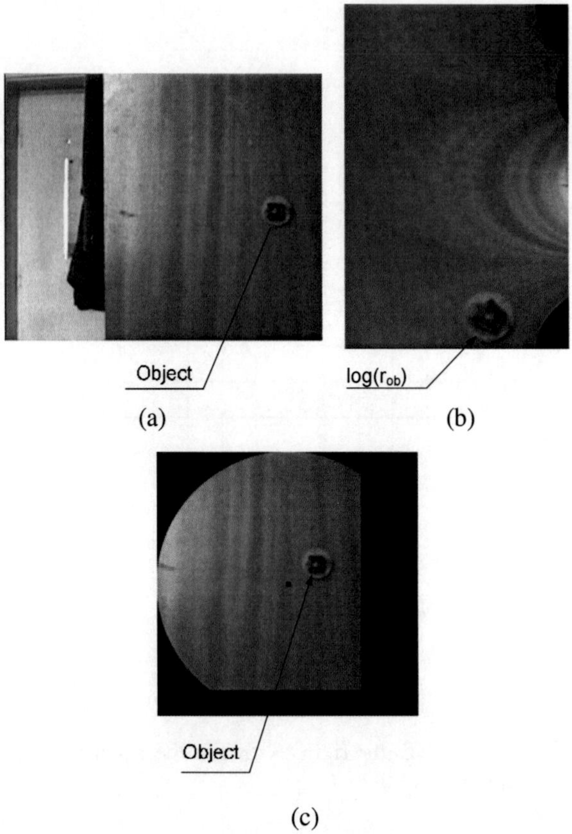

Fig. 7.10. The result of translation and scaling of an object concerning the optic axis of visual perception system in real images captured from mobile robot visual perception sensors: (a) is input image without translation and scaling; (b) log-polar image without translation and scaling; (c) – restored image without translation and scaling; (d) – input image with translation and scaling; (e) – log-polar image with translation and scaling; (f) – restored image with translation and scaling

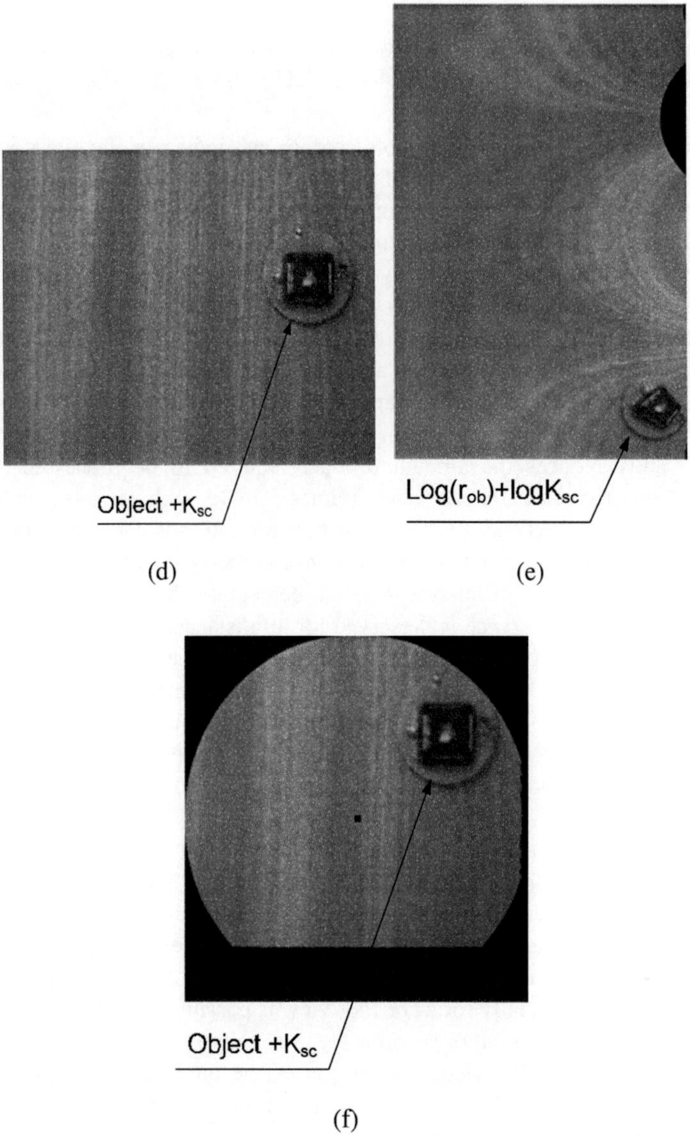

Fig. 7.10 *(continued)*

7.3 Algorithm for Motion Control with Log-Polar Visual Mobile Robot Perception

7.3.1 The Basic Principles and Steps of the Algorithm for Motion Control with Log-Polar Visual Mobile Robot Perception

There are many cases in mobile robots outdoor and indoor applications when the mobile robot motion is performed and controlled with an appropriate tracking algorithm based on the processing of information from all mobile robot visual, audio, ultrasound, laser or other sensors mounted on the mobile robot platform. Most of these tracking algorithms give priority to visual information capturing with mobile robot visual perception sensor or video cameras.

The log-polar visual perception properties and advantages in rotation, translation and scaling of objects described above are used in the proposed mobile robot tracking algorithm. The log-polar transformation provides two major benefits to tracking from a mobile robot platform moving along outdoor or indoor road. First, it significantly reduces the amount of data required to be processed since it collapses the original Cartesian video frames into log-polar images with much smaller dimensions. Second, the log-polar transformation is capable of mitigating perspective distortion due to its scale invariance property.

This second aspect is of interest in visual perception for mobile robot tracking because the target appearance is preserved for all distances from the mobile robot video camera in video perception system. This works however only if the center of log-polar transformation is coincident with the vanishing point of perspective view. Therefore, in the proposed tracking algorithm is included the procedure to keep the center of log-polar transform on the vanishing point (center of perspective view) at every video frame compensating for the carrying mobile robot movements.

To the development of this algorithm is supposed on some prior knowledge about the outdoor or indoor environment. For example, the outdoor perspective view of a road edges and painted lane separation lines can be used in visual mobile robot perception and tracking algorithm an also in estimating the vanishing point location of perspective view. In the similar way, for the indoor perspective view of a room of especially for a corridor view is possible to exploit the existence of lines at two side of a room or corridor.

Log-polar mobile robot visual perception of the outdoor road edges, painted lane separation lines or indoor two side existing room or corridor lines possess the following important features:

- if the road or corridor lines converge at the center of vanishing point of perspective view, they are transformed as parallel lines in case of log-polar mobile robot visual perception;

- if there is a shift of the center of road or corridor lines convergence from the center of vanishing point of perspective view, these lines are perceived from log-polar mobile robot visual perception system as bended lines or curves instead of parallel lines;
- the bend of road or corridor lines in the log-polar images is mostly in the left region (fovea) of log-polar image plane, while the peripheral sections (retina) of these lines stay parallel withstanding larger shifts of the central point toward the center of vanishing point of perspective view, because the fovea (left) region of the log-polar image is more sensitive to the center point shift in opposition to the retina or peripheral (right) region.

The last property is based on the geometry of the log-polar transformation:

- the angular shifts corresponding to the center point displacement are larger with regards to a point in the fovea region opposed to a point in the retina or periphery because the latter is more distant from the log-polar mapping central point;
- also because the size of receptive fields in the fovea are smaller then in the periphery.

These important features of mobile robot visual perception in log-polar images of the outdoor road edges or indoor corridor lines are demonstrated in Fig.7.11. It is chosen an example of simplified graphical representation of a corridor perspective view with convergence (left down in Fig.7.11.) and without convergence (left up in Fig.7.11.) of the corridor lines with the center of vanishing point of perspective view. After the log-polar transformation the mobile robot visual perception of these lines seem parallel (right up in Fig.7.11.) or bend (right down in Fig.7.11.) lines, respectively.

The above mentioned features of mobile robot visual perception in the log-polar images of the outdoor road edges or indoor corridor lines and are chosen as the base of the proposed log-polar visual perception mobile robot tracking algorithm shown in Fig. 7.12.

Briefly the basic steps of the proposed algorithm for mobile robot tracking of outdoor road edges or indoor corridor lines using log-polar visual perception are listed in the flow chart on Fig. 7.12.

The first step describe the necessary image capture procedure from the mounted on the mobile robot platform visual perception mobile robot sensors like mono or stereo video cameras with or without pan-tilt device. The images defined in Cartesian coordinate plane with equation (7.1) are captured as static frames or frames separated from a continuous video stream given from video cameras.

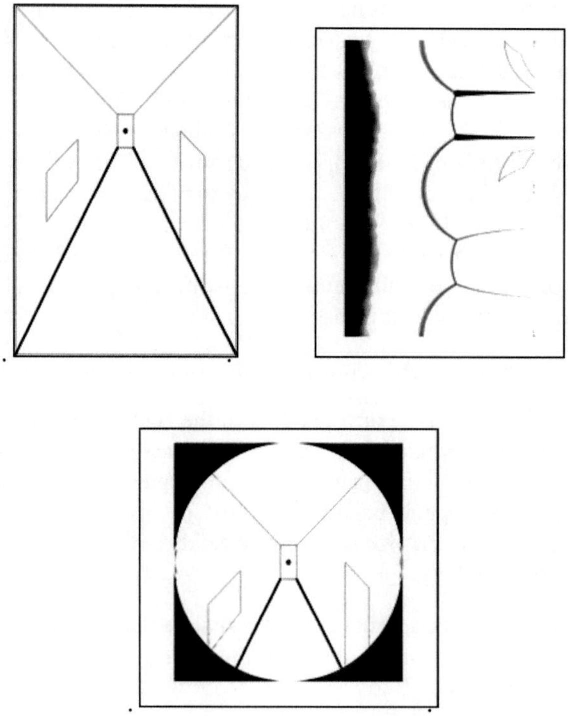

Fig. 7.11 Simplified graphical representation of a corridor perspective view with convergence (these lines seem parallel in log-polar mobile robot visual perception) and without convergence (these lines seem bend in log-polar mobile robot visual perception) of the corridor lines with the center of vanishing point of perspective view

Each image frame is transformed in next step of the algorithm using the equations from (7.4) to (7.8) for the conversion of mobile robot Cartesian visual perception in log-polar visual perception. The initial coordinates x_{ip}, y_{ip} of centerof log-polar transformation are chosen equal to coordinates x_{vp}, y_{vp} of the center or vanishing point of perspective view:

$$x_{ip}, y_{ip} = x_{vp}, y_{vp} \tag{7.33}$$

The condition (7.33) is necessary to satisfy the above mentioned log-polar mobile robot visual perception feature, that if the road or corridor lines converge at the center of vanishing point of perspective view, they are transformed as parallel lines in case of log-polar mobile robot visual perception. When this is done, it can be used the capability of log-polar visual perception of points or objects translation in direction of radial log-polar axis, which is determined with the equations (7.30), (7.31) and (7.32) and is demonstrate on Fig. 7.8.

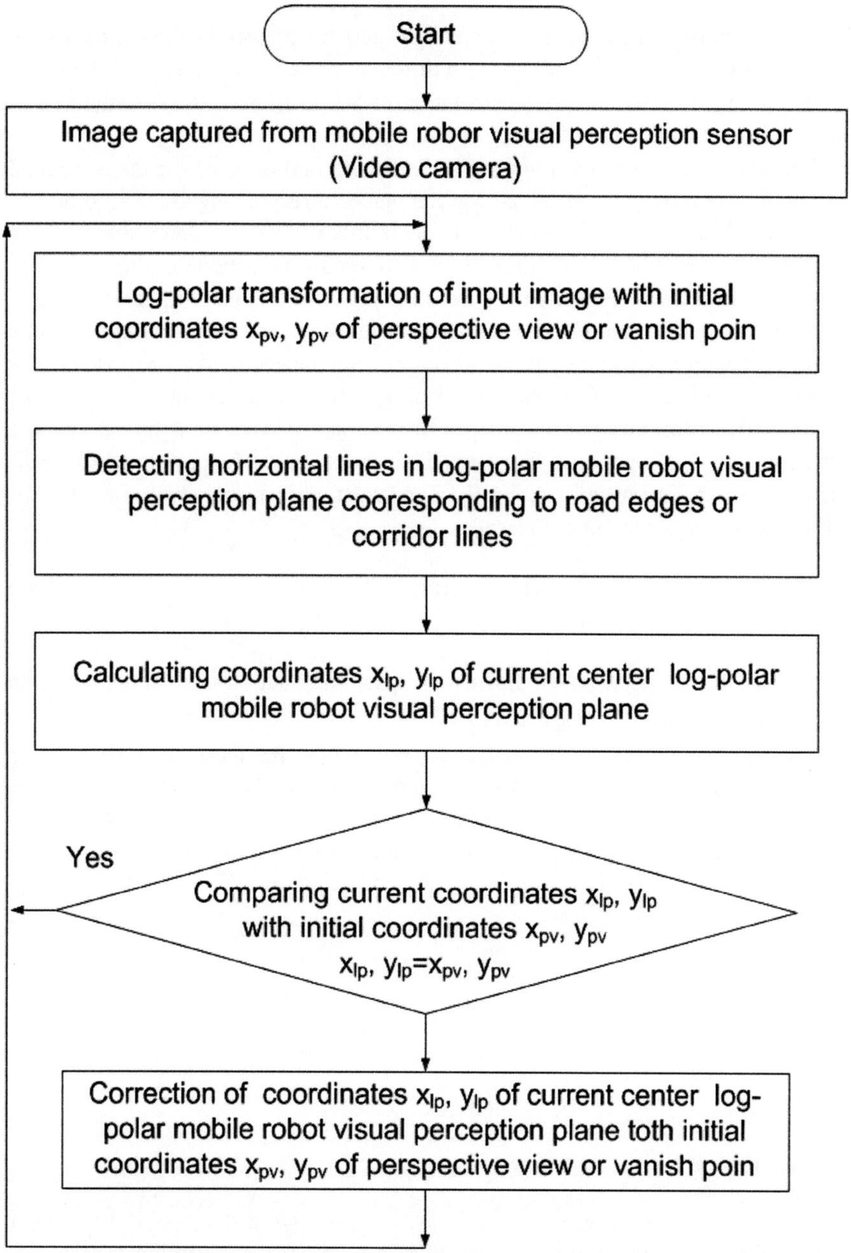

Fig. 7.12. The basic steps of the proposed algorithm for mobile robot tracking of outdoor road edges or indoor corridor lines using log-polar visual perception 26

The log-polar image feature to converge the road or corridor lines of perspective view as parallel horizontal lines (in direction of radial log-polar coordinate direction u) is proposed to use in calculation of the positions of these parallel lines in direction v of angle log-polar coordinate v.

The next step of the algorithm in Fig. 7.12 is to find or detecting the horizontal parallel lines in log-polar visual perception plane corresponding to the road or corridor lines. Here is possible to use the popular and well known algorithms for lines detection in images performing for example first edge detection, or other local operators to separate the existing lines from the log-polar image of the road or corridor. The choices of one of these methods for lines detection or separation depend upon the concrete content of the road or corridor images. If the road or corridor images are simple with little number of objects like in the example on Fig. 7.11 the lines detection in log-polar images of the road or corridor is an easy or not time consuming computation operation. This operation can be performed using directly the log-polar image from Fig. 7.11 and calculating the sum s_r(v) of values of all image pixels in each image row:

$$s_r(v) = \sum_{u=1}^{nu} p(u,v), \quad \text{for v = 1,2,3,.......nv;} \tag{7.34}$$

where

nu, nv are the log-polar image dimensions in u and v coordinate direction of the log-polar image $p(u,v)$, respectively.

The equation (7.34) can be analyzed for searching the local minimums (if the road or corridor lines are black like in example of Fig. 7.11) or local maximums (if the road or corridor lines are white like in example of Fig. 7.11) in the sum $s_r(v)$ of values of all image pixels in each image row to find the log-polar coordinate v_{li} for the existing in the log-polar images parallel lines:

$$v_{li} = \min_{v=1,2,3,...nv}(s_r(v)) \ , \tag{7.35}$$

if the road or corridor lines are black,

or

$$v_{li} = \max_{v=1,2,3,...nv}(s_r(v)) \, , \tag{7.36}$$

if the road or corridor lines are white,

where

v_{li} is the angle coordinate v_l of the *i-th* parallel line in the analyzed log-polar image of the road or corridor in area of mobile robot visual perception.

The using of equation (7.34) in calculations of the sum s_r *(v)* of values of all image pixels in each image row to find the log-polar coordinate v_{li} for the existing in the log-polar images parallel lines is shown in Fig.7.13 for the example of mobile robot visual perception in log-polar image of the indoor corridor lines in Fig. 7.11. It is seen from the Fig.7.13 two minimums in sum s_r*(v)* from which is 27.

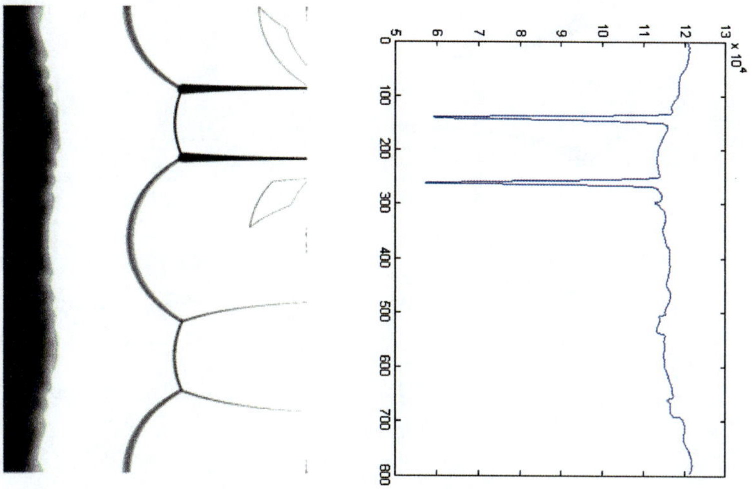

Fig. 7.13 In the sum $s_r(v)$ are seen two minimums from which is very easy to calculating the log-polar coordinates v_{li} for the perceived in the log-polar image two parallel lines corresponding to the indoor corridor lines

The example shown in Fig.7.13 is of the case when the condition (7.33) is satisfied, i.e. the coordinates x_{ip}, y_{ip} of center of log-polar transformation are chosen equal to coordinates x_{vp}, y_{vp} of the center or vanishing point of perspective view. If there is some difference between the initial coordinates x_{ip}, y_{ip} of center of log-polar transformation from the coordinates x_{vp}, y_{vp} of the center or vanishing point of perspective view, then the parallel horizontal lines (in direction of radial log-polar coordinate direction u) are changed as curves and they seem in this case not as parallel lines. These changes are show on Fig. 7.14 for the case when the condition (7.33) is not satisfied - the coordinates x_{ip}, y_{ip} of center of log-polar transformation are not equal to coordinates x_{vp}, y_{vp} of the center or vanishing point of perspective view.

The difference of the sum $s_r(v)$ shown on Fig. 7.13 and Fig. 7.14 are used in the next step of proposed algorithm in Fig. 7.12 to compare the log-polar coordinates v_{li} of the parallel lines calculated in log-polar images in the initial step of algorithm with the current values of coordinates v_{li} of lines calculated in current step of the algorithm shown in Fig.7.12. This comparison give as a final result the possibility to transform the comparison of initial and current coordinates v_{li} in log-polar plane as a comparison of the initial and current coordinates x_{ip}, y_{ip} in Cartesian plane of visual mobile robot perception sensors.

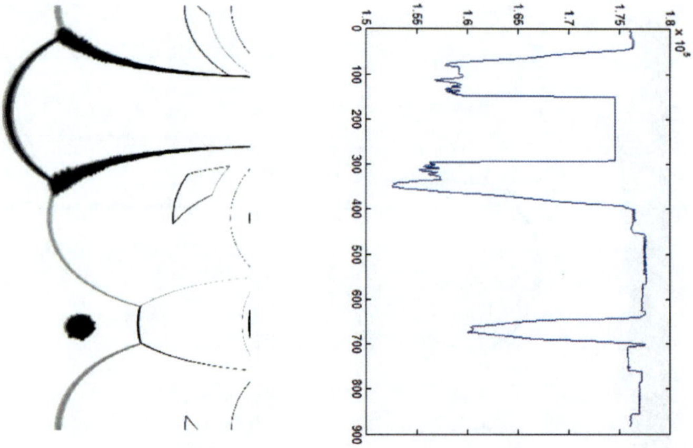

Fig. 7.14 The parallel horizontal lines (in direction of radial log-polar coordinate direction u) are changed as curves they seem in this case not as parallel lines because of differences between the initial coordinates x_{ip}, y_{ip} of center of log-polar transformation from the coordinates x_{vp},y_{vp} of the center or vanishing point of perspective view

The result of these comparisons in log-polar and then in Cartesian image planes are applied in the last step of the presented in Fig. 7.12 algorithm when the differences of the initial and current coordinates x_{ip}, y_{ip} are used doing the correction of the center of log-polar transformation coordinates x_{vp}, y_{vp} to be in concordance or equal to the center or vanishing point of perspective view in the firs step of the next cycle of the proposed algorithm and to satisfy the condition (7.33). If the images are a sequence of frames in a video stream from mobile robot visual perception sensors, it is usually to treat the time of execution of all steps of each cycle in the proposed algorithm equivalent to the duration of an image frame.

7.3.2 Simulation and Test Results for the Algorithm of Motion Control with Log-Polar Visual Mobile Robot Perception

The briefly described algorithm proposed for motion control with log-polar visual mobile robot perception is simulated and tested in the following ways:

- using preliminary recorded images from mobile robot visual perception sensors in form of static images or video streams;
- creating Matlab programs and Simulink [33] models for simulation of the proposed algorithm;

- extending and embedding the created Simulink models of the proposed algorithm as real time working application with digital signal processor of the Texas Instruments Development Kit TMS320C6416T [34];
- simulations and real tests with Microsoft Robotic Studio [35], MobileSim and MobileEyes [36] for some existing models of the mobile robots.

The simulations of the algorithm proposed for motion control with log-polar visual mobile robot perception are prepared with the following created Simulink model shown in Fig. 7.15.

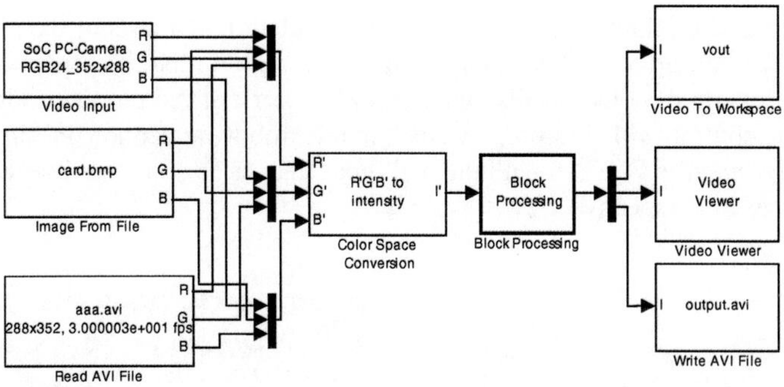

Fig. 7.15 The Simulink model of the algorithm proposed for motion control with log-polar visual mobile robot perception

The possibilities of input color images perceived with mobile robot visual sensors are shown in the model on Fig. 7.14 with the following blocks:

- Video input block - from video camera (block SoC PC-Camera);
- Image From File - from file (for example card.bmp in the block);
- Read AVI File - from preliminary recorded video streaming file (for example aaa.avi in the block).

The input color image for a simulation can be chosen enabling the corresponding RGB component of the chosen input color image from the block of multiplexing. The outputs of the multiplexed RGB component are converted as grayscale image (I) in the Color Space Conversion block.

The main block in the model on Fig. 7.15 is Block Processing. This is a Simulink block specially developed to perform the proposed algorithm for motion control with log-polar visual mobile robot perception. It is simulated first as Matlab program, which then is converted as a Simulink block.

The visualization of the input, log-polar and output images is prepared in three ways with the following three Simulink blocks:

- Video to Workspace block - to use the results of the proposed algorithm for motion control with log-polar visual mobile robot perception for the next or post-processing in Matlab Workspace;
- Video Viewer block – to direct visualization of the input or resultantlog-polar images on the computer monitor or display;
- Write AVI File block – to record in a video streaming file the results of the proposed algorithm for motion control with log-polar visual mobile robot perception.

The created Simulink model is used in the simulation and tests of the proposed algorithm for motion control with log-polar visual mobile robot perception.

An impression of some results taken from the work of the proposed algorithm for motion control with log-polar visual mobile robot perception in simulation mode is presented in Fig.7.16 with the real input images of an indoor corridor with existing lines in perspective view.

Input image of indoor corridor

Log-polar image of indoor corridor

Restored image from log-polar image of indoor corridor

Fig. 7.16 The input, log-polar and restored log-polar test images of the indoor corridor with the calculated sum $s_r(v)$ from which can be determined the log-polar coordinates v_{li} for the perceived in the log-polar image parallel lines corresponding to the indoor corridor lines

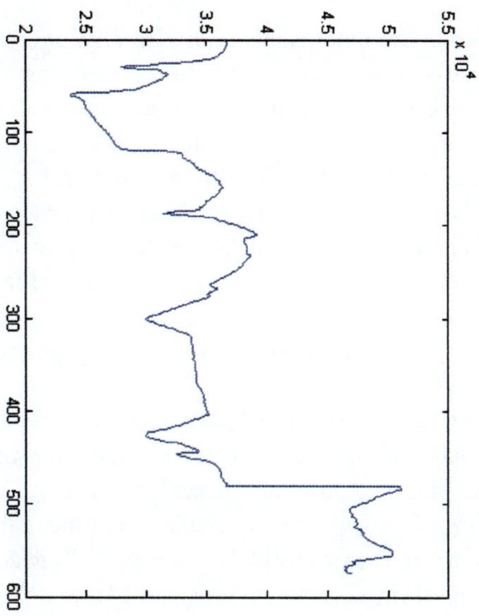

The sum $s_r(v)$ (equation 7.34) for the log-polar image

Fig. 7.16 *(continued)*

The calculated sum $s_r(v)$ is used in post-processing stage of the algorithm to determine the log-polar coordinates v_{li} for the perceived in the log-polar image parallel lines corresponding to the indoor corridor lines. The comparison of the graphical representation of this sum $s_r(v)$ in simulation with a real indoor corridor image (Fig. 7.16) with the corresponding sum $s_r(v)$ in simulation with an example of simplified graphical representation of a corridor perspective view (Fig. 7.11 and Fig. 7.13) show in the case of real images of the corridor a more complicated and difficult situation for the determination of the log-polar coordinates v_{li} for the perceived in the log-polar image parallel lines corresponding to the indoor corridor lines. This lead to the need of more difficult post-processing algorithm or more precise processing of log-polar image before calculating sum $s_r(v)$. This is mentioned in the § 7.3.1 as the need to use some effective and popular well known algorithms for lines detection in log-polar images performing for example first edge detection, or other local operators to separate the existing lines from the log-polar image of the road or corridor. The choices of one of these methods for lines detection or separation depend upon the concrete content of the real corridor images using in the simulations.

After calculating and post-processing of the sum $s_r(v)$ are determined the log-polar coordinates v_{li} for the perceived in the log-polar image parallel lines corresponding to the indoor corridor lines. These current coordinates v_{li} are used to evaluate the existence of the differences between the log-polar initial coordinates v_{li} of the parallel lines calculated in log-polar images in the initial step of algorithm, as is shown in the block schema in Fig. 7.12 of the algorithm. The results from this comparison are used to do, in the last step of the presented in Fig. 7.12 algorithm, the correction of the current coordinates x_{ip}, y_{ip} in Cartesian plane to be in concordance or equal to the center or vanishing point of perspective view and to satisfy the condition (7.33). The information for correction of the of the current coordinates x_{ip}, y_{ip} in Cartesian plane of visual mobile robot perception sensors pass from output of the Block Processing in Simulink model onl Fig. 7.15 through the block for demultiplexing and is entered in the block Video to Workspace block, when is possible to use the results of the proposed algorithm for motion control with log-polar visual mobile robot perception for the next or post-processing in Matlab Workspace.

The results of using the information for correction of the current coordinates x_{ip}, y_{ip} in Cartesian plane of visual mobile robot perception sensors and finally to using for mobile robot motion control are presented in Fig. 7.17.

The image on the Fig. 7.17 is graphical simulation of the real image of the corridor in Fig. 7.16.with the trajectory (solid line in Fig. 7.17) necessary to follow the mobile robot to the end of the corridor. With small circles are pointed some of the real positions in the steps of mobile robot motion control using the information taken as the result of the proposed algorithm for correction of the of the current coordinates x_{ip}, y_{ip} in Cartesian plane of visual mobile robot perception sensors and mobile robot current position. It is seen the trend to achieved a more precise motion control of the mobile robot with comparison to same task of mobile robot following the trajectory to the end of the corridor, but using only the visual information and image processing in Cartesian mobile robot visual perception plane. These results are marked in graphical simulation in Fig. 7.17 as small black rectangles.

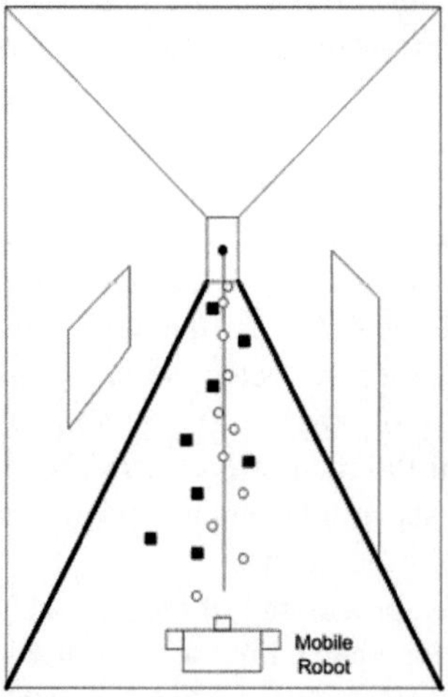

Fig. 7.17. The results of motion control when mobile robot executes the task to a target moving to the end of the corridor.

For achievement of some improvements in calculation in the proposed algorithm and to realize the real or near real time work of the mobile robot motion control and tracking when the robot follow the corridor to a given target or object in the end of the corridor is proposed to create Simulink model with the embedded in Matlab Toolbox to real time programs execution with digital signal processor of the Texas Instruments Development Kit TMS320C6416T [34]. The proposed Simulink model is presented in Fig. 7.18.

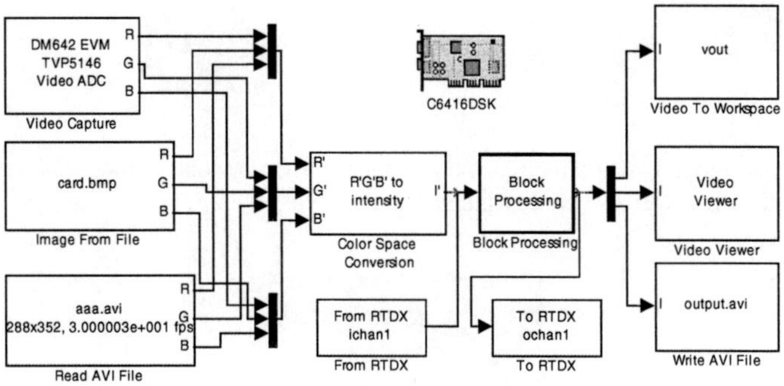

Fig. 7.18 The proposed Simulink model with the embedded in Matlab Toolbox for real time programs execution of the algorithm with digital signal processor of the Texas Instruments Development Kit TMS320C6416T [34]

The new block in the extended from Fig. 7.15 Simulink model are the block to connect the Development Kit TMS320C6416T with Simulink embedded blocks:

- Video Capture – from Daughter Card DM642 EVM of mobile robot input images;
- the blocks To RTDX and From RTDX – to real time execution and connection of the module Development Kit TMS320C6416T with host computer;
- initialization block C6416DSK.

Some of the above mentioned experiments and results (Fig. 7.16 and Fig. 7.17) are performed with the presented in Fig.7.18 Simulink model.

In the simulations and tests are used the following models of the mobile robots Pioneer 3-DX (Mobil Robots INC) and Lego Robot Mindstorms NXT shown in Fig. 7.19.

An impression of the possibility to direct capturing of images direct in log-polar plane and using this visual information in the mobile robots visual perception sensors to reduce the time consuming operation in Cartesian to log-polar transformation is presented in Fig. 7.20 with the developed model of log-polar visual sensor [37].

Fig. 7.19 The models of the mobile robots Pioneer 3-DX (Mobil Robots INC) and Lego Robot Mindstorms NXT used in the simulation and tests

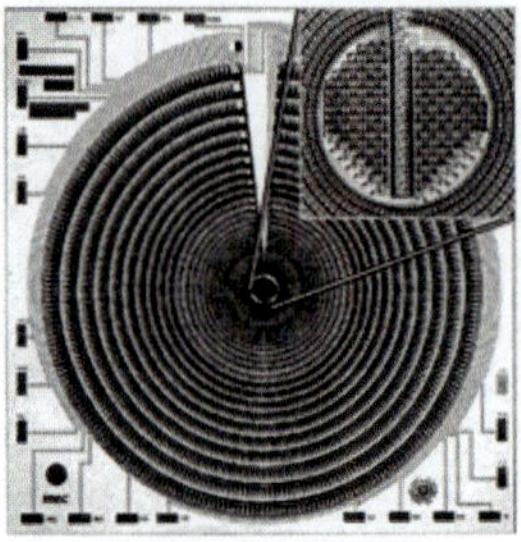

Fig. 7.20 The developed model of log-polar visual sensor [37].

7.4 Conclusion

After the development and testing of the proposed algorithm for motion control with log-polar visual mobile robot perception is possible to conclude and resume of the main results achieved in the algorithm development and to outline the possible future improvements of this algorithm:

- the features and advantages of visual mobile robot perception in log-polar image plane are used to development of an algorithm for mobile robot motion control suitable in popular for mobile robots application tasks like outdoor road lanes or indoor corridor lines tracking;
- the results from simulations and near real time tests carried with the proposed algorithm for chosen examples of mobile robot following the corridor to an object in the end of the corridor given a satisfactory precision in mobile robot motion and tracking control in comparison with some well known methods using visual or other mobile robot perception sensors in motion control and tracking algorithms;

- the achieved positive theoretical and experimental results reassure to think for the future improvements of the proposed algorithm in more realistic situations an more difficult mobile robot motion control and tracking in direction of time consuming minimization, using effective calculation algorithms, digital signal processors (DSP) mentioned above, programmable logic arrays (FPGA) and visual mobile robot perception sensor capturing images direct in log-polar representation, show above as developed experimental exemplars.

References

1. Bigun, J.: Vision with direction. Springer, Heidelberg (2006)
2. Ahle, E., Söffker, D.: A cognitive-oriented architecture to realize autonomous behavior – part II: application to mobile robots. In: Proc. 2006 IEEE Conf. on Systems, Man, and Cybernetics, Taipei, Taiwan, October 8-11, pp. 2221–2227 (2006)
3. Ciftcioglu, Ö., Bittermann, M.S., Sariyildiz, I.S.: Towards computer-based perception by modeling visual perception: a probabilistic theory. In: Proc. 2006 IEEE Int. Conf. on Systems, Man, and Cybernetics, Taipei, Taiwan, October 8-11, pp. 5152–5159 (2006)
4. Bundsen, C.: A theory of visual attention. Psychological Review 97(4), 523–547 (1990)
5. Itti, L., Koch, C., Niebur, E.: A model of saliency-based visual attention for rapid scene analysis. IEEE Trans. on Pattern Analysis and Machine Intelligence 20(11), 1254–1259 (1998)
6. Oriolio, G., Ulivi, G., Vendittelli, M.: Real-time map building and navigation for autonomous robots in unknown environments. IEEE Trans. on Systems, Man and Cybernetics – Part B: Cybernetics 28(3), 316–333 (1998)
7. Foster, J.: The Nature of Perception. Oxford University Press (2000)
8. Bertero, M., Poggio, T.A., Torre, V.: Ill-posed problems in early vision. Proceedings of the IEEE 76(8), 869–889 (1988)
9. Hecht-Nielsen, R.: The mechanism of thought Proc. IEEE World Congress on Computational Intelligence WCCI 2006. Int. Joint Conf. on Neural Networks, Vancouver, Canada, July 16-21, pp. 1146–1153 (2006)
10. Bundsen, C.: A theory of visual attention. Psychological Review 97(4), 23–547 (1990)
11. Ciftcioglu, Ö., Bittermann, M.S., Sariyildiz, I.S.: Autonomous robotics by perception. In: Proc. ISCIS & ISIS 2006, Joint 3rd Int. Conf. on Soft Computing and Intelligent Systems and 7th Int. Symp. on Advanced Intelligent Systems, Tokyo, Japan, September 20-24, pp. 1963–1970 (2006)
12. Eckmiller, R., Baruth, O., Neumann, D.: On human factors for interactive man-machine vision: requirements of the neural visual system to transform objects into percepts. In: Proc. IEEE World Congress on Computational Intelligence WCCI 2006 - Int. Joint Conf. on Neural Networks, Vancouver, Canada, July 16-21, pp. 699–703 (2006)
13. Plumert, J.M., Kearney, J.K., Cremer, J.F., Recker, K.: Distance perception in real and virtual environments. ACM Trans. Appl. Percept. 2(3), 216–233 (2005)

14. Beetz, M., Arbuckle, T., Belker, T., Cremers, A.B., Schulz, D., Bennewitz, M., Burgard, W., Hähnel, D., Fox, D., Grosskreutz, H.: Integrated, plan-based control of autonomous robots in human environments. IEEE Intelligent Systems 16(5), 56–65 (2001)
15. Hachour, O.: Path planning of Autonomous Mobile robot. International Journal of Systems Applications, Engineering & Development 2(4), 178–190 (2008)
16. Wang, M., Liu, J.N.K.: Online path searching for autonomous robot navigation. In: Proc. IEEE Conf. on Robotics, Automation and Mechatronics, Singapore, December 1-3, vol. 2, pp. 746–751 (2004)
17. Bekiarski, A., Pleshkova-Bekiarska, S.: Visual Design of Mobile Robot Audio and Video System in 2D Space of Observation. In: International Conference on Communications, Electromagnetic and Medical applications (CEMA), Athens, vol. 6-9 XI, pp. 14–18 (2008)
18. Bekiarski, A., Pleshkova-Bekiarska, S.: Neural Network for Audio Visual Moving Robot Tracking to Speaking Person. In: 10th WSEAS Neural Network, Praha, pp. 92–95 (2009)
19. Bekiarski, A.: Audio Visual System with Cascade-Correlation Neural Network for Moving Audio Visual Robot. In: 10th WSEAS Neural Network, Praha, pp. 96–99 (2009)
20. Bekiarski, A., Pleshkova-Bekiarska, S.: Simulation of Audio Visual Robot Perception of Speech Signals and Visual Information. In: International Conference on Communications, Electromagnetic and Medical applications (CEMA), vol. 6-9 XI, pp. 19–24 (2008)
21. Ahle, E., Söffker, D.: A cognitive-oriented architecture to realize autonomous behavior – part I: theoretical background. In: Proc. 2006 IEEE Conf. on Systems, Man, and Cybernetics, Taipei, Taiwan, October 8-11, pp. 2215–2220 (2006)
22. Ahle, E., Söffker, D.: A cognitive-oriented architecture to realize autonomous behavior – part II: application to mobile robots. In: Proc. 2006 IEEE Conf. on Systems, Man, and Cybernetics, Taipei, Taiwan, October 8-11, pp. 2221–2227 (2006)
23. Adams, B., Breazeal, C., Brooks, R.A., Scassellati, B.: Humanoid robots: a new kind of tool. IEEE Intelligent Systems and Their Applications 15(4), 25–31 (2000)
24. Zitova, B., Flusser, J.: Image registration methods: A survey. IVC 21(11), 977–1000 (2003)
25. Traver, V.J., Pla, F.: The log-polar image representation in pattern recognition tasks. In: Proceedings of Pattern Recognition and Image Analysis, vol. 2652, pp. 1032–1040 (2003)
26. Zokai, S., Wolberg, G.: Image registration using log-polar mappings for recovery of large-scale similarity and projective transformations. IEEE Transactions on Image Processing 14, 1422–1434 (2005)
27. Luengo-Oroz, M.A., Angulo, J., Flandrin, G., Klossa, J.: Mathematical Morphology in Polar-Logarithmic Coordinates. Application to Erythrocyte Shape Analysis. In: Marques, J.S., Pérez de la Blanca, N., Pina, P. (eds.) IbPRIA 2005. LNCS, vol. 3523, pp. 199–206. Springer, Heidelberg (2005)
28. Jain, R., Bartlett, S.L., O'Brien, N.: Motion stereo using ego-motion complex logarithmic mapping. IEEE Trans. Pattern Analys. Machine Intell. 9(3) (May 1987)

29. Massimo, T., Sandini, G.: On the advantages of the polar and log-polar mapping for direct estimation of time-to-impact from optical flow. IEEE Trans. Pattern Analys. Machine Intell. 15(4) (April 1993)
30. Schwartz, E.L.: Computational anatomy and functional architecture of the striate cortex: a spatial mapping approach to perceptual coding. Vision Res. 20, 645–669 (1980)
31. Schwartz, E.L.: Spatial mapping in the primate sensory projection: Analytic structure and relevance to perception. Biological Cybernetics 25, 181–194 (1977)
32. Shah, S., Levine, M.D.: Visual information processing in primate cone pathways. I. A model. IEEE Transactions on Systems, Man and Cybernetics, Part B 26, 259–274 (1996)
33. Matlab & Simulink R2011a, `http://www.mathworks.com/products/matlab/`
34. TMS320C6416T, D.S.P.: Starter Kit (Starter Kits), `http://focus.ti.com/dsp/`
35. Microsoft Robotic Studio (2008), `http://msdn.microsoft.com/en-us/robotics/`
36. MobileSim & MobileEyes, `http://www.mobilerobots.com/`
37. Pardo, F., Dierickx, B., Scheffer, D.: Space-Variant Non-Orthogonal Structure CMOS Image Sensor Design. IEEE Journal of Solid State Circuits 33(6), 842–849 (1998)

Chapter 8
Motion Estimation for Objects Analysis and Detection in Videos

Margarita Favorskaya

Siberian State Aerospace University, 31 Krasnoyarsky Rabochy, Krasnoyarsk, 660014 Russia
favorskaya@sibsau.ru

Abstract. The motion estimation methods are used for modeling of various physical processes, the behavior of objects, and prediction of events. In this chapter the moving objects in videos are generally considered. Such motion estimation methods are classified as comparative methods and gradient methods. The comparative motion estimation methods are usually used in real-time applications. Many aspects of block-matching modifications are discussed including Gaussian mixture model, Lie operators, bilinear deformations, multi-level motion model, etc. The gradient motion estimation methods assist to realize the motion segmentation in complex dynamic scenes because only they provide a required accuracy. Application of the 2D tensors (in spatial domain) or the 3D tensors (in spatio-temporal domain) depends from the solved problem. Development of the gradient motion estimation methods is necessary for intelligent recognition of objects and events in complex scenes, video indexing in multimedia databases.

Keywords: Motion estimation, block matching, optical flow, structural tensor, flow tensor, visual imagery, infrared imagery.

8.1 Introduction

The motion estimation plays a key role in inner and outdoor surveillance, technological controllers, video coding, video editing, etc. Here we'll consider methods of motion detection in videos for recognition systems generally. Such methods permit to realize the segmentation of dynamic visual objects more accurately; also they are used for recognition of events at the highest stage of the intelligent video processing [15] [53].

The receiving of information about various types of motion and the automatic forming of motion classes are the complex tasks [13] [14]. Motions in videos have a different repeatability in space and time domains; so they are divided into temporal textures, active motions, events, and composite motions. The short description of motion classes and their applications are presented in Table 8.1. In surveillance

R. Kountchev et al. (Eds.): Advances in Reasoning-Based Image Processing, ISRL 29, pp. 211–253.
springerlink.com

systems some temporal textures such as leaves movement under a wind, sea wave's fluctuations, clouds movement, and so on are usually suppressed. Others motion classes are needed in the intelligent methods and algorithms for objects detection and recognition.

Table 8.1 Classification of motion classes

Motion class	Short description	Applications
Temporal textures	Statistical regularities in space and time domains	Analysis of the turbulence of liquids and gases, landscape images recognition, motion analysis of small homogeneous objects
Active motions	Structures that are repeatable in space and time domains	People surveillance, interactive systems "user–computer", navigation systems of robots
Events	Motions that are non-repeatable in space and time domains	People surveillance, retrieval in digital libraries, analysis of sport's competitions, surveillance of emergency events and incidents
Composite motions	Multi-level motions including all previous motion classes	Analysis of visual imagery obtained from a moving camera (people surveillance, robots navigation, surveillance of emergency events and incidents)

Images of complex scenes obtained from a stationary camera may be interpreted as a set of static and dynamic regions, which are further classified as foreground and background objects. If camera is maintained on a moving platform then all dynamic regions in scene possess multilevel motion features. In both cases it is important to determine the motion features of regions, additionally color, texture, geometric and others descriptors. Object analysis will be more complete if we find the rigid and the non-rigid dynamic regions in time domain. Such accumulated information permits to advance hypothesis about a set of closely related regions to be a single whole object.

Visual object has a global motion vector which is calculated as a sum of local motion vectors of regions [48]. One of perspective modifications of the optical flow method is based on a joint application of the 3D structure tensor received from the visual imagery and the 3D flow tensor extracted from the infrared imagery. At that, motion estimation is realized by calculating features of geometric primitives (points, corners, lines) only in moving regions. Such intelligent technology permits to decrease the computing cost due to the dropping of geometric primitives in static regions. The new estimation method based on visual and infrared imageries finds the moving periodical structures which are later included into the connected video graph of a scene.

In Section 8.2 we'll discuss two main categories of motion estimation methods: the comparative methods and the gradient methods. Section 8.3 is dedicated to the tensor approach for motion estimation in videos. In section8.4 you'll find the experimental results which were received for some motion estimation methods.

Tasks for self testing (Section 8.5), conclusion (Section 8.6) and recommend references (Section 8.7) are at the end of this chapter.

8.2 Classification of Motion Estimation Methods

At present many motion estimation methods in videos exist, and they are used for a motion definition of different physical objects. All physical objects are classified as physical processes and phenomenon, solid state objects (with finite and infinite sets of projections), situations and events in temporal progress. Usually the motion detection methods are divided into two categories: the comparative methods and the gradient methods. More often used methods of motion detection and/or motion estimation are presented in Table 8.2. They estimate a motion in spatio-temporal domain and characterize various groups of physical processes, objects and events [6].

Table 8.2 Classification of motion estimation methods

Groups of objects	Comparative methods	Gradient methods
Dynamic textures		Method of spatio-temporal fractal analysis
		Analysis based on autoregression functions
Objects with finite sets of projections	Background subtraction method*	Edge points tracking
	Block-matching method*	Feature points tracking
	Calculation of density motion functions*	Building of motion trajectories of objects
	Motion patterns of optical flow	
Objects with infinite sets of projections	Background subtraction method*	Edge points tracking
	Block-matching method*	Feature points tracking
		Optical flow method
		Building of global and local motion trajectories of objects and their parts
Actions and events		Detection of relative motions
		Prediction of motion trajectories
		Building of actions graph
		Building of events graph

High speed and less accurate methods are labeled by symbol "*".

Let's discuss some well-known motion estimation methods which are used for analysis and detection of objects in videos. In section 8.2.1 we'll analyze the comparative motion estimation methods, generally modifications of the basic block-matching method. In section 8.2.2 the gradient motion estimation methods will be considered. They are more complex than the comparative motion estimation methods, but provide more exact results.

8.2.1 Comparative Motion Estimation Methods

The main assumption of the comparative motion estimation methods consists in small displacements of objects in scene between two sequential frames. We may substitute such displacement by a parallel transition of environment in any point of frame by some vector with a sufficiently high precision. Usually frames in visual imagery satisfy to such restriction excepted by the areas of sharp changes in scene. Let's assume that a motion of objects is described by an almost continuous function. We'll discuss the most applied comparative methods of motion estimation in videos – a background subtraction and a block-matching method including its modifications.

8.2.1.1 Background Subtraction Method

The simplest motion estimation technique is the method of background subtraction which is based on following assumption. During initial n frames, scene is not changed, and starting from (n+1) frame the objects of interest (pedestrians, vehicles and others moving objects) can appear and disappear from a visual field. For each current frame, such parameters as values of brightness, color components in each pixel are compared with corresponding values in each pixel of initial averaged (etalon) frame of a visual imagery. Such method is a noise-dependent method, that's why median filter or mathematical morphological operations are applied for a received binary image. Filter parameters determine the method sensitivity and degree of errors. The singular realization simplicity and the high algorithm speed are the advantages of this method [12]. In spite of these advantages, following problems exist which transform this method to the non-used approach in practice:

- Shadow occurrence from moving objects.
- Dynamic background of scene.
- Quick or slow luminance changes.
- Camera inaccuracy.

More accurate methods of motion estimation in videos can overcome such disadvantages.

8.2.1.2 The Basic Block-Matching Method

The entity of block-matching method consists in a choice of some region in a current frame and in the search of the similar region in a following frame. If the location of the detected region is differed from the location of the initial region then we assume that the movement occurred, and the motion vector can be calculated. Firstly the current frame is divided on the non-crossed blocks with similar sizes $N \times N$ (usually 16×16 pixels) which are defined by the brightness function $f_{t-1}(x,y)$ where (x,y) are coordinates in space domain, t is a discrete time in temporal domain. Secondly for each block in small neighborhood $-S_x<d_x<+S_x$ and $-S_y<d_y<+S_y$ (where S_x, S_y are the sizes of a neighborhood along *OX*, *OY* axes correspondingly) we try to find the most similar block in following frame $f_t(x+d_x,y+d_y)$. The similarity is determined by the used metric. The decision about correspondence of regions is based on the minimization of an error functional *e* according to the used metric. Usually three metrics are applied:

- The sum of absolute differences (*SAD*, *Sum of Absolute Differences*)

$$SAD(d_x,d_y)=\sum_{x=1}^{N}\sum_{y=1}^{N}\left|f_{t-1}(x,y)-f_t(x+d_x,y+d_x)\right| \quad , \tag{8.1}$$

- The sum of squared differences (*SSD*, *Sum of Squared Differences*)

$$SSD(d_x,d_y)=\sum_{x=1}^{N}\sum_{y=1}^{N}\left(f_t(x,y)-f_{t-1}(x+d_x,y+d_y)\right)^2 \quad , \tag{8.2}$$

- The mean of squared differences (*MSD*, *Mean of Squared Differences*)

$$MSD(d_x,d_y)=\frac{1}{N\times N}\sum_{x=1}^{N}\sum_{y=1}^{N}\left(f_t(x,y)-f_{t-1}(x+d_x,y+d_y)\right)^2 \quad . \tag{8.3}$$

Vector $\mathbf{MV}(d_x,d_y)$ for which the error functional *e* ($e_{SAD(dx,dy)}$, $e_{SSD(dx,dy)}$ or $e_{MSD(dx,dy)}$) has the minimum value, is considered as the displacement vector for given block (Fig. 8.1). On fig 8.1 vector $\mathbf{MV}(d_x,d_y)$ shows the displacement of left and top corner of marked block from a frame $(t-1)$ to a frame t.

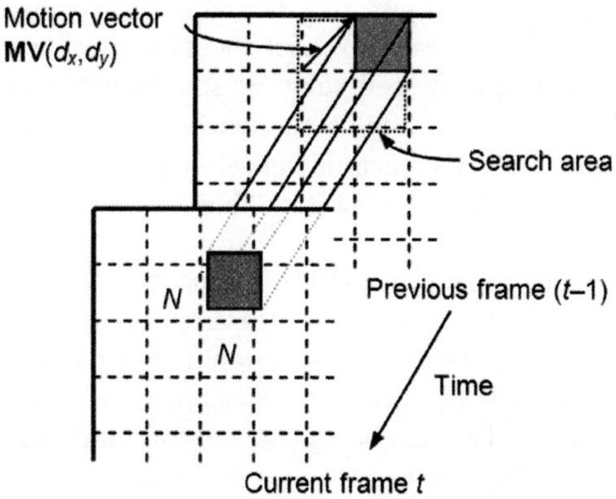

Fig. 8.1 Scheme of the basic block-matching method

The basic block-matching method has various interpretations such as Full Search (FS), Pattern Search (PS), Recursive Search (RS), and Block Search (BS) [19]. Full Search is based on the calculation of all values of motion vectors from an acceptable search area and the definition of a minimum value of the error functional according to expressions (8.1) – (8.3). Such procedure has a high computational complexity, and it can not be used in real-time systems. Some modifications of Full Search exist such as Three-Step Search (TSS), Conjugate Direction Search (CDS), Dynamic Window Search (DSW), Cross-Search Algorithm (CSA), and Two-Dimensional Logarithmic Search (TDLS). All these methods direct on an increased calculation speed of FS.

Pattern Search assumes that a function $SAD(d_x,d_y)$ of analyzed surrounding blocks has a monotone convergence to its minimum into a search area $-S_x<d_x<+S_x$ and $-S_y<d_y<+S_y$. The check of points from such area permits to localize this minimum. The orthogonal pattern minimizes the check numbers by a reduction of the search area sizes twice for each procedure step. Rectangular 4-points and 8-points patterns are usually used. Pattern Search demonstrates a good system response speed, but it finds often local minimums instead of the global minimum. The main advantage of such approach consists in calculation of a motion vector for each separate block regardless of the search results in adjacent blocks in current and previous frames. This fact makes Pattern Search more efficient for analysis of intensive and complex motion in scene in comparison with Recursive Search.

During Recursive Search, the checking subset of a search area is separately formed for each block based on the results of adjacent processed blocks in current and previous frames. The main idea consists in the assumption of the smoothness of a motion vectors field that reduces a search time. According to experiments, a motion vectors field is an enough smooth field in spatio-temporal domain with the exception of discontinuities of a brightness function which are appeared near boundaries of visual objects during irregular motions or scene changes. That's why in the most cases it is necessary to check the motion vectors of some adjacent blocks in search area and specify their values in small neighborhood for finding of global minimum. Such method is the fastest realization from concerned modifications of the basic block-matching method, and shows good results for a weak motion. However an imposed constraint on smoothness of a motion vectors field is a very strong restriction.

Block Search method works under following assumptions. The calculation of a motion vector of each block is realized regardless of adjacent blocks, and we know the number of an object for which a current image block is concerned. Such approach is one of the most perspective realizations but at present it has a high computational complexity. Potentially it is the most accurate and a noise-safe method but it is needs in a previous well-realized segmentation.

8.2.1.3 The Motion Estimation in Noisy Videos

The noise occurrence changes the models (8.1) – (8.3). Let's assume that frames f_k and f_{k-1} are noisy and that a frame f_{k-1} is less noisy than a frame f_k. Two cases are possible: (1) a noise presents only on the current frame in the imagery whereas the previous frame is not affected, and (2) a noise presents on both frames. If only the current frame is affected by noise then the following degradation model may be used:

$$g_t(x,y) = f_t(x,y) + n_t(x,y) \ , \tag{8.4}$$

where $g_t(x,y)$ is a brightness function in spatial location (x,y) of a noisy current frame; $f_t(x,y)$ is a brightness function of an original frame without noise; $n_t(x,y)$ denotes an additive white Gaussian noise.

Let's denote the matching criterion calculated between a noisy frame and an original frame in MSD-metric as $MSD^n(d_x,d_y)$ (noisy MSD) and calculate $MSD^n(d_x,d_y)$ using expressions (8.3) and (8.4):

$$\begin{aligned} MSD^n(d_x,d_y) &= \frac{1}{N\times N}\sum_{x=1}^{N}\sum_{y=1}^{N}\left(g_t(x,y) - f_{t-1}(x+d_x, y+d_y)\right)^2 = \\ &= \frac{1}{N\times N}\sum_{x=1}^{N}\sum_{y=1}^{N}\left(\left(f_t(x,y)+n_t(x,y)\right) - f_{t-1}(x+d_x, y+d_y)\right)^2 \ , \end{aligned} \tag{8.5}$$

After developing the squared term in (8.5) we'll receive:

$$\begin{aligned} MSD^n(d_x,d_y) &= MSD(d_x,d_y) + \frac{1}{N\times N}\sum_{x=1}^{N}\sum_{y=1}^{N} n_t^{\,2}(x,y) + \\ &+ \frac{2}{N\times N}\sum_{x=1}^{N}\sum_{y=1}^{N} n_t(x,y)\left(f_t(x,y) - f_{t-1}(x+d_x, y+d_y)\right) \ . \end{aligned} \tag{8.6}$$

The minimum of function (8.6) is conditioned only by a reduction of the term

$$B(d_x,d_y) = \frac{2}{N\times N}\sum_{x=1}^{N}\sum_{y=1}^{N} n_t(x,y)\left(f_t(x,y) - f_{t-1}(x+d_x, y+d_y)\right) \ ,$$

because the term

$$\frac{1}{N\times N}\sum_{x=1}^{N}\sum_{y=1}^{N} n_t^2(x,y)$$

changes only the $MSD^n(d_x,d_y)$ minimum value without modifying its localization.

According to a noise nature influencing on the current frame, two cases are available. (1) The simple case:

$$\begin{aligned} \mu &= \frac{1}{N\times N}\sum_{x=1}^{N}\sum_{y=1}^{N} n_t(x,y) \cong 0 \ , \\ \sigma^2 &= \frac{1}{N\times N}\sum_{x=1}^{N}\sum_{y=1}^{N} n_t^2(x,y) \ . \end{aligned} \tag{8.7}$$

(2) The complex case: if we have a non-zero mean ($\mu \neq 0$) of variables n_t then value $B(d_x,d_y)$ increases and gives more variations of displacement (d_x,d_y). This leads to a modification of $MSD^n(d_x,d_y)$ minimum localization. Since the previous frame is not affected by a noise, the considered frame will be filtered from a noise but it would suffer from motion artifacts. In this case expression (8.5) will be rewritten in following manner:

$$MSD^n(d_x,d_y) = \frac{1}{N \times N}\sum_{x=1}^{N}\sum_{y=1}^{N}\left(g_t(x,y) - g_{t-1}(x+d_x, y+d_y)\right)^2 = \\ = \frac{1}{N \times N}\sum_{x=1}^{N}\sum_{y=1}^{N}\left((f_t(x,y) + n_t(x,y)) - \right. \\ \left. -\left(f_{t-1}(x+d_x, y+d_y) + n_{t-1}(x+d_x, y+d_y)\right)\right)^2 . \tag{8.8}$$

Note that a noise in an imagery can be stationary (the means for adjacent frames have the same values; and also variances are equal), and not stationary. In last case a filtering performance is applied [3].

Let's consider modifications of the basic block-matching method which are used for digital image stabilization, reconstruction of missing frames or missing fragments in frames in visual imagery, video inner-frame coding, and many others applications [10] [19] [21] [31] [43].

8.2.1.4 Fast Block-Matching Motion Estimation

For reduction of a computation cost involved in block-matching motion estimation some adaptive thresholding algorithms were suggested [24] [27] [28] [45]. Such algorithms use an adaptive threshold for each frame to prejudge if results of a block matching computation are worthwhile [39]. Let's discuss two adaptive fast block-matching algorithms which are differed by a threshold sampling. The first algorithm determines an optimal trade-off threshold (based on the difference between a current frame and a previous frame) between run-time and distortion, and the second algorithm determines a threshold according to a user specified percentage of computed blocks.

A direct prediction uses the same block in a previous frame and finds a motion vector in the search area. The prediction error termed in the original SAD-metric (OSAD) for each block is defined as

$$OSAD = \sum_{x=1}^{N}\sum_{y=1}^{N}\left|f_{t-1}(x,y) - f_t(x,y)\right| \ , \tag{8.9}$$

where $f_{t-1}(x,y)$, denotes a brightness pixel value at the (x,y) position in a current frame and $f_t(x,y)$ represents a brightness pixel value at the same position in a previous frame. To measure the gain from motion estimation, the reduced distortion is defined as

$$\Delta SAD = OSAD - SAD \ . \tag{8.10}$$

An observation regarding the correlation between the *OSAD* and the *ΔSAD* demonstrates that blocks with high *OSAD* values can reduce a high distortion under the motion estimation. The two-step procedure was suggested in [17]. In the first algorithm, the *OSAD* values for all blocks are calculated by expression (8.9) and pushed in a heap in decreasing order. If the relation *ΔSAD*/*SAD*<10% then this block is considered. The sequence of the motion vector computation is ordered by the *OSAD* values; and the motion vector estimation process stops when a number of consecutive small *ΔSAD* values occurs. In the second algorithm, the *p*-percentile of the *OSAD* values in the current frame is calculated as the desired threshold for the frame. The *OSAD* values are obtained by randomly sampling a small set of blocks. Then the *p*-percentile is estimated using the sampling data, and part of the large *OSAD* values is removed.

Using the adaptive threshold for each frame we can select effective blocks and remove ineffective blocks from the motion estimation procedure. Therefore such adaptive thresholding algorithms of motion estimation reduce a computation cost. The experiments have shown that the proposed algorithms outperform the non-thresholding approach and the constant-thresholding approach.

8.2.1.5 Gaussian Mixture Model

The most of motion estimation algorithms are based on the following assumptions:

- Illumination in scene has a spatially and temporally uniform distribution.
- Objects move on the frontal plane relative to camera; that's why the scale changes and rotations are absent.
- Occlusions of several objects are not enabled.

However local variations of illumination (such as shadows) are usually located in scene; and one of approaches is a statistical method of modeling of image blocks as a mixture of Gaussian distributions [4]. The main statistical parameters are the priori probabilities, means, and covariance matrices; they may be used for optimization of received functions by Expectation–Maximization algorithm (EM algorithm). The similarity measure of the best matching between adjacent blocks in sequential frames is calculated by Mahalanobis distance or Extended Mahalanobis distance. Gaussian mixture model is described by following expression:

$$p(\mathbf{X}/\mathbf{\Theta}_k)=\sum_{i=1}^{k}\alpha_i p(X_i/\theta_i)=\sum_{i=1}^{k}\alpha_i p(X_i/\mu_i,\Sigma_i) \ , \tag{8.11}$$

where $\mathbf{\Theta}_k$ is the collection of all parameters in the mixture, $\mathbf{\Theta}_k=(\theta_1,\ldots,\theta_k,\alpha_1,\ldots,\alpha_k)$; k is the number of components in mixtures; $\theta_i=(\mu_i,\Sigma_i)$ are the parameters of a Gaussian distribution; α_i, $i\in 1,k$ are the weighting coefficients, $\alpha_i\geq 0$, and

$$\sum_{i=1}^{k}\alpha_i=1 \ ;$$

each component density $p(X/\theta_i)$ is a Gaussian probability density function calculated by formula:

$$p(X_i/\mu_i,\Sigma_i)=\frac{1}{(2\pi)^{n/2}|\Sigma_i|^{1/2}}e^{-(1/2)(\mathbf{X}-\boldsymbol{\mu})^T\Sigma_i^{-1}(\mathbf{X}-\boldsymbol{\mu})}\ , \quad (8.12)$$

where n is a dimension of of "mixture" function $\mathbf{X}$ which describes a set of brightness functions in considered block $BL(x,y)$; $\boldsymbol{\mu}$ is a set of mean values, and $\boldsymbol{\Sigma}_i$ is a positive defined covariance matrix.

Given a set of N samples, $\mathbf{X}=\{X_t\}$, t=1,..,N, the logarithmic likelihood function for the Gaussian mixture model is expressed as follows:

$$\log\ p(\mathbf{X}/\boldsymbol{\Theta}_k)=\log\prod_{t=1}^{N}p(X_t/\boldsymbol{\Theta}_k)=\sum_{t=1}^{N}\log\sum_{i=1}^{k}\alpha_i p(X_t/\theta_i)\ , \quad (8.13)$$

which is maximized a Maximum Likelihood (ML) estimate of vector $\boldsymbol{\Theta}_k$ via the EM algorithm.

The EM algorithm highly depends from the initial set of parameters. If the initial parameters are not selected well then the EM algorithm may converge into a local maximum. The similarity between a block in the current image and the more resembling one in a search window on the reference image is measured by the minimization of Extended Mahalanobis distance between the clusters of the Gaussian mixture.

It is well known that the Kullback-Leibler divergence and the Extended Mahalanobis distance are the popular measures of statistical fields. The Kullback-Leibler divergence indicates a dissimilarity measure between two functions of probabilities. However it has not a symmetric distribution and it is not a true metric. The Mahalanobis distance metric may be extended to a distance measure between two Gaussian distributions $GD_1(\mu_1,\Sigma_1)$ and $GD_2(\mu_2,\Sigma_2)$ by combining of their covariance matrices in such manner:

$$D_{EMhn}(GD_1,GD_2)=\sqrt{(\mu_1-\mu_2)^T(\Sigma_1+\Sigma_2)^{-1}(\mu_1-\mu_2)}\ . \quad (8.14)$$

Based on expression (8.14) three distances d_1, d_2 and d_3 for the components of strong weights, the components of medium weights, and the components of weak weights correspondingly are used:

$$\begin{aligned} d_1&=(\alpha_1^1\mu_1^1-\alpha_2^1\mu_2^1)^T(\alpha_1^1\Sigma_1^1+\alpha_2^1\Sigma_2^1)^{-1}(\alpha_1^1\mu_1^1-\alpha_2^1\mu_2^1)\ ,\\ d_2&=(\alpha_1^2\mu_1^2-\alpha_2^2\mu_2^2)^T(\alpha_1^2\Sigma_1^2+\alpha_2^2\Sigma_2^2)^{-1}(\alpha_1^2\mu_1^2-\alpha_2^2\mu_2^2)\ ,\\ d_3&=(\alpha_1^3\mu_1^3-\alpha_2^3\mu_2^3)^T(\alpha_1^3\Sigma_1^3+\alpha_2^3\Sigma_2^3)^{-1}(\alpha_1^3\mu_1^3-\alpha_2^3\mu_2^3)\ . \end{aligned} \quad (8.15)$$

During modeling of two Gaussian distributions, often the cost function is defined by the Extended Mahalanobis distance between the components of strong weights (d_1) and the components of weak weights (d_3). Full Search realization of the block-matching method is used for such statistical approach [36].

8.2.1.6 Lie Operators in Motion Estimation

Often the motion of video objects in imagery can not be characterized as a linear motion. That's why it is important to considerate the affine motion model which is based on Lie operators and Lie algebra [16]. Several Lie operators corresponding to various motions may be employed in the context of motion estimation to detect small degrees of rotation (L_R), scaling (L_S), horizontal scaling (L_{Sx}), vertical scaling (L_{Sy}), parallel deformation (S_P), and diagonal deformation (L_D) [37]. The definitions of these Lie operators are given in Table 8.3.

Table 8.3 Several types of non-translational motions and their Lie operators

Lie operation	Corresponding transform	Block-matching result	
Rotation	$t_{\theta_R} = \begin{pmatrix} x \\ y \end{pmatrix} \leftrightarrow \begin{pmatrix} x\cos\theta_R - y\sin\theta_R \\ x\sin\theta_R + y\cos\theta_R \end{pmatrix}$	$L_R = y\frac{\partial}{\partial x} - x\frac{\partial}{\partial y}$,	$f_R^{\theta_R} = f + \theta_R \times L_R(f)$
Scaling	$t_{\theta_S} = \begin{pmatrix} x \\ y \end{pmatrix} \leftrightarrow \begin{pmatrix} x+\theta_S \cdot x \\ y+\theta_S \cdot y \end{pmatrix}$	$L_S = x\frac{\partial}{\partial x} + y\frac{\partial}{\partial y}$,	$f_S^{\theta_S} = f + \theta_S \times L_S(f)$
Vertical scaling	$t_{\theta_{Sx}} = \begin{pmatrix} x \\ y \end{pmatrix} \leftrightarrow \begin{pmatrix} x+\theta_S \cdot x \\ y \end{pmatrix}$	$L_{Sx} = x\frac{\partial}{\partial x}$,	$f_{Sx}^{\theta_{Sx}} = f + \theta_{Sx} \times L_{Sx}(f)$
Horizontal scaling	$t_{\theta_{Sy}} = \begin{pmatrix} x \\ y \end{pmatrix} \leftrightarrow \begin{pmatrix} x \\ y+\theta_S \cdot y \end{pmatrix}$	$L_{Sy} = y\frac{\partial}{\partial y}$,	$f_{Sy}^{\theta_{Sy}} = f + \theta_{Sy} \times L_{Sy}(f)$
Parallel deformation	$t_{\theta_P} = \begin{pmatrix} x \\ y \end{pmatrix} \leftrightarrow \begin{pmatrix} x+\theta_P \cdot x \\ y-\theta_P \cdot y \end{pmatrix}$	$L_P = x\frac{\partial}{\partial x} - y\frac{\partial}{\partial y}$,	$f_P^{\theta_P} = f + \theta_P \times L_P(f)$
Diagonal deformation	$t_{\theta_D} = \begin{pmatrix} x \\ y \end{pmatrix} \leftrightarrow \begin{pmatrix} x+\theta_D \cdot x \\ y+\theta_D \cdot y \end{pmatrix}$	$L_D = y\frac{\partial}{\partial x} + x\frac{\partial}{\partial y}$,	$f_D^{\theta_D} = f + \theta_D \times L_D(f)$

For example, if function f describing some block is rotated by a very small degree θ_R, then the rotation block $f_R^{\theta_R}$ can be approximated by applying the corresponding operator L_R as

$$f_R^{\theta_R} = f + \theta_R \times L_R(f) \ , \tag{8.16}$$

where

$$L_R(f) = \left(y\frac{\partial}{\partial x} - x\frac{\partial}{\partial y} \right) f = y\left(\frac{\partial f}{\partial x}\right) - x\left(\frac{\partial f}{\partial y}\right) \tag{8.17}$$

and

$$\frac{\partial f}{\partial x} \approx \frac{1}{2}[f(x+1, y) - f(x-1, y)] ,$$
$$\frac{\partial f}{\partial y} \approx \frac{1}{2}[f(x, y+1) - f(x, y-1)] . \tag{8.18}$$

In practice, Lie operators would be economically applied to the blocks in the "moving" regions. These moving regions can be detected by finding the difference between the adjacent frames using the thresholding background subtraction. However even through the background of the scene does not change significantly, some background blocks can be skipped when we apply Lie operators. Thus, the applying of Lie operators would not be effective if objects are unlikely founded in the previous frame, but any scaling or deformation from lighting condition would be detectable by using Lie operators [38].

Further development of the motion estimation accuracy connects with applying different Lie operators in combination. One can use firstly the Full Search method, and only then try to find the best combination of four types of Lie operators (L_R, L_S, L_P, and L_D). In order to reduce the high computational complexity associated with the Full Search method, then one apply the following parameter-search methods: Dynamic Programming (DP-like) Search, Iterative Search, and Serial Search. They combine Lie operators in different ways, with varying accuracy-complexity tradeoffs.

8.2.1.7 Bilinear Deformable Block-Matching Method

A bilinear deformable block matching (BDBM) method was developed for ultrasound applications to improve a motion tracking [2]. This iterative approach uses a local bilinear model with eight parameters for controlling the local mesh deformation:

$$u(x, y) = a_u \cdot x + b_u \cdot y + c_u \cdot x \cdot y + d_u \quad ,$$
$$v(x, y) = a_v \cdot x + b_v \cdot y + c_v \cdot x \cdot y + d_v \quad , \tag{8.19}$$

where u and v are respectively the lateral and axial displacements in each position (x,y). In this context, one needs to find the eight parameters of the bilinear model to estimate a local displacement (Fig. 8.2, RROI is an abbreviation of Rectangular Region of Interest).

The estimation of four translations of corners shown on Fig. 8.2 allows estimate the parameters of the bilinear model. The algorithm works iteratively with a multi-scale approach [46]. At each resolution level, the computation grid is refined by a bilinear interpolation, and the current image region is deformed using the bilinear parameters estimated at the previous iteration. The next iteration starts with four deformed blocks which allow a better estimation of corner translations of the current region of interest. The iterative multi-scale approach has the advantage of

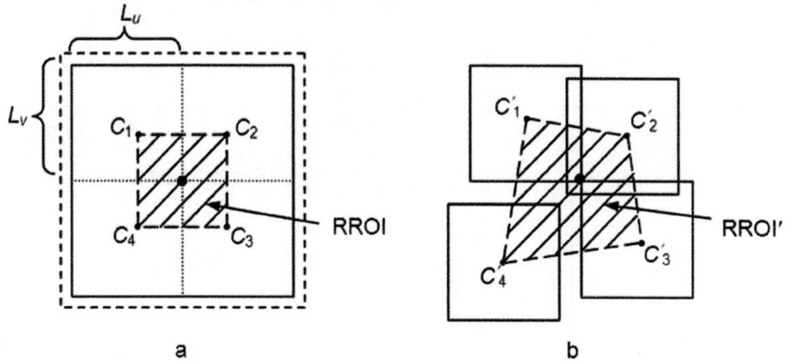

Fig. 8.2 Scheme of the bilinear spatial transformation: (a) initial block; (b) deformable block

a decreasing motion error with the advancement of iterations [7]. Experiments demonstrated that an interpolation factor of 3 (at two iterations) provides a better compromise between a motion estimation accuracy and an estimation time.

8.2.1.8 Fuzzy Logic Interpretation

It is known that a fuzzy logic has been successfully applied in many control systems [34] [40]. The inter-frame fuzzy search (IIFS) algorithm for block-matching estimation is the most famous realization. Using the inter-block and inter-frame correlations, IIFS algorithm can determine the motion vectors of image blocks quickly and correctly. However, the IIFS algorithm is not suitable for hardware implementation due to very irregular data flow. A modified inter-block/inter-frame fuzzy search (MIIFS) algorithm also was suggested with more easily realization by VLSI technology (Very-Large Scale Integration). SAD metric must be adapted to the block distortion measure (BDM) as shown on Fig. 8.3. Function (8.1) is changed, and becomes

$$SAD\left(d_x, d_y, v, u\right)= \\ =\sum_{x=1}^{N}\sum_{y=1}^{N}\left|f_t\left(x+d_x, y+d_y\right)-f_{t-1}\left(x+d_x+u, y+d_x+v\right)\right| \quad , \tag{8.20}$$

where u, v are additional fuzzy displacements along Cartesian axis, $|u|\le w$; $|v|\le w$, w is a size of maximum extension of search area which is not multiply N generally.

The $SAD(d_x,d_y,u,v)$ values are computed for each candidate block within the search area. A block with the minimum SAD value is considered the best-matched block, and a motion vector $\mathbf{MV}(d_x,d_y)$ is built on the value (u,v) for the best-matched block:

$$\mathbf{MV}\left(d_x, d_y\right)=\left(u,v\right)\Big|_{\min SAD\left(d_x, d_y, u, v\right)} . \tag{8.21}$$

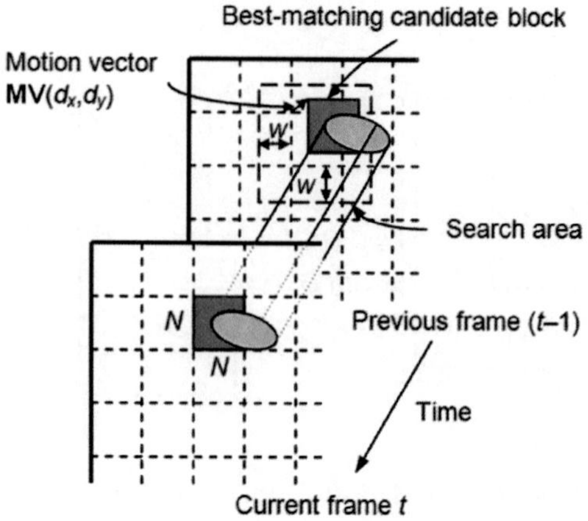

Fig. 8.3 Scheme of the block distortion measure

Let's suppose that an image frame is processed from the top-left block to the bottom-right block. MIIFS algorithm applies the two main working processes for calculation of a motion vector. Firstly, it defines the initial search center using the predicted displacement (with the help of a fuzzy reasoning), and then it finds a motion vector within the search area. Secondly, MIIFS algorithm searches the whole search area with a 3×3 movable search window either until the local minimum point lies in the center of the current search window, or until the number of search iterations reaches 3. The experiments have shown the better results of MIIFS algorithm applying in real-working systems relatively others fast block-matching search algorithms.

8.2.1.9 Multi-level Motion Model

Sometimes one can use the multi-level motion model in perspective scenes [35]. So, the traditional single-level block matching (SLBM) method is extended to the multi-level block matching (MLBM) algorithm. MLBM uses variable matching blocks and search windows. A large block initially is used to provide a coarse-resolution estimation of the whole motion field. Each subsequent level uses a smaller block as a search window for improvement the spatial resolution without the influence of the prior levels.

Such multi-level scheme is applied for processing of ultrasound images. The ultrasound images are characterized by discrete speckle patterns, and excessive filtering transforms the speckle statistics from a Rayleigh distribution to pseudo-Gaussian distribution. The spatial motion model-based block matching algorithm combines the multi-level scheme with a smoothness constraint based on the assumption that a motion field is continuous. This assumption is generally valid when the motion vectors are contained within a single moving or deforming object. The center of the search window at each subsequent level Lv_i is displaced by the motion vector estimated in the previous level (Fig. 8.4). The coarser-resolution motion estimations (from the previous level) are reduced to the same resolution by interpolation. This procedure is repeated until the bottom level is reached. Since the motion estimations from coarser levels are passed to the finer levels as offsets of the center of the search windows, the final motion estimation is then the sum of the vectors found in all stages of the hierarchy.

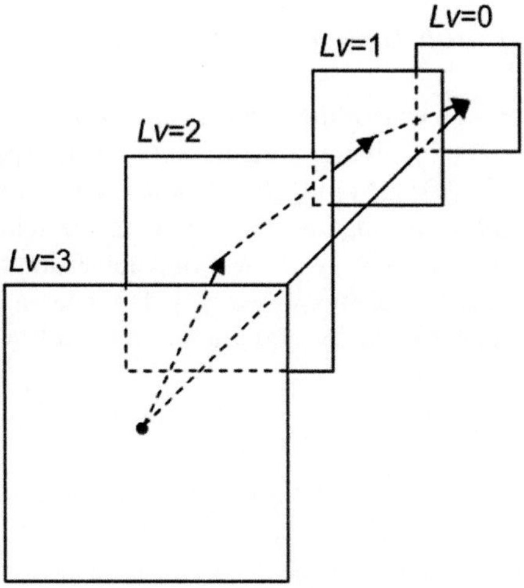

Fig. 8.4 Global vector estimation in MLBM algorithm

8.2.1.10 Global Block-Matching Method

When we analyze a complex scene, we need to find the local motion vectors (according to the traditional block-matching method), and estimate a global motion in sequential frames. Such tasks appear in video coding and video editing systems. The global matching algorithms are also based on the correlative functions calculating absolute differences, mean differences, or root mean square differences of each pixel [1]. In addition these algorithms have the matching functions including additional and subtraction operations, and cumulative computing. In these cases speed of algorithms is high but the matching accuracy is low.

Others global matching algorithms are based on the location of the ranks in the matching matrix when the correlative parameters between original frames and the entire matching matrix are composed into the array of vectors in the directions of columns or rows. However the larger location coefficients influence on the result of matching operation, and we have a situation when some pixels with high location coefficients determine the whole result of matching operation. The approach from paper [49] compensates such disadvantages by usage the oblique vectors in the direction of 45° tilt angle.

Thereby we considered the comparative motion estimation methods, discussed their advantages and disadvantages. Their specialties are connected with a less computation complexity and a less accuracy relatively the group of the gradient motion estimation methods.

8.2.2 Gradient Motion Estimation Methods

Usually the gradient motion estimation methods are used for the motion segmentation in dynamic scenes and receiving of the odometry data (speed and acceleration parameters) [5] [20] [22] [23] [25] [32] [50]. These tasks can not be accomplished by the comparative motion estimation methods with the required accuracy. The motion segmentation is a key step in many computer vision applications such as video surveillance, traffic monitoring, robotics, video indexing, etc. [41] [42]. The motion segmentation strategies can be classified into the following groups [51]:

- Image difference techniques (the multiple independent motions of non-rigid and articulated objects).
- Statistical techniques (the multiple motions with occlusions and temporary stopping).
- Wavelet methods (the analysis of the different frequency image components in a simple motion cases).
- Optical flow (the apparent motions in visual imagery).
- Layer-based techniques (the most natural solution for occlusions).
- Manifold clustering techniques (trajectories in motion segmentation based on feature points tracking).

Fig. 8.5 shows the scheme of a features extraction and motion estimation in videos. The specific procedures based on various gradient methods are greatly differed and as a result define the accuracy and computational cost [8].

Let's discuss the gradient motion estimation methods which realizations depend from groups of objects as shown in Table 8.1. In this chapter we'll consider the optical flow methods for motion estimation of visual objects based on spatio-temporal 3D gradients, oriented filters in spatio-temporal volume, feature tracking, and tensor approach.

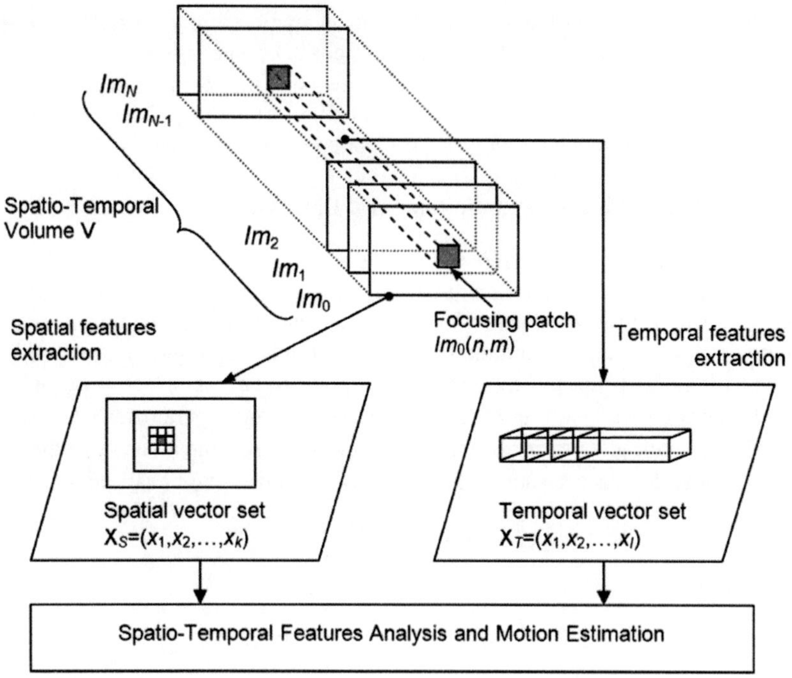

Fig. 8.5 The scheme of spatio-temporal features extraction and motion estimation in videos

8.2.2.1 Spatio-Temporal 3D Gradients

The motion estimation method and its modifications may be based on 3D local spatio-temporal 3D gradients. For some image point **s** the Bag-Of-Features (BoF) vector is built. Such features are retrievable from a spatio-temporal volume $N_x \times N_y \times N_t$. The invariance to spatial and temporal scales is achieved by adding in feature vector $\mathbf{f}=(x_s,y_s,t_s,\sigma_{xy},\sigma_t)^T$ the standard deviations σ_{xy} and σ_t of pixels brightness in local region $\mathbf{r}_s=(x_r,y_r,t_r,w_r,h_r,l_r)^T$, where w_r,h_r,l_r are a width, a height and a length of local region correspondingly. At first, a full feature vector near a point of interest is formed, moreover surrounding region in divided on squares which consist from 2×2×2 cells. Then 3D histograms for each cell in a polar coordinate system are built; the axes of such polar coordinate system are the length of gradient vector, and two orientation angles φ and ψ in space. As a result 3D space near a point of interest is partitioned on some numbers of polyhedrons (in this case, polyhedrons are looked like as tetrahedral pyramids). A pyramid center is situated in a point of interest; and the amount of such pyramids is equal 32. Then the histograms approximation and normalization in the space of such polyhedrons are realized, a summarized histogram is built, and a direction which conforms to the maximum of gradients changes, is calculated by usage of a threshold function. Experiments show good calculating abilities. One may denominate such approach as the motion estimation method based on local spatio-temporal pseudo-3D gradients because we analyze a set of images but not 3D scene space. In this case the back transitions into 3D space and polyhedrons construction are not a realizable process because one uses the circle approximation around a point of interest by sectors and find a motion direction as a projection on image plane.

8.2.2.2 Oriented Filters in Spatio-temporal Data Volume

In many researches only aperiodic objects translations are considered [44]. The question about a quick calculation of periodic translation features is not decided at present. For description of local structures and local motion estimation in spatio-temporal volume I. Laptev, Schmid C., earlier T. Lindeberg suggested to apply the oriented filters [26]. Let $\mathbf{p}=(x,y,t)^T$ denotes a point in (2+1)D space-time, and let f:$R^3 \rightarrow R$ represents a spatio-temporal image. Then a multi-parameter scale-space L:$R^3 \times G \rightarrow R$ is defined as a convolution between function f and a family of the spatio-temporal scale-space kernels h: $R^3 \times G \rightarrow R$ (G is a Gaussian in various scales)

$$L(\cdot;\Sigma) = h(\cdot;\Sigma) * f(\cdot)$$

parameterized by covariance matrices $\mathbf{\Sigma}$ in a group G:

$$\mathbf{\Sigma} = \begin{pmatrix} \lambda_1 c^2 + \lambda_2 s^2 + u^2\lambda_t & (\lambda_2 - \lambda_1)cs + uv\lambda_t & u\lambda_t \\ (\lambda_2 - \lambda_1)cs + uv\lambda_t & \lambda_1 s^2 + \lambda_2 c^2 + v^2\lambda_t & v\lambda_t \\ u\lambda_t & v\lambda_t & \lambda_t \end{pmatrix}, \quad (8.22)$$

where parameters λ_1, λ_2, c, s characterize a spatial anisotropic smoothness: λ_1, λ_2 are eigenvalues of matrix describing a local region along axes OX and OY with angle orientation α, c=cos α, s=sin α; λ_t is a eigenvalue of matrix describing a local region along axis OZ (temporal axis); parameters (u,v) show a filter orientation in the spatio-temporal volume (if $\lambda_1=\lambda_2$ and $(u,v)=(0,0)$ then standard deviations $\sigma_{x,y}^2=\lambda_1=\lambda_2$, and $\sigma_t^2=\lambda_t$).

The second-moment matrix is a good descriptor which estimates the local image deformations (relatively $\mathbf{p}$ and $\mathbf{q}$ points). It can be defined as

$$\mu(\mathbf{p},\Sigma) = \int_{q \in R^3} (\nabla L(\mathbf{q}))(\nabla L(\mathbf{q}))^T w(\mathbf{p}-\mathbf{q};\Sigma) d\mathbf{q}, \quad (8.23)$$

where $\nabla L(\cdot)=(L_x,L_y,L_t)^T$ is a spatio-temporal gradient in point $\mathbf{p}(x,y)$ at the instant t; L_x, Ly are spatial gradients along spatial axes OX, OY correspondingly; L_t is a temporal gradient along temporal axis OZ; w is a spatio-temporal window function.

For simplicity, the smoothing operation is modeled by 3D Gaussian kernel with covariance matrix $\mathbf{\Sigma}$

$$h(\mathbf{p};\Sigma) = g(\mathbf{p};\Sigma) = \exp\left(\left(-\mathbf{p}^T \Sigma^{-1} \mathbf{p}/2\right) / (2\pi)^{3/2} \sqrt{\det\Sigma}\right).$$

In a space-time separable case matrix $\mathbf{\Sigma}$ can be represented as a convolution between 2D Gaussian in a space domain

$$g_{2D}(x, y, \sigma^2) = 1/(2\pi\sigma^2)\exp(-(x^2 + y^2)/2\sigma^2)$$

and 1D Gaussian over a time domain

$$g_{1D}(t, \tau^2) = 1/(\sqrt{2\pi\tau})\exp(-t^2/2\tau^2)\,.$$

So, expression (8.23) one can rewritten in matrix form as

$$\begin{pmatrix} \mu_{xx} & \mu_{xy} & \mu_{xt} \\ \mu_{xy} & \mu_{yy} & \mu_{yt} \\ \mu_{xt} & \mu_{yt} & \mu_{tt} \end{pmatrix} = \int_{\Omega_w} \begin{pmatrix} L_x^2 & L_xL_y & L_xL_t \\ L_xL_y & L_y^2 & L_xL_t \\ L_xL_t & L_yL_t & L_t^2 \end{pmatrix} dw\,, \qquad (8.24)$$

where Ω_w is an integration region.

8.2.2.3 Feature Tracking

Usually a feature tracking method is used at the higher-level motion estimation; it permits to built visual objects trajectories, calculate parameters of dynamic structures, and find particular objects motions [30]. Also if restrictions on an observable scene are known then one can calculate the parameters of inner and outer camera calibration using the feature tracking method.

Feature tracking method includes two steps: the detection (a finding of feature points on an initial frame and a limitation of amount of feature points), and the tracking (a calculation of coordinates of new feature points for each following frame). However a part of feature points set disappears from time to time owing to the camera movements or the scene changes. Also the points neighborhoods may have such great distortions that feature points became usual points, otherwise a feature points set is changed in temporal domain.

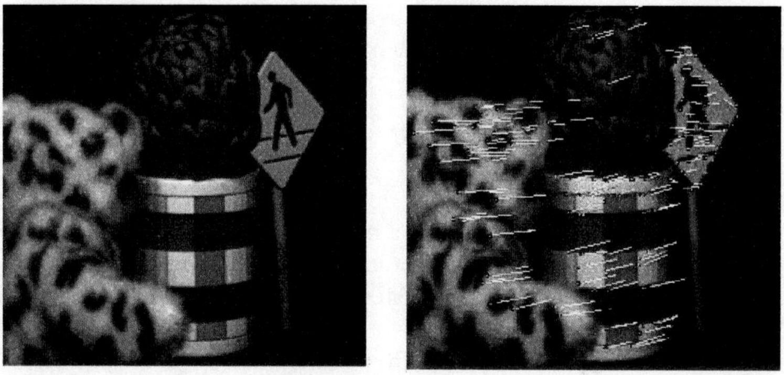

a b

Fig. 8.6. Example of the Lucas-Canade algorithm work: (a) initial image, (b) result image

All modern feature points tracking algorithms are based on the works of B. D. Lucas and T. Canade (1981–1991). Later the extensions of feature points tracking algorithms were designed including affine transformations and brightness changes in feature points' neighborhoods. Any modified algorithm transforms to the basic Lucas-Canade algorithm by replacement corresponding variables on constant values. Fig. 8.6 b demonstrates the work of the Lucas-Canade algorithm.

Let's consider some modifications of the Lucas-Canade algorithm:

– The Lucas-Canade algorithm assumes that a feature point translates relatively the coordinate axes; any others distortions are absent. This algorithm is applied for functions with any dimension n. Let $I(\mathbf{P}_J, t_J)$ is a function describing a brightness in feature point $\mathbf{P}_J$ with coordinates (x,y) in image J (in movement t_J). A brightness function of following frame K in the same feature point $\mathbf{P}_K$ in movement t_K (for pixels situated far from the frame boundaries) is described by following equation

$$I(\mathbf{P}_J, t_J) = I(\mathbf{P}_K, t_K) = \text{const} . \tag{8.25}$$

For small image distortions (from frame to frame) we consider that a window near a feature point is simply shifted, and the position of the point $\mathbf{P}_K$ is defined as

$$\mathbf{P}_K = \mathbf{P}_J + \mathbf{d} \quad ,$$

where $\mathbf{d}$ is a displacement vector.

It is necessary to find a function $I(\mathbf{P}_J+\mathbf{d})$ describing the following image so that "subtraction" of these points neighborhoods will have a minimum value according to some metric.

– The Tomasi-Canade algorithm refines the Lucas-Canade algorithm. It is assumed that a motion is a displacement which is calculated by iterative solution of a system of linear equations. This algorithm searches the point in which the minimum of some function is achieved by using the gradient descent method. A translation along gradient direction occurs during a current iteration.

– The Shi-Tomasi-Canade algorithm considers the affine distortions of a feature point's displacement, that's why a pixels motion near a feature point is defined as

$$\mathbf{P}_k = \mathbf{A}\mathbf{P}_j + \mathbf{d} \quad ,$$

where $\mathbf{A}$ is a affine matrix (2×2).

– The Jin-Favaro-Soatto algorithm is a modification of the Shi-Tomasi-Canade algorithm with affine distortions of brightness in considered feature point.

Let's consider one of a practical realization of the Lucas-Canade algorithm based on the following steps:

Step 1. The static frame is extracted from videos with an object of interest (pedestrian, vehicle, etc.).

Step 2. The grid 16×16 pixels is covered an image with a following calculation of a response function for each pixel from neighborhood. Let's note that a response function is different for various methods of feature point tracking.

Step 3. The maximum value of a response function for each quadrant is defined.

Step 4. Feature points are selected by defined thresholding value.

In practice expression (8.25) can not be strongly hold, so we try to find such motion for which the brightness difference between a current features position and a future features position would be minimized. So, we find such value $\mathbf{P}_K$ for which the function minimum E is obtained:

$$E = \left| I\left(P_k, t_k\right) - I\left(P_j, t_j\right) \right| \quad ,$$

or as the function minimum E considering a neighborhood:

$$E = \min \sum_{i=1}^{W} \left(I_i\left(P_K, t_K\right) - I_i\left(P_J, t_J\right) \right)^2 \quad .$$

Often for a feature points extraction the operator suggested by H. Moravec, and also C. Harris detector are used. Moravec operators detect an angle measure for a pixel. Harris detector is the enhanced Moravec operator. The measure is a brightness minimal difference finding along 8 directions. The applying of Moravec operator for each pixel in image creates an angle map. Harris detector for each pixel in image calculates values of a corners response function estimated a correlation degree between point neighborhood image and angle image:

$$M_c = \det(\mathbf{A}) - k \cdot \mathrm{trace}^2(\mathbf{A}) \quad , \tag{8.26}$$

where trace $(\cdot)$ is a spur of matrix **A**; parameter k is equal 0.04 (suggested by C. Harris); matrix **A** is calculated as

$$\mathbf{A} = \sum_u \sum_v w(u,v) \begin{vmatrix} \dfrac{\partial^2 I}{\partial x^2} & \dfrac{\partial I}{\partial x} \cdot \dfrac{\partial I}{\partial y} \\ \dfrac{\partial I}{\partial x} \cdot \dfrac{\partial I}{\partial y} & \dfrac{\partial^2 I}{\partial y^2} \end{vmatrix} \quad , \tag{8.27}$$

where $I(x,y)$ is an image brightness in point with coordinates (x,y); $w(u,v)$ is a windowing function describing a neighborhood (u,v) of point with coordinates (x,y) (the weighing sum is equal 1 in the simplest case); $\partial I/\partial x$ and $\partial I/\partial y$ are partial derivatives along axes OX and OY.

Image points corresponding the local minimums of function (8.26) are considered as the feature points. The calculation of the first-derivative of a digital image is based on various discrete approximations of 2D gradients in a neighborhood of some pixel. The direction of a gradient vector matches the direction of a maximum change speed of brightness function in point (x,y). We can receive discrete approximations of partial derivatives in each point using Sobel gradient operator. For noise reduction in obtained feature points we use the smoothing by a Gauss filter in partial derivatives maps. In most cases the amount of obtained feature points is very large; it means that their tracking will be a complex issue. That's why the constraints on a minimal distance between obtained feature points are determined, and extraneous feature points are rejected.

Let's discuss the algorithm of feature points tracking. Let J and K are two adjacent frames from a visual imagery. We may consider such images as continuous 2D functions. It is necessary to track a point feature $\mathbf{P}_0=[x_0,y_0]$ from image J on image K by finding its displacement $\mathbf{d}=[d_x,d_y]^T$. The difference ε between locations of a feature point on two adjacent frames considering a spatial neighborhood W is calculated as:

$$\varepsilon = \int_W [K(\mathbf{P}_0) - J(\mathbf{P}_0 - \mathbf{d})]^2 \, dx\, dy \, . \tag{8.28}$$

Let's replace $\mathbf{P}_0$ with $(\mathbf{P}+\mathbf{d}/2)$ in expression (8.28)

$$\varepsilon = \int_W \left[K\left(\mathbf{P}+\frac{\mathbf{d}}{2}\right) - J\left(\mathbf{P}-\frac{\mathbf{d}}{2}\right)\right]^2 dx\, dy \tag{8.29}$$

and find a displacement $\mathbf{d}$ which minimizes the difference ε:

$$\frac{\partial \varepsilon}{\partial d} = 2\int_W \left[K\left(\mathbf{P}+\frac{\mathbf{d}}{2}\right) - J\left(\mathbf{P}-\frac{\mathbf{d}}{2}\right)\right] \cdot \left[\frac{\partial K\left(\mathbf{P}+\frac{\mathbf{d}}{2}\right)}{\partial d} - \frac{\partial J\left(\mathbf{P}-\frac{\mathbf{d}}{2}\right)}{\partial d} \right] dx\, dy = 0 \, . \tag{8.30}$$

For solution of equation (8.29) the displacement values of images J and K can be performed as low order Taylor's series:

$$\begin{aligned} K\left(\mathbf{P}+\frac{\mathbf{d}}{2}\right) &\approx K(\mathbf{P}) + \frac{\partial_x}{2}\frac{\partial K}{\partial d_x}(\mathbf{P}) + \frac{\partial_y}{2}\frac{\partial K}{\partial d_y}(\mathbf{P}) \, , \\ J\left(\mathbf{P}-\frac{\mathbf{d}}{2}\right) &\approx J(\mathbf{P}) - \frac{\partial_x}{2}\frac{\partial J}{\partial d_x}(\mathbf{P}) - \frac{\partial_y}{2}\frac{\partial J}{\partial d_y}(\mathbf{P}) \, . \end{aligned} \tag{8.31}$$

Using expressions (8.31), equation (8.30) takes an approximate form

$$\frac{\partial \varepsilon}{\partial d} \approx \int_W \left[K(\mathbf{P}) - J(\mathbf{P}) + \frac{\mathbf{g}^T(\mathbf{P})\cdot \mathbf{d}}{2} \right] \mathbf{g}(\mathbf{P})\, dx\, dy = 0 \, , \tag{8.32}$$

where

$$\mathbf{g}(\mathbf{P}) = \begin{vmatrix} \frac{\partial}{\partial d_x}(J(\mathbf{P}) + K(\mathbf{P})) \\ \frac{\partial}{\partial d_y}(J(\mathbf{P}) + K(\mathbf{P})) \end{vmatrix} \, . \tag{8.33}$$

After formula (8.32) transformation we come to a matrix equation

$$\mathbf{Z} \cdot \mathbf{d} = \mathbf{e} \,, \tag{8.34}$$

where $\mathbf{Z}$ is a matrix with dimension 2×2 and module

$$\mathbf{Z} = \int_W \mathbf{g}(\mathbf{P}) \mathbf{g}^T(\mathbf{P}) \, dx \, dy$$

and $\mathbf{e}$ is a column vector with dimension 2×1 and module

$$\mathbf{e} = \int_W [J(\mathbf{P}) - K(\mathbf{P})] \mathbf{g}(\mathbf{P}) \, dx \, dy \ .$$

The received equation permits to define an approximate displacement of feature point between two frames. If a displacement does not a zero-converge after some iterations then such feature point is considered as a lost point. General scheme of the algorithm for a feature points tracking is shown on Fig. 8.7, and on Fig. 8.8 one can see a scheme of the algorithm for a calculation of a feature point displacement.

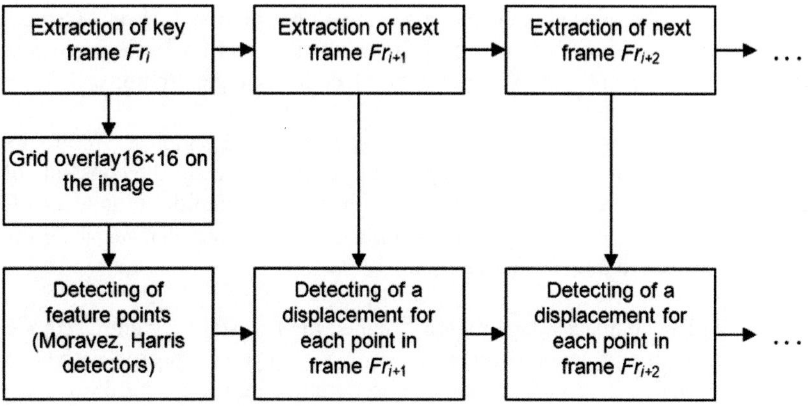

Fig. 8.7. General scheme of the algorithm for a feature points tracking

We assume that a key frame contains an object of interest, and then a grid 16×16 is overlaid on a key frame. Feature points are founded with using the chosen detector, and for each following frame the displacement of each feature point is defined. Using displacements we can build the field of local motion vectors; the lengths of such motion vectors are usually normalized, and show a degree of displacements. The directions of local motion vectors visualize the optical flow field in a graphic form.

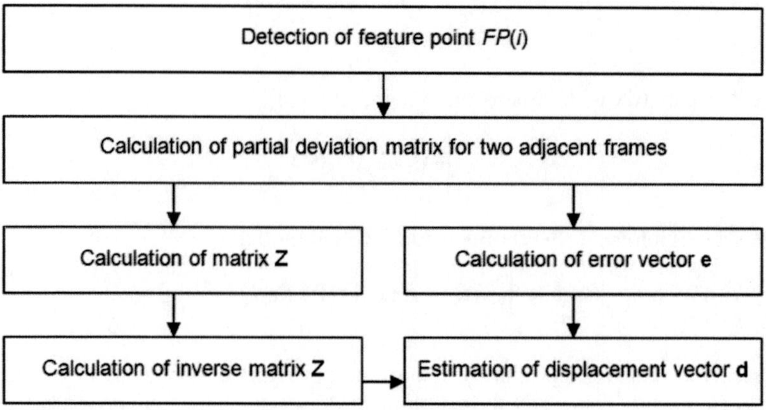

Fig. 8.8 A scheme of the algorithm for a calculation of a feature point displacement

8.3 Local Motion Estimation Based on Tensor Approach

Any visual object in videos is characterized by global motion features. They are defined from analysis of local motion features of rigid and non-rigid regions situated closely in sequential frames. Let's introduce following definitions.

Definition 1. The compact image region is called the rigid region R_r (the region with a finite projections set) if the compact image area possesses the constant color $\mathbf{F}_c$ and texture $\mathbf{F}_t$ features under determined luminance conditions, and has a finite projection set $\mathbf{P}_s$, $|\mathbf{P}_s|\rightarrow$const in frontal plane with contour changes according to affine or projective transformation group.

Definition 2. The compact image region is called the non-rigid region R_{nr} (the region with an infinite projections set) if the compact image area possesses the constant color $\mathbf{F}_c$ and texture $\mathbf{F}_t$ features under determined luminance conditions, and has an infinite projections set $\mathbf{P}_d$, $|\mathbf{P}_d|\rightarrow\infty$ in frontal plane with various contour changes.

In practice the non-rigid regions have various but periodic contour changes that facilitates the forming of following hypotheses about video objects. Let's assume that the sizes of such regions are well less the sizes of image

$$\int_{\Omega_{R_r,R_{nr}}} f(x,y,t)dx\,dy\,dt << \int_{\Omega} f(x,y,t)dx\,dy\,dt \quad , \tag{8.35}$$

where (x,y) are spatial coordinates; t is a time; $f(x,y,t)=1$ is a normalized square functional; $\Omega_{R_r,R_{nr}}$ is a set of points owned to dynamic regions; Ω is a set of points owned to observable image.

8.3.1 The Initiation Stage

At the initiation stage it is required to find the local dynamic regions which pretend on visual objects. If we have a moving camera then all objects in scene are characterized by their motion features. In this case the initiation stage (a detection of a stationary background) we are missing, and going direct to estimation of a spatio-temporal data with following finding the motion levels in scene. Let's discuss the case with a stationary camera.

Let some moving objects of interest are in scene, and others static objects are covered to a background. The procedure of finding the objects of interest is periodically repeated as far as visual objects can appear and disappear in scene; also meteorological conditions may change. The initial stage supposes a fast but rough motion estimation of the dynamic regions. We may use the simplest method of background subtraction and the complex method of mixture composition based on Gauss distributions [52]. Also surveillance systems are working at night, and they use an infrared camera which forms an infrared image of scene with less resolution then a visual image [47]. An infrared camera does not map the shadows; so the common usage of two imageries permits to compensate artifacts and achieves a more efficiency of processing algorithms.

Statistical descriptors of an enhanced background model deal with a simultaneous sample of N frames and a receiving of average images I_{med} for both types of imageries. For each pixel with a brightness function $I_t(x,y)$, a mean value $\mu(x,y)$, and a deviation $\sigma^2(x,y)$ for N frames are calculated in following manner:

$$\mu(x,y)=\frac{\sum_{t=1}^{N} w_t(x,y)\cdot I_t(x,y)}{\sum_{t=1}^{N} w_t(x,y)}\,, \tag{8.36}$$

$$\sigma^2(x,y)=\frac{\sum_{t=1}^{N} w_t(x,y)\cdot\left(I_t(x,y)-\mu(x,y)\right)^2}{1-\frac{1}{N}\sum_{t=1}^{N} w_t(x,y)}\,, \tag{8.37}$$

where $w_t(x,y)$ are the weighting coefficients. They are calculated by normal distribution and used for minimization of spikes which are maximally removed from average I_{med}:

$$w_t(x,y)=\frac{1}{\sqrt{2\pi\sigma_{ex}^2}}\exp\left(-\frac{\left(I_t(x,y)-I_{med}(x,y)\right)^2}{2\sigma_{es}^2}\right)\,. \tag{8.38}$$

Standard deviation σ^2_{ex} is calculated for N adjacent frames. The usage of the weighting values in the statistical background model permits to receive a robust background model without a training imagery.

On basis of the background statistics (expressions (8.36), (8.37)) from infrared imagery we can receive the masks of the moving regions D^{IS} by criteria

$$D^{IS}(x,y)=\begin{cases}1\ , & \dfrac{(I(x,y)-\mu(x,y))^2}{\sigma^2(x,y)}>Z^2\ ,\\ 0\ , & \text{in other cases}\ ,\end{cases} \quad (8.39)$$

where Z is a thresholding value.

For detection of moving regions from an infrared imagery we can apply the operator of morphological erosion for a mask D^{IS} (with sizes 5×5 elements) and the algorithm of a regions link. Any region with a square less 0.1% from a whole image is rejected. The similar mask we can build for a visual imagery D^{VS} but it will contain shadows and noisy. On Fig. 8.9 one see examples of input infrared (Fig. 8.9 a) and visual (Fig.8.9 b) images, also received masks D^{IS} and D^{VS} (Fig. 8.9 c and 8.10 d correspondingly), and a result of a background subtraction method applied for two adjacent frames (Fig. 8.9 e) from test base OTCBVS'07, Thermal "Sequence 1 a" and Color "Sequence 1 b".

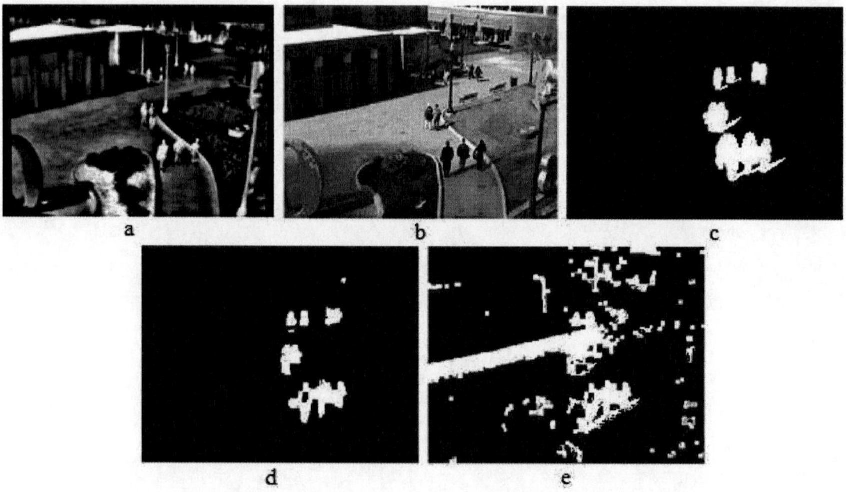

Fig. 8.9 Examples of a moving regions detection: (a) an input infrared frame, (b) an input visual frame; (c) a mask D^{IS}; (d) a mask D^{VS}; (e) a background subtraction

8.3.2 Motion Estimation in Visual Imagery

The motion estimation of the regions displacements in a visual imagery is well described by a 3D structural tensor, and for an infrared imagery one can use a 3D flow tensor. Such estimations are based on gradient frames differences and possesses the most accuracy in comparison with others motion estimation methods in videos.

Let's define a discrete function $\mathbf{I}^{\Delta}$ determined in points of 3D grid $\{(x_i,y_i,z_k)=\mathbf{p}_{ijk}\}$, where x_i,y_i,z_k are grid coordinates along axes OX, OY, OZ correspondingly, $x_i\in[a_{OX},b_{OX}]$, $y_j\in[a_{OY},b_{OY}]$, $z_k\in[a_{OZ},b_{OZ}]$, $1\le i\le i_0$, $1\le j\le j_0$, $1\le k\le k_0$, $\mathbf{I}^{\Delta}(\mathbf{p}_{ijk})=\mathbf{I}^{\Delta}_{ijk}$. Let's build the mathematical model of motion estimation of regions in a visual imagery [9]. Let's find a differentiable function $\mathbf{I}(x,y,z)$ for which $\mathbf{I}^{\Delta}(\mathbf{p}_{ijk})=\mathbf{I}^{\Delta}_{ijk}$ in points of grid, where an arbitrary point is denoted as $\mathbf{p}=[x,y,t]$, (x,y) are a spatial coordinates of frame along axes OX, OY correspondingly, and t is a temporal coordinate considering a sequence of frames appearance along axis OZ. Then the expression for motion estimation of spatio-temporal data in visual imagery $\mathbf{I}(\mathbf{p})$ (under the condition of the persistent scene luminance) relatively the position of a local point $\mathbf{p}$ has a form:

$$\frac{d\mathbf{I}(\mathbf{p})}{dt}=\frac{\partial I(\mathbf{p})}{\partial x}v_x+\frac{\partial I(\mathbf{p})}{\partial y}v_y+\frac{\partial I(\mathbf{p})}{\partial t}v_t=\nabla\mathbf{I}^T(\mathbf{p})\mathbf{v}(\mathbf{p}) \quad , \qquad (8.40)$$

where $\mathbf{v}(\mathbf{p})=[v_x,v_y,v_t]$ is an error vector in a local neighborhood $\nabla\mathbf{I}(\mathbf{q})$ relatively a point $\mathbf{p}$. The module of a speed vector $\mathbf{v}(\mathbf{p})$ in a visual imagery is determined under the condition of a minimum function (8.40) using a local 3D volume $\mathbf{\Omega}(\mathbf{p},\mathbf{q})$ centered relatively a vector $\mathbf{p}$, where $\mathbf{q}$ is a local point with coordinates $\mathbf{q}=[x_n,y_n,t_n]$. Let's denote the error vector between gradient $\nabla\mathbf{I}(\mathbf{q})$ for point $\mathbf{q}$ and a speed vector $\mathbf{v}(\mathbf{p})$ as $\mathbf{e}(\mathbf{p},\mathbf{q})$. So a module of an error vector $\mathbf{e}(\mathbf{p},\mathbf{q})$ is written as (Fig. 8.10):

$$\|\mathbf{e}(\mathbf{p},\mathbf{q})\|=\left\|\nabla\mathbf{I}(\mathbf{q})-\left(\nabla\mathbf{I}^T(\mathbf{q})\mathbf{v}(\mathbf{p})\right)\mathbf{v}(\mathbf{p})\right\| \quad . \qquad (8.41)$$

Usually for motion estimation of a point translation quadratic functional is chosen:

$$\rho\left(\|\mathbf{e}(\mathbf{p},\mathbf{q})\|\right)=\|\mathbf{e}(\mathbf{p},\mathbf{q})\|^2 \quad . \qquad (8.42)$$

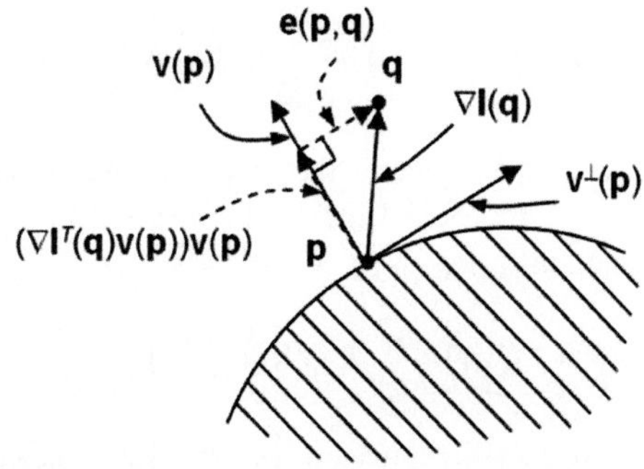

Fig. 8.10 Scheme of a point **p** motion in a frame

Let's find the speed estimation $e_{ls}(\mathbf{p})$ of a point displacement by the least-squares method in a local volume $\boldsymbol{\Omega}(\mathbf{p},\mathbf{q})$ with conditions $\|\mathbf{v}(\mathbf{p})\|=1$ and $\|\nabla\mathbf{I}(\mathbf{q})\|=1$:

$$e_{ls}^{s}(\mathbf{p}) = \int_{\Omega_{\mathbf{p},\mathbf{q}}} \|\mathbf{e}(\mathbf{p},\mathbf{q})\|^{2} W(\mathbf{p},\mathbf{q})d\mathbf{q} \quad , \tag{8.43}$$

where $W(\mathbf{p},\mathbf{q})$ is a spatial invariant function having a Gauss distribution and increasing a gradient value in central pixel relatively a neighborhood. Simplifying the expression (8.43) we'll receive

$$\begin{aligned} e_{ls}^{s}(\mathbf{p}) = & \int_{\Omega(\mathbf{p},\mathbf{q})} \left(\nabla\mathbf{I}^{T}(\mathbf{q})\nabla\mathbf{I}(\mathbf{q})\right) W(\mathbf{p},\mathbf{q})d\mathbf{q} - \\ & - \int_{\Omega(\mathbf{p},\mathbf{q})} \left(\mathbf{v}^{T}(\mathbf{p})\left(\nabla\mathbf{I}(\mathbf{q})\nabla\mathbf{I}^{T}(\mathbf{q})\right)\mathbf{v}(\mathbf{p})\right) W(\mathbf{p},\mathbf{q})d\mathbf{q} \quad . \end{aligned} \tag{8.44}$$

The minimization of value e^{s}_{ls} relatively a speed $\mathbf{v}$, $\|\mathbf{v}(\mathbf{p})\|=1$, is equivalent to finding a maximum of a secondary member in expression (8.43). Using te method of Lagrange multipliers (with one restriction in our case), let's rewrite the criteria (8.44) in following manner:

$$\begin{aligned} e_{ls}^{S}(\mathbf{p}) = \mathbf{v}^{T}(\mathbf{p}) & \left(\int_{\Omega(\mathbf{p},\mathbf{q})} \left(\nabla\mathbf{I}(\mathbf{q})\nabla\mathbf{I}^{T}(\mathbf{q})\right) W(\mathbf{p},\mathbf{q})d\mathbf{q} \right) \mathbf{v}(\mathbf{p}) + \\ & + \alpha\left[1 - \mathbf{v}(\mathbf{p})^{T}\mathbf{v}(\mathbf{p})\right] \quad , \end{aligned} \tag{8.45}$$

where α is a Lagrange multiplier.

Let's suppose that a speed vector $\mathbf{v}(\mathbf{p})$ has the constant value into a spatial volume $\boldsymbol{\Omega}(\mathbf{p},\mathbf{q})$ and differentiate the expression (9.45) performing an integrand product $(\nabla\mathbf{I}(\mathbf{q})\nabla\mathbf{I}^{T}(\mathbf{q}))$ as a kernel of a scalar product

$$K(\mathbf{q}) = \varphi(\mathbf{q})\varphi^{T}(\mathbf{q}) = \sum_{j=0}^{m_1} \varphi_j(\mathbf{q})\varphi_j^{T}(\mathbf{q}) \quad ,$$

which is a particular case of the Mercer theorem [33]. According to the Mercer's theorem, a continuous symmetrical kernel which is defined on closed interval $\mathbf{a}\le\mathbf{p}\le\mathbf{b}$ may be performed in series

$$K(\mathbf{q}) = \sum_{i=0}^{\infty} \lambda_i \varphi_i(\mathbf{q})\varphi_i^{T}(\mathbf{q})$$

with positive coefficients $\lambda_i>0$ for all values i. Functions $\varphi_i(\mathbf{q})$ are called the eigenfunctions of expansion, and numbers λ_i are eigenvalues of these functions. Thereby we receive an approximate expression for estimation of a vector $\mathbf{v}_{es}(\mathbf{p})$

$$\mathbf{J}_{S}(\mathbf{p},\mathbf{W}(\mathbf{p}))\mathbf{v}_{es}(\mathbf{p}) = \alpha\mathbf{v}_{es}(\mathbf{p}) \quad , \tag{8.46}$$

where $\mathbf{J}_S(\mathbf{p},\mathbf{W})$ is a 3D structural tensor of a spatio-temporal volume centered relatively a vector $\mathbf{p}$. In expression (8.46) the term

$$\mathbf{J}_s(\mathbf{p},\mathbf{W}(\mathbf{p})) = \int\limits_{\Omega(\mathbf{p},\mathbf{q})} \left(\nabla\mathbf{I}(\mathbf{q})\nabla\mathbf{I}^T(\mathbf{q})\right) W(\mathbf{p},\mathbf{q})\,d\mathbf{q} \tag{8.47}$$

is a structural tensor in a point **p** which uses a kernel **W**, and is optimized by the least-squares method. The maximum eigenvalue gives a gradient estimation in a point **p** which is a more robust estimation. For ideal edge a minimal eigenvalue is equal zero.

However a robust error estimation method ought to be a noisy-insensitive that the least-squares method (8.42) is not provided. The Gaussian robust error function is a special case of the Leclerc robust error function

$$\rho\left(\|\mathbf{e}(\mathbf{p},\mathbf{q})\|, m, \eta\right) = 1 - e^{-\frac{\|\mathbf{e}(\mathbf{p},\mathbf{q})\|^2}{\phi^2(m,\eta)}} , \tag{8.48}$$

where a functional $\phi^2(m,\eta)$ is an analog of a dispersion σ^2 in a Gauss function, m is a penalty parameter for high error spikes with additional parameter η^2=2. Also one can use the Geman-McClure criteria

$$\rho\left(\|\mathbf{e}(\mathbf{p},\mathbf{q})\|, m\right) = \frac{\|\mathbf{e}(\mathbf{p},\mathbf{q})\|^2}{m^2 + \|\mathbf{e}(\mathbf{p},\mathbf{q})\|^2} = 1 - \frac{m^2}{m^2 + \|\mathbf{e}(\mathbf{p},\mathbf{q})\|^2} , \tag{8.49}$$

where m is a penalty parameter.

Function (8.49) limits the influence of high error spikes. We can find the minimum of an error function for criteria (8.49) and a robust structural Geman-McClure tensor calculating by an iterative procedure

$$\mathbf{J}_S(\mathbf{p},\mathbf{v}_i(\mathbf{p}),\mathbf{W}(\mathbf{p}))\mathbf{v}_{i+1}(\mathbf{p}) = \alpha\mathbf{v}_{i+1}(\mathbf{p}) , \tag{8.50}$$

where $\mathbf{J}_S(\mathbf{p},\mathbf{v}(\mathbf{p}),\mathbf{W})$ is a 3D structural Geman-McClure tensor

$$\mathbf{J}_s(\mathbf{p},\mathbf{v}(\mathbf{p}),\mathbf{W}(\mathbf{p})) = \int\limits_{\Omega(\mathbf{p},\mathbf{q})} \frac{m^2}{\left(\nabla\mathbf{I}^T(\mathbf{q})\nabla\mathbf{I}(\mathbf{q}) - \mathbf{v}^T(\mathbf{p})\left(\nabla\mathbf{I}(\mathbf{q})\nabla\mathbf{I}^T(\mathbf{q})\right)\mathbf{v}(\mathbf{p}) + m^2\right)} \cdot \left(\nabla\mathbf{I}(\mathbf{q})\nabla\mathbf{I}^T(\mathbf{q})\right) W(\mathbf{p},\mathbf{q})\,d\mathbf{q} \tag{8.51}$$

An iterative procedure (8.50) is usually come to a local minimum; it is well known some convergence criterions, for example,

$$\|\mathbf{v}_{i+1} - \mathbf{v}_i\| < \varepsilon , \quad \text{trace}\left(\mathbf{J}(\mathbf{p},\mathbf{v}_i(\mathbf{p}),W)\right) < k_{trace} , \tag{8.52}$$

and also the sizes of a kernel $W(\mathbf{p},\mathbf{q})$.

A structural tensor is appropriated for a gradient enhancement near a central pixel and for a gradient influence decreasing in a neighborhood $\Omega(\mathbf{p},\mathbf{q})$. Usually for these purposes a scale-invariant Gaussian with a kernel $W(\mathbf{p},\mathbf{q})$ is used, and a uniform scale change is assumed [18]. Let's use a spatial alterable kernel $W(\mathbf{p},\mathbf{q})$, that is a Gaussian function which is adaptive for sizes and orientations of a neighborhood

$\Omega(\mathbf{p},\mathbf{q})$. Let's assume that a neighborhood $\Omega(\mathbf{p},\mathbf{q})$ has a form of circle, and then it adapts in a oriented ellipse according to eigenvalues λ_i^1 and λ_i^2, $\lambda_i^1>\lambda_i^2$, of a structure tensor $\mathbf{J}(\mathbf{p},\mathbf{v}_{i-1}(\mathbf{p}),W_i)$ at the ith iteration. Using these eigenvalues one can update the semi-major and semi-minor axes at the $(i+1)^{th}$ iteration minimizing a difference between a motion vector and a received vector at the ith iteration. Such adaptation permits to enhance the estimation of the oriented structures in the image (particularly the gradient structures during a regions motion). From some number of iterations (experiments show that number of iterations does not exceed 2–3) a local ellipse neighborhood can be associated with each moving pixel. The influence of such local neighborhood will be minimally on a neighborhood moving pixels. A neighborhood with two and more edges is interpreted as angles. For localization of such areas let's choose very small regions. A suggested adaptive structural tensor may find such regions on an image. In this case a function calculating spatial changeable Gaussian kernel is modified by the following iterative procedure

$$\mathbf{J}_S(\mathbf{p},\mathbf{v}_i(\mathbf{p}),\mathbf{W}_{i+1}(\mathbf{p}))\mathbf{v}_{i+1}(\mathbf{p})=\alpha\mathbf{v}_{i+1}(\mathbf{p}) \quad . \tag{8.53}$$

So, it is enough 2–3 iterations to achieve a convergence criterion (8.52).

The expression of a 3D structural tensor centered relatively a vector $\mathbf{p}$ we may rewrite in a matrix's form (without a spatial filter $\mathbf{W}(\mathbf{p},\mathbf{q})$):

$$\mathbf{J}_S(\mathbf{p})=\begin{bmatrix} \int_\Omega \frac{\partial I}{\partial x}\frac{\partial I}{\partial x}d\mathbf{q} & \int_\Omega \frac{\partial I}{\partial x}\frac{\partial I}{\partial y}d\mathbf{q} & \int_\Omega \frac{\partial I}{\partial x}\frac{\partial I}{\partial t}d\mathbf{q} \\ \int_\Omega \frac{\partial I}{\partial y}\frac{\partial I}{\partial x}d\mathbf{q} & \int_\Omega \frac{\partial I}{\partial y}\frac{\partial I}{\partial y}d\mathbf{q} & \int_\Omega \frac{\partial I}{\partial y}\frac{\partial I}{\partial t}d\mathbf{q} \\ \int_\Omega \frac{\partial I}{\partial t}\frac{\partial I}{\partial x}d\mathbf{q} & \int_\Omega \frac{\partial I}{\partial t}\frac{\partial I}{\partial y}d\mathbf{q} & \int_\Omega \frac{\partial I}{\partial t}\frac{\partial I}{\partial t}d\mathbf{q} \end{bmatrix} \quad . \tag{8.54}$$

The simplest method of a motion estimation is to compute a spur of matrix trace($\mathbf{J}_S(\mathbf{p})$), and compare it with a thresholding value. However, it is required to remember that the expression

$$\mathrm{trace}(\mathbf{J}_S(\mathbf{p}))=\int_\Omega \|\nabla\mathbf{I}\|^2 d\mathbf{q} \tag{8.55}$$

includes the full changes of the gradient only in a spatial domain.

8.3.3 Motion Estimation in Infrared Imagery

The motion estimation in an infrared imagery containing a thermal energy distribution of objects is based on the applying a flow tensor which permits to estimate a motion without calculation eigenvalues of a 3D structural tensor. Let's find the second derivative of expression (8.40) relative to a variable t. Then we receive

$$\frac{d}{dt}\left(\frac{\partial \mathbf{I}(\mathbf{p})}{\partial t}\right)=\frac{\partial^2 I(\mathbf{p})}{\partial x \partial t}v_x+\frac{\partial^2 I(\mathbf{p})}{\partial y \partial t}v_y+\frac{\partial^2 I(\mathbf{p})}{\partial t^2}v_t+ \\ +\frac{\partial I(\mathbf{p})}{\partial x}a_x+\frac{\partial I(\mathbf{p})}{\partial y}a_y+\frac{\partial I(\mathbf{p})}{\partial t}a_t \quad , \tag{8.56}$$

or in a vector form

$$\frac{d}{dt}\left(\nabla\mathbf{I}^T(\mathbf{p})\mathbf{v}(\mathbf{p})\right)=\frac{\partial \nabla\mathbf{I}^T(\mathbf{p})}{\partial t}\mathbf{v}(\mathbf{p})+\nabla\mathbf{I}^T(\mathbf{p})\mathbf{a}(\mathbf{p}) \quad ,$$

where $\mathbf{a}(\mathbf{p})$ is an acceleration vector.

Let's use the similar approach from section 8.3.2 for finding an error $e^F_{ls}(\mathbf{p})$ assuming that a speed is a constant, and $\|\mathbf{v}(\mathbf{p})\|=1$. We may write

$$e^F_{ls}(\mathbf{p})=\mathbf{v}^T(\mathbf{p})\left(\int\limits_{\Omega(\mathbf{p},\mathbf{q})}\left(\frac{\partial\nabla\mathbf{I}(\mathbf{p})}{\partial t}\cdot\frac{\partial\nabla\mathbf{I}^T(\mathbf{p})}{\partial t}\right)W(\mathbf{p},\mathbf{q})d\,\mathbf{q}\right)\mathbf{v}(\mathbf{p})+ \\ +\alpha\left[1-\mathbf{v}(\mathbf{p})^T\mathbf{v}(\mathbf{p})\right] \quad . \tag{8.57}$$

Under the condition of a constant speed in a spatial volume $\mathbf{\Omega}(\mathbf{p},\mathbf{q})$ the expression for a 3D flow tensor $\mathbf{J}_F(\mathbf{p},\mathbf{W}(\mathbf{p}))$ takes a form

$$\mathbf{J}_F(\mathbf{p},\mathbf{W}(\mathbf{p}))=\int\limits_{\Omega(\mathbf{p},\mathbf{q})}W(\mathbf{p},\mathbf{q})\frac{\partial\nabla\mathbf{I}(\mathbf{p})}{\partial t}\cdot\frac{\partial\nabla\mathbf{I}^T(\mathbf{p})}{\partial t}d\,\mathbf{q} \quad ,$$

or in an extension matrix form (without taking into account a spatial filter $\mathbf{W}(\mathbf{p},\mathbf{q})$)

$$\mathbf{J}_F(\mathbf{p})=\begin{bmatrix} \int\limits_\Omega\left\{\frac{\partial^2 I}{\partial x\partial t}\right\}^2 d\mathbf{q} & \int\limits_\Omega\frac{\partial^2 I}{\partial x\partial t}\frac{\partial^2 I}{\partial y\partial t}d\mathbf{q} & \int\limits_\Omega\frac{\partial^2 I}{\partial x\partial t}\frac{\partial^2 I}{\partial t^2}d\mathbf{q} \\ \int\limits_\Omega\frac{\partial^2 I}{\partial y\partial t}\frac{\partial^2 I}{\partial x\partial t}d\mathbf{q} & \int\limits_\Omega\left\{\frac{\partial^2 I}{\partial y\partial t}\right\}^2 d\mathbf{q} & \int\limits_\Omega\frac{\partial^2 I}{\partial y\partial t}\frac{\partial^2 I}{\partial t^2}d\mathbf{q} \\ \int\limits_\Omega\frac{\partial^2 I}{\partial t^2}\frac{\partial^2 I}{\partial x\partial t}d\mathbf{q} & \int\limits_\Omega\frac{\partial^2 I}{\partial t^2}\frac{\partial^2 I}{\partial y\partial t}d\mathbf{q} & \int\limits_\Omega\left\{\frac{\partial^2 I}{\partial t^2}\right\}^2 d\mathbf{q} \end{bmatrix} . \tag{8.58}$$

As one can see from expression (8.58), the elements of flow tensor contain the values of temporal gradients changes that permit to separate efficiently static and dynamic image regions. A trace of flow tensor matrix can be expressed as:

$$\operatorname{trace}(\mathbf{J}_F(\mathbf{p}))=\int\limits_{\Omega(\mathbf{p},\mathbf{q})}\left\|\frac{\partial\nabla\mathbf{I}(\mathbf{p})}{\partial t}\right\|^2 d\mathbf{q} \quad . \tag{8.59}$$

The expression (8.59) can be directly used for a motion regions classification without calculation eigenvalues of static regions matrixes. If it is necessary, the expression (8.57) one can minimize for receiving an estimation of a motion vector $\mathbf{v}_{es}(\mathbf{p})$ using the equation

$$\mathbf{J}_F(\mathbf{p},\mathbf{W}(\mathbf{p}))\mathbf{v}_{es}(\mathbf{p}) = \alpha\mathbf{v}_{es}(\mathbf{p}) \quad .$$

Thereby for motion estimation it is ought to calculate derivatives

$$I_{xt}(\mathbf{p}) = \frac{\partial^2 I(\mathbf{p})}{\partial x \partial t} \ , \quad I_{yt}(\mathbf{p}) = \frac{\partial^2 I(\mathbf{p})}{\partial y \partial t} \ , \quad I_{tt}(\mathbf{p}) = \frac{\partial^2 I(\mathbf{p})}{\partial t \, \partial t} \quad ,$$

and execute an integration on the area $\mathbf{\Omega}(\mathbf{p},\mathbf{q})$. These derivatives are calculated as the convolutions of infrared frames with kernels of the separable filters with a possibility of cascade 1D convolutions. The noise reduction is realized by usage any smoothing filter. For a calculation of derivatives I_{xt}, I_{yt}, and I_{tt}, the spatial convolutions I_{xs}, I_{ys}, and I_{ss} are computed where s is a smoother filter. Thus, for each input frame the sets I_{xs}, I_{ys}, and I_{ss} are calculated and memorized. Then after accumulation of enough data the sum of derivatives $I^2_{xt}+I^2_{yt}+I^2_{tt}$ are founded for some input infrared frames, and a motion mask MF_M is created. The mask MF_M which is received by a 3D flow tensor one can use for a building of a mask MS_M of visual frames. In this case we find and analyze eigenvalues of a matrix $\mathbf{J}_S(\mathbf{p},\mathbf{W}(\mathbf{p}))$ in local area of a visual frame in compliance with a received mask MF_M.

8.3.4 Elaboration of Boundaries of Moving Regions

There are two problems of creating a mask MF_M in infrared imagery connecting with (1) availability of "empties" into slowly moving regions, and (2) inaccurate region boundaries when sizes of dynamic regions are more then real sizes of dynamic video objects. Inaccurate boundaries lead to merge the masks of the adjacent objects images and false interpretation of the objects trajectories. The first problem is solved by the mathematical morphology methods; the second problem-solving suggests the application of active contours method for a segmentation of moving regions [29]. Active contours are divided on the parametric active contours (classic "snakes") based on the Lagrange functions and the geometrical active contours (levels) based on the Euler functions which possess a low computer cost and a topological flexibility [54]. In last case a curve C is determined by a Lipschitz function ϕ as

$$C = \{(x, y) | \phi(x, y) = 0\} \ ,$$

and a curve of zero level is determined by a function $\phi(x,y,t)$. Propagation of a curve C in a normal direction with a speed F is defined from a solution of the differential equation

$$\frac{\partial \phi}{\partial t} = |\nabla \phi| F \quad , \qquad \phi(x, y, 0) = \phi_0(x, y) \quad .$$

Propagation of an active contour is occurred by expression

$$\frac{\partial \phi}{\partial t} = g_F(\mathbf{I})(c + \mathrm{K}(\phi))|\nabla\phi| + \nabla\phi \cdot \nabla g_F(\mathbf{I}) \quad , \tag{8.60}$$

where $g_F(\mathbf{I})$ is a halt function of a contour generation; c is a speed constant; $\mathrm{K}(\phi)$ is a curvature function

$$\mathrm{K}(\phi) = \mathrm{div}\left(\frac{\nabla\phi}{|\nabla\phi|}\right) = \frac{\phi_{xx}\phi_y^2 - 2\phi_x\phi_y\phi_{xy} + \phi_{yy}\phi_x^2}{\left(\phi_x^2 + \phi_y^2\right)^{\frac{3}{2}}} \quad .$$

A sign of a speed constant c determines a compression or an expansion of a curve. Function $\mathrm{K}(\phi)$ makes the smoother boundaries. The terms $\nabla\phi$ and $\nabla g_F(\mathbf{I})$ from equation (8.60) decrease a deviation of a curve from an object boundary. The following elaboration of regions boundaries is based on the edge and corner detectors. The use of such detectors or feature tracking in local regions permits to estimate a relative speed and acceleration value of a moving region.

It is well known three methods of boundaries detection in color images. In generalized methods a detection of boundaries is realized in each RGB-channel, transformed into the grey levels and calculated as a weighting sum. Sometimes color boundaries are presented as a set of vectors based on the order statistics. One of perspective solutions is a tensor approach relating to the multi-dimensional gradient methods which build the 1D estimations along all dimensions in each boundary point. Two types of tensors are the most significant – 2D color structural tensor and Beltrami color metric tensor. The expression for 2D color structural tensor $\mathbf{J}_c(\mathbf{I})$ has a form:

$$\mathbf{J}_C(\mathbf{I}) = \begin{bmatrix} \sum\limits_{i=R,G,B}\left(\frac{\partial I_i}{\partial x}\right)^2 & \sum\limits_{i=R,G,B}\frac{\partial I_i}{\partial x}\frac{\partial I_i}{\partial y} \\ \sum\limits_{i=R,G,B}\frac{\partial I_i}{\partial y}\frac{\partial I_i}{\partial x} & \sum\limits_{i=R,G,B}\left(\frac{\partial I_i}{\partial y}\right)^2 \end{bmatrix} \quad . \tag{8.61}$$

Eigenvalues of tensor (8.61) are defined as

$$\lambda_{1,2} = \frac{1}{2}\left(J_c(1,1) + J_c(2,2) \pm \sqrt{(J_c(1,1) - J_c(2,2))^2 + (2J_c(1,2))^2}\right) \quad .$$

Beltrami color metric tensor $\mathbf{J}_B(\mathbf{I})$ is determined in 5D color space $\{x,y,R,G,B\}$ by following expression

$$\mathbf{J}_B(\mathbf{I}) = \begin{bmatrix} 1 + \sum\limits_{i=R,G,B}\left(\frac{\partial I_i}{\partial x}\right)^2 & \sum\limits_{i=R,G,B}\frac{\partial I_i}{\partial x}\frac{\partial I_i}{\partial y} \\ \sum\limits_{i=R,G,B}\frac{\partial I_i}{\partial y}\frac{\partial I_i}{\partial x} & 1 + \sum\limits_{i=R,G,B}\left(\frac{\partial I_i}{\partial y}\right)^2 \end{bmatrix} \quad , \tag{8.62}$$

which determinant is calculated as

$$\det(\mathbf{J}_B(\mathbf{I})) = 1 + \mathrm{trace}(\mathbf{J}_C(\mathbf{I})) + \det(\mathbf{J}_C(\mathbf{I})) \approx 1 + (\lambda_1 + \lambda_2) + \lambda_1\lambda_2 \quad ,$$

where λ_1, λ_2 are eigenvalues of tensor $\mathbf{J}_C(\mathbf{I})$).

Let's remind other color edge and angle detectors:

– Harris detector $HR(\mathbf{I}_{RGB})$ uses an adaptive parameter k (the condition $k \to 0$ indicates an angle)

$$HR(\mathbf{I}_{RGB}) = \det(\mathbf{J}_c(\mathbf{I})) - k \cdot \mathrm{trace}^2(\mathbf{J}_c(\mathbf{I})) \approx \lambda_1\lambda_2 - k \cdot (\lambda_1 + \lambda_2)^2 \quad . \quad (8.63)$$

– Shi-Tomasi detector $ST(\mathbf{I}_{RGB})$ reinforces angles and cuts boundaries (one of eigenvalues is closed to zero near a boundary) but it can not be used in active contours method

$$ST(\mathbf{I}_{RGB}) = \min(\lambda_1, \lambda_2) \quad . \quad (8.64)$$

– Cumani detector $CU(\mathbf{I}_{RGB})$ finds edges and boundaries on image successfully

$$CU(\mathbf{I}_{RGB}) = \max(\lambda_1, \lambda_2) \quad . \quad (8.65)$$

Beltrami and Cumani detectors show good results for boundaries detection in color images; Beltrami detector is a less sensitive operator for variable boundary in visual imagery.

After contour detection from the both types of imageries the contours superposition is occurred by the analysis of a 6D space $\{x,y,R,G,B,\mathbf{I}_R\}$. The superposed function g_E is determined as a minimum value from two functions $g_{RGB}(x,y)$ and $g_{IF}(x,y)$ normalized in range [0…1]

$$g_E(x, y, R, G, B, \mathbf{I}_R) = \min\{g_{\mathrm{RGB}}(x, y), g_{IF}(x, y)\} \quad .$$

A repetitive procedure of a region motion estimation and a contour elaboration for sequential frames (possibly with frames resampling) builds a set of local vectors of a dynamic region motion and calculates the relative values of vectors modules during surveillance.

8.3.5 Classification of Dynamic Regions

A classification of dynamic regions is realized by eigenvalues analysis of symmetric covariance 3D structure tensor $\mathbf{J}_S(\mathbf{p})$ with size 3×3 pixels (expression (8.54)). Eigenvalues $\mathbf{\Lambda}=\{\lambda_k\}$, ($k$=1,2,3) are characterized the local brightness displacements along three axes in Euclidean space. We can use such brightness displacements for estimation of a local orientation of dynamic regions. According to the principal components analysis (PCA) eigenvalues λ_k of such matrix are sorted in the decreased order $\lambda_1 \geq \lambda_2 \geq \lambda_3 \geq 0$. The first eigenvector corresponds to the largest eigenvalue and points a direction of the most data change. Ratio of each eigenvalue to the sum of all three eigenvalues is defined the energy concentration along corresponding axis. Such eigenvalues are used for detection of local changes in imagery. The least eigenvalue is used for determination of differences in frames. Then we can build the brightness maps $\lambda_1(I)$, $\lambda_2(I)$, $\lambda_3(I)$ based on eigenvalues $\lambda_1(x,y,t)$, $\lambda_2(x,y,t)$, $\lambda_3(x,y,t)$ of a local 3D structure tensor. A map $\lambda_1(I)$ shows the motion of dynamic objects and sometimes the isolated texture in background regions. A map $\lambda_2(I)$ is a less informative map. A map $\lambda_3(I)$ generates small "empties" into masks of video objects but permits to estimate a degree of variability of dynamic regions shapes. The correlation coefficient R (by Pearson criterion) between neighbor frames Fr_t и Fr_{t+1} is calculated as

$$R = \frac{\sum_{i=1}^{n}(v_i - \bar{v})(u_i - \bar{u})}{\sqrt{\sum_{i=1}^{n}(v_i - \bar{v})^2 \sum_{i=1}^{n}(u_i - \bar{u})^2}} , \tag{8.66}$$

where n is a total amount of pixels in a frame,

$$\bar{v} = \frac{1}{n}\sum_{i=1}^{n} v_i , \quad \bar{u} = \frac{1}{n}\sum_{i=1}^{n} u_i , \quad v_i \in \lambda_3(I_t) , \quad u_i \in \lambda_3(I_{t+1}) .$$

A spread of correlation coefficients will be a substantially smaller value for rigid regions relatively non-rigid regions. A value of such spread S one can calculate using formula of standard deviation:

$$S = \sqrt{\frac{1}{N-1}\sum_{i=1}^{N}(R_i - R_{avg})^2} , \tag{8.67}$$

where N is amount of frames in scene, R_i is a value of a correlation coefficient for frame i, R_{avg} is average of values R_i, $i \in [1\ldots N]$.

If value S exceeds a thesholding value then we may consider that the region underwent the intense changes. A detection of rigid and non-rigid regions permits to extract the man-caused objects and the anthropogenic objects in scene for the following recognition procedure [11].

8.4 Experimental Researches

All experiments were executed by the experimental software SPOER v. 2.06 (Software for Processing, Objects and Events Recognition) which was developed in Laboratory of Image and Videos Processing (Department of Computer Science of Siberian State Aerospace University). Software SPOER is oriented on a system hierarchical processing of imageries, up to the highest recognition level of objects and events. This software requires a teacher learning and tuning coefficients for detectors, graphs, nets and classifiers. Some low-level modules are working in automatic modes. Software SPOER is the Windows application, includes the program modules performing as *.dll files. Such realization permits to modify any system module without influence on others modules. (Data protocol was previously designed.) Application of *.dll files permits to use various program languages such as C, C++, Object Pascal. Each module is a complex program realizing various research algorithms and allowing the iterative parameters tuning except of Configuration Module.

Motion Estimation Module assigns for analysis of imageries and includes following main functions:

- Block-Matching method.
- Feature Points Tracking.
- Optical Flow method.
- Combined methods.

Each method has the own tuning parameters which are selected in an interactive mode. The main parameters of three basic methods are shown in Table 8.4., and the main screen of Motion Estimation Module is presented on Fig. 8.11. Motion Estimation Module builds the zoomed images of visual objects (for example, pedestrians) on which the directions of the normalized local motion vectors are well visible. (These vectors were built according to the optical flow method applied to a visual imagery.) The fields of the local vectors for a linear motion are presented on Fig. 8.12.

Table 8.4 Main parameters of basic motion estimation methods

Method	Parameters
Block-matching method	Sizes of a neighborhood
	Sizes of an object detection area
	Frames count between keyframes
Feature points tracking	Sizes of a neighborhood
	Sizes of an object detection area
	Frames count between keyframes
	Count of feature points in frame
Optical flow based on tensor approach	Sizes of a neighborhood
	Sizes of an object detection area
	Frames count between keyframes
	Eigenvalues thresholding
	Thresholding interval
	Object displacement between keyframes
	Smoother filtration of vector fields
	Maximum distance for binding of moving regions
	Minimal sizes of moving regions

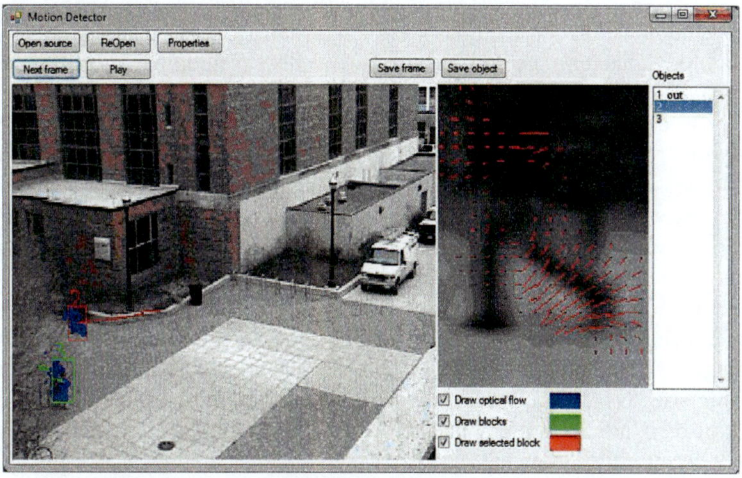

Fig. 8.11 The main screen on Motion Estimation Module

Fig. 8.12 Rendering of the local motion vectors in the visual imagery Color "Sequence 6 b" (OTCBVS'07) frames 2–25

Five motion estimation methods were tested during experiments. It was shown that the block-matching method and the optical flow method based on flow tensor (an infrared imagery) have closed results with a low precision. The optical flow method based on structural tensor (a visual imagery) and the feature points tracking demonstrate results with the highest precision. The optical flow method based on structural tensor is characterized by a less computation cost in comparison with the feature points tracking. A joint use of visual and infrared imageries is actually for definition the modules of the speed local vectors, but directions of the speed local vectors are not exact. However in conditions of lower brightness such approach based on the joint use of the both types of imageries is the most intelligent decision.

Generalized motion estimations of regions concerning to objects of interest for test imageries "Hamburg taxi", "Rubic cube", "Silent", visual and infrared imageries from base "OTCBVS'07" and own video materials are presented in Table 8.5. Values of recognition accuracy and error of moving regions are better in infrared imagery because the "moving" shadows were not considered by algorithm.

Table 8.5 Generalized motion estimations

Imagery	Recognition accuracy, %	Recognition error, %
"Hamburg taxi"	93.82	2.76
"Rubic cube"	82.61	3.81
"Silent"	92.87	2.54
Infrared "Sequence 6 a"	98.15	2.18
Color "Sequence 6 b"	90.29	2.63
"Sequence 6 a" and "Sequence 6 b"	94.91	2.51
Video 1*	90.12	2.94
Video 2**	91.28	3.29
Video 3***	73.04	2.12

* – Video with dynamic structures less 10 % of frame square
** – Video with dynamic structures less 40 % of frame square
*** – Video with dynamic structures up to 100 % of frame square.

8.5 Tasks for Self-testing

1. Design a program which is based on a background subtraction method. Use for processing a set of sequential frames from a test visual imagery. Investigate the influence of (1) "white" noise; (2) dynamic background of scene; (3) quick and slow luminance changes; (4) shadow occurrence. Realize possible compensations of shadows using a brightness component in HSV-, HSB-, HLS-color spaces. Build graphics of a noise influence on amount of "moving" pixels. You may use a pseudo-color representation of processed frames.
2. Realize the basic block-matching algorithm (by one of modifications of Full Search (FS)) and estimate the motion vectors according to SAD-, SSD-, and

MSD-metrics in test videos. Investigate the influence of (1) non-crossed blocks sizes; (2) noise.

3. Realize the basic block-matching algorithm by Pattern Search (PS), Recursive Search (RS), and Block Search (BS) modifications. Investigate a time distribution for these modifications using different types of video (a simple scene, a complex scene, a slow motion, a quick motion).
4. Simplify the expression (8.8) and analyze the remaining terms in error contribution. Use the basic block-matching algorithm for learning dependences between a motion estimation accuracy and a stationary/non-stationary noise in whole imagery.
5. Realize a fast block-matching method with various procedures of threshold calculation: the non-thresholding approach, the constant-thresholding approach (a user selection), and the adaptive-thresholding approach (a spatial frame content).
6. Calculate a Gaussian mixture model of moving image blocks by expressions (8.11), (8.13) and use a Full Search realization of the block-matching method applying Mahalanobis or Extended Mahalanobis distances.
7. Create a program for motion estimation in videos using Lie operators from Table 8.3 and their combinations for simple test imageries with the geometrical regular shapes objects.
8. Design a program for ultrasound imagery processing and apply a bilinear deformable block matching method. Realize an iterative multi-scale approach for decreasing a motion error in static scenes with moving elements.
9. Create a program based on the inter-frame fuzzy search algorithm (for a block-matching estimation) and develop its modification using a Very-Large Scale Integration technology. Compare experimental results of a basic block-matching method and a fuzzy logic interpretation using the given test imagery.
10. Create a program for calculation the local spatio-temporal 3D gradients previously extracted a feature vector from a moving local region. Use a summarized histogram for detection of a motion direction of a local region.
11. Design a program for calculation eigenvalues of a matrix 3×3 pixels by one of known standard mathematical direct or iterative procedures. Calculate descriptions of local structures and their local motion estimations in spatio-temporal volume using frame-to-frame difference.
12. Realize the Moravec and Harris detectors for a color visual imagery and an infrared imagery. Create a program for feature points tracking in imagery. Compare results of tracking in visual and infrared imageries (from OTCBVS'07) separately and by a joint use decision.
13. Realize a simple background model (based on calculation a mean values in each pixel of N frames) and enhanced background model (based on formulas (8.36)–(8.39)).
14. Realize a motion estimation method for visual imagery based on 3D structural tensor in moving regions. Don't calculate 3D structural tensor in every pixel in each frame but only in "moving" regions and their surroundings. You may receive robust motion estimation by a block-matching algorithm or by

image pyramid applications for detection of "moving" regions. Use a local ellipse adaptive kernel for a linear object motion. Apply expressions (8.53), (8.54) for motion estimation in scene. Create video summarized masks for sequential frames. Define rigid and non-rigid regions in your test imagery according to expressions (8.65)–(8.66).

15. Accomplish a previous task but for an infrared imagery (from OTCBVS'07) using 3D flow tensor. Calculate derivatives along axes *OX*, *OY* (spatial domain) and axis *OZ* (temporal domain) for creating the infrared summarized masks for sequential frames. Match visual and infrared masks, estimate the received results.
16. Create a program for boundaries detection in color images. Realize the parametric active contours based on Lagrange functions and the geometrical active contours based on Euler functions. Investigate the edge and corner detectors based on expressions (8.60)–(8.64) for various types of images (smooth image, image with pronounced isotropic and anisotropic texture).

8.6 Conclusion

We considered the main motion estimation methods for objects analysis and detection in videos. More simple comparative methods may be used separately in real-time applications. Also they give the robust estimations for following use of more accurate and complex in realization the gradient methods in detected moving regions. Such two-level procedure creates a basis for rigid and non-rigid regions labeling. Later, static regions (finding by well-known segmentation methods) and dynamic regions joint into the video spatial structures which possess the motion features according to a compact hypothesis. Such approach provides more accurate dynamic features extractions and shows good results of dynamic objects recognition in comparison with existing methods of pattern recognition based on templates, group transformations and others approaches.

References

1. Alzoubi, H., Pan, W.: Fast and accurate global motion estimation algorithm using pixel subsampling. Information Sciences 178(17), 3415–3425 (2008), doi:10.1016/j.ins.2008.05.004
2. Basarab, A., Liebgott, H., Morestin, F., Lyshchik, A., Higashi, T., Asato, R., Delachartre, P.: A method for vector displacement estimation with ultrasound images and its application for thyroid nodular disease. Med. Image Analysis 12(3), 259–274 (2008), doi:10.1016/j.media.2007.10.007
3. Benmoussat, N., Belbachir, M.F., Benamar, B.: Motion estimation and compensation from noisy image sequences: A new filtering scheme. Image and Vision Computing 25(5), 686–694 (2007), doi:10.1016/j.imavis.2006.05.010
4. Boudlal, A., Nsiri, B., Aboutajdine, D.: Modeling of Video Sequences by Gaussian Mixture: Application in Motion Estimation by Block Matching Method . EURASIP J. on Advances in Signal Processing (2010), doi:10.1155/2010/210937

5. Bugeau, A., Perez, P.: Detection and segmentation of moving objects in complex scenes. Computer Vision and Image Understanding 113(4), 459–476 (2009), doi:10.1016/j.cviu.2008.11.005
6. Denman, S., Fookes, C., Sridharan, S.: Improved simultaneous computation of motion detection and optical flow for object tracking. Digital Image Computing: Techniques and Applications (2009), doi:10.1109/DICTA.2009.35
7. Dikbas, S., Arici, T., Altunbasak, Y.: Fast motion estimation with interpolation-free sub-sample accuracy. IEEE Transactions on Circuits and Systems for Video Technology 20(7), 1047–1051 (2010), doi:10.1109/TCSVT.2010.2051283.
8. Doshi, A., Bors, A.G.: Smoothing of optical flow using robustified diffusion kernels. Image and Vision Computing 28(12), 1575–1589 (2010), doi:10.1016/j.imavis.2010.04.001
9. Favorskaya, M.: Estimation of Objects Motion Based on Tensor Approach. J. Digital Signal Processing 1, 2–9 (2010)
10. Favorskaya, M., Zotin, A., Damov, M.: Intelligent Inpainting System for Texture Reconstruction in Videos with Text Removal. In: International Congress on Ultra Modern Telecommunications and Control Systems (2010), doi:10.1109/ICUMT.2010.5676476
11. Favorskaya, M.: Recognition of dynamic visual patterns based on group transformations. In: International Conference on Pattern Recognition and Image Analysis: New Information Technologies, vol. 1, pp. 185–188 (2010)
12. Fernandez-Caballero, A., Castillo, J.C., Martínez-Cantos, J., Martinez-Tomas, R.: Optical flow or image subtraction in human detection from infrared camera on mobile robot. Robotics and Autonomous Systems (2011), doi:10.1016/j.robot.2010.06.002
13. Gao, X., Li, X., Feng, J., Tao, D.: Shot-based video retrieval with optical flow tensor and HMMs. Pattern Recognition Letters 30(2), 140–147 (2009), doi:10.1016/j.patrec.2008.02.009
14. Gao, X., Yang, Y., Tao, D., Li, X.: Discriminative optical flow tensor for video semantic analysis. Computer Vision and Image Understanding 113(3), 372–383 (2009), doi:10.1016/j.cviu.2008.08.007
15. Hannuksela, J., Sangi, P., Heikkila, J.: Vision-based motion estimation for interaction with mobile devices. Computer Vision and Image Understanding 108(1/2), 188–195 (2007), doi:10.1016/j.cviu.2006.10.014
16. Jang, S.W., Pomplun, M., Kim, G.Y., Choi, H.I.: Adaptive robust estimation of affine parameters from block motion vectors. Image and Vision Computing 23(14), 1250–1263 (2005), doi:10.1016/j.imavis.2005.09.003
17. Jayaswal, D.J., Zaveri, M.A., Chaudhari, R.E.: Multi step motion estimation algorithm. In: International Conference and Workshop on Emerging Trends in Technology (2010), doi:10.1145/1741906.1742012
18. Kemouche, M.S., Aouf, N.: A Gaussian mixture based optical flow modeling for object detection. In: International Conference on Crime Detection and Prevention, vol. 2, pp. 1–6 (2009), doi:10.1049/ic.2009.0256
19. Kim, B.G., Song, S.K., Mah, P.S.: Enhanced block motion estimation based on distortion-directional search patterns. Pattern Recognition Letters 27(12), 1325–1335 (2006), doi:10.1016/j.patrec.2006.01.004
20. Klappstein, J., Vaudrey, T., Rabe, C., Wedel, A., Klette, R.: Moving object segmentation using optical flow and depth information. LNCS (2009), doi:10.1007/978-3-540-92957-4_53
21. Lee, H., Jeong, J.: Content adaptive binary block matching motion estimation algorithm. In: Midwest Symposium on Circuits and Systems (2010), doi:10.1109/MWSCAS.2010.5548850

22. Lee, K.J., Kwon, D., Yun, I.D., Lee, S.U.: Optical flow estimation with adaptive convolution kernel prior on discrete framework. Computer Vision and Pattern Recognition (2010), doi:0.1109/CVPR.2010.5539953
23. Liao, B., Du, M., Hu, J.: Color optical flow estimation based on gradient fields with extended constraints. In: International Conference on Networking and Information Technology (2010), doi:10.1109/ICNIT.2010.5508511
24. Liawa, Y.C., Lai, J.Z.C., Hong, Z.C.: Fast block matching using prediction and rejection criteria. Signal Processing 89(6), 1115–1120 (2009), doi:10.1016/j.sigpro.2008.12.012
25. Lin, D., Grimson, E., Fisher, J.: Modeling and estimating persistent motion with geometric flows. Computer Vision and Pattern Recognition (2010), doi:10.1109/CVPR.2010.5539848
26. Lindeberg, T., Akbarzadeh, A., Laptev, I.: Galilean-diagonalized spatio-temporal interest operators. International Conference on Pattern Recognition 1, 57–62 (2004), doi:10.1109/ICPR.2004.1334004
27. Liu, K., Qian, J., Yang, R.: Block matching algorithm based on RANSAC algorithm. In: International Conference on Image Analysis and Signal Processing (2010), doi:10.1109/IASP.2010.5476127
28. Liu, X., Cong, W.: Hybrid-template adaptive motion estimation algorithm based on block matching. In: International Conference on Computer and Communication Technologies in Agriculture Engineering (2010), doi:10.1109/CCTAE.2010.5543459
29. Liu, P.R., Meng, M.Q.H., Liu, P.X., Tong, F.F.L., Wang, X.: Optical flow and active contour for moving object segmentation and detection in monocular robot. In: International Conference on Robotics and Automation (2006), doi:10.1109/ROBOT.2006.1642328
30. Lowe, D.G.: Distinctive Image Features from Scale-Invariant Keypoints. International Journal of Computer Vision 60(2), 91–110 (2004), doi:10.1023/B:VISI.0000029664.99615.94
31. Lu, X., Manduchi, R.: Fast image motion segmentation for surveillance applications. Image and Vision Computing (2010), doi:10.1016/j.imavis.2010.08.001
32. Mahmoud, H.A., Muhaya, F.B., Hafez, A.: Lip reading based surveillance system. In: International Conference on Future Information Technology (2010), doi:10.1109/FUTURETECH.2010.5482688
33. Mercer, J.: Functions of positive and negative type and their connection with the theory of integral equations. Phil. Trans. R Soc. Lond. 209(441-4589), 415–446 (1909), doi:10.1098/rsta.1909.0016
34. Moreno-Garcia, J., Rodriguez-Benitez, L., Fernandez-Caballero, A., Lopez, M.T.: Video sequence motion tracking by fuzzification techniques. Applied Soft Computing 10(1), 318–331 (2010), doi:10.1016/j.asoc.2009.08.002
35. Nisar, H., Choi, T.S.: Multiple initial point prediction based search pattern selection for fast motion estimation. Pattern Recognition 42(3), 475–486 (2009), doi:10.1016/j.patcog.2008.08.010
36. Nsiri, B., Boudlal, A., Aboutajdine, D.: Modeling of video sequences by Gaussian mixture: Application in motion estimation by block matching method. Eurasip J. on Advances in Signal Processing (2010), doi:10.1155/2010/210937
37. Pan, W.D., Yoo, S.M., Park, C.H.: Complexity accuracy tradeoffs of Lie operators in motion estimation. Pattern Recognition Letters 28(7), 778–787 (2007), doi:10.1016/j.patrec.2006.11.006
38. Park, H., Martin, G.R., Bhalerao, A.: Local affine image matching and synthesis based on structural patterns. IEEE Transactions on Image Processing 19(8), 1968–1977 (2010), doi:10.1109/TIP.2010.2045704

39. Park, S.J., Jeon, G., Kim, H., Jeong, J., Kim, S.N., Lim, J.: Adaptive partial block matching algorithm for fast motion estimation. In: Digest of Technical Papers International Conference on Consumer Electronics (2010), doi:10.1109/ICCE.2010.5418765
40. Parrilla, E., Riera, J., Torregrosa, J.R.: Fuzzy control for obstacle detection in object tracking. Mathematical and Computer Modelling 52(7/8), 1228–1236 (2010), doi:10.1016/j.mcm.2010.02.014
41. Pers, J., Sulic, V., Kristan, M., Perse, M., Polanec, K., Kovacic, S.: Histograms of optical flow for efficient representation of body motion. Pattern Recognition Letters 31(11), 1369–1376 (2010), doi:10.1016/j.patrec.2010.03.024
42. Quan, H.: A new method of dynamic texture segmentation based on optical flow and level set combination. In: International Conference on Information Science and Engineering (2009), doi:10.1109/ICISE.2009.95
43. Saha, A., Mukherjee, J., Sural, S.: New pixel-decimation patterns for block matching in motion estimation. Signal Processing: Image Communication 23(10), 725–738 (2008), doi:10.1016/j.image.2008.08.004
44. Scharr, H.: Optimal Filters for Extended Optical Flow. In: Jähne, B., Mester, R., Barth, E., Scharr, H. (eds.) IWCM 2004. LNCS, vol. 3417, pp. 14–29. Springer, Heidelberg (2007), doi:10.1007/978-3-540-69866-1_2
45. Soroushmehr, S.M.R., Samavi, S., Shirani, S.: Block matching algorithm based on local codirectionality of blocks. In: IEEE International Conference on Multimedia and Expo., pp. 201–204 (2009), doi:10.1109/ICME.2009.5202471
46. Touil, B., Basarab, A., Delachartre, P., Bernard, O., Friboulet, D.: Analysis of motion tracking in echocardiographic image sequences: Influence of system geometry and point-spread function. Ultrasonics 50(3), 373–386 (2010), doi:10.1016/j.ultras.2009.09.001
47. Wang, X., Tang, Z.: Modified particle filter-based infrared pedestrian tracking. Infrared Physics & Technology 53(4), 280–287 (2010), doi:10.1016/j.infrared.2010.04.002
48. Werlberger, M., Pock, T., Bischof, H.: Motion estimation with non-local total variation regularization. Computer Vision and Pattern Recognition (2010), doi:10.1109/CVPR.2010.5539945
49. Yu, F., Hui, M., Han, W., Wang, P., Dong, L., Zhao, Y.: The application of improved block-matching method and block search method for the image motion estimation. Optics Communications 283(23), 4619–4625 (2010), doi:10.1016/j.optcom.2010.07.006
50. Yupeng, X., Xin, W., Feng, H.: Application of optical flow field for intelligent tracking system. In: International Symposium on Intelligent Information Technology Application (2008), doi:10.1109/IITA.2008.475
51. Zappella, L., Llado, X., Provenzi, E., Salvi, J.: Enhanced Local Subspace Affinity for feature-based motion segmentation. Pattern Recognition 44(2), 454–470 (2011), doi:10.1016/j.patcog.2010.08.015
52. Zhang, W., Fang, X., Yang, X., Wu, Q.M.J.: Spatiotemporal Gaussian mixture model to detect moving objects in dynamic scenes. J. Electron Imaging 16(2) (2007), doi:10.1117/1.2731329
53. Zhang, W., Wua, Q.M.J., Yin, H.: Moving vehicles detection based on adaptive motion histogram. Digital Signal Processing 20(3), 793–805 (2010), doi:10.1016/j.dsp.2009.10.006
54. Zinbi, Y., Chahir, Y., Elmoataz, A.: Moving object segmentation using optical flow with active contour model. In: International Conference on Information and Communication Technologies: From Theory to Applications (2008), doi:10.1109/ICTTA.2008.4530112

Chapter 9
Shape-Based Invariant Feature Extraction for Object Recognition

Mingqiang Yang[1], Kidiyo Kpalma[2], and Joseph Ronsin[2]

[1] ISE, Shandong University, 250100, Jinan, China
[2] Université Européenne de Bretagne, France - INSA, IETR, UMR 6164, F-35708 RENNES
yangmq@sdu.edu.cn,
{kidiyo.kpalma,joseph.ronsin}@insa-rennes.fr

Abstract. The emergence of new technologies enables generating large quantity of digital information including images; this leads to an increasing number of generated digital images. Therefore it appears a necessity for automatic systems for image retrieval. These systems consist of techniques used for query specification and retrieval of images from an image collection. The most frequent and the most common means for image retrieval is the indexing using textual keywords. But for some special application domains and face to the huge quantity of images, keywords are no more sufficient or unpractical. Moreover, images are rich in content; so in order to overcome these mentioned difficulties, some approaches are proposed based on visual features derived directly from the content of the image: these are the content-based image retrieval (CBIR) approaches. They allow users to search the desired image by specifying image queries: a query can be an example, a sketch or visual features (e.g., colour, texture and shape). Once the features have been defined and extracted, the retrieval becomes a task of measuring similarity between image features. An important property of these features is to be invariant under various deformations that the observed image could undergo.

In this chapter, we will present a number of existing methods for CBIR applications. We will also describe some measures that are usually used for similarity measurement. At the end, and as an application example, we present a specific approach, that we are developing, to illustrate the topic by providing experimental results.

9.1 Introduction

Pattern recognition is the ultimate goal of most computer vision research. Shape feature extraction and representation are the bases of object recognition. It is also a research domain which plays an important role in many applications ranging from image analysis and pattern recognition, to computer graphics and computer animation. The feature extraction stage produces a representation of the content that is

R. Kountchev et al. (Eds.): Advances in Reasoning-Based Image Processing, ISRL 29, pp. 255–314.
springerlink.com

useful for shape matching. Usually the shape representation is kept as compact as possible for the purposes of efficient storage and retrieval and it integrates perceptual features that allow the human brain to discriminate between shapes. Efficient shape features must present some essential properties such as:

- **identifiability:** shapes which are found perceptually similar by human have the same features but different from the others,
- **translation, rotation and scale invariance:** the location, rotation and scaling changing of the shape must not affect the extracted features,
- **affine invariance:** the affine transform performs a linear mapping from 2D coordinates to other 2D coordinates that preserves the "straightness" and "parallelism" of lines. Affine transform can be constructed using sequences of translations, scales, flips, rotations and shears. The extracted features must be as invariant as possible with affine transforms.
- **noise resistance:** features must be as robust as possible against noise, i.e. they must be the same, in a given range, whichever be the strength of the noise that affects the pattern,
- **occultation resistance:** when some parts of a shape are occulted by other objects, the feature of the remaining part must not change, in a given range, compared to the original shape,
- **statistical independence:** two features must be statistically independent. This represents compactness of the representation,
- **reliability**: as long as one deals with the same pattern, the extracted features must remain the same.

In general, shape descriptor is some set of numbers that are produced to describe a given shape feature. A descriptor attempts to quantify shape in ways that agree with human intuition (or task-specific requirements). Good retrieval accuracy requires a shape descriptor to be able to effectively find perceptually similar shapes from a database. Usually, the descriptors are gathered under the form of a vector. Shape descriptors should meet the following requirements:

- **completeness:** the descriptors should be as complete as possible to represent the content of the information items,
- **compactness:** the descriptors should be represented and stored compactly. The size of descriptor vector must not be too large,
- **simplicity:** the computation of distance between descriptors should be simple; otherwise the execution time would be too long,
- **accessibility:** it describes how easy (or difficult) it is to compute a shape descriptor in terms of memory requirements and computation time,
- **large scope:** it indicates the extent of the class of shapes that can be described by the method,
- **uniqueness:** it indicates whether a one-to-one mapping exists between shapes and shape descriptors,
- **stability:** this describes how stable a shape descriptor is to “small” changes in shape.

Shape feature extraction and representation plays an important role in the following categories of applications:

- **shape retrieval:** searching for all shapes in a typically large database of shapes that are similar to a query shape. Usually all shapes within a given distance from the query are determined or at least the first few shapes that have the smallest distance.
- **shape recognition and classification:** determining whether a given shape matches a model sufficiently, or which one of representative class is the most similar,
- **shape alignment and registration:** transforming or translating one shape so that it best matches another shape, in whole or in part,
- **shape approximation and simplification:** constructing a shape from fewer elements (points, segments, triangles, etc.), that is still similar to the original.

To this end, many shape description and similarity measurement techniques have been developed in the past. A number of new techniques have been proposed in recent years, leading to three main classification methods:

- **contour-based methods and region-based methods:** this is the most common and general classification and it is proposed by MPEG-7 which is a multimedia content description standard. It is based on the use of shape boundary points as opposed to shape interior points. Under each class, different methods are further divided into structural approaches and global approaches. This sub-class is based on whether the shape is represented as a whole or represented by segments/sections (primitives).
- **space domain and feature domain:** methods in space domain match shapes on point (or point feature) basis, while feature domain techniques match shapes on feature (vector) basis.
- **information preserving (IP) and non-information preserving (NIP):** IP methods allow an accurate reconstruction of a shape from its descriptor, while NIP methods are only capable of partial ambiguous reconstruction. For object recognition purpose, IP is not a requirement.

Various algorithms and methods are documented in a vast literature. In this chapter, for sake of application conveniences, we reclassify them according to the **processing methods** i.e. the way the data of the shape are mathematically modelled and processed. The whole hierarchy of the classification is shown in figure 9.1.

Without being complete, we will describe and group a number of these methods together. So this chapter is organized as follows: section 2 presents 1D functions used in shape description. Section 3 presents some approaches for polygonal approximation of contours. Section 4 is dedicated to spatial interrelation features and section 5 presents shape moments. Sections 6 and 7 are, respectively, dedicated to scale space features and transform domain feature. Section 8 presents a summary table showing the properties of the methods. In order to illustrate this study, a practical example, based on a new shape descriptor, is presented in section 9.

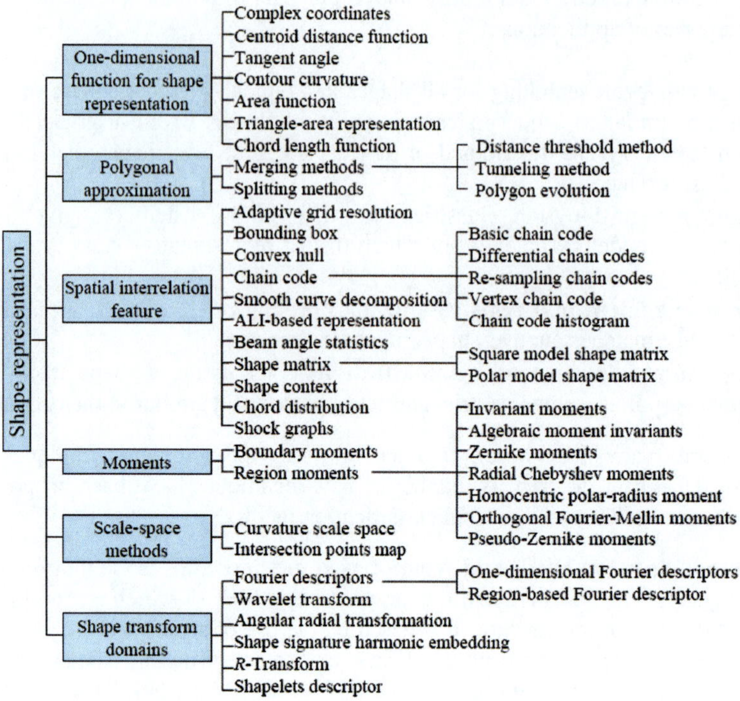

Fig. 9.1 An overview of shape description techniques

9.2 One-Dimensional Function for Shape Representation

The one-dimensional function which is derived from shape boundary coordinates is also often called shape signature [22, 53]. The shape signature usually captures the perceptual feature of the shape [48]. Complex coordinates, centroid distance function, tangent angle (turning angles), curvature function, area function, triangle-area representation and chord length function are the commonly used shape signatures.

Shape signature can describe a shape all alone; it is also often used as a preprocessing to other feature extraction algorithms, for example, Fourier descriptors, wavelet description. In this section, the shape signatures are introduced.

9.2.1 Complex Coordinates

A complex coordinates function is simply the complex number generated from the coordinates of boundary points, $P_n(x(n),y(n))$, $n\in[1,N]$:

$$z(n) = [x(n) - g_x] + i[y(n) - g_y] \tag{9.1}$$

where (g_x, g_y) is the centroid of the shape.

9.2.2 Centroid Distance Function

The centroid distance function r(n) is expressed by the distance of the boundary points from the centroid (g_x, g_y) of a shape, so that

$$r(n) = \sqrt{(x(n) - g_x)^2 + (y(n) - g_y)^2} \quad (9.2)$$

Due to the subtraction of centroid, which represents the position of the shape, from boundary coordinates, both complex coordinates and centroid distance representation are invariant to translation.

9.2.3 Tangent Angle

The tangent angle function at a point $P_n(x(n); y(n))$ is defined by a tangential direction of a contour [54]:

$$\theta(n) = \theta_n = \arctan\frac{y(n) - y(n-\omega)}{x(n) - x(n-\omega)} \quad (9.3)$$

where ω represents a small window to calculate $\theta(n)$ more accurately, since every contour is a digital curve .

Tangent angle function has two problems. One is noise sensitivity. To decrease the effect of noise, the contour is filtered by a low-pass filter with appropriate bandwidth before calculating the tangent angle function. The other is discontinuity, due to the fact that the tangent angle function assumes values in a range of length 2π, usually in the interval of $[-\pi,\pi]$ or $[0,2\pi]$. Therefore θ_n in general contains discontinuities of size 2π. To overcome the discontinuity problem, with an arbitrary starting point, the cumulative angular function φ_n is defined as the angle differences between the tangent at any point P_n along the curve and the tangent at the starting point P_0 [30, 50]:

$$\varphi(n)=[\theta(n)-\theta(0)] \quad (9.4)$$

In order to be in accordance with human intuition that a circle is "shapeless", assume $t=2\pi n/N$, then $\varphi(n)=\varphi(tN/2\pi)$. A periodic function is termed as the cumulative angular deviant function $\psi(t)$ and is defined as

$$y(t) = j(\frac{N}{2\pi}t) - t, \ t \in [0,2\pi] \quad (9.5)$$

where N is the total number of contour points.

In [25], the authors proposed a method based on tangent angle. It is called tangent space representation. A digital curve C simplified by polygon evolution is represented in the tangent space by the graph of a step function, where the x-axis represents the arc length coordinates of points in C and the y-axis represents the direction of the line segments in the decomposition of C. For example, figure 9.2 shows a digital curve and its step function representation in the tangent space.

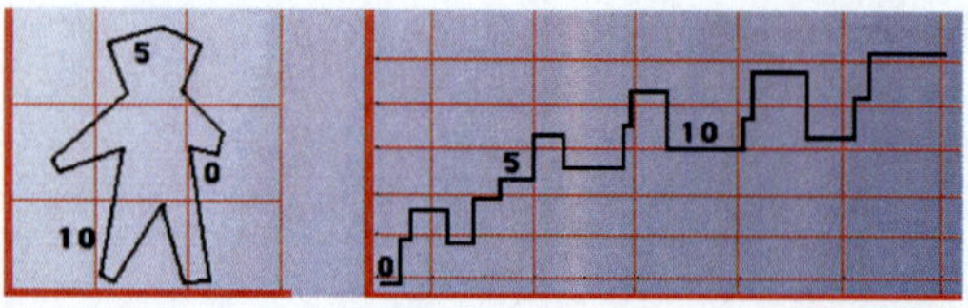

Fig. 9.2 Digital curve and its step function representation in the tangent space

9.2.4 Contour Curvature

Curvature is a very important boundary feature for human being to judge similarity between shapes. It also has salient perceptual characteristics and has proven to be very useful for shape recognition [47]. In order to use *K(n)* for shape representation, we quote the curvature function, *K(n)*, from [19, 32] as:

$$K(n) = \frac{\dot{x}(n)\ddot{y}(n) - \dot{y}(n)\ddot{x}(n)}{(\dot{x}(n)^2 - \dot{y}(n)^2)^{3/2}} \tag{9.6}$$

where $\dot{x}$ (or $\dot{y}$) and $\ddot{x}$ (or $\ddot{y}$) are, respectively, the first and second order derivatives of x (or y). Therefore, it is possible to compute the curvature of a planar curve from its parametric representation. If *n* is the normalized arc-length parameter *s*, then equation (9.6) can be written as:

$$K(s) = \dot{x}(s)\ddot{y}(s) - \dot{y}(s)\ddot{x}(s) \tag{9.7}$$

As given in equation (9.7), the curvature function is computed only from parametric derivatives, and, therefore, it is invariant under rotations and translations. However, the curvature measure is scale dependent, i.e., inversely proportional to the scale. A possible way to achieve scale independence is to normalize this measure by the mean absolute curvature, i.e.,

$$K'(s) = \frac{K(s)}{\frac{1}{N}\sum_{s=1}^{N}|K(s)|} \tag{9.8}$$

where *N* is the number of points on the normalized contour.

When the size of the curve is an important discriminative feature, the curvature should be used without the normalization; otherwise, for the purpose of scale-invariant shape analysis, the normalization should be performed by the following algorithm.

Let $P = \sum_{n=1}^{N} d_n$ be the perimeter of the curve and $L = \sum_{n=1}^{N} \sqrt{d_n}$, where d_n is the length of the chord between points p_n and p_{n+1}, $n=1, 2, \ldots, N-1$. An approximate arc-length parameterization based on the centripetal method is given by the following [19]:

$$s_k = s_{k-1} + \frac{P\sqrt{d_{k-1}}}{L}, k = 2,3,...,N \quad (9.9)$$

with s_1=0. Starting from an arbitrary point and following the contour clockwise, we compute the curvature at each interpolated point using equation (9.7). Figure 9.3 is an example of curvature function. Clearly, as a descriptor, the curvature function can distinguish different shapes.

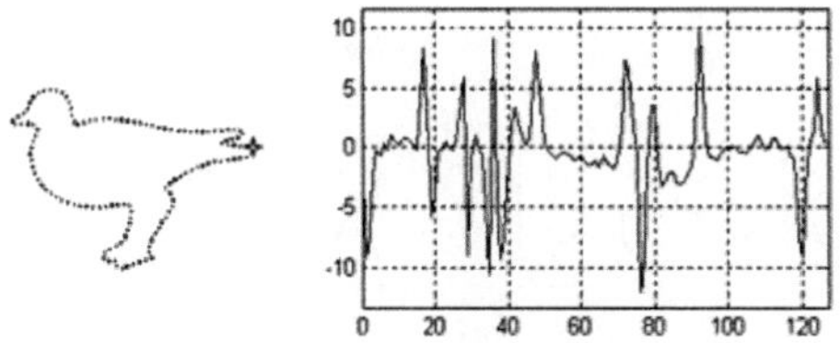

(a) (b)

(a) Contours normalized to 128 points; the dots marked star are the starting points on the contours; (b) curvature functions; the curvature is computed clockwise.

Fig. 9.3 Curvature function

Convex and concave vertices will imply negative and positive values, respectively (the opposite is verified for counter clockwise sense).

9.2.5 Area Function

When the boundary points change along the shape boundary, the area of the triangle formed by two successive boundary points and the centre of gravity also changes. This forms an area function which can be exploited as shape representation. Figure 9.4 shows an example where *S*(*n*) is the area between the successive boundary points P_n, P_{n+1} and centre of gravity *G*.

The area function is linear under affine transform. However, this linearity only works for shape sampled at its same vertices.

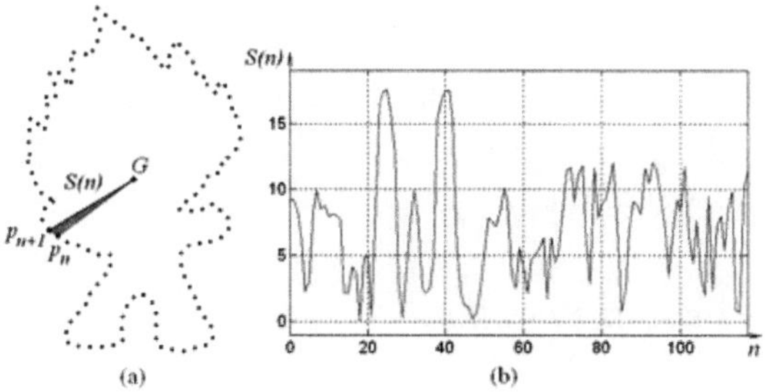

(a) (b)

(a) Original contour; (b) the area function of (a).

Fig. 9.4 Area function

9.2.6 Triangle-Area Representation

The triangle-area representation (TAR) signature is computed from the area of the triangles formed by the points on the shape boundary [2, 3]. The curvature of the contour point (x_n, y_n) is measured using the *TAR* function defined as follows:

For each three consecutive points $P_{n-t_s}(x_{n-t_s}, y_{n-t_s})$, $P_n(x_n, y_n)$, and $P_{n+t_s}(x_{n+t_s}, y_{n+t_s})$, where $n \in [1, N]$ and $t_s \in [1, N/2-1]$, N is even the signed area of the triangle formed by these points is given by:

$$TAR(n,t_s) = \frac{1}{2}\begin{vmatrix} x_{n-t_s} & y_{n-t_s} & 1 \\ x_n & y_n & 1 \\ x_{n+t_s} & y_{n+t_s} & 1 \end{vmatrix} \tag{9.10}$$

when the contour is traversed in counter clockwise direction, positive, negative and zero values of TAR mean convex, concave and straight-line points, respectively. Figure 9.5 shows these three types of the triangle areas and the complete TAR signature for the hammer shape.

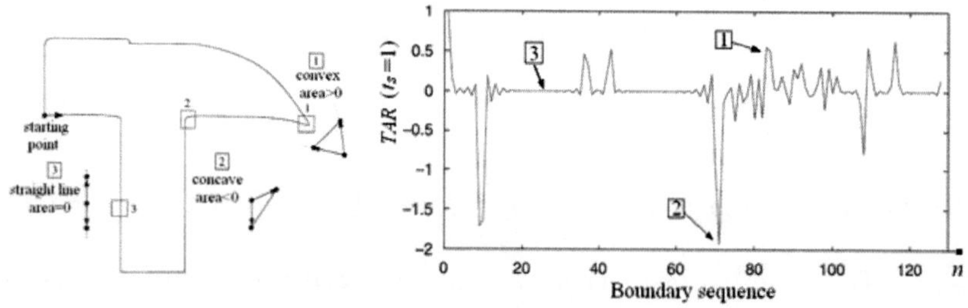

Fig. 9.5 Three different types of the triangle-area values and the TAR signature for the hammer shape

By increasing the length of the triangle sides, i.e., considering farther points, the equation 9.10 will represent longer variations along the contour. The TARs with different triangle sides can be regarded as different scale space functions. The total TARs, $t_s \in [1, N/2-1]$, compose a multi-scale space TAR. In [3], authors show that the multi-scale space TAR is relatively invariant to the affine transform and robust to non-rigid transform.

9.2.7 Chord Length Function

The chord length function is derived from shape boundary without using any reference point. For each boundary point p, its chord length function is the shortest distance between p and another boundary point p' such that line pp' is perpendicular to the tangent vector at p [53].

The chord length function is invariant to translation and it overcomes the biased reference point (which means the centroid is often biased by boundary noise or defections) problems. However, it is very sensitive to noise, there may be drastic burst in the signature of even smoothed shape boundary.

9.2.8 Discussions

A shape signature represents a shape by a 1-D function derived from shape contour. To obtain the translation invariant property, they are usually defined by relative values. To obtain the scale invariant property, normalization is necessary. In order to compensate for orientation changes, shift matching is needed to find the best matching between two shapes. Having regard to occultation, Tangent angle, Contour curvature and Triangle-area representation have invariance property. In addition, shape signatures are computationally simple.

Shape signatures are sensitive to noise, and slight changes in the boundary can cause large errors in matching. Therefore, it is undesirable to directly describe shape using a shape signature. Further processing is necessary to increase its robustness and reduce the matching load. For example, a shape signature can be simplified by quantizing the signature into a signature histogram, which is rotationally invariant.

9.3 Polygonal Approximation

Polygonal approximation can be set to ignore the minor variations along the edge, and instead capture the overall shape. This is useful because it reduces the effects of discrete pixelization of the contour. In general, there are two methods to realize it. One is merging, the other is splitting [18].

9.3.1 Merging Methods

Merging methods add successive pixels to a line segment if each new pixel that is added does not cause the segment to deviate too much from a straight line.

9.3.1.1 Distance Threshold Method

Choose one point as a starting point, on the contour. For each new point that we add, let a line go from the starting point to this new point. Then, we compute the squared error for every point along the segment/line. If the error exceeds some threshold, we keep the line from the start point to the previous point and start a new line. In practice, the most of practical error measures in use are based on distance between vertices of the input curve and the approximated linear segment [62]. The distance $d_k(i,j)$ from curve vertex $P_k(x_k,y_k)$ to the corresponding approximated linear segment defined by $P_i(x_i,y_i)$ and $P_j(x_j,y_j)$ is as follows (and illustrated in figure 9.6):

$$d_k(i,j)=\frac{\left|(x_j-x_i)(y_i-y_k)-(x_i-x_k)(y_j-y_i)\right|}{\sqrt{(x_j-x_i)^2+(y_j-y_i)^2}} \tag{9.11}$$

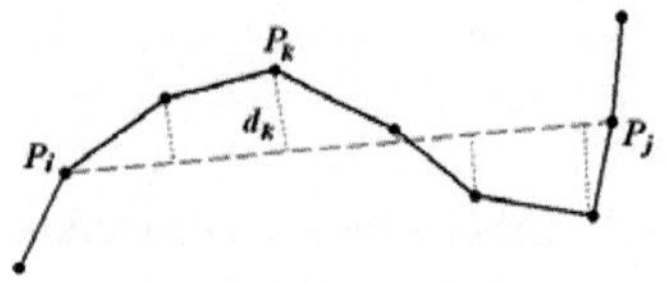

Fig. 9.6 Illustration of the distance from a point on the boundary to a linear segment

9.3.1.2 Tunnelling Method

If we have thick boundaries rather than single-pixel thick ones, we can still use a similar approach called tunnelling. Imagine that we are trying to lay straight rods along a curved tunnel, and that we want to use as few as possible. We can start at one point and lay a straight rod as long as possible. Eventually, the curvature of the "tunnel" won't let us go any further, so we lay one rod after another until we reach the end.

Both the distance threshold and tunnelling methods efficiently can do polygonal approximation. However, the great disadvantage is that the position of starting point will affect greatly the approximate polygon.

9.3.1.3 Polygon Evolution

The basic idea of polygons evolution presented in [26] is very simple: in every evolution step, a pair of consecutive line segments (the line segment is the line between two consecutive vertices) s_1 and s_2 is substituted with a single line segment joining the endpoints of s_1 and s_2. The key property of this evolution is the order of the substitution. The substitution is done according to a relevance measure K given by

$$K(s_1,s_2)=\frac{\beta(s_1,s_2)l(s_1)l(s_2)}{l(s_1)+l(s_2)} \tag{9.12}$$

where $\beta(s_1,s_2)$ is the turn angle at the common vertex of segments s_1, s_2 and $l(\alpha)$ is the length of α, α=s_1 or s_2, normalized with respect to the total length of a polygonal curve. The evolution algorithm assumes that vertices which are surrounded by segments with high values of $K(s_1,s_2)$ are more important than those with a low values (see figure 9.7 for illustration).

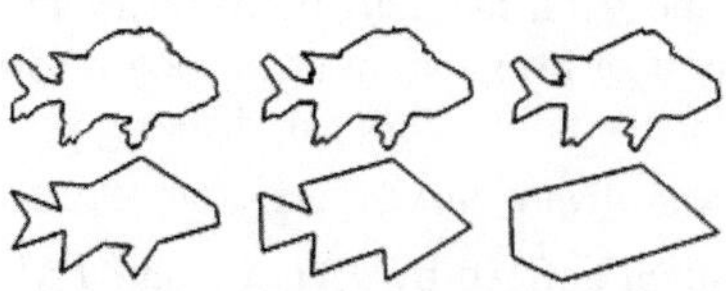

Fig. 9.7 A few stages of polygon evolution according to a relevant measure

The curve evolution method achieves the task of shape simplification, i.e., the process of evolution compares the significance of vertices of the contour based on a relevance measure. Since any digital curve can be seen as a polygon without loss of information (with possibly a large number of vertices), it is sufficient to study evolutions of polygonal shapes for shape feature extraction.

9.3.2 Splitting Methods

Splitting methods work by first drawing a line from one point on the boundary to another. Then, we compute the perpendicular distance from each point along the boundary segment to the line. If this exceeds some threshold, we break the line at the point of greatest distance. We then repeat the process recursively for each of the two new lines until we don't need to break any more. See figure 9.8 for an example.

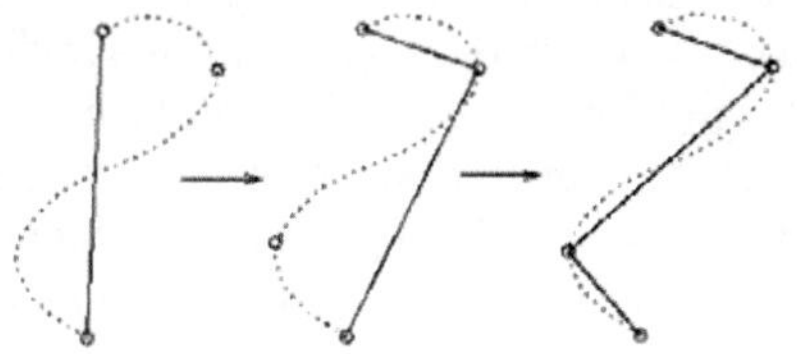

Fig. 9.8 Splitting methods for polygonal approximation

This is sometimes known as the "fit and split" algorithm. For a closed contour, we can find the two points that lie farthest apart and fit two lines between them, one for one side and one for the other. Then, we can apply the recursive splitting procedure to each side.

9.3.3 Discussions

Polygonal approximation technique can be used as a simple method for contour representation and description. The polygon approximation has some interesting properties:

- it leads to simplification of shape complexity with no blurring effects,
- it leads to noise elimination,
- although irrelevant features vanish after polygonal approximation, there is no dislocation of relevant features,
- the remaining vertices on a contour do not change their positions after polygonal approximation.

Polygonal approximation technique can also be used as pre-processing method for further features extracting methods from a shape.

9.4 Spatial Interrelation Feature

Spatial interrelation feature describes the region or the contour of shapes by observing and featuring the relations between their pixels or curves. In general, the representation is done by observing their geometric features: length, curvature, relative orientation and location, area, distance and so on.

9.4.1 Adaptive Grid Resolution

The adaptive grid resolution (AGR) scheme was proposed by [11]. In the AGR, a square grid that is just big enough to cover the entire shape is overlaid on it. A resolution of the grid cells varies from one portion to another according to the content of the portion of the shape. On the borders or the detail portion on the shape, the highest resolution, i.e. the smallest grid cells, are applied; on the other hand, in the coarse regions of the shape, lower resolution, i.e. the biggest grid cells, are applied.

To guarantee rotation invariance, it needs to reorient the shape into a unique common orientation. First, one has to find the major axis of the shape. The major axis defined as is the straight line segment joining the two points on the boundary farthest away from each other. Then rotate the shape so that its major axis is parallel to the *x*-axis.

One method to compute the AGR representation of a shape relies on a quad-tree decomposition on the bitmap representation of the shape [11]. The decomposition is based on successive subdivision of the bitmap into four equal-sized quadrants. If a bitmap-quadrant does not consist entirely of part of shape, it is recursively subdivided into smaller quadrants until we reach bitmap-quadrants, i.e., termination condition of the recursion is that the resolution reaches that one predefined: figure 9.9(a) shows an example of AGR.

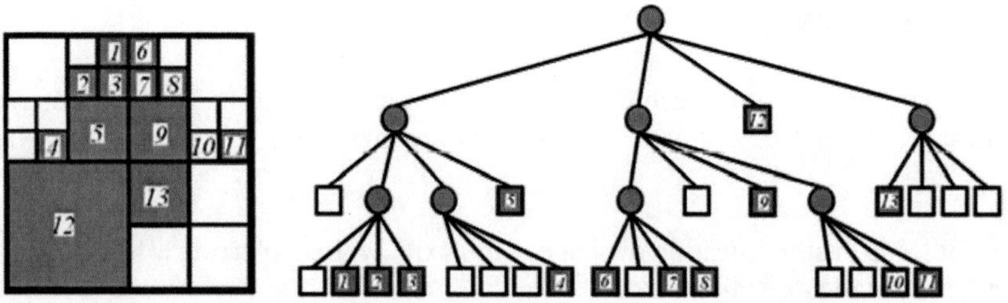

(a) (b)

(a) Adaptive Grid Resolution (AGR) image; (b) quad-tree decomposition of AGR.

Fig. 9.9 Adaptive resolution representations

Each node in the quad-tree covers a square region of the bitmap. The level of the node in the quad-tree determines the size of the square. The internal nodes (shown by grey circles) represent "partially covered" regions; the leaf nodes shown by white boxes represent regions with all 0*s* while the leaf nodes shown by black boxes represent regions with all 1*s*. The "all 1*s*" regions are used to represent the shape as shown on figure 9.9(b). Each rectangle can be described by 3 numbers: its centre coordinates $C=(C_x, C_y)$ and its size (i.e. side length) *S*. So each shape can be mapped to a point in 3*n*-dimensional space, where *n* is the number of the rectangles occupied by the shape region. Due to prior normalization, AGR representation is invariant under rotation, scaling and translation. It is also computationally simple.

9.4.2 Bounding Box

Bounding box computes homeomorphisms between 2D lattices and its shapes. Unlike many other methods, this mapping is not restricted to simply connected shapes but applies to arbitrary topologies [7].

The minimum bounding rectangle or bounding box of *S* is denoted by *B(S)*; its width and height, are called *w* and *h*, respectively. An illustration of this procedure and its result is shown in figure 9.10.

(a) Compute the bounding box B(S) of a pixel set S; (b) subdivide S into n vertical slices; (c) compute the bounding box B(Sj) of each resulting pixel set S_j, where j=1, 2,..., n; (d) subdivide each B(Sj) into m horizontal slices; (e) compute the bounding box $B(S_{ij})$ of each resulting pixel set Sij, where i = 1, 2,..., m.

Fig. 9.10 The five steps of bounding box splitting

Figure 9.11 shows the algorithm flowchart based on bounding box that divides a shape S into m (row)×n (column) parts. The output B is a set of bounding boxes.

If $v=(v_x, v_y)^T$ denotes the location of the bottom left corner of the initial bounding box of *S*, and $u_{ij}=(u_x^{ij}, u_y^{ij})^T$ denotes the centre of sample box $\mathbf{B}_{ij}$, then the coordinates

$$\begin{pmatrix} \mu_x^{ij} \\ \mu_y^{ij} \end{pmatrix} = \begin{pmatrix} (u_x^{ij} - v_x)/w \\ (u_y^{ij} - v_y)/h \end{pmatrix} \tag{9.13}$$

provide a scale invariant representation of *S*. Sampling *k* points of an *m*×*n* lattice therefore allows to represent *S* as a vector

$$r = [\mu_x^{i(1)j(1)}, \mu_y^{i(1)j(1)}, \ldots, \mu_x^{i(k)j(k)}, \mu_y^{i(k)j(k)}] \tag{9.14}$$

where $i(\alpha)<i(\beta)$ if $\alpha<\beta$ and likewise for the index *j*.

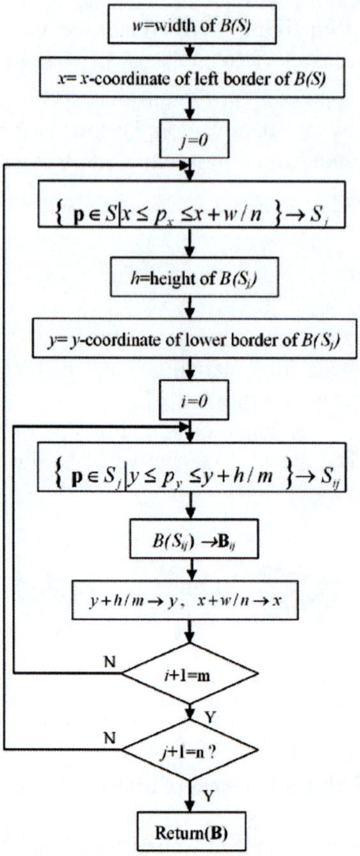

Fig. 9.11 Flowchart of shape divided by bounding box

To represent each bounding box, one method consists of sampling partial points of the set of bounding boxes (see figure 9.12).

Fig. 9.12 A sample points on lattice and examples of how it is mapped onto different shapes

Bounding box representation is a simple computational geometry approach to compute homeomorphisms between shapes and lattices. It is storage and time efficient. It is invariant to rotation, scaling and translation and also robust against noisy shape boundaries.

9.4.3 Convex Hull

The approach is based on the fact that the shape is represented by a series of convex hulls. The convex hull *H* of a region consists of its smallest convex region including it. In other words, for a region *S*, the convex hull *conv*(*S*) is defined as the smallest convex set in R^2 containing *S*. In order to decrease the effect of noise, common practice is to first smooth a boundary prior to partitioning it.

The representation of the shape may then be obtained by a recursive process which results in a concavity tree (see figure 9.13). Each concavity can be described by its area, chord (the line connects the cut of the concavity) length, maximum curvature, distance from maximum curvature point to the chord. The matching between shapes becomes a string or a graph matching.

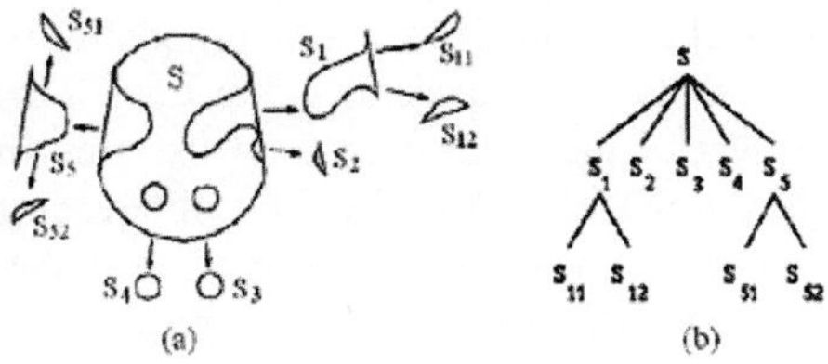

(a) Convex hull and its concavities; (b) concavity tree representation of convex hull.

Fig. 9.13 Illustration of recursive process of convex hull

Convex hull representation has a high storage efficiency. It is invariant to rotation, scaling and translation and also robust against noisy shape boundaries (after filtering). However, extracting the robust convex hulls from the shape is where the shoe pinches. [14, 16] and [41] gave the boundary tracing method and morphological methods to achieve convex hulls respectively.

9.4.4 Chain Code

Chain code is a common approach for representing different rasterized shapes as line-drawings, planar curves, or contours. Chain code describes an object by a sequence of unit-size line segments with a given orientation [51]. Chain code can be viewed as a connected sequence of straight-line segments with specified lengths and directions [28].

9.4.4.1 Basic Chain Code

Freeman [57] first introduced a chain code that describes the movement along a digital curve or a sequence of border pixels by using so-called 8-connectivity or 4-connectivity. The direction of each movement is encoded by the numbering scheme i=0, 1, …,7 or i=0, 1, 2, 3 denoting a counter-clockwise angle of $45^{0}\times i$ or $90^{0}\times i$ regarding the positive *x*-axis, as shown in figure 9.14.

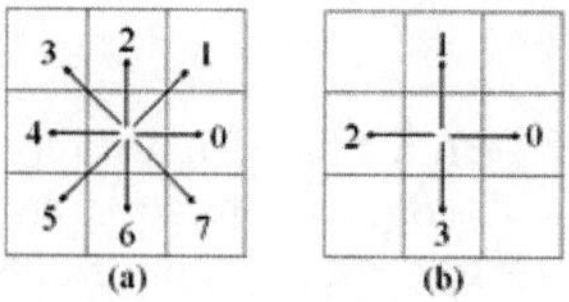

(a) Chain code in eight directions (8-connectivity); (b) chain code in four directions (4-connectivity).

Fig. 9.14 Basic chain code direction

By encoding relative, rather than absolute position of the contour, the basic chain code is translation invariant. We can match boundaries by comparing their chain codes, but with the two main problems: 1) it is very sensitive to noise; 2) it is not rotationally invariant. To solve these problems, differential chain codes (DCC) and resampling chain codes (RCC) were proposed.

DCC encodes differences in the successive directions. This can be computed by subtracting each element of the chain code from the previous one and taking the result modulo n, where n is the connectivity. This differencing process allows us to rotate the object in 90-degree increments and still compare the objects, but it doesn't get around the inherent sensitivity of chain codes to rotation on the discrete pixel grid.

RCC consists of re-sampling the boundary onto a coarser grid and then computing the chain codes of this coarser representation. This smoothes out small variations and noise but can help compensate for differences in chain-code length due to the pixel grid.

9.4.4.2 Vertex Chain Code (VCC)

To improve chain code efficiency, in [28] the authors proposed a chain code for shape representation according to VCC. An element of the VCC indicates the number of cell vertices, which are in touch with the bounding contour of the shape in that element's position. Only three elements "1", "2" and "3" can be used to represent the bounding contour of a shape composed of pixels in the rectangular grid. Figure 9.15 shows the elements of the VCC to represent a shape.

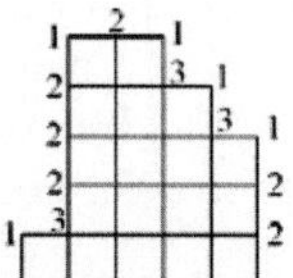

Fig. 9.15 Vertex chain code

9.4.4.3 Chain Code Histogram (CCH)

Iivarinen and Visa have derived a CCH for object recognition [58]. The CCH is computed as h_i=#{$i \in M$, M is the range of chain code}, #{α} denotes getting the number of the value α. The CCH reflects the probabilities of different directions present in a contour. If the chain code is used for matching it must be independent of the choice of the starting pixel in the sequence. The chain code usually has high dimensions and is sensitive to noise and any distortion. So, except for the CCH, the other chain code approaches are often used as contour representations, but not as contour attributes.

9.4.5 Smooth Curve Decomposition

In [9], the authors proposed smooth curve decomposition as shape descriptor. The segment between the curvature zero-crossing points from a Gaussian smoothed boundary are used to obtain primitives, called tokens. The feature for each token corresponds to its maximum curvature and its orientation. In figure 9.16, the first number in the parentheses is its maximum curvature and the second is its orientation.

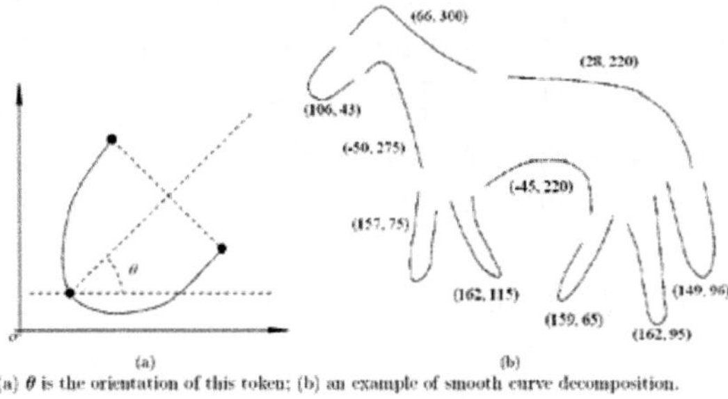

Fig. 9.16 Smooth curve decomposition

The similarity between two tokens is measured by the weighted Euclidean distance. The shape similarity is measured according to a non-metric distance. Shape retrieval based on token representation has shown to be robust in the presence of partially occulted objects, translation, scaling and rotation.

9.4.6 Symbolic Representation Based on the Axis of Least Inertia

In [17], a method of representing a shape in terms of multi-interval valued type data is proposed. The proposed shape representation scheme extracts symbolic features with reference to the axis of least inertia, which is unique to the shape. The axis of least inertia (ALI) of a shape is defined as the line for which the integral of the square of the distances to points on the shape boundary is a minimum.
Once the ALI is calculated, each point on the shape curve is projected on to ALI. The two farthest projected points say $E1$ and $E2$ on ALI are chosen as the extreme points as shown in figure 9.17. The Euclidean distance between these two extreme points defines the length of ALI. The length of ALI is divided uniformly by a fixed number n; the equidistant points are called feature points. At every feature point chosen, an imaginary line perpendicular to the ALI is drawn. It is interesting to note that these perpendicular lines may intersect the shape curve at several

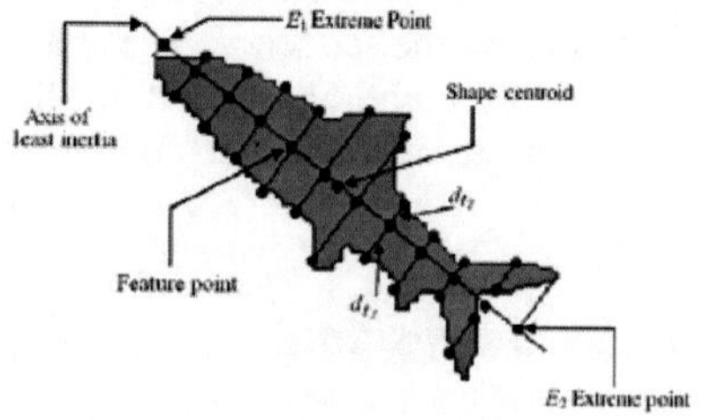

Fig. 9.17 Symbolic features based axis of least inertia

points. The length of each imaginary line in shape region is computed and the collection of these lengths in an ascending order defines the value of the feature at the respective feature point.

Let *S* be a shape to be represented and *n* the number of feature points chosen on its ALI. Then the feature vector *F* representing the shape *S*, is in general of the form $F=[\mathbf{f}_1,\mathbf{f}_2,...,\mathbf{f}_t,...,\mathbf{f}_n]$, where $f_t=\{d_{t_1},d_{t_2},\cdots,d_{t_k}\}$ for some $t_k \geq 1$.

The feature vector *F* representing the shape *S* is then invariant to image transformations viz., uniform scaling, rotation, translation and flipping (reflection).

9.4.7 Beam Angle Statistics

Beam angle statistics (BAS) shape descriptor is based on the beams originated from a boundary point, which are defined as lines connecting that point with the rest of the points on the boundary [5].

Let *B* be the shape boundary. $B=\{P_1,P_2,\cdots,P_N\}$ is represented by a connected sequence of points, $P_i=(x_i, y_i)$, $i=1,2,\cdots,N$, where *N* is the number of boundary points. For each point P_i, the beam angle between the forward beam vector $V_{i+k}=\overrightarrow{P_iP_{i+k}}$ and backward beam vector $V_{i-k}=\overrightarrow{P_iP_{i-k}}$ in the kth order neighbourhood system, is then computed as (see figure 9.18, *k*=5 for example)

$$C_k(i)=\theta_{V_{i+k}}-\theta_{V_{i-k}} \tag{9.15}$$

where $\theta_{V_{i+k}}=\arctan\frac{y_{i+k}-y_i}{x_{i+k}-x_i}$, $\theta_{V_{i-k}}=\arctan\frac{y_{i-k}-y_i}{x_{i-k}-x_i}$

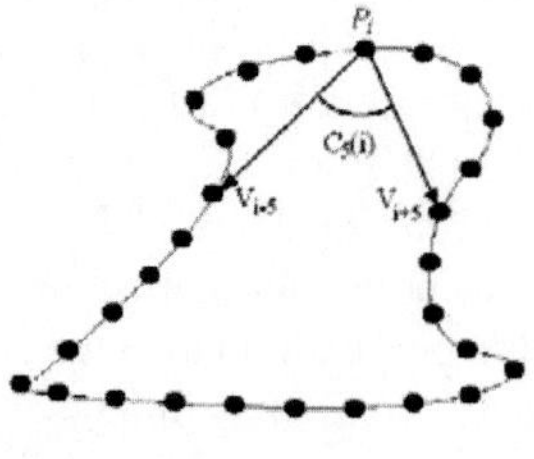

Fig. 9.18 Beam angle at the neighbourhood system 5 for a boundary point

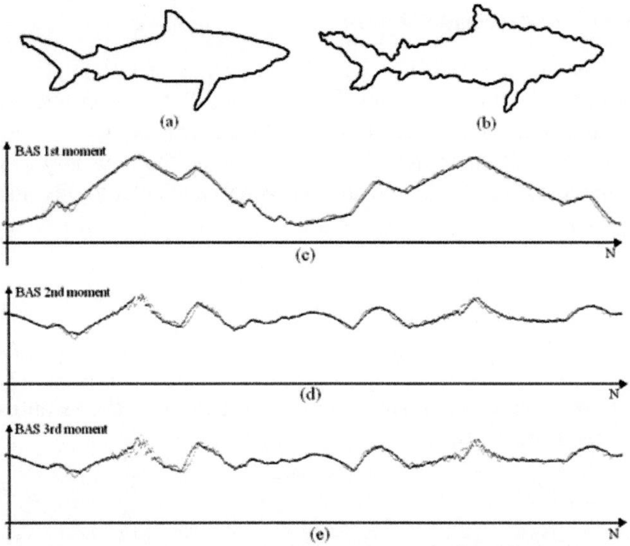

(a) Original contour; (b) noisy contour; (c), (d) and (e) are the BAS plot 1st, 2nd and 3rd moment, respectively.

Fig. 9.19 The BAS descriptor for original and noisy contour

For each boundary point P_i of the contour, the beam angle $C_k(i)$ can be taken as a random variable with the probability density function $P(C_k(i))$. Therefore, beam angle statistics (BAS), may provide a compact representation for a shape descriptor. For this purpose, m^{th} moment of the random variable $C_k(i)$ is defined as follows:

$$E[(C(i))^m] = \sum_{k=1}^{(N/2)-1} (C_k(i))^m \cdot P_k(C_k(i)), \quad m=1, 2, \cdots \tag{9.16}$$

In the above formula E indicates the expected value. Figure 9.19 shows an example of this descriptor.

Beam angle statistics shape descriptor captures the perceptual information using the statistical information based on the beams of individual points. It gives globally discriminative features to each boundary point by using all other boundary points. BAS descriptor is also quite stable under distortions and is invariant to translation, rotation and scaling.

9.4.8 Shape Matrix

Shape matrix descriptor requires an $M \times N$ matrix to present a region shape. There are two basic modes of shape matrix: Square model [59] and Polar model [44].

9.4.8.1 Square Model Shape Matrix

Square model of shape matrix, also called grid descriptor [29, 59], is constructed according to the following algorithm: for the shape *S*, construct a square centred on the centre of gravity *G* of *S*. The size of each side is equal to 2*L*, where *L* is the maximum Euclidean distance from *G* to a point *M* on the boundary of the shape. Point *M* lies in the centre of one side and *GM* is perpendicular to this side.

Divide the square into *N*×*N* subsquares and denote S_{kj}, $k,j=1,\cdots,N$, the sub-squares of the grid. Define the shape matrix $SM=[B_{kj}]$,

$$B_{kj} = \begin{cases} 1 \Leftrightarrow \mu(S_{kj} \cap S) \geq \mu(S_{kj})/2 \\ 0 \text{ otherwise} \end{cases} \tag{9.17}$$

where μ(*F*) is the area of the planar region *F*. Figure 9.20 shows an example of square model of shape matrix.

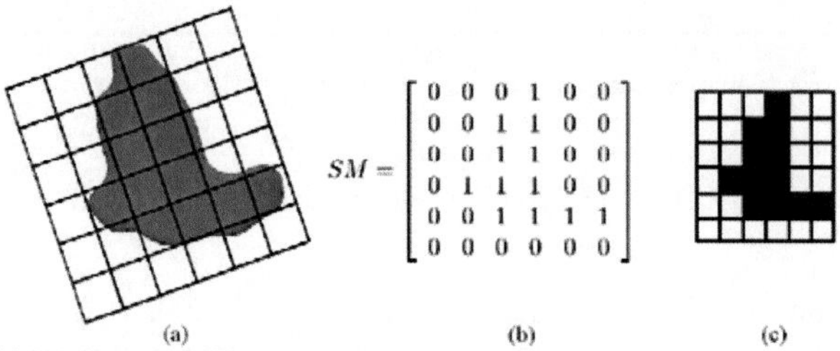

(a) Original shape region; (b) square model shape matrix; (c) reconstruction of the shape region.

Fig. 9.20 Square model shape matrix

For a shape with more than one maximum radius, it can be described by several shape matrices and the similarity distance is the minimum distance between these matrices. In [59], authors gave a method to choose the appropriate shape matrix dimension.

9.4.8.2 Polar Model Shape Matrix

Polar model of shape matrix is constructed by the following steps. Let *G* be the centre of gravity of the shape, and *GA* be the maximum radius of the shape. Using *G* as centre, draw *n* circles with radii equally spaced. Starting from *GA*, and counter clockwise, draw radii that divide each circle into *m* equal arcs. The values of the matrix are the same as those in square model shape matrix. Figure 9.21 shows an example, where n = 5 and m =12. Its polar model of shape matrix is

$$PSM = \begin{bmatrix} 1 & 1 & 1 & 1 & 1 & 1 & 1 & 1 & 1 & 1 & 1 & 1 \\ 1 & 1 & 1 & 1 & 1 & 1 & 1 & 1 & 1 & 1 & 1 & 1 \\ 1 & 1 & 1 & 1 & 1 & 1 & 1 & 1 & 0 & 1 & 1 & 1 \\ 1 & 0 & 0 & 0 & 0 & 1 & 1 & 0 & 0 & 0 & 0 & 1 \\ 1 & 0 & 0 & 0 & 0 & 1 & 1 & 0 & 0 & 0 & 0 & 0 \end{bmatrix} \tag{9.18}$$

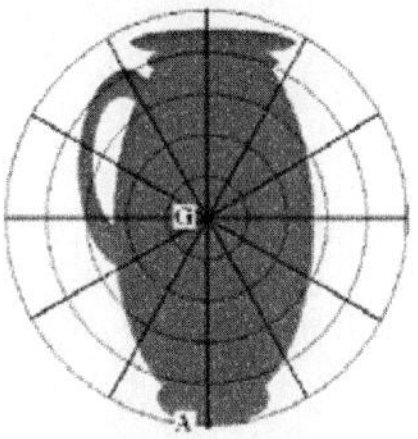

Fig. 9.21 Polar model shape

Polar model of shape matrix is simpler than square model because it only uses one matrix no matter how many maximum radii are on the shape. However, since the sampling density is not constant with the polar sampling raster, a weighed shape matrix is necessary. For the detail, refer to [44].

The shape matrix exists for every compact shape. There is no limit to the scope of the shapes that the shape matrix can represent. It can describe even shapes with holes. Shape matrix is also invariant under translation, rotation and scaling of the object. The shape of the object can be reconstructed from the shape matrix; the accuracy is given by the size of the grid cells.

9.4.9 *Shape Context*

In [8], the shape context has been shown to be a powerful tool for object recognition tasks. It is used to find corresponding features between model and image.

Shape contexts analysis begins by taking N samples from the edge elements on the shape. These points can be on internal or external contours. Consider the vectors originating from a point to all other sample points on the shape. These vectors express the appearance of the entire shape relative to the reference point. This descriptor is the histogram of the relative polar coordinates of all other points:

$$h_i(k) = \#\{ Q \neq P_i : (Q - P_i) \in bin(k) \} \tag{9.19}$$

An example is shown in figure 9.22 where (c) is the diagram of log-polar histogram that has 5 bins for the polar direction and 12 bins for the angular direction. The histogram of a point P_i is formed by the following steps: putting the center of the histogram bins diagram on the point P_i, each bin of this histogram contains a count of all other sample points on the shape falling into that bin. Note that on this figure, the shape contexts (histograms) for the points marked by 'o' (in (a)), '◊' (in (b)) and '◁' (in (a)) are shown in (d), (e) and (f), respectively. It is clear that the shape contexts for the points marked by 'o' and '◊', which are computed for relatively similar points on the two shapes, have visual similarity. By contrast, the shape context for '◁' is quite different from the others. Obviously, this descriptor is a rich description, since as N gets large, the representation of the shape becomes exact.

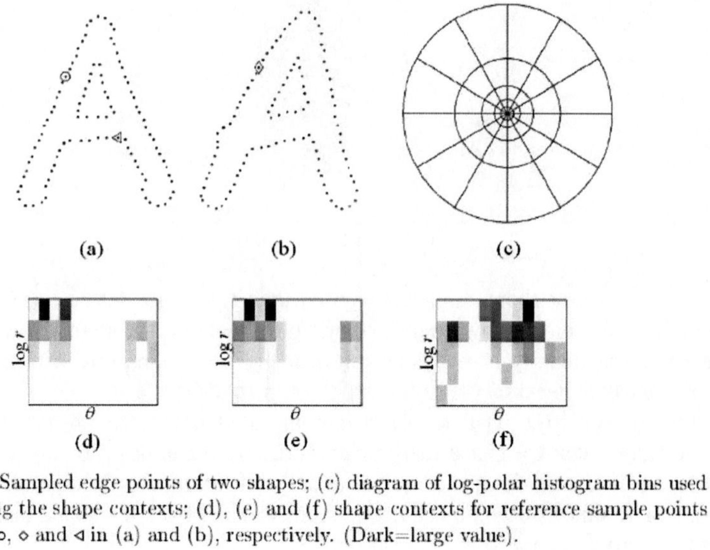

Fig. 9.22 Shape context computation and graph matching

Shape context matching is often used to find the corresponding points on two shapes. It has been applied to a variety of object recognition problems [8, 33, 45, 55]. The shape context descriptor has the following invariance properties:

- translation: the shape context descriptor is inherently translation invariant as it is based on relative point locations.
- scaling: for clutter-free images the descriptor can be made scale invariant by normalizing the radial distances by the mean (or median) distance between all point pairs.
- rotation: it can be made rotation invariant by rotating the coordinate system at each point so that the positive x-axis is aligned with the tangent vector.
- shape variation: the shape context is robust against slight shape variations.
- few outliers: points with a final matching cost larger than a threshold value are classified as outliers. Additional 'dummy' points are introduced to decrease the effects of outliers.

9.4.10 Chord Distribution

The basic idea of chord distribution is to calculate the lengths of all chords in the shape (all pair-wise distances between boundary points) and to build a histogram of their lengths and orientations [40]. The "lengths" histogram is invariant to rotation and scales linearly with the size of the object. The "angles" histogram is invariant to object size and shifts relative to object rotation. Figure 9.23 gives an example of chord distribution.

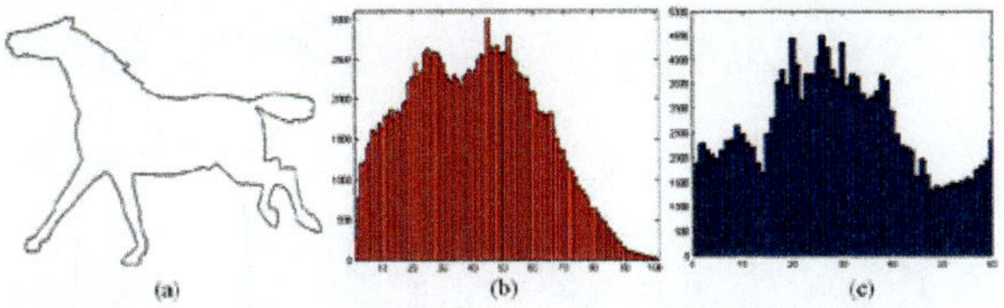

(a) Original contour; (b) chord lengths histogram; (c) chord angles histogram (each stem covers 3)

Fig. 9.23 Chord distribution

9.4.11 Shock Graphs

Shock graphs is a descriptor based on the medial axis. The medial axis is the most popular method that has been proposed as a useful shape abstraction tool for the representation and modelling of animate shapes. Skeleton and medial axes have been extensively used for characterizing objects satisfactorily using structures that are composed of line or arc patterns. Medial axis is an image processing operation which reduces input shapes to axial stick-like representations. It is as the loci of centres of bi-tangent circles that fit entirely within the foreground region being considered. Figure 9.24 illustrates the medial axis for a rectangular shape.

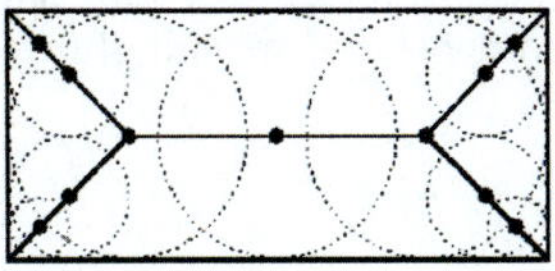

Fig. 9.24 Medial axis of a rectangle defined in terms of bi-tangent circles

We notice that the radius of each circle is variable. This variable is a function of the loci of points on the medial axis. We call this function as the radius function.

A shock graph is a shape abstraction that decomposes a shape into a set of hierarchically organized primitive parts. Siddiqi and Kimia define the concept of a shock graph [39] as an abstraction of the medial axis of a shape onto a directed acyclic graph (DAG). Shock segments are curve segments of the medial axis with monotonic flow, and give a more refined partition of the medial axis segments (see figure 9.25).

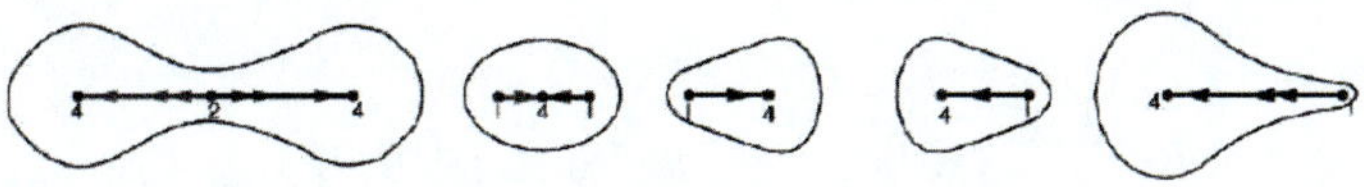

Fig. 9.25 Shock segments

The skeleton points are first labelled according to the local variation of the radius function at each point. Shock graph can distinguish the shapes but the medial axis cannot. Figure 9.26 shows two examples of shapes and their shock graphs.

Fig. 9.26 Examples of shapes and their shock graphs

To calculate the distance between two shock graphs, in [38], the authors employ a polynomial-time edit-distance algorithm. It shows that this algorithm has good performance against boundary perturbations, articulation and deformation of parts, segmentation errors, scale variations, viewpoint variations and partial occultation.

9.4.12 Discussions

Spatial feature descriptor is a direct method to describe a shape. These descriptors can apply tree-based theory (Adaptive grid resolution and Convex hull), statistic (Chain code histogram, Beam angle statistics, Shape context and Chord distribution) or syntactic analysis (Smooth curve decomposition) to extract or represent the feature of a shape. This description scheme not only compresses the data of a shape, but also provides a compact and meaningful form to facilitate further recognition operations.

9.5 Moments

This concept is issued from the concept of moments in mechanics where mass repartition of objects are observed. It is an integrated theory system. For both contour and region of a shape, one can use moment's theory to analyze the object.

9.5.1 Boundary Moments

Boundary moments, analysis of a contour, can be used to reduce the dimension of boundary representation [41]. Assume shape boundary has been represented as a 1-D shape representation $z(i)$ as introduced in Section 2, the rth moment m_r and central moment μ_r can be estimated as

$$m_r = \frac{1}{N} \sum_{i=1}^{N} [z(i)]^r \quad and \quad \mu_r = \frac{1}{N} \sum_{i=1}^{N} [z(i) - m_1]^r \tag{9.20}$$

where N is the number of boundary points.

The normalized moments $\overline{m}_r = m_r / (\mu_2)^{r/2}$ and $\overline{\mu}_r = \mu_r / (\mu_2)^{r/2}$ are invariant to shape translation, rotation and scaling. Less noise-sensitive shape descriptors can be obtained from

$$F_1 = \frac{(\mu_2)^{1/2}}{m_1}, \quad F_2 = \frac{\mu_3}{(\mu_2)^{3/2}} \quad and \quad F_3 = \frac{\mu_4}{(\mu_2)^2} \tag{9.21}$$

The other boundary moments method treats the 1-D shape feature function $z(i)$ as a random variable **v** and creates a K bins histogram $p(v_i)$ from $z(i)$. Then, the rth central moment is obtained by

$$\mu_r = \sum_{i=1}^{K} (v_i - m)^r p(v_i) \quad and \quad m = \sum_{i=1}^{K} v_i p(v_i) \tag{9.22}$$

The advantage of boundary moment descriptors is that they are easy to implement. However, it is difficult to associate higher order moments with physical interpretation.

9.5.2 Region Moments

Among the region-based descriptors, moments are very popular. These include moment invariants, Zernike moments, Radial Chebyshev moments, etc.

The general form of a moment function m_{pq} of order $(p+q)$ of a shape region can be given as:

$$m_{pq} = \sum_x \sum_y \Psi_{pq}(x,y) f(x,y) \quad p,q = 0,1,2\cdots \tag{9.23}$$

where Ψ_{pq} is known as the *moment weighting kernel* or the *basis set*; $f(x,y)$ is the shape region defined as follows

$$f(x,y) = \begin{cases} 1 & \text{if } (x,y) \in D \\ 0 & \text{otherwise} \end{cases} \tag{9.24}$$

where D represents the image domain.

9.5.2.1 Invariant Moments (IM)

Invariant moments (IM) are also called geometric moment invariants. Geometric moments, are the simplest of the moment functions with basis $\Psi_{pq} = x^p y^q$, while complete, is not orthogonal [57]. Geometric moment function m_{pq} of order $(p+q)$ is given as:

$$m_{pq} = \sum_x \sum_y x^p y^q f(x,y) \quad p,q = 0,1,2\cdots \tag{9.25}$$

The geometric central moments, which are invariant to translation, are defined as

$$\mu_{pq} = \sum_x \sum_y (x - \bar{x})^p (y - \bar{y})^q f(x,y) \text{ with } p,q = 0,1,2\cdots \tag{9.26}$$

where $\bar{x} = m_{10} / m_{00}$ and $\bar{y} = m_{01} / m_{00}$

A set of 7 invariant moments (IM) is given by [57]:

$$\phi_1=\eta_{20}+\eta_{02} \tag{9.27}$$

$$\phi_2=(\eta_{20}-\eta_{02})^2+4\eta_{11}^{\ 2} \tag{9.28}$$

$$\phi_3=(\eta_{30}-3\eta_{12})^2+(3\eta_{21}-\eta_{03})^2 \tag{9.29}$$

$$\phi_4=(\eta_{30}+\eta_{12})^2+(\eta_{21}+\eta_{03})^2 \tag{9.30}$$

$$\phi_5=(\eta_{30}-3\eta_{12})(\eta_{30}+\eta_{12})\left[(\eta_{30}+\eta_{12})^2-3(\eta_{21}+\eta_{03})^2\right]+(3\eta_{21}-\eta_{03})(\eta_{21}+\eta_{03}) \cdot\left[3(\eta_{30}+\eta_{12})^2-(\eta_{21}+\eta_{03})^2\right] \tag{9.31}$$

$$\phi_6=(\eta_{20}-\eta_{02})\left[(\eta_{30}+\eta_{12})^2-(\eta_{21}+\eta_{03})^2\right]+4\eta_{11}^{\ 2}(\eta_{30}+\eta_{12})(\eta_{21}+\eta_{03}) \tag{9.32}$$

$$\phi_7=(3\eta_{21}-\eta_{03})(\eta_{30}+\eta_{12})\left[(\eta_{30}+\eta_{12})^2-3(\eta_{21}+\eta_{03})^2\right]+(3\eta_{12}-\eta_{03})(\eta_{21}+\eta_{03}) \cdot\left[3(\eta_{30}+\eta_{12})^2-(\eta_{21}+\eta_{03})^2\right] \tag{9.33}$$

where $\eta_{pq}=\mu_{pq}/\mu_{00}^{\ \gamma}$ and $\gamma=1+(p+q)/2$ for $p+q=2,3,\cdots$

IM are computationally simple. Moreover, they are invariant to rotation, scaling and translation. However, they have several drawbacks [10]:

- information redundancy: since the basis is not orthogonal, these moments suffer from a high degree of information redundancy.
- noise sensitivity: higher order moments are very sensitive to noise.
- large variation in the dynamic range of values: since the basis involves powers of p and q, the moments computed have large variation in the dynamic range of values for different orders. This may cause numerical instability when the image size is large.

9.5.2.2 Algebraic Moment Invariants

The algebraic moment invariants are computed from the first m central moments and are given as the eigenvalues of predefined matrices, $M_{[j,k]}$, whose elements are scaled factors of the central moments [43]. The algebraic moment invariants can be constructed up to arbitrary order and are invariant to affine transformations. However, algebraic moment invariants performed either very well or very poorly on the objects with different configuration of outlines.

9.5.2.3 Zernike Moments (ZM)

Zernike Moments (ZM) are orthogonal moments [10]. The complex Zernike moments are derived from orthogonal Zernike polynomials:

$$V_{nm}(x,y)=V_{nm}(r\cos\theta,r\sin\theta)=R_{nm}(r)\exp(jm\theta) \tag{9.34}$$

where $R_{nm}(r)$is the orthogonal radial polynomial:

$$R_{nm}(r)=\sum_{s=0}^{(n-|m|)/2}(-1)^s\frac{(n-s)!}{s!\times\left(\frac{n-2s+|m|}{2}\right)!\left(\frac{n-2s-|m|}{2}\right)!}r^{n-2s} \quad (9.35)$$

n=0,1,2,⋯; 0≤ $|m|$≤n; and $n-|m|$ is even.

Zernike polynomials are a complete set of complex valued functions that are orthogonal over the unit disk, i.e., $x^2+y^2\leq 1$. The Zernike moment of order n with repetition m of shape region $f(x,y)$ is given by:

$$Z_{nm}=\frac{n+1}{\pi}\sum_{r}\sum_{\theta}f(r\cos\theta,r\sin\theta)\cdot R_{nm}(r)\cdot\exp(jm\theta) \quad r\leq 1 \quad (9.36)$$

Zernike moments (ZM) have the following advantages [35]:

- rotation invariance: the magnitudes of Zernike moments are invariant to rotation.
- robustness: they are robust to noise and minor variations in shape.
- expressiveness: since the basis is orthogonal, they have minimum information redundancy.

However, the computation of ZM (in general, continuous orthogonal moments) pose several problems:

- coordinate space normalization: the image coordinate space must be transformed to the domain where the orthogonal polynomial is defined (unit circle for the Zernike polynomial).
- numerical approximation of continuous integrals: the continuous integrals must be approximated by discrete summations. This approximation not only leads to numerical errors in the computed moments, but also severely affects the analytical properties such as rotational invariance and orthogonality.
- computational complexity: computational complexity of the radial Zernike polynomial increases as the order becomes large.

9.5.2.4 Radial Chebyshev Moments (RCM)

The radial Chebyshev moment of order p and repetition q is defined as [34]:

$$S_{pq}=\frac{1}{2\pi\rho(p,m)}\sum_{r=0}^{m-1}\sum_{\theta=0}^{2\pi}t_p(r)\cdot\exp(-jq\theta)\cdot f(r,\theta) \quad (9.37)$$

where $t_p(r)$ is the scaled orthogonal Chebyshev polynomials for an $N\times N$ image such that

$$t_p(x)=\frac{(2p-1)t_1(x)t_{p-1}(x)-(p-1)\left\{1-\frac{(p-1)^2}{N^2}\right\}t_{p-2}(x)}{p},\quad p>1 \tag{9.38}$$

with $t_0(x)=1$, $t_1(x)=(2x-N+1)/N$ and where $\rho(p,N)$ is the squared-norm:

$$\rho(p,N)=\frac{N\left(1-\frac{1}{N^2}\right)\left(1-\frac{2^2}{N^2}\right)\cdots\left(1-\frac{p^2}{N^2}\right)}{2p+1},\quad p=0,1,\cdots,N-1 \tag{9.39}$$

and $m=(N/2)+1$.

The mapping between (r,θ) and image coordinates (x,y) is given by:

$$x=\frac{rN}{2(m-1)}\cos(\theta)+\frac{N}{2} \text{ and } y=\frac{rN}{2(m-1)}\sin(\theta)+\frac{N}{2} \tag{9.40}$$

Compared to Chebyshev moments, radial Chebyshev moments possess rotational invariance property.

9.5.3 Discussions

Besides the previous moments, there are other moments for shape representation, for example, homocentric polar-radius moment [20], orthogonal Fourier-Mellin moments (OFMMs) [21], pseudo-Zernike Moments [31], etc. The study shows that the moment-based shape descriptors are usually concise, robust and easy to compute. They are also invariant to scaling, rotation and translation of the object. However, because of their global nature, the disadvantage of moment-based methods is that it is difficult to correlate high order moments with a shape's salient features.

9.6 Scale Space Approaches

Scale space approaches are issued from multiscale representation that allows handling shape structure at different scales. In scale space theory a curve is embedded into a continuous family $\{\Gamma_\sigma:\sigma\geq 0\}$ of gradually simplified versions. The main idea of scale spaces is that the original curve $\Gamma=\Gamma_0$ should get more and more simplified, and so small structures should vanish as parameter σ increases. Thus due to different scales (values of σ), it is possible to separate small details from relevant shape properties. The ordered sequence $\{\Gamma_\sigma:\sigma\geq 0\}$ is referred to as evolution of Γ.

A lot of shape features can be analyzed in scale-space theory to get more information about shapes. Here we introduced 2 scale-space approaches: curvature scale-space (CSS) and intersection points map (IPM).

9.6.1 Curvature Scale-Space

The curvature scale-space (CSS) method, proposed by F. Mokhtarian in 1988, was selected as a contour shape descriptor for MPEG-7. This approach is based on multi-scale representation and curvature to represent planar curves. For convenience, a contour is defined with a discrete parameterization as following:

$$\Gamma(\mu)=(x(\mu),y(\mu)) \tag{9.41}$$

An evolved version of that curve is defined by

$$\Gamma_\sigma(\mu)=(X(\mu,\sigma),Y(\mu,\sigma)) \tag{9.42}$$

where $X(\mu,\sigma)=x(\mu)*g(\mu,\sigma)$ and $Y(\mu,\sigma)=y(\mu)*g(\mu,\sigma)$, * is the convolution operator, and $g(\mu,\sigma)$ denotes a Gaussian filter with standard deviation σ defined by

$$g(\mu,\sigma)=\frac{1}{\sigma\sqrt{2\pi}}\exp(\frac{-\mu^2}{2\sigma^2}) \tag{9.43}$$

Functions $X(\mu,\sigma)$ and $Y(\mu,\sigma)$ are given explicitly by

$$X(\mu,\sigma)=\int_{-\infty}^{\infty} x(v)\frac{1}{\sigma\sqrt{,2\pi}}\exp(\frac{-(\mu-v)^2}{2\sigma^2})dv \tag{9.44}$$

$$Y(\mu,\sigma)=\int_{-\infty}^{\infty} y(v)\frac{1}{\sigma\sqrt{,2\pi}}\exp(\frac{-(\mu-v)^2}{2\sigma^2})dv \tag{9.45}$$

The curvature of the contour is given by

$$k(\mu,\sigma)=\frac{X_\mu(\mu,\sigma)Y_{\mu\mu}(\mu,\sigma)-X_{\mu\mu}(\mu,\sigma)Y_\mu(\mu,\sigma)}{(X_\mu(\mu,\sigma)^2-Y_\mu(\mu,\sigma)^2)^{3/2}} \tag{9.46}$$

where

$$X_\mu(\mu,\sigma)=\frac{\partial}{\partial\mu}(x(\mu)*g(\mu,\sigma))=x(\mu)*g_\mu(\mu,\sigma) \tag{9.47}$$

$$X_{\mu\mu}(\mu,\sigma)=\frac{\partial^2}{\partial\mu^2}(x(\mu)*g(\mu,\sigma))=x(\mu)*g_{\mu\mu}(\mu,\sigma) \tag{9.48}$$

$$Y_\mu(\mu,\sigma)=\frac{\partial}{\partial\mu}(y(\mu)*g(\mu,\sigma))=y(\mu)*g_\mu(\mu,\sigma) \tag{9.49}$$

$$Y_{\mu\mu}(\mu,\sigma)=\frac{\partial^2}{\partial\mu^2}(y(\mu)*g(\mu,\sigma))=y(\mu)*g_{\mu\mu}(\mu,\sigma) \tag{9.50}$$

Note that σ is also referred to as a scale parameter. The process of generating evolved versions of Γ_σ as σ increases from 0 to ∞ is referred to as the evolution of Γ_σ. This technique is suitable for removing noise and smoothing a planar curve as well as gradual simplification of a shape.

The function defined by $k(\mu,\sigma)=0$ is the CSS image of Γ. Figure 9.27 is a CSS image examples.

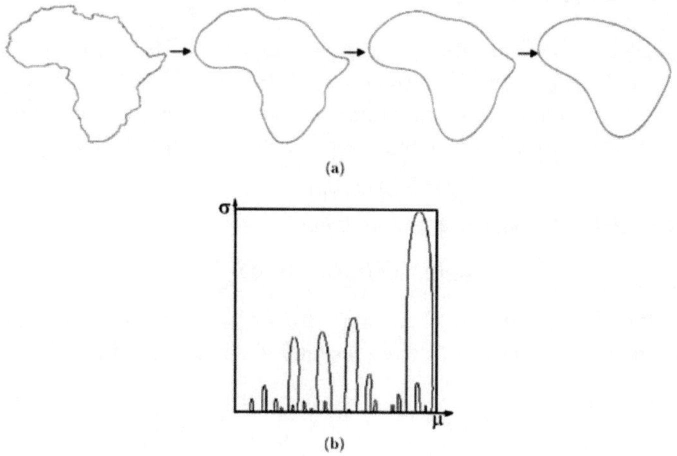

(a) Evolution of Africa: from left to right σ=0(original), σ=4, σ=8 and σ=16, respectively; (b) CSS image of Africa.

Fig. 9.27 Curvature scale-space image

The representation of CSS is the maxima of CSS contour of an image. Many methods for representing the maxima of CSS exist in the literatures [19, 36, 52] and the CSS technique has been shown to be robust contour-based shape representation technique. The basic properties of the CSS representation are as follows:

- it captures the main features of a shape, enabling similarity-based retrieval;
- it is robust to noise, changes in scale and orientation of objects;
- it is compact, reliable and fast;
- It retains the local information of a shape. Every concavity or convexity on the shape has its own corresponding contour on the CSS image.

Although CSS has a lot of advantages, it does not always give results in accordance with human vision system. The main drawbacks of this description are due to the problem of shallow concavities/convexities on a shape. It can be shown that the shallow and deep concavities/convexities may create the same large contours on the CSS image. In [1, 49], the authors gave some methods to alleviate these effects.

9.6.2 Intersection Points Map

Similarly to the CSS, many methods also use a Gaussian kernel to progressively smooth the curve relatively to the varying bandwidth. In [24], the authors proposed a new algorithm, intersection points map (IPM), based on this principle. Instead of characterizing the curve with its curvature involving 2nd order derivatives, it uses the intersection points between the smoothed curve and the original. As the

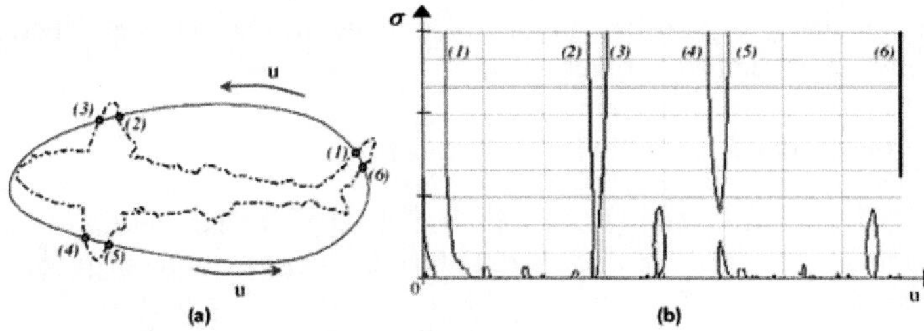

(a) An original contour; (b) an IPM image in the (u,σ) plane. The IPM points indicated by (1)-(6) refer to the corresponding intersection points in (a).

Fig. 9.28 Example of the IPM

standard deviation of the Gaussian kernel increases, the number of the intersection points decreases. By analyzing these remaining points, features for a pattern can be defined. Figure 9.28 represents an example of IPM.

The IPM pattern can be identified regardless of its orientation, translation and scale change. It is also resistant to noise for a range of noise energy. The main weakness of this approach is that it fails to handle occulted contours and those having undergone a non-rigid deformation. Since this method deals only with curve smoothing, it needs only the convolution operation in the smoothing process. So this method is faster than the CSS one with equivalent performances.

9.6.3 Discussions

As multi-resolution analysis in signal processing, scale-space theory can obtain abundant information about a contour with different scales. In scale-space, global pattern information can be interpreted from higher scales, while detailed pattern information can be interpreted from lower scales. Scale-space algorithm benefits from the boundary information redundancy in the new image, making it less sensitive to errors in the alignment or contour extraction algorithms. The great advantages are the high robustness to noise and the great coherence with human perception.

9.7 Shape Transform Domains

With operators transforming data pixels into frequency domain, a description of shape can be obtained with respect to its frequency content. The transform domain class includes methods which are formed by the transform of the detected object or the transform of the whole image. Transforms can therefore be used to characterize the appearance of images. The shape feature is represented by all or partial coefficients of a transform.

9.7.1 Fourier Descriptors

Although, Fourier descriptor (FD) is a 40-year-old technique, it is still considered as a valid description tool. The shape description and classification using FD either in contours or regions are simple to compute, robust to noise and compact. It has many applications in different areas.

9.7.1.1 One-Dimensional Fourier Descriptors

In general, Fourier descriptor (FD) is obtained by applying Fourier transform on a shape signature that is a one-dimensional function derived from shape boundary coordinates (cf. Section 2). The normalized Fourier transformed coefficients are called the Fourier descriptor of the shape. FD derived from different signatures has significant different performance on shape retrieval. As shown in [52, 53], FD derived from centroid distance function *r(t)* outperforms FD derived from other shape signatures in terms of overall performance. The discrete Fourier transform of *r(t)* is then given by

$$a_n = \frac{1}{N} \sum_{t=0}^{N-1} r(t) \exp\left(\frac{-j2\pi nt}{N}\right), \quad n=0,1,\cdots,N-1 \tag{9.51}$$

Since the centroid distance function *r(t)* is only invariant to rotation and translation, the acquired Fourier coefficients have to be further normalized so that they are scaling and starting point independent shape descriptors. From Fourier transform theory, the general form of the Fourier coefficients of a contour centroid distance function $r(t)$ transformed through scaling and change of start point from the original function $r(t)^{(o)}$ is given by

$$a_n = \exp(jn\tau) \cdot s \cdot a^{(o),n} \tag{9.52}$$

where a_n and $a^{(o)n}$ are the Fourier coefficients of the transformed shape and the original shape, respectively, τ is the angle incurred by the change of starting point and *s* is the scale factor. Now considering the following expression:

$$b_n = \frac{a_n}{a_1} = \frac{\exp(jn\tau) \cdot s \cdot a_n^{(o)}}{\exp(j\tau) \cdot s \cdot a_1^{(o)}} = \frac{a_n^{(o)}}{a_1^{(o)}} \exp[j(n-1)\tau] = b_n^{(o)} \exp[j(n-1)\tau] \tag{9.53}$$

where b_n and $b^{(o)n}$ are the normalized Fourier coefficients of the transformed shape and the original shape, respectively. If we ignore the phase information and only use magnitude of the coefficients, then $|b_n|$ and $|b^{(o)n}|$ are the same. In other words, $|b_n|$ is invariant to translation, rotation, scaling and change of start point.

The set of magnitudes of the normalized Fourier coefficients of the shape $\left\{ |b_n|, 0<n<N \right\}$ are used as shape descriptors, denoted as

$$\left\{ FD_n, 0<n<N \right\}. \tag{9.54}$$

One-dimensional FD has several interesting characteristics such as simple derivation, simple normalization and simple to do matching. As indicated in [52], for efficient retrieval, 10 FDs are sufficient for shape description.

9.7.1.2 Region-Based Fourier Descriptor

The region-based FD is referred to as generic FD (GFD), which can be used for general applications. Basically, GFD is derived by applying a modified polar Fourier transform (MPFT) on shape image [48, 54]. In order to apply MPFT, the polar shape image is treated as a normal rectangular image. The steps are as follows

1. the approximated normalized image is rotated counter clockwise by an angular step sufficiently small.
2. the pixel values along positive x-direction starting from the image center are copied and pasted into a new matrix as row elements.
3. the steps 1 and 2 are repeated until the image is rotated by 360°.

The result of these steps is that an image in polar space plots into Cartesian space. Figure 9.29 shows the polar shape image turning into normal rectangular image.

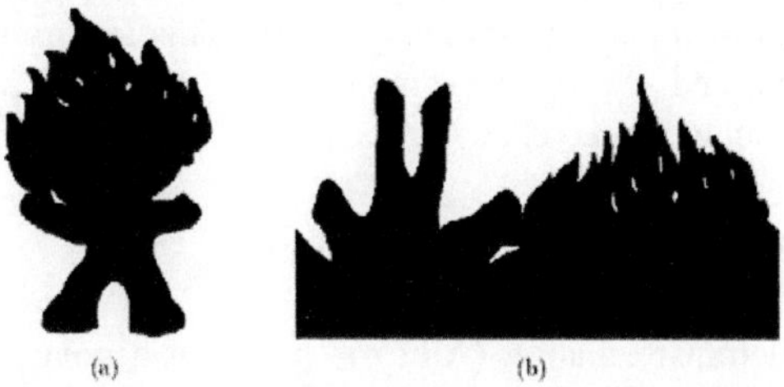

(a) Original shape image in polar space; (b) polar image of (a) plotted into Cartesian space.

Fig. 9.29 The polar shape image turns into normal rectangular image.

The Fourier transform is obtained by applying a discrete 2D Fourier transform on this shape image, so that

$$pf(\rho,\phi)=\sum_r \sum_i f(r,\theta_i) exp[j2\pi(\frac{r}{R}\rho+\frac{2\pi i}{T}\phi)] \tag{9.55}$$

where $0\leq r=\sqrt{[(x-g_x)^2+(y-g_y)^2]}<R$, and $\theta_i=i(2\pi/T)$; $0\leq\rho<R$, $0\leq\phi<T$ with (g_x,g_y) being the centre of mass of the shape; R and T are the radial and angular resolutions. The acquired Fourier coefficients are translation invariant. Rotation and scaling invariance are achieved by the following:

$$GFD=\left\{\frac{|pf(0,0)|}{area}, \frac{|pf(0,1)|}{|pf(0,0)|}, \ldots, \frac{|pf(0,n)|}{|pf(0,0)|}, \ldots, \frac{|pf(m,0)|}{|pf(0,0)|}, \ldots, \frac{|pf(m,n)|}{|pf(0,0)|}\right\} \tag{9.56}$$

where *area* is the area of the bounding circle in which the polar image resides. m is the maximum number of the radial frequencies selected and n is the maximum number of selected angular frequencies. m and n can be adjusted to achieve hierarchical coarse to fine representation requirement.

For efficient shape description, following the implementation of [54], 36 GFD features reflecting m=4 and n=9 are selected to index the shape. The experimental results have shown GFD as invariant to translation, rotation, and scaling. For obtaining the affine and general minor distortions invariance, in [54], the authors proposed Enhanced Generic Fourier Descriptor (EGFD) to improve the GFD properties.

9.7.2 Wavelet Transform

A hierarchical planar curve descriptor is developed by using the wavelet transform [13]. This descriptor decomposes a curve into components of different scales so that the coarsest scale components carry the global approximation information while the finer scale components contain the local detailed information. The wavelet descriptor has many desirable properties such as multi-resolution representation, invariance, uniqueness, stability, and spatial localization. In [23], the authors use dyadic wavelet transform deriving an affine invariant function. In [12], a descriptor is obtained by applying the Fourier transform along the axis of polar angle and the wavelet transform along the axis of radius. This feature is also invariant to translation, rotation, and scaling. At same time, the matching process of wavelet descriptor can be accomplished cheaply.

9.7.3 Angular Radial Transformation

The angular radial transformation (ART) is based in a polar coordinate system where the sinusoidal basis functions are defined on a unit disc. Given an image function in polar coordinates, $f(\rho,\theta)$, an ART coefficient F_{nm} (radial order n, angular order m) can be defined as [37]:

$$F_{nm}=\int_0^{2\pi}\int_0^1 V_{nm}(\rho,\theta)f(\rho,\theta)\rho d\rho d\theta \tag{9.57}$$

where $V_{nm}(\rho,\theta)$ is the ART basis function and is separable in the angular and radial directions so that:

$$V_{nm}(\rho,\theta)=A_m(\theta)R_n(\rho) \tag{9.58}$$

The angular basis function, A_m, is an exponential function used to obtain orientation invariance. This function is defined as:

$$A_m(\theta)=\frac{1}{2\pi}e^{jm\theta} \tag{9.59}$$

where R_n, the radial basis function, is defined as:

$$R_n(\rho)=\begin{cases}1 & \text{if n}=0\\ 2\cos(\pi n\rho) & \text{if n}\neq 0\end{cases} \tag{9.60}$$

In MPEG-7, twelve angular and three radial functions are used (n<3,m<12). Real parts of the 2-D basis functions are shown in figure 9.30.

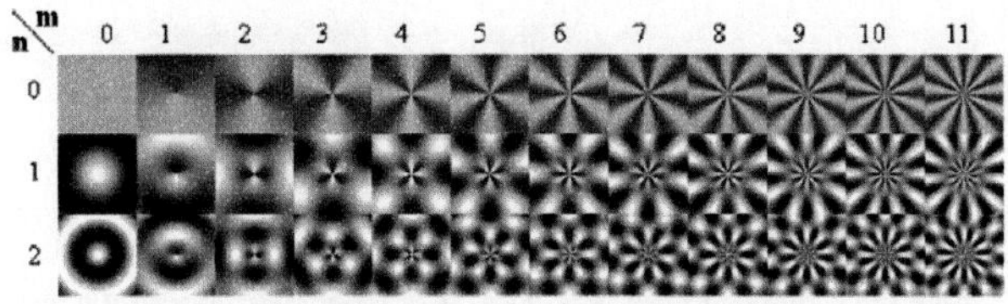

Fig. 9.30 Real parts of the ART basis functions

For scale normalization, the ART coefficients are divided by the magnitude of ART coefficient of order n=0,m=0. MPEG-7 standardization process showed the efficiency of 2-D angular radial transformation. This descriptor is robust against translation, scaling, multi-representation (remeshing, weak distortions) and noises.

9.7.4 Shape Signature Harmonic Embedding

A harmonic function is obtained by a convolution between the Poisson kernel $P_R(r,\theta)$ and a given boundary function $u(Re^{j\phi})$. Poisson kernel is defined by

$$P_R(r,\theta)=\frac{R^2-r^2}{R^2-2Rr\cos(\theta)+r^2} \tag{9.61}$$

The boundary function could be any real- or complex-valued function, but here we choose shape signature functions for the purpose of shape representation. For any shape signature $s[n],n=0,1,\cdots,N-1$, the boundary values for a unit disk can be set as

$$u(Re^{j\phi})=u(Re^{j\omega_0 n})=s[n] \tag{9.62}$$

where $\omega_0=2\pi/N$, $\phi=\omega_0 n$.

So the harmonic function u can be written as

$$u(re^{j\theta})=\frac{1}{2\pi}\int_0^{2\pi} u(Re^{j\phi})P_R(r,\phi-\theta)d\phi \tag{9.63}$$

The Poisson kernel $P_R(r,\theta)$ has a low-pass filter characteristic, where the radius r is inversely related to the bandwidth of the filter. The radius r is considered as the scale parameter of a multi-scale representation [27]. Another important property is $P_R(0,\theta)=1$, indicating $u(0)$ is the mean value of boundary function $u(Re^{j\phi})$.

In [27], the authors proposed a formulation of a discrete closed-form solution for the Poisson's integral formula of equation (9.63), so that one can avoid the need for approximation or numerical calculation of the Poisson summation form.

As in Subsection 7.1.2, the harmonic function inside the disk can be mapped to a rectilinear space for a better illustration. Figure 9.31 shows an example for a star shape. Here, we used curvature as the signature to provide boundary values.

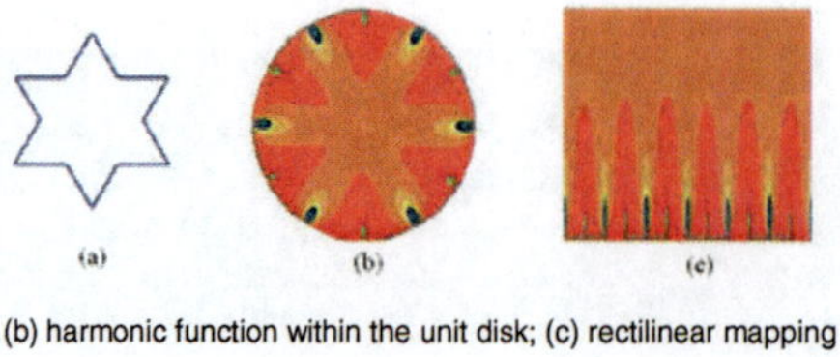

Fig. 9.31 Harmonic embedding of curvature signature

The zero-crossing image of the harmonic functions is extracted as a shape feature. This shape descriptor is invariant to translation, rotation and scaling. It is also robust to noise. Figure 9.32 is an example. The original curve is corrupted with different noise levels, and the harmonic embeddings show robustness to the noise.

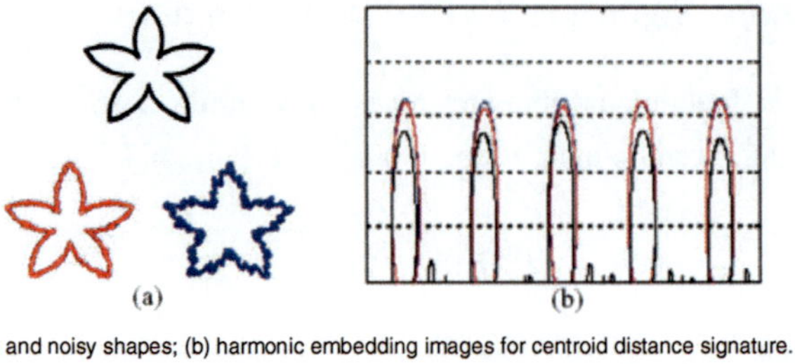

Fig. 9.32 Centroid distance signature harmonic embedding that is robust to noisy boundaries

In addition, it is more efficient than CSS descriptor. However, it is not suitable for similarity retrieval, because it is inconsistent with non-rigid transform.

9.7.5 ℛ-Transform

The ℛ-Transform to represent a shape is based on the Radon transform. The approach is presented as follows. We assume that the function f is the domain of a shape. Its Radon transform is defined by:

$$T_R(\rho,\theta)=\int_{-\infty}^{\infty}\int_{-\infty}^{\infty} f(x,y)\delta(x\cos\theta+y\sin\theta-\rho)dxdy \tag{9.64}$$

where δ(.) is the Dirac delta-function such that:

$$\delta(x)=\begin{cases}1 \text{ if } x=0\\ 0 \text{ otherwise}\end{cases} \tag{9.65}$$

$\theta\in[0,\pi]$ and $\rho\in(-\infty,\infty)$. In other words, Radon transform $T_R(\rho,\theta)$ is the integral of f over the line $L_{(\rho,\theta)}$ defined by $\rho=x\cos\theta+y\sin\theta$.

Figure 9.33 is an example of a shape and its Radon transform.

(a) Shape (b) Radon transform of (a)

Fig. 9.33 A shape and its Radon transform

The following transform is defined as $\mathcal{R}$-transform:

$$\mathrm{R}_{f}(\theta) = \int_{-\infty}^{-\infty} T_R^2(\rho,\theta)d\rho \tag{9.66}$$

where $T_R(\rho,\theta)$ is the Radon transform of the domain function f. In [42], the authors show the following properties of $\mathcal{R}_f(\theta)$:

- periodicity: $\mathcal{R}_f(\theta \pm \pi) = \mathcal{R}_f(\theta)$
- rotation: a rotation of the image by an angle θ_0 implies a translation of the $\mathfrak{R}$-transform of θ_0: $\mathcal{R}_f(\theta+\theta_0)$.
- translation: the $\mathfrak{R}$-transform is invariant under a translation of the shape f by a vector $\vec{u} = (x_0, y_0)$.
- scaling: a change of the scaling of the shape f induces a scaling only in the amplitude of the $\mathcal{R}$-transform.

Given a large collection of shapes, one $\mathcal{R}$-transform per shape is not efficient to distinguish from the others because the $\mathcal{R}$-transform provides a highly compact shape representation. In this perspective, to improve the description, each shape is projected in the Radon space for different segmentation levels of the Chamfer distance transform. Chamfer distance transform is introduced in [60, 61].

Given the distance transform of a shape, the distance image is segmented into N equidistant levels to keep the segmentation isotropic. For each distance level, pixels having a distance value superior to that level are selected and at each level of segmentation, an $\mathcal{R}$-transform is computed. In this manner, both the internal structure and the boundaries of the shape are captured. Since a rotation of the shape implies a corresponding shift of the $\mathcal{R}$-transform. Therefore, a one-dimensional Fourier transform is applied on this function to obtain the rotation invariance. After the one-dimensional discrete Fourier transform F, $\mathcal{R}$-transform descriptor vector is defined as follows:

$$RTD = \left(\frac{F\mathcal{R}^1(\frac{\pi}{M})}{F\mathcal{R}^1(0)}, \ldots, \frac{F\mathcal{R}^1(\frac{i\pi}{M})}{F\mathcal{R}^1(0)}, \ldots, \frac{F\mathcal{R}^1(\pi)}{F\mathcal{R}^1(0)}, \ldots, \frac{F\mathcal{R}^N(\frac{\pi}{M})}{F\mathcal{R}^N(0)}, \ldots, \frac{F\mathcal{R}^N(\frac{i\pi}{M})}{F\mathcal{R}^N(0)}, \ldots, \frac{F\mathcal{R}^N(\pi)}{F\mathcal{R}^N(0)} \right) \tag{9.67}$$

where $i \in [1,M]$, M is the angular resolution, $F\mathcal{R}^{\alpha}$ is the magnitude of Fourier transform to $\mathcal{R}$-transform and $\alpha \in [1,N]$, is the segmentation level of Chamfer distance transform.

9.7.6 Shapelet Descriptor

Shapelet descriptor was proposed to present a model for animate shapes and for extracting meaningful parts of objects. The model assumes that animate shapes (2D simple closed curves) are formed by a linear superposition of a number of shape bases. A basis function $\psi(s;\mu,\sigma)$ is defined in [15] so that $\mu\in[0,1]$ indicates the location of the basis function relative to the domain of the observed curve, and σ is the scale of the function ψ. Figure 9.34 shows the shape of the basis function ψ at different σ values. It displays variety with different parameter and transforms.

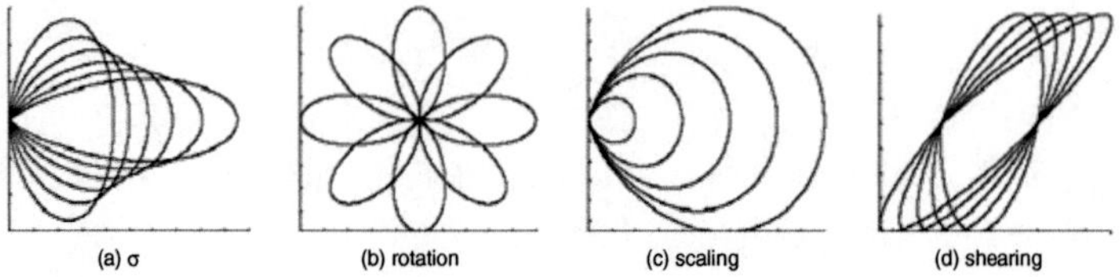

Fig. 9.34 Each shape base is a lobe-shaped curve

The basis functions are subject to affine transformations by a 2×2 matrix of basis coefficients:

$$A_k=[\ a_k;b_k;c_k;d_k\ ] \tag{9.68}$$

The variables for describing a base are denoted by $\mathbf{b}_k=(A_k,\mu_k,\sigma_k)$ and are termed basis elements. The shapelet is defined by

$$\gamma(s;\mathbf{b}_k)=A_k\psi(s;\mu_k,\sigma_k) \tag{9.69}$$

Figure 9.34(b,c,d) demonstrates shapelets obtained from the basis functions ψ by the affine transformations of rotation, scaling, and shearing respectively, as indicated by the basis coefficient A_k. By collecting all the shapelets at various μ, σ, A and discretizing them at multiple levels, a dictionary is obtained

$$\Delta=\{\gamma(s;b):\forall b;a\gamma_0,a>0\}. \tag{9.70}$$

A special shapelet γ_0 is defined as an ellipse. Shapelets are the building blocks for shape contours, and they form closed curves by linear addition:

$$\Gamma(s)=\begin{bmatrix} x_0 \\ y_0 \end{bmatrix}+\sum_{k=1}^{K}\begin{bmatrix} a_k b_k \\ c_k d_k \end{bmatrix}\psi(s;\mu_k,\sigma_k)+\mathbf{n}(s) \tag{9.71}$$

where (x_0,y_0) is the centroid of the contour and $\mathbf{n}$ is residue.

A discrete representation $\mathbf{B}=(K,\mathbf{b}_1,\mathbf{b}_2,\cdots,\mathbf{b}_K)$, shown by the dots in second row of figure 9.35, represents a shape. $\mathbf{B}$ is called the "shape script" by analogy to music scripts, where each shapelet is represented by a dot in the (μ,σ) domain. The horizontal axis is $\mu\in[0,1]$ and the vertical axis is the σ. Large dots correspond to big coefficient matrix

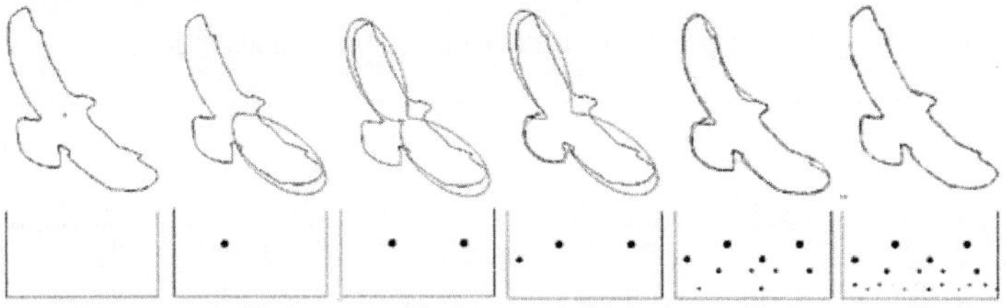

Fig. 9.35 Pursuit of shape bases for an eagle contour

$$A_k^2 = a_k^2 + b_k^2 + c_k^2 + d_k^2 \tag{9.72}$$

Clearly, computing the shape script **B** is a non-trivial task, since Δ is over-complete and there will be multiple sets of bases that reconstruct the curve with equal precision. [15] gave some pursuit algorithms to use shapelets representing a shape.

9.7.7 Discussions

As a kind of global shape description technique, shape analysis in transform domains takes the whole shape as the shape representation. The description scheme is designed for this representation. Unlike the spatial interrelation feature analysis, shape transform projects a shape contour or region into an other domain to obtain some of its intrinsic features. For shape description, there is always a trade-off between accuracy and efficiency. On one hand, shape should be described as accurate as possible; on the other hand, shape description should be as compact as possible to simplify indexing and retrieval. For a shape transform analysis algorithm, it is very flexible to accomplish a shape description with different accuracy and efficiency by choosing the number of transform coefficients.

9.8 Summary Table

For convenience, to compare these shape feature extraction approaches in this chapter, we summarize their properties in Table 9.1.

Frankly speaking, it is not equitable to affirm a property of an approach by rudely speaking "good" or "bad" because certain approaches have great differences in performances under different conditions.

For example, the method area function is invariant with affine transform under the condition of the contours sampled at its same vertices; whereas it is not robust to affine transform if the condition can't be contented. In addition, some approaches have good properties for certain type shapes; however it is not for the others. For example, the method shapelets representation is especially suitable for blobby objects, and it has shortcomings in representing elongated objects.

So the simple evaluations in this table are only as a reference. These evaluations are drawn by assuming that all the necessary conditions have been contented for each approach.

Table 9.1 Properties of shape feature extraction approaches

Shape representation			Recon-structure	Invariance				Resistance			Computational complexity
				Translation	Rotation	Scale	Affine transform	Noise	Occultation	Non-rigid deformation	
Shape signatures	Complex coordinates		Yes	Bad	Bad	Bad	Bad	Average	Good	Bad	Low
Shape signatures	Central distance		No	Good	Good	Good	Bad	Average	Good	Bad	Low
Shape signatures	Tangent angle		No	Good	Good	Good	Bad	Bad	Good	Average	Low
Shape signatures	Curvature function		No	Good	Good	Good	Bad	Bad	Good	Average	Low
Shape signatures	Area function		No	Good	Good	Good	Good	Good	Good	Bad	Low
Shape signatures	Triangle-area representation		No	Good	Good	Good	Good	Good	Average	Average	Low
Shape signatures	Chord length function		No	Good	Good	Good	Bad	Bad	Bad	Bad	Low
Polygonal approximation	Merging	Distance threshold	No	Good	Good	Good	Bad	Good	Bad	Bad	Average
Polygonal approximation	Merging	Tunneling	No	Good	Good	Good	Bad	Good	Bad	Bad	Average
Polygonal approximation	Merging	Polygon evolution	No	Good	Good	Good	Bad	Good	Bad	Bad	Average
Polygonal approximation	Splitting		No	Good	Good	Good	Bad	Good	Bad	Bad	Average
Space interrelation Feature	Adaptive grid resolution		Yes	Good	Good	Good	Bad	Good	Good	Bad	Low
Space interrelation Feature	Bounding box		Yes	Good	Good	Good	Average	Good	Good	Average	Average
Space interrelation Feature	Convex hull		No	Good	Good	Good	Good	Average	Bad	Bad	High
Space interrelation Feature	Chain code	Basic chain code	Yes	Good	Bad	Bad	Bad	Bad	Good	Bad	Low
Space interrelation Feature	Chain code	Vertex chain code	Yes	Good	Bad	Bad	Bad	Bad	Good	Bad	Low
Space interrelation Feature	Chain code	Statistic chain code	No	Good	Bad	Bad	Bad	Bad	Bad	Bad	Low
Space interrelation Feature	Smooth curve decomposition		No	Good	Good	Good	Bad	Good	Good	Average	Average
Space interrelation Feature	ALI-based representation		No	Good	Good	Good	Average	Good	Average	Bad	Average
Space interrelation Feature	Beam angle statistics		No	Good	Good	Good	Bad	Good	Bad	Bad	Low
Space interrelation Feature	Shape matrix	Square model	Yes	Good	Good	Good	Bad	Bad	Good	Bad	Average
Space interrelation Feature	Shape matrix	Polar model	Yes	Good	Good	Good	Bad	Bad	Good	Bad	Low
Space interrelation Feature	Shape context		No	Good	Good	Good	Bad	Bad	Average	Average	Average
Space interrelation Feature	Chord distribution		No	Good	Good	Good	Bad	Good	Bad	Bad	Low
Space interrelation Feature	Shock graphs		Yes	Good	Good	Good	Good	Good	Good	Good	High
Moments	Boundary moments		No	Good	Good	Good	Bad	Average	Bad	Bad	Low
Moments	Region moments	Invariant moments	No	Good	Good	Good	Bad	Bad	Bad	Bad	Average
Moments	Region moments	Algebraic Moment	No	Good	Good	Good	Good	Average	Bad	Bad	Average
Moments	Region moments	Zernike Moments	No	Good	Good	Good	Bad	Good	Average	Average	High
Moments	Region moments	Radial Chebyshev Moments	No	Good	Good	Good	Bad	Good	Average	Average	High
Scale-space methods	Curvature scale space		No	Good	Good	Good	Average	Good	Good	Average	Average
Scale-space methods	Intersection points map		No	Good	Good	Good	Average	Good	Good	Bad	Average
Shape transform domains	Fourier descriptors	1-D Fourier descriptor	No	Good	Good	Good	Bad	Bad	Bad	Bad	Average
Shape transform domains	Fourier descriptors	Region-based Fourier descriptor	No	Good	Good	Good	Good	Good	Average	Average	High
Shape transform domains	Wavelet transform		No	Good	Good	Good	Good	Average	Average	Bad	Average
Shape transform domains	Angular radial transformation		No	Good	Good	Good	Bad	Good	Bad	Bad	High
Shape transform domains	Signature harmonic embedding		No	Good	Good	Good	Average	Good	Average	Bad	High
Shape transform domains	$\Re$ -Transform		No	Good	Good	Good	Bad	Good	Average	Average	High
Shape transform domains	Shapelets descriptor		No	Good	Good	Good	Bad	Good	Bad	Bad	High

9.9 Illustrative Example: A Contour-Based Shape Descriptor

In this section is presented a new contour-based shape descriptor we are developing: it belongs to the class of scale-space methods. Fundamental concepts about affine transforms are introduced, the method and its properties are presented and the method is then evaluated by applying it to shape retrieval from the MPEG-7 CE-Shape-1 database that consists of 1 400 contours.

9.9.1 Fundamental Concepts

Thereafter fundamental concepts are introduced and defined: the affine transform and 2 parameters which are linear (affine invariant) under affine transforms.

9.9.1.1 Closed Curve

Let us consider the discrete parametric equation of a closed curve Γ:

$$\Gamma(\mu) = (x(\mu), y(\mu)) \tag{9.73}$$

where $\mu \in \{0, ..., N-1\}$; an application curve may be parameterized with any number of vertices N.

9.9.1.2 Affine Transforms

The affine transformed version of a shape can be represented by the following equations:

$$\begin{bmatrix} x_a(\mu) \\ y_a(\mu) \end{bmatrix} = \begin{bmatrix} a & b \\ c & d \end{bmatrix} \begin{bmatrix} x(\mu) \\ y(\mu) \end{bmatrix} + \begin{bmatrix} e \\ f \end{bmatrix} = A \begin{bmatrix} x(\mu) \\ y(\mu) \end{bmatrix} + B \tag{9.74}$$

where $x_a(\mu)$ and $y_a(\mu)$ represent the coordinates of the transformed shape. Translation is represented by matrix B, while scaling, rotation and shear are reflected in matrix A. Corresponding values of coefficients of A can be found in the following matrices:

$$A_{Scaling} = \begin{bmatrix} S_x & 0 \\ 0 & S_y \end{bmatrix}, A_{Rotation} = \begin{bmatrix} \cos\theta & -\sin\theta \\ \sin\theta & \cos\theta \end{bmatrix}, A_{Shear} = \begin{bmatrix} 1 & k \\ 0 & 1 \end{bmatrix} \tag{9.75}$$

If S_x is equal to S_y, $A_{Scaling}$ represents uniform scaling and shape is not deformed under rotation, uniform scaling and translation. However, non-uniform scaling and shear contribute to shape deformation under general affine transforms.

9.9.1.3 Affine Invariant Parameters

The *arc length* parameter observed on a closed contour transforms linearly under any linear transformation up to the similarity transform. Translation and rotation do not affect the *arc length*; scaling scales the parameter by the same amount. An arbitrary choice of a starting point only introduces a shift in the parameter. However, the *arc length* is nonlinearly transformed under an affine transform and would not be a suitable parameter in this situation [46].

There are two parameters which are linear under affine transforms. They are the *affine arc length*, and the *enclosed area.*

The first parameter can be derived from the properties of determinants. It is defined as follows:

$$\tau = \int_{\alpha}^{\beta} [x'(s)y''(s) - x''(s)y'(s)]^{1/3} ds \tag{9.76}$$

where $x(s)$ and $y(s)$ are the coordinates of points on the contour and α and β are the curvilinear abscissa of 2 points on it.

The second affine invariant parameter is *enclosed area*, which is based on the property of affine transforms: under affine mapping, all areas are changed in the same ratio. Based on this property, Arbter et al.[4] defined a parameter ψ , which is linear under a general affine transform, so that:

$$\psi = \frac{1}{2} \int_{\alpha}^{\beta} |x(s)y'(s) - y(s)x'(s)| ds \tag{9.77}$$

where $x(s)$ and $y(s)$ are the coordinates of points on the contour with the origin of the system located at the centroid of the contour and α and β the curvilinear abscissa of 2 points on it. The parameter ψ is essentially the cumulative sum of triangular areas produced by connecting the centroid to pairs of successive vertices on the contour.

9.9.2 Equal Area Normalization

All points on a contour could be expressed in terms of the parameter of index points along the contour curve from a specified starting point. With affine transforms, the position of each point changes and it is possible that the number of points between two specified points changes too. So if we parameterize the contour using the equidistant vertices, the index point along the contour curve will change under affine transforms. For example, figure 9.36(a) is the top view of a plane, and (e) is its rear top view, so (e) is one of possible affine transforms of image (a). Via region segmentation or edge following, we obtain the contours of the two planes (b) and (f). (c) and (g) are parts of the contours (b) and (f) normalized by equidistant vertices respectively. In figure 9.36(c), the number of points on the segment between the points A and B is 21; however, the number is 14 in the same segment in figure 9.36(g). So the contour normalised by equidistant vertices is variant under possible affine transforms.

In order to make it be invariant under affine transforms, a novel curve normalization approach is proposed, which provides an affine invariant description of object curves at low computational cost, while at the same time preserving all information on curve shapes. We call this approach "equal area normalization" (EAN).

All points on a shape contour could be expressed in terms of two functions $\hat{\Gamma}(m) = (\hat{x}(m), \hat{y}(m))$, $m \in [0, M-1]$, where variable m is measured along the contour curve from a specified starting point and M is the total number of points on the contour. The steps of EAN are presented as follows:

1) Normalize $\hat{\Gamma}(m)$ to N points with equidistant vertices. The new functions are denoted $\overline{\Gamma}(\mu) = (\overline{x}(\mu), \overline{y}(\mu))$ and all the points on the contour are $\overline{P}_{\mu}$, where $\mu \in [0, N-1]$.

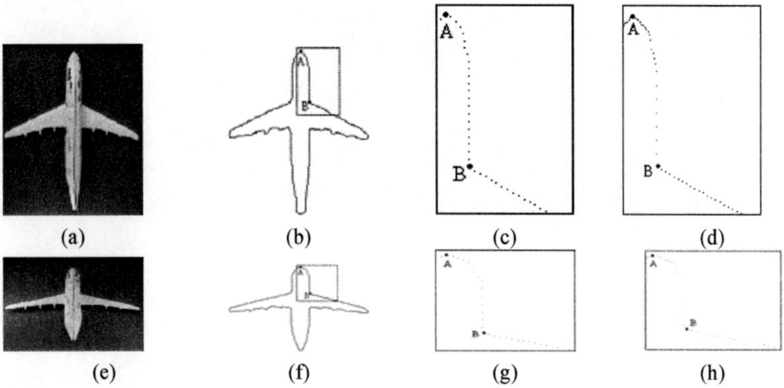

Fig. 9.36 The comparison of equidistant vertices normalization and equal area normalization. (a) is the image of the top view of a plane. (b) is the contour of image (a). (c) is a part of contour (b) normalized by equidistant vertices. (d) is a part of contour (b) normalized by equal area. (e) is the image of rear top view of the plane. (f) is the contour of image (e). (g) is a part of contour (f) normalized by equidistant vertices. (h) is a part of contour (f) normalized by equal area.

2) Calculate the second order moments of the contour at its centroid G.
3) Transfer the contour to make its centroid G be the origin of the system.
4) Point $\overline{P}_N(\overline{x}(N),\overline{y}(N))$ is assumed to be the same as the first point $\overline{P}_0(\overline{x}(0),\overline{y}(0))$.

Compute the area of the contour using the formula:

$$S=\frac{1}{2}\sum_{\mu=0}^{N-1}\left|\ \overline{x}(\mu)\overline{y}(\mu+1)-\overline{x}(\mu+1)\overline{y}(\mu)\ \right| \tag{9.78}$$

where $\frac{1}{2}\left|\overline{x}(\mu)\overline{y}(\mu+1)-\overline{x}(\mu+1)\overline{y}(\mu)\right|$ is the area of the triangle whose vertices are $\overline{P}_\mu(\overline{x}(\mu),\overline{y}(\mu))$, $\overline{P}_{\mu+1}(\overline{x}(\mu+1),\overline{y}(\mu+1))$, and centroid G (see figure 9.37).

5) Let the number of points on the contour after EAN be N. Of course, any other number of points could be chosen. Therefore, after EAN, each enclosed area S_{part} defined by any two successive points on the contour and the centroid G is equal to $S_{part}=S/N$.
6) Suppose all the points on the contour after EAN are P_t. Let $\Gamma(t)=(x(t),y(t))$ represent the contour, where $t\in[0,N-1]$. Select point $\overline{P}_0(\overline{x}(0),\overline{y}(0))$ on the equidistant vertices normalization as the starting point $P_0\ (x(0),y(0))$ of the EAN. On segment $P_0\overline{P}_1$, we seek a point $P_1(x(1),y(1))$, so that the area $S(0)$ of the triangle whose vertices are $P_0\ (x(0),y(0))$, $P_1(x(1),y(1))$ and $G(0,0)$ is equal to S_{part}. If there is no point to satisfy this condition, we seek the point P_1 on the segment $\overline{P}_1\overline{P}_2$. So area $S(0)$, which is the sum of the areas of triangle $\overline{P}_0\overline{P}_1G$ and triangle $\overline{P}_1P_1G$, is equal to S_{part}. If again there is yet no

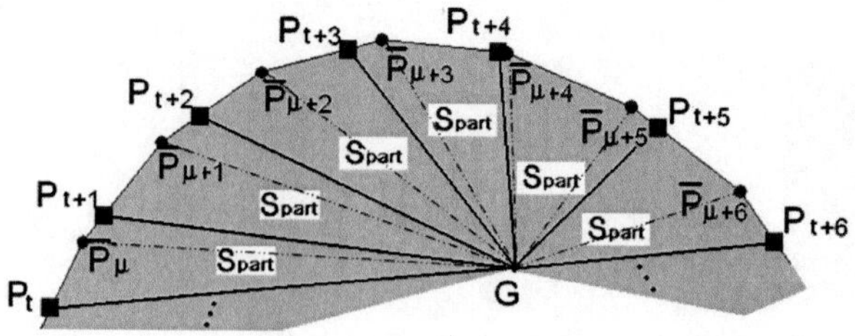

Fig. 9.37 The method of equal area normalization. "•" is the vertex $\overline{P}$ of equidistant vertices normalization, and "■" is the point P of equal area normalization. G is the centroid of the contour.

point to satisfy the condition, we continue to seek for the point in the next segment until the condition is satisfied. This point P_1 is the second point on the normalized contour.

7) From point $P_1(x(1), y(1))$, we use the same method to calculate all the other points $P_t(x(t), y(t))$, $t \in [2, N-1]$ along the contour. Because the area of each closed zone, e.g. the polygon $P_t[\overline{P}_\mu \overline{P}_{\mu+1} \cdots]P_{t+1}G$ where $t \in [0, N-2]$ is equal to S_{part}, the total area of $N-1$ polygon is equal to $(N-1)\cdot S_{part}$. According to the step 5, the area of the last zone $\overline{P}_{N-1}[\overline{P}_\mu \overline{P}_{\mu+1} \cdots \overline{P}_{N-1}]P_0 G$ is exactly equal to:

$$S-(N-1)\cdot S_{part} = N \cdot S_{part} - (N-1)\cdot S_{part} = S_{part}$$

From figure 9.37 we know that the area of triangle $P_t P_{t+1} G$ is approximately equal to the area S_{part} of polygon $P_t[\overline{P}_\mu \overline{P}_{\mu+1} \cdots]P_{t+1}G$ if the two points $\overline{P}_\mu$ and $\overline{P}_{\mu+1}$ are close enough or the number N of the points on the contour is large enough. Therefore, we can use the points P_t, $t \in [0, N-1]$ to replace the points $\overline{P}_\mu$, $\mu \in [0, N-1]$; the EAN process is then complete.

After this normalization, the number of vertices on the segment between the two appointed points is invariant under affine transforms. Figure 9.36(d) and (h) are the same parts of figure 9.36(c) and (g) respectively. We notice that the distance between the consecutive points is not uniform. In figure 9.36(d), the number of points between points A and B is 23, the number is also 23 in figure 9.36(g). Therefore, after applying EAN, the index of the points on a contour can remain stable with their positions under affine transforms. This property will be very advantageous when extracting the robust attributes of a contour and decreasing complexity in the measurement of similarity. We can also use EAN with the other algorithms, to improve their robustness with affine transforms. For example, before applying the curvature scale space (CSS) algorithm [32], the contour can be normalized by EAN: none of the maximum points in the CSS image will change under affine transforms. This is beneficial when calculating the similarity between two CSS attributes.

9.9.3 Normalized Part Area Vector

In this section, we look for the existing relations between the part area S_{part}, affine transforms and low-pass filtering.

THEOREM 1

Let $\Gamma_a(\mu)=(x_a(\mu), y_a(\mu))$ be the transformed version of a curve $\Gamma(\mu)= (x(\mu), y(\mu))$ under an affine transform A, where μ is an arbitrary parameter, $\Gamma_{af}(\mu)=(x_{af}(\mu), y_{af}(\mu))$ notes that $\Gamma_a(\mu)$ is filtered by a linear low-pass filter F. Let $\Gamma_f(\mu)=(x_f(\mu), y_f(\mu))$ note that $\Gamma(\mu)$ is filtered by the same low-pass filter F, and $\Gamma_{fa}(\mu)=(x_{fa}(\mu), y_{fa}(\mu))$ refers to the transformed version of $\Gamma_f(\mu)$ under the same affine transform A. The curve $\Gamma_{af}(\mu)$ is then the same as curve $\Gamma_{fa}(\mu)$. In other words: $F(A(\Gamma(\mu)))=A(F(\Gamma(\mu)))$. The following figure illustrates this theorem.

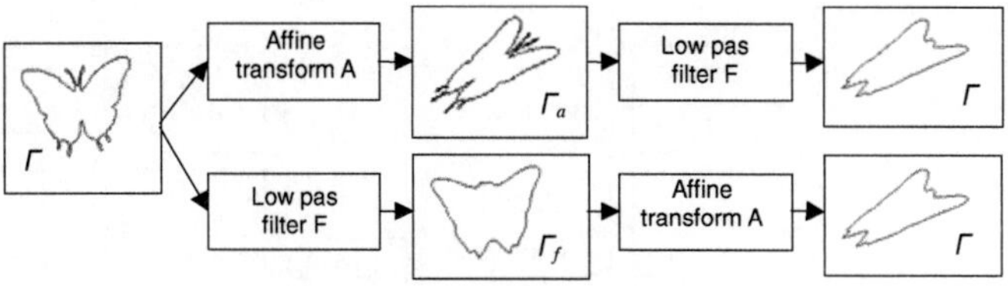

Fig. 9.38 Illustration of theorem1

PROOF

From (9.74) we have

$$x_a(\mu) = ax(\mu) + by(\mu) + e \tag{9.79}$$

$$y_a(\mu) = cx(\mu) + dy(\mu) + f \tag{9.80}$$

For the entire contour, we transfer its centre of gravity to the origin of the system, so that the translation e and f can be removed. Therefore, the affine transform can be represented by two simple formulae:

$$x_a(\mu) = ax(\mu) + by(\mu) \tag{9.81}$$

$$y_a(\mu) = cx(\mu) + dy(\mu) \tag{9.82}$$

The computation starts by convolving each coordinate of the curve $\Gamma_a(\mu)$ with a linear low-pass filter F whose impulse response is $g(\mu)$. In the continuous form this leads to:

$$x_{af}(\mu) = x_a(\mu) * g(\mu) \quad = [ax(\mu) + by(\mu)] * g(\mu)$$

$$= ax(\mu) * g(\mu) + by(\mu) * g(\mu) \quad = ax_f(\mu) + by_f(\mu) \tag{9.83}$$

where $*$ denotes the convolution. Likewise,

$$y_{af}(\mu) = cx_f(\mu) + dy_f(\mu) \tag{9.84}$$

By comparison of equations (9.79-9.84), it is clear that point $(x_{af}(\mu), y_{af}(\mu))$ is the same as point $(x_f(\mu), y_f(\mu))$ transformed by the affine transform A. So curve $\Gamma_{af}(\mu)$ is the same as curve $\Gamma_{fa}(\mu)$. Theorem1 indicates that exchanging the computation order between affine transform and filtering does not change the result.

THEOREM 2

For any affine transform of a closed contour, using EAN sets parameter t to produce the curve $\Gamma_a(t)=(x_a(t), y_a(t))$. If area $s_p(t)$ is the area of an enclosed sector whose vertices are a pair of successive points and the centroid of the contour and if $\Gamma_{af}(t)=(x_{af}(t), y_{af}(t))$ indicates that $\Gamma_a(t)$ is filtered by a low-pass filter F, then the changes in enclosed areas $s_p(t)$ on the $\Gamma_{af}(t)$ are linear with affine mapping as illustrated on figure 9.39.

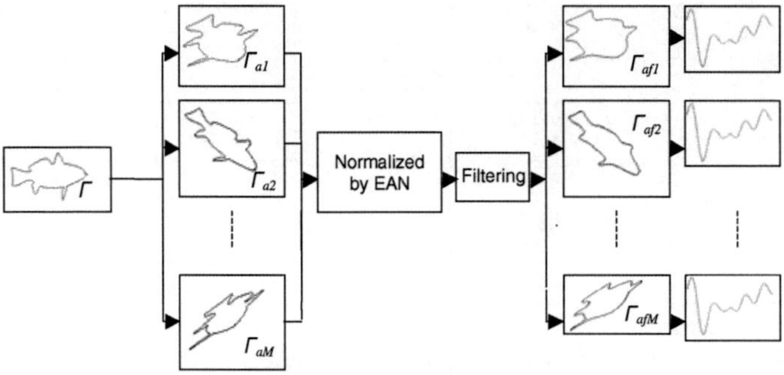

Fig. 9.39 Illustration of theorem2

PROOF

In section 3, we know the enclosed area $s_p(t)$ of the triangle on the filtered affine contour whose vertices are $(x_{af}(t), y_{af}(t))$, $(x_{af}(t+1), y_{af}(t+1))$ and the centroid G is

$$s_p(t) = \frac{1}{2}\left| x_{af}(t) y_{af}(t+1) - x_{af}(t+1) y_{af}(t) \right| \tag{9.85}$$

Due to THEOREM1,

$$x_{af}(t) = x_{fa}(t) = ax_f(t) + by_f(t) \tag{9.86}$$

$$y_{af}(t) = y_{fa}(t) = cx_f(t) + dy_f(t) \tag{9.87}$$

and

$$x_{af}(t+1) = x_{fa}(t+1) = ax_f(t+1) + by_f(t+1) \tag{9.88}$$

$$y_{af}(t+1) = y_{fa}(t+1) = cx_f(t+1) + dy_f(t+1) \tag{9.89}$$

Therefore from equation (9.85), we can write

$$\begin{aligned} s_p(t) &= \frac{1}{2}\big|[ax_f(t)+by_f(t)][cx_f(t+1)+dy_f(t+1)] \\ &\quad -[ax_f(t+1)+by_f(t+1)][cx_f(t)+dy_f(t)]\big| \\ &= \frac{1}{2}\big|adx_f(t)y_f(t+1)+bcx_f(t+1)y_f(t) \\ &\quad -adx_f(t+1)y_f(t)-bcy_f(t+1)x_f(t) \\ &= \frac{1}{2}|ad-bc|\cdot|x_f(t)y_f(t+1)-x_f(t+1)y_f(t)| \end{aligned} \tag{9.90}$$

Observing equation (9.90), $s_p(t)$ is just linearly proportional by a scale factor $|ad-bc|$. Accordingly we have proved that enclosed areas $s_p(t)$ are linear with affine mapping.

DEDUCTION

The proportion $v'(t)$ of closed areas $s_p(t)$ with the total area S of the filtered contour is preserved under general affine transforms.

PROOF

According to relation (9.90), the total area S of the filtered contour is:

$$S = \frac{1}{2}|ad-bc|\cdot\sum_{t=0}^{N-1}|x_f(t)y_f(t+1)-x_f(t+1)y_f(t)| \tag{9.91}$$

so that

$$\begin{aligned} v'(t) &= s_p(t)/S \\ &= |x_f(t)y_f(t+1)-x_f(t+1)y_f(t)|/\sum_{t=0}^{N-1}|x_f(t)y_f(t+1)-x_f(t+1)y_f(t)| \end{aligned} \tag{9.92}$$

Equation (9.92) indicates that $v'(t)$ is not related to the affine parameters a, b, c and d. Therefore $v'(t)$ is preserved under general affine transforms. In addition, we can deduce a major property of $v'(t)$: the integration of $v(t)=[v'(t)-1/N]$ is equal to zero as shown by equation (9.93).

$$\sum_{t=0}^{N-1} v(t) = \sum_{t=0}^{N-1}[v'(t)-\frac{1}{N}] = \sum_{t=0}^{N-1}\frac{s_p(t)}{S} - \sum_{t=0}^{N-1}\frac{1}{N} = \frac{S}{S}-\frac{N}{N} = 0 \tag{9.93}$$

We refer to vector $v(t)$ as the normalized part area vector (NPAV). Figure 9.40 shows an example of NPAV. The contour is normalized to 512 points by EAN.

As THEOREM2 and its deduction show, in all cases, even those with severe deformations, the function $s_p(t)$ is also preserved. Only the amplitude changes

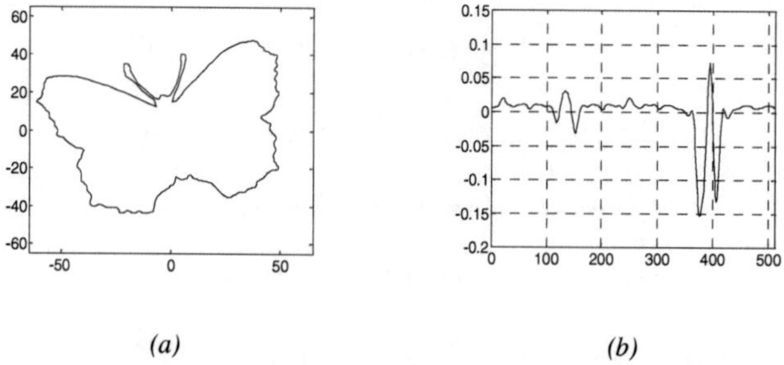

Fig. 9.40 An example of NPAV. (a) The contour of a butterfly. (b) The NPAV of the contour (a).

under general affine transforms; the NPAV $v(t)$ has an affine-invariant feature. In the following section, we will present the results of our experiments which evaluate the property of the proposed algorithm.

9.9.4 Experimental Results

In this section, we will evaluate the behaviour of NPAV $v(t)$ in relation to the affine transforms, EAN, filtering and noise by presenting various experimental results. We consider the results of our experiments on the MPEG-7 CE-shape-1 database containing 1400 images of shapes. The contours in this database yield a great range of shape variation. The framework of these experiments is shown in figure 9.41.

The input signal of the experiment is an original contour from the MPEG-7 database, and the output is the maximum linear correlation coefficient between the NPAV $v(t)$ of the upper pathway and that of the lower pathway. The upper pathway includes the affine transforms, noise power control, equal area normalization and the

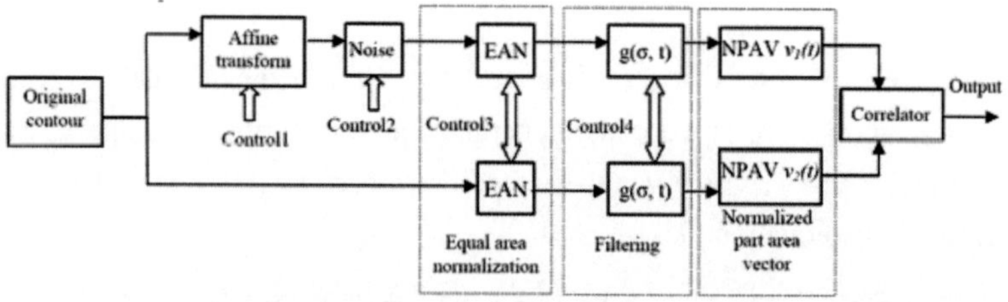

Fig. 9.41 The framework of experiments

low-pass filter. The output signal of the upper pathway is the NPAV $v_1(t)$ of the affine contour. The lower pathway includes the equal area normalization and low-pass filter. The output signal of the lower pathway is the NPAV $v_2(t)$ of the original contour. The correlator is then applied to calculate the maximum linear correlation coefficient between NPAV $v_1(t)$ and NPAV $v_2(t)$ by shifting the vector $v_2(t)$ circularly.

The four control points are presented as follows:

- **Control1** is applied to control the parameters of affine transforms. We can control the affine contour by respectively scaling, rotating and shearing or mixing the transforms.
- **Control2** controls the power of the noise.
- **Control3** controls the parameters of equal area normalization in the upper and lower pathways. The parameters are the number of normalized points and the position of the starting point.
- **Control4** is applied to simultaneously control the bandwidth of the low-pass filters in the upper and lower pathways. Here, the filter is a Gaussian kernel with a scale parameter σ given by the standard deviation of the filter.

9.9.4.1 The NPAV and Scaling Transforms

The scaling transform is one of the affine transform modes. It is obtained by applying the matrix $A_{scaling}$ to the contour coordinates. For a uniform scaling transformation, we choose the parametric matrix

$$A_{Scaling1} = \begin{bmatrix} \gamma_1 & 0 \\ 0 & \gamma_1 \end{bmatrix} \tag{9.94}$$

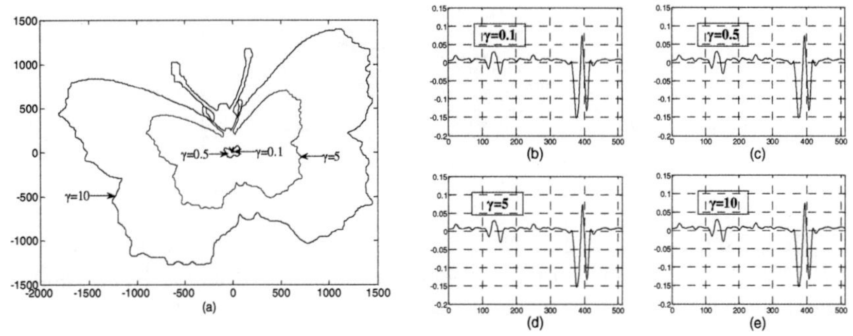

Fig. 9.42 Illustration of the robustness of NPAV under uniform scaling transforms. (a) is 4 different scale contours. (b), (c), (d) and (e) are the NPAVs of each contour in (a) respectively.

Suppose γ_1 is successively equal to 0.1, 0.5, 5, 10, for the different observed contours of figure 9.42 (a): we obtain figures 9.42 (b)-(e) respectively. For the non-uniform scaling transformation, we choose the parametric matrix

$$A_{Scaling2} = \begin{bmatrix} \gamma_2 & 0 \\ 0 & 1 \end{bmatrix} \tag{9.95}$$

Suppose γ_2 is successively equal to 0.1, 0.5, 5, 10, for the figure 9.43 (a): we obtain figures 9.43 (b)-(e) respectively.

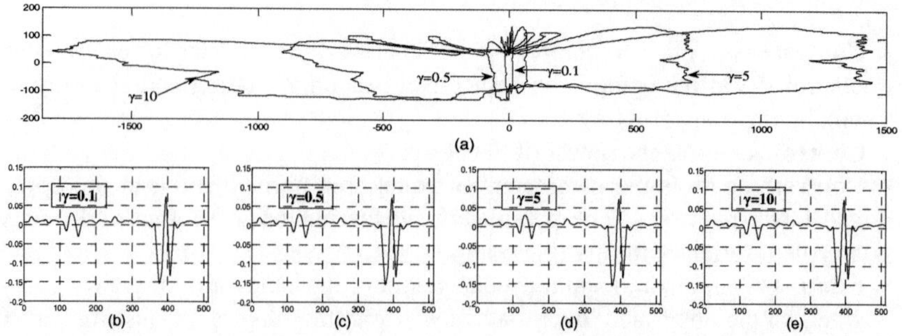

Fig. 9.43 Illustration of the robustness of NPAV under non-uniform scaling transforms. (a) is 4 different scale contours. (b), (c), (d) and (e) are the NPAVs of each contour in (a) respectively.

Figures 9.42 and 9.43 show that although the shape of the butterfly changes with scaling, the NPAV remains identical.

Furthermore, we calculate statistical results. For all the shapes in the database, the average correlation between the NPAV of the original shape and that of its scaling transforms under matrixes $A_{scaling1}$ and $A_{scaling2}$ is presented in Table 9.2.

Table 9.2 Correlation with the Scaling Transforms

	Correlation coefficient	
γ_1/γ_2	Uniform	Non-uniform
0.1	0.999	0.972
0.5	0.999	0.985
5	0.999	0.980
10	0.999	0.976

Table 9.2 shows that the uniform scaling transform has practically no effect on the NPAV of the contour and that non-uniform scaling affects the NPAV only a little. So the NPAV is very robust under scaling transforms.

9.9.4.2 The NPAV and Rotation Transforms

As stated before, a rotation transform is obtained by applying the matrix $A_{Rotation}$ to the contour coordinates. Let the rotation angles θ be 60°, 120°, 180°, 240° and 300°. The position of the starting point of a contour is unchanged. Figure 9.44 shows the rotated copies of the pattern on Figure 9.40(a) and their NPAVs. As can be seen in figure 9.44, all the NPAVs are identical. Furthermore, we present the statistical results. For all the shapes in the database, the average correlation between the NPAV of the original shape and that of its rotated transforms under the rotation angles θ is presented in Table 9.3.

In this way, as the position of the starting point does not change so that the NPAVs maintain their invariance.

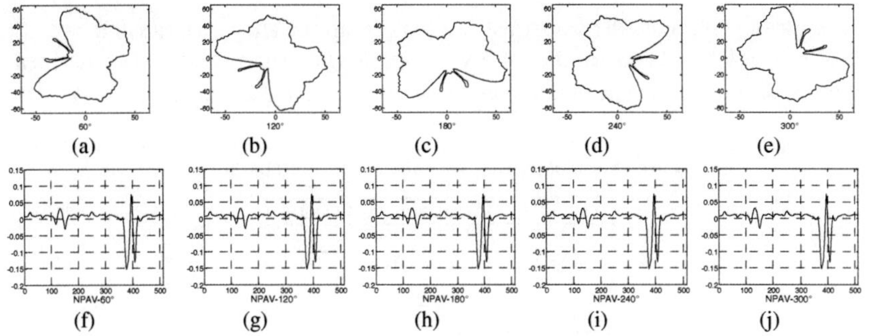

Fig. 9.44 Illustration of the robustness of NPAV under rotation: (a)-(e) are the same contour with 5 different orientations. (f)-(j) are the NPAVs of each contour in (a)-(e) respectively.

Table 9.3 Correlation with the rotation transforms

θ	Correlation coefficient
60°	1.000
120°	1.000
180°	1.000
240°	1.000
300°	1.000

9.9.4.3 The NPAV and Shearing Transforms

As introduced in subsection 9.1.2, a shearing is obtained by applying the matrix A_{Shear} to the contour coordinates. Let the shearing parameter k be successively equal to 0.1, 0.5, 5 and 10. Suppose the position of the starting point of the contour remains unchanged. Figure 9.45 shows the shearing copies of pattern figure 9.40(a) and their NPAVs.

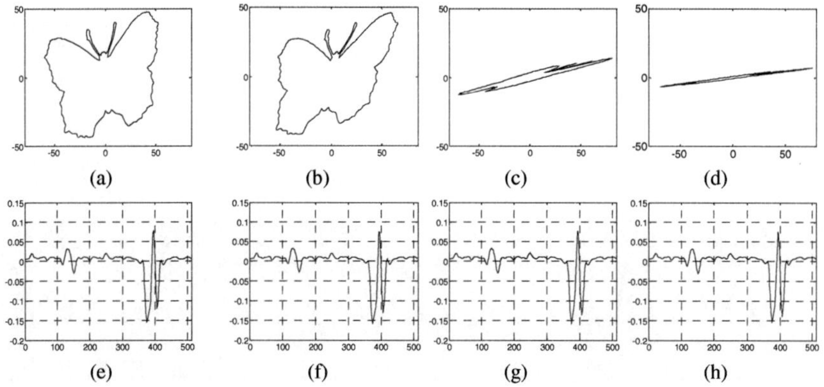

Fig. 9.45 Illustration of the robustness of NPAV under shearing transforms. (a)-(d) are the same contour with 4 different shearing transforms. (e)-(h) are the NPAVs of each contour in (a)-(d) respectively.

We further calculate the statistical results. For all the shapes in the database, the average correlation between the NPAV of the original shape and that of its shearing transform under shearing parameter k is presented in the Table 9.4.

Table 9.4 Correlation with the shearing transforms

k	Correlation coefficient
0.1	0.989
0.5	0.987
5	0.980
10	0.972

The three above experiments indicate that the descriptor NPAV is relatively affine-invariant. As shown below, we will evaluate the effect of equal area normalization (EAN) under the same affine transform. Suppose the contour rotates 60° counter clockwise and the shearing parameter k is 1.5. The affine matrix is

$$A = \begin{bmatrix} \frac{2+3\sqrt{3}}{4} & -\frac{3-2\sqrt{3}}{4} \\ \frac{\sqrt{3}}{2} & \frac{1}{2} \end{bmatrix} \tag{9.96}$$

Adjust 'Control2' to let the power of noise be 0. Adjust 'Control4' to let $\sigma = 10$. The contents of the experimental set include the two following aspects: the relation between the NPAV and the number of points normalized by EAN and the relation between the NPAV and the position of the starting points on a contour. The results are presented in subsections *9.4.4* and *9.4.5.*

9.9.4.4 The NPAV and the Number of Points Normalized by EAN

The number of points on the contour is normalized to 64, 128 and 256. Figure 9.47 shows the original and the affined contours of the pattern in figure 9.41(a) with various numbers of points and their NPAVs.

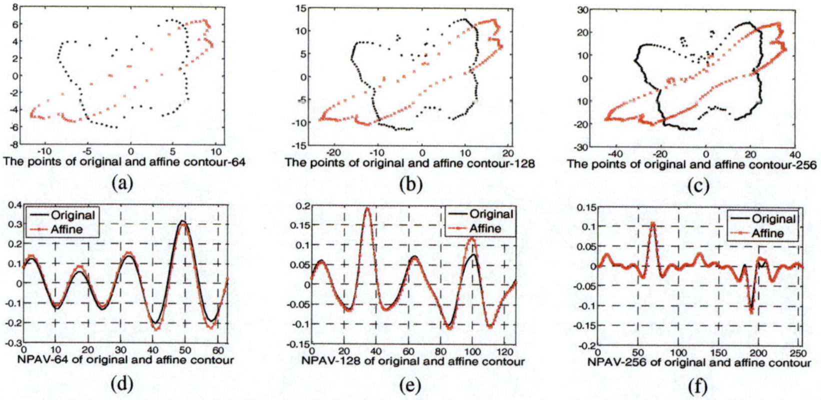

Fig. 9.46 Illustration of the robustness of NPAV under the contour normalized to different number of points. (a)-(c) are the original and affine contours normalized to 64, 128 and 256 respectively. (d)-(f) are the NPAVs of the two contours in (a)-(c) respectively.

We notice that although the NPAVs of the contour with different numbers of points are very different from each other, the NPAV of the affine contour and that of original contour are almost identical under the same number of points normalized by EAN. Table 9.5 presents the statistical results. It shows that under the different number of points normalized by EAN, the NPAV changes slightly under affine transforms.

Table 9.5 Correlation under different normalized number of points

Number of points	Correlation coefficient
64	0.985
128	0.992
256	0.993

9.9.4.5 The NPAV and the Position of the Starting Point

Suppose the position of the starting point (SP) is located on the different positions '1', '2', '3' or '4', as illustrated on figure 9.47(a)-(d). These positions are located at 12.5%, 25%, 37.5% and 50% of the original starting with 100% corresponding to the total number of points. Suppose the contour is normalized to 256 points. figure 9.47 shows the effect of different starting points to the NPAV of the original and affine contours.

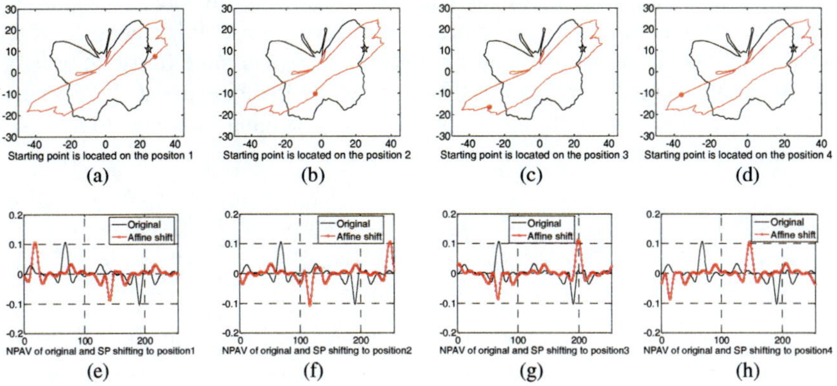

Fig. 9.47 Illustration of the robustness of a NPAV with various positions of the starting point. (a)-(d) are the original and the affine contours with the starting point located at position 1-4 respectively. '☆'is the position of the starting point on the original contour; ' • ''is the position of the starting point on the affine contour . (e)-(h) are the NPAVs of the two contours with different positions for the starting point in (a)-(d) respectively.

As obvious from figure 9.47, the NPAV of various starting point positions are topologically identical except for a 'circular' delay. In this way, a shift in the starting point is equivalent to a circular delay in the NPAV.

We further calculate the statistical results. To calculate the correlation between the NPAV of the original contour and that of its affine transforms with a shift of starting point, we move the NPAV of the original shape point by point and search for the highest value of the correlation. For all the shapes in the database, the average correlations of the various starting point positions under the same number of points normalized by EAN are presented in the Table 9.6.

As can be seen in the table, the position of the starting point on the contour does not affect the robustness of the NPAV.

Table 9.6 Correlation under different position of starting point

Starting point shift	Correlation coefficient
12.5%	0.989
25%	0.987
37.5%	0.974
50%	0.972

9.9.4.6 The NPAV and Noise

For different reasons, it happens that the curve undergoes perturbations so that it becomes noisy. To reduce the effect of noise, the curve is first smoothened by applying a low-pass Gaussian filter of standard deviation set to $\sigma = 2$. The NPAVs and their shape contaminated by the random uniform noise with different *SNR* are presented in figure 9.48. It reveals that the NPAV is quite robust to boundary noise and irregularities, even in the presence of severe noise. It is clear that, as the noise amplitude increases, the contours become more and more fuzzy. In order to calculate the average correlation coefficient, we do the experiments by contaminating the test contours with random uniform noise ranging from high to low *SNR* affecting the database.

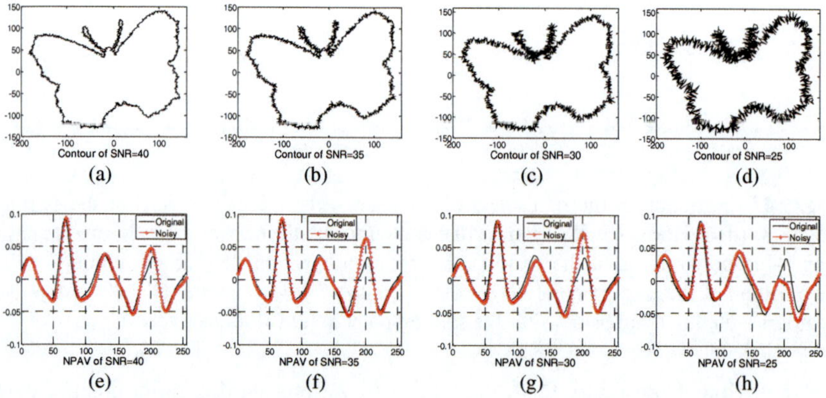

Fig. 9.48 Demonstration of NPAV under the condition of different SNR. (a)-(d) are the contour contaminated by different noise power. (e)-(h) are the NPAVs of contours in (a)-(d) respectively.

Table 9.7 Correlation under different *SNR*

SNR	Correlation coefficient
40dB	0.964
35dB	0.963
30dB	0.949
25dB	0.898

Table 9.7 shows the average correlation coefficient of all the NPAVs of shapes in the database under different *SNR*.

This shows the NPAV's suitability for use in noisy conditions. By analyzing the experimental results, we notice that NPAV is quite robust to scale, orientation, shearing of objects, noise and the position of starting point. Therefore, NPAV can be used to characterize a pattern for recognition purposes.

9.9.4.7 Evaluation on Pattern Retrieval

In order to assess the retrieval performance, we create affine transformed versions of our existing shape contours. Suppose the contour rotates θ counter-clockwise, and the shearing parameter is k. The matrix A is then constructed as follows.

$$A = \begin{bmatrix} k \cdot \sin\theta + \cos\theta & k\cos\theta - \sin\theta \\ \sin\theta & \cos\theta \end{bmatrix} \tag{9.97}$$

Let us consider $\theta = 2n\pi/5$ with n=0, 1, 2 and 4, and k=1 and 2. Therefore, 10 affine transforms are applied to each contour so that the new database consists of $1400 \times 10 = 14000$ transformed contours. The similarity measure between two NPAV attributes, $v_1(i)$ *and* $v_2(i)$, *i=0, 1, …, N-1* $v_1(i), v_2(i), i \in [0, N-1]$ can be represented by the following functions:

$$d_1 = \min_{n=0}^{N-1}\left[\sum_{i=0}^{N-1} |v_1(i) - v_2(j)|\right] \qquad d_2 = \min_{n=0}^{N-1}\left[\sum_{i=0}^{N-1} |v_1(N-1-i) - v_2(j)|\right] \tag{9.98}$$

where $j = \begin{cases} i+n, & i+n < N \\ i+n-N, & i+n > N-1 \end{cases}$

Then similarity is given by:

$$d = \min(d_1, d_2) \tag{9.99}$$

Figure 9.49 gives examples of the retrieval results. Each example is presented in two rows, starting with the input query on the left-hand side and followed by the outputs from the system in response to that query. It denotes the first 20 retrieved contours and their similarity distance *d*.

We notice that all 10 affine transforms of the query contour appear in the first 10 images. And the similarity distance of the first non-relevant contour is much greater than that of related contours. So, we can retrieve the similar contours easily. For the 1400 query original contours, the statistical average distance of the first 10 related contours is only 31.5% of the average distance of the first non-relevant contour.

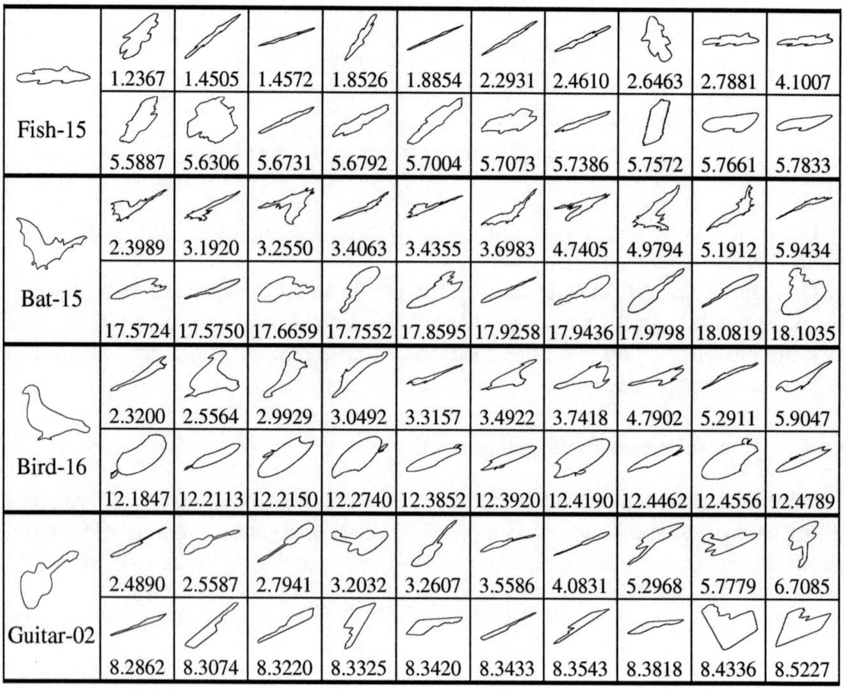

Fig. 9.49 Illustrative retrieval results obtained by the multi-scale NPAV.

All these previous results indicate that NPAV stays robust under affine transforms.

9.10 Conclusion

In this chapter we have studied and compared the methods of shape-based feature extraction and representation. About 40 techniques for extraction of shape features have been shortly described, referenced in a bibliography and synthetically compared. Unlike the traditional classification, the approaches of shape-based feature extraction and representation were classified by their processing approaches. These processing approaches included shape signatures, polygonal approximation methods, spatial interrelation feature, moments approaches, scale-space methods and shape transform domains: in such way, one can easily select the appropriate processing approach. A synthetic table has been established for a fast and global comparison of the performances of these approaches.

To go more deeply in shape based feature extraction we have also described and evaluated a new method designed for extracting invariants of a shape under affine transform. Our representation is based on the association of two parameters: the affine arc length and the enclosed area, viz. we normalize a contour to affine-invariant length by the affine enclosed area. For the needs of this new approach, we proved two theorems and a deduction. They revealed that, for a filtered contour, the part enclosed area is linear under affine transforms. We further defined the affine-invariance vector: the normalized part area vector (NPAV). After a number of experiments applied to the MPEG-7 CE-shape-1 database, we demonstrated that NPAV is quite robust with respect to affine transforms and noise, even in the presence of severe noise.

References

[1] Abbasi, S., Mokhtarian, F., Kittler, J.: Enhancing CSS-based shape retrieval for objects with shallow concavities. Image and Vision Computing 18(3), 199–211 (2000)

[2] Alajlan, N., Kamel, M.S., Freeman, G.: Multi-object image retrieval based on shape and topology. Signal Processing: Image Communication 21, 904–918 (2006)

[3] Alajlan, N., Rube, I.E., Kamel, M.S., Freeman, G.: Shape retrieval using triangle-area representation and dynamic space warping. Pattern Recognition 40(7), 1911–1920 (2007)

[4] Arbter, K., Snyder, W., Burkhardt, H., Hirzinger, G.: Applications of affine-invariant Fourier descriptors to recognition of 3-D objects. IEEE Trans. Pattern Analysis and Machine Intelligence 12(7), 640–646 (1990)

[5] Arica, N., Vural, F.: BAS: a perceptual shape descriptor based on the beam angle statistics. Pattern Recognition Letters 24(9-10) (2003)

[6] Badawy, O.E., Kamel, M.: Shape Retrieval using Concavity Trees. In: Proceedings of the 17th International Conference on Pattern Recognition, pp. 111–114 (2004)

[7] Bauckhage, C., Tsotsos, J.K.: Bounding box splitting for robust shape classification. In: Proc. IEEE International Conference on Image Processing, pp. 478–481 (2005)

[8] Belongie, S., Malik, J., Puzicha, J.: Shape Matching and Object Recognition Using Shape Context. IEEE Trans. Pattern Analysis and Machine Intelligence 24(4), 509–522 (2002)

[9] Berretti, S., Bimbo, A.D., Pala, P.: Retrieval by shape similarity with perceptual distance and effective indexing. IEEE Trans. on Multimedia 2(4), 225–239 (2000)

[10] Celebi, M.E., Aslandogan, Y.A.: A Comparative Study of Three Moment-Based Shape Descriptors. In: Proc. of the International Conference of Information Technology: Codingand Computing, pp. 788–793 (2005)

[11] Chakrabarti, K., Binderberger, M., Porkaew, K., Mehrotra, S.: Similar shape retrieval in MARS. In: Proc. IEEE International Conference on Multimedia and Expo. (2000)

[12] Chen, G., Bui, T.D.: Invariant Fourier-wavelet descriptor for pattern recognition. Pattern Recognition 32, 1083–1088 (1999)

[13] Chuang, C.-H., Kuo, C.-C.: Wavelet Descriptor of Planar Curves: Theory and Applications. IEEE Trans. Image Processing 5(1), 56–70 (1996)

[14] Davies, E.: Machine Vision: Theory, Algorithms, Practicalities. Academic Press, New York (1997)

[15] Dubinskiy, A., Zhu, S.C.: A Multi-scale Generative Model for Animate Shapes and Parts. In: Proc. Ninth IEEE International Conference on Computer Vision, ICCV (2003)

[16] Gonzalez, R., Woods, R.: Digital image processing, 2nd edn. Pearson Education North Asia Limited and Publishing House of Electronics Industry (2002)

[17] Guru, D., Nagendraswam, H.: Symbolic representation of two-dimensional shapes. Pattern Recognition Letters 28, 144–155 (2007)

[18] Han, S., Yang, S.: An Invariant Feature Representation for shape Retrieval. In: Proc. Sixth International Conference on Parallel and Distributed Computing, Applications and Technologies (2005)

[19] Jalba, A., Wilkinson, M., Roerdink, J.: Shape representation and recognition through morphological curvature scale spaces. IEEE Trans. Image Processing 15(2), 331–341 (2006)

[20] Jin, K., Cao, M., Kong, S., Lu, Y.: Homocentric Polar-Radius Moment for Shape Classification. In: The 8th International Conference on Proc. Signal Processing (2006)

[21] Kan, C., Srinath, M.D.: Invariant character recognition with Zernike and orthogonal Fourier-Mellin moments. Pattern Recognition 35, 143–154 (2002)

[22] Kauppinen, H., Seppanen, T., Pietikainen, M.: An Experimental Comparison of Autoregressive and Fourier-Based Descriptors in 2-D Shape Classification. IEEE Trans. Pattern Analysis and Machine Intelligence 17(2), 201–207 (1995)

[23] Khalil, M., Bayoumi, M.: A Dyadic Wavelet Affine Invariant Function for 2D Shape Recognition. IEEE Trans. Pattern Analysis and Machine Intelligence 25(10), 1152–1164 (2001)

[24] Kpalma, K., Ronsin, J.: Multiscale contour description for pattern recognition. Pattern Recognition Letters 27, 1545–1559 (2006)

[25] Latecki, L.J., Lakamper, R.: Shape Similarity Measure Based on Correspondence of Visual Parts. IEEE Trans. Pattern Analysis and Machine Intelligence 22(10), 1185–1190 (2000)

[26] Latecki, L.J., Lakamper, R.: Convexity rule for shape decomposition based on discrete Contour Evolution. Computer Vision and Image Understanding 73(3), 441–454 (1999)

[27] Lee, S.-M., Abbott, A.L., Clark, N.A., Araman, P.A.: A Shape Representation for Planar Curves by Shape Signature Harmonic Embedding. In: Proc. IEEE Computer Society Conference on Computer Vision and Pattern Recognition (2006)

[28] Liu, Y.K., Wei, W., Wang, P.J., Zalik, B.: Compressed vertex chain codes. Pattern Recognition 40(11), 2908–2913 (2007)

[29] Lu, G., Sajjanhar, A.: Region-based shape representation and similarity measure suitable for content based image retrieval. ACM Multimedia System Journal 7(2), 165–174 (1999)

[30] Lu, K.-J., Kota, S.: Compliant Mechanism Synthesis for Shape-Change Applications: Preliminary Results. In: Proceedings of SPIE Modelling, Signal Processing, and Control Conference, pp. 161–172 (2002)

[31] Mehtre, B.M., Kankanhalli, M.S., Lee, W.F.: Shape Measures for Content Based Image Retrieval: A Comparison. Pattern Recognition 33(3), 319–337 (1997)

[32] Mokhtarian, F., Mackworth, A.K.: A Theory of Multiscale, Curvature-Based Shape Representation for Planar Curves. IEEE Trans. Pattern Analysis and Machine Intelligence 14(8), 789–805 (1992)

[33] Mori, G., Malik, J.: Estimating Human Body Configurations Using Shape Context Matching. In: Heyden, A., Sparr, G., Nielsen, M., Johansen, P. (eds.) ECCV 2002. LNCS, vol. 2352, pp. 666–680. Springer, Heidelberg (2002)

[34] Mukundan, R.: A new class of rotational invariants using discrete orthogonal moments. In: Sixth IASTED International Conference on Signal and Image Processing, pp. 80–84 (2004)

[35] Mukundan, R., Ong, S., Lee, P.: Image Analysis by Tchebichef Moments. IEEE Trans. Image Processing 10(9), 1357–1364 (2001)

[36] Peng, J., Yang, W., Cao, Z.: A Symbolic Representation for Shape Retrieval in Curvature Scale Space. In: Proc. International Conference on Computational Intelligence for Modelling Control and Automation and International Conference on Intelligent Agents Web Technologies and International Commerce (2006)

[37] Ricard, J., Coeurjolly, D., Baskurt, A.: Generalizations of angular radial transform for 2D and 3D shape retrieval. Pattern Recognition Letters 26(14) (2005)

[38] Sebastian, T., Klein, P., Kimia, B.: Recognition of Shapes by Editing Their Shock Graphs. IEEE Trans. Pattern Analysis and Machine Intelligence 26(5), 550–571 (2004)

[39] Siddiqi, K., Kimia, B.: A Shock Grammar for Recognition. In: Proceedings of the IEEE Conference Computer Vision and Pattern Recognition, pp. 507–513 (1996)

[40] Smith, S.P., Jain, A.K.: Chord distribution for shape matching. Computer Graphics and Image Processing 20, 259–271 (1982)

[41] Sonka, M., Hlavac, V., Boyle, R.: Image Processing, Analysis and Machine Vision. Chapman and Hall, London (1993)

[42] Tabbone, S., Wendling, L., Salmon, J.-P.: A new shape descriptor defined on the Radon transform. Computer Vision and Image Understanding 102(1), 42–51 (2006)

[43] Taubin, G., Cooper, D.: Recognition and positioning of rigid objects using algebraic moment invariants. In: SPIE Conference on Geometric Methods in Computer Vision, pp. 175–186 (1991)

[44] Taza, A., Suen, C.: Discrimination of planar shapes using shape matrices. IEEE Trans. System, Man, and Cybernetics 19(5), 1281–1289 (1989)

[45] Thayananthan, A., Stenger, B., Torr, P.H.S., Cipolla, R.: Shape Context and Chamfer Matching in Cluttered Scenes. In: Proc. IEEE Computer Society Conference on Computer Vision and Pattern Recognition (2003)

[46] Tieng, Q.M., Boles, W.W.: Wavelet-Based Affine Invariant Representation: A Tool for Recognizing PlanarObjects in 3D Space. IEEE Trans. Pattern Analysis and Machine Intelligence 19(8), 846–857 (1997)

[47] Wang, Y.P., Lee, K.T.: Multiscale curvature-based shape representation using B-spline wavelets. IEEE Trans. Image Process 8(10), 1586–1592 (1999)

[48] Yadava, R.B., Nishchala, N.K., Gupta, A.K.: Retrieval and classification of shape-based objects using Fourier, generic Fourier, and wavelet-Fourier descriptors technique: A comparative study. Optics and Lasers in Engineering 45(6), 695–708 (2007)

[49] Yang, M., Kpalma, K., Ronsin, J.: Scale-controlled area difference shape descriptor. In: Proc. SPIE, Electronic Imaging science and Technology (2007)

[50] Zahn, C.T., Roskies, R.Z.: Fourier Descriptors for Plane closed Curves. IEEE Trans. Computer c-21(3), 269–281 (1972)

[51] Zhang, D., Lu, G.: Review of shape representation and description techniques. Pattern Recognition 37, 1–19 (2004)

[52] Zhang, D., Lu, G.: A comparative study of curvature scale space and Fourier descriptors for shape-based image retrieval. Visual Communication and Image Representation 14(1) (2003)

[53] Zhang, D., Lu, G.: A Comparative Study of Fourier Descriptors for Shape Representation and Retrieval. In: Proc. 5th Asian Conference on Computer Vision (2002)

[54] Zhang, D.S., Lu, G.: A Comparative Study on Shape Retrieval Using Fourier Descriptors with DifferentShape Signatures. In: Proc. International Conference on Intelligent Multimedia and Distance Education (ICIMADE 2001) (2001)

[55] Zhang, H., Malik, J.: Learning a discriminative classifier using shape context distances. In: Proc. IEEE Computer Society Conference on Computer Vision and Pattern Recognition (2003)

[56] ISO/IEC JTC1/SC29/WG11, MPEG-7 Overview (version 10), Technical report (2004)

[57] Hu, M.-K.: Visual Pattern Recognition by Moment Invariants. IRE Trans. Information Theory IT-8, 179–187 (1962)

[58] Iivarinen, J., Visa, A.: Shape recognition of irregular objects. In: Proc. SPIE, Intelligent Robots and Computer Vision XV: Algorithms, Techniques, Active Vision, and Materials Handling, pp. 25–32 (1996)

[59] Flusser, J.: Invariant Shape Description and Measure of Object Similarity. In: Proc. 4th International Conference on Image Processing and its Applications, pp. 139–142 (1992)

[60] di Baja, G.S., Thiel, E.: Skeletonization algorithm running on path-based distance maps. Image Vision Computer 14, 47–57 (1996)

[61] Borgefors, G.: Distance Transformations in Digital Images. Computer Vision, Graphics, and Image Processing, 344–371 (1986)

[62] Kolesnikov, A.: Efficient algorithms for vectorization and polygonal approximation, Ph.D thesis, University of Joensu, Finland (2003)

Chapter 10
Object-Based Image Retrieval System Using Rough Set Approach

Neveen I. Ghali[1], Wafaa G. Abd-Elmonim[1], and Aboul Ella Hassanien[2]

[1] Faculty of Science, Al-Azhar University, Cairo Egypt
nev_ghali@yahoo.com
[2] Faculty of Computers and Information, Cairo University, Cairo, Egypt
aboitcairo@gmail.com

Abstract. In this chapter, we present an object-based image retrieval system using the rough set theory. The system incorporates two major modules: Preprocessing and Object-based image retrieval. In preprocessing, an image-based object segmentation algorithm in the context of the rough set theory is used to segment the images into meaningful semantic regions. A new object similarity measure is proposed for the image retrieval. Performance is evaluated on an image database and the effectiveness of proposed image retrieval system is demonstrated. Experimental results show that the proposed system performs well in terms of speed and accuracy.

10.1 Introduction

The growth of the size of data and number of existing databases far exceeds the ability of humans to analyze this data, which creates both a need and an opportunity to extract knowledge from databases. There is a pressing need for efficient information management and mining of the huge quantities of image data that are routinely being used in databases [16,17,18,19,20]. These data are potentially an extremely valuable source of information, but their value is limited unless they can be effectively explored and retrieved, and it is becoming increasingly clear that in order to be efficient, data mining must be based on semantics. However, the extraction of semantically rich meta-data from computationally accessible low-level features poses tremendous scientific challenges [16,17,21].

Content-based image retrieval (CBIR) systems are needed to effectively and efficiently use the information that is intrinsically stored in these image databases. This image retrieval system has gained considerable attention, especially during the last decade. CBIR is an attractive research area due to the rapid growth in image databases in a variety of application domains, such as entertainment, commerce, education, biomedicine, military, document archives,

R. Kountchev et al. (Eds.): Advances in Reasoning-Based Image Processing, ISRL 29, pp. 315–329.
springerlink.com

art collections, geographical information systems and image classification and searching [5,6,7,8,10].

In a typical CBIR system, queries are normally formulated either by query by example or similarity retrieval, selecting from a color, shape, skeleton, and texture features or a combination of two or more features. The system then compares the query with a database representing the stored images. The output from a CBIR system is usually a ranked list of images in order of their similarity to the query.

The most commonly used techniques in content-based image retrieval and classification are neural networks, genetic algorithms, decision trees, fuzzy theory and rough set theory [3,24]. Rough set concept was introduced by Polish logician, Professor Zdzislaw Pawlak in early eighties [11,12,13]. This theory become very popular among scientists around the world and the rough sets are now one of the most developing tools for data analysis [25]. Rough set data analysis was used for the discovery of data dependencies, data reduction, approximate set classification, and rule induction from databases. In the case of image processing the generated rules represent the underlying semantic content of the images [24].

In gray scale images, boundaries between objects are often ill defined because of grayness and/or spatial ambiguities [9,23]. This uncertainty can be handled by describing the different objects as rough sets. The work introduced in this chapter demonstrates an application of rough set theory for object extraction from gray scale images. This has been done by defining an image as a collection of pixels. The equivalence relation induced partition of pixels lying within each non-overlapping window over the image. With this definition the roughness of various transforms or partitions, which are known as rough set equivalence classes of the image can be computed using image windows of different sizes, (i.e., granules).

This chapter has the following organization. Section (2) provides an explanation of the basic framework of rough set theory, along with some of the key definitions and gives an introduction to rough image processing including rough images, rough representation of a region of interest, applications including content-based image retrieval categories. Section (3) provides the proposed object-based image retrieval scheme and its components. Experimental results are discussed in Section (4). Finally, conclusions are discussed in Section (5).

10.2 Basic Concepts

10.2.1 Rough Sets: Short Description

Due to space limitations we provide only a brief explanation of the basic framework of rough set theory, along with some of the key definitions. A more comprehensive review can be found in sources such as [11,12,13].

Rough sets theory provides a novel approach to knowledge description and to approximation of sets. Rough theory was introduced by Pawlak during the early eighties [11] and is based on an approximation space-based approach to classifying sets of objects. In rough sets theory, feature values of sample objects are collected in what are known as information tables. Rows of a such a table correspond to objects and columns correspond to object features.

Let $\mathcal{O}, \mathcal{F}$ denote a set of sample objects and a set of functions representing object features, respectively. Assume that $B \subseteq \mathcal{F}, x \in \mathcal{O}$. Further, let $x_{\sim_B}$ denote

$$x_{/\sim_B} = \{y \in \mathcal{O} \mid \forall \phi \in B, \phi(x) = \phi(y)\},$$

i.e., $x_{/\sim_B}$ (description of x matches the description of y). Rough sets theory defines three regions based on the equivalent classes induced by the feature values: lower approximation $\underline{B}X$, upper approximation $\overline{B}X$ and boundary $BND_B(X)$. A lower approximation of a set X contains all equivalence classes $x_{/\sim_B}$ that are proper subsets of X, and upper approximation $\overline{B}X$ contains all equivalence classes $x_{/\sim_B}$ that have objects in common with X, while the boundary $BND_B(X)$ is the set $\overline{B}X \setminus \underline{B}X$, *i.e.*, the set of all objects in $\overline{B}X$ that are not contained in $\underline{B}X$. Any set X with a non-empty boundary is *roughly* known relative, *i.e.*, X is an example of a rough set.

The indiscernibility relation $\sim_B$ (also written as Ind_B) is a mainstay of rough set theory. Informally, $\sim_B$ is a set of all classes of objects that have matching descriptions. Based on the selection of B (*i.e.*, set of functions representing object features), $\sim_B$ is an equivalence relation that partitions a set of objects $\mathcal{O}$ into classes (also called elementary sets [11]). The set of all classes in a partition is denoted by $\mathcal{O}_{/\sim_B}$ (also by $\mathcal{O}/Ind_B$). The set $\mathcal{O}/Ind_B$ is called the quotient set. Affinities between objects of interest in the set $X \subseteq \mathcal{O}$ and classes in a partition can be discovered by identifying those classes that have objects in common with X. Approximation of the set X begins by determining which elementary sets $x_{/\sim_B} \in \mathcal{O}_{/\sim_B}$ are subsets of X.

10.2.2 Rough Image Processing

Various rough image processing methodologies have been applied to handle the different challenges posed by multimedia imaging. We can define rough image processing as the collection of all approaches and techniques that understand, represent and process the images, their segments and features as rough sets (see, *e.g.* [26,27]). In this section, we first describe the ability of rough sets to handle and represent images, followed by the description of the rough representation of a region of interest.

10.2.2.1 The Ability of Rough Sets to Handle Images

Rough sets provide reasonable structures for the overlap boundary given domain knowledge. Research involving color images appears in [28]. Histones (*i.e.*, encrustations of a histogram) are used as the primary measure and as a visualization of multi-dimensional color information. The basic idea of a Histone is to build a histogram on top of the histograms of the primary color components red, green, and blue. We can consider the base histogram correlates with the lower approximation, whereas the encrustation correlates with the upper approximation. The problem of a machine vision application where an object is imaged by a camera system is considered in [29]. The object space can be modeled as a finite subset of the Euclidean space when the objects image is captured via an imaging system. Rough sets can bound such sets and provide a mechanism for modeling the spatial uncertainty in the object's image.

In grayscale images boundaries between object regions are often ill defined because of grayness or spatial ambiguities [30]. This uncertainty can be effectively handled by describing the different objects as rough sets with upper (or outer) and lower (or inner) approximations. Here the concepts of upper and lower approximation can be viewed, respectively, as outer and inner approximations of an image region in terms of granules [30].

Definition 1. *(Rough image [31]) Let the universe U be an image consisting of a collection of pixels. Then if we partition U into a collection of non-overlapping windows of size $m \times n$, each window can be considered as a granule G. Given this granulation, object regions in the image can be approximated by rough sets.*

A rough image *is a collection of pixels along with the equivalence relation induced partition of an image into sets of pixels lying within each non-overlapping window over the image. With this definition, the roughness of various transforms (or partitions) of an image can be computed using image granules for windows of different sizes.*

10.2.2.2 Rough Representation of a Region of Interest

A region of interest (ROI), is a selected subset of samples within an image identified for a particular purpose. The concept of ROI is commonly used in medical imaging. For example, the boundaries of a tumor may be defined on an image or in a volume, for the purpose of measuring its size. The endocardial border may be defined on an image, perhaps during different phases of the cardiac cycle, say end-systole and end-diastole, for the purpose of assessing cardiac function.

Hirano and Tsumoto [2] introduced the rough direct representation of ROIs in medical images. The main advantage of this method is its ability to represent inconsistency between the knowledge-driven shape and image-driven shape of a ROI using rough approximations. The method consists of three steps. First, they derive discretized feature values that describe the characteristics of a ROI. Secondly, using all features, they build up the basic regions (granules) in the image so that each region contains voxels that are indiscernible on all features. Finally, according to the given knowledge about the ROI, they construct an ideal shape of the ROI and approximate it by the basic categories. Then the image is split into three regions: a set of voxels that are:

(1) certainly included in the ROI (Positive region),
(2) certainly excluded from the ROI (Negative region),
(3) possibly included in the ROI (Boundary region).

The ROI is consequently represented by the positive region associated with some boundary regions.

The authors in [2] described the procedures for rough representation of ROIs under single and multiple types of classification knowledge. Usually, the constant variables defined in the prior knowledge, for example some threshold values, do not meet the exact boundary of images due to inter-image variances of the intensity. The approach tries to roughly represent the shape of the ROI by approximating the given shapes of the ROI by the primitive regions derived from feature of the image itself. It is reported that the simplest case occurs when we have only information about the intensity range of the ROI. In this case intensity thresholding is a conventional approach to obtain the voxels that fall into the given range. Let us denote the lower and upper thresholds by Th_L and Th_H, respectively. Then the ROI can be represented by:

$$ROI = \{x(p) \mid Th_L \leq I(x(p)) \leq Th_H\}, \tag{10.1}$$

where $x(p)$ denotes a voxel at location p and $I(x(p))$ denotes intensity of voxel $x(p)$.

Figure (10.1) illustrates the concept of rough ROI representation. The left image is an original grayscale image. Assume that the ROIs are three black circular regions: ROI_1, ROI_2, and ROI_3. Also assume that we are given a prior knowledge about the ROIs, that is, the lower threshold value Th_L of the ROIs, derived from some knowledge base. With this knowledge we can segment an ideal ROI $X_{\hat{ROI}}$ as follows:

$$X_{\hat{ROI}} = \{x(p) | Th_L \leq I(p)\}. \tag{10.2}$$

However, $X_{\hat{ROI}}$ does not correctly match the expected ROIs. This is because Th_L was too small to separate the ROIs. Th_L is a global threshold determined on the other sets, therefore, it should not be directly applied to this image.

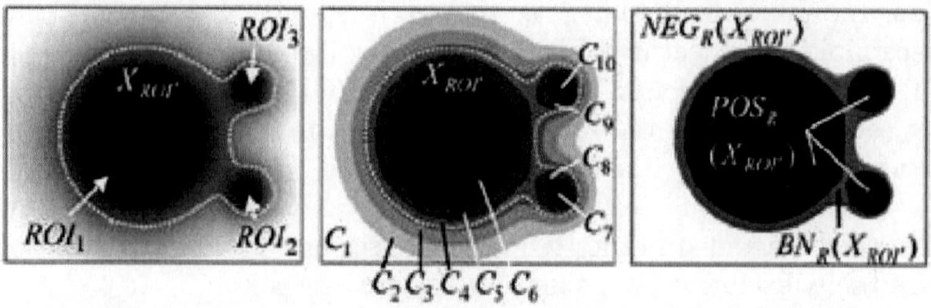

Fig. 10.1 Rough ROI representation. Left: an original image. Middle: elementary categories. Right: roughly segmented ROI [2]

10.2.3 Image Retrieval Systems: Problem Definition and Categories

Image retrieval systems attempt to search through a collection of images stored in an organized database to find the images that are perceptually similar to a query image.

Problem Definition: Assume that we have an image database that contains a collection of images $IDB = \{I_1, I_2, \ldots I_n\}$. The user can specify a query to retrieve a number of relevant images. Let Q be a query image and $D(I_i, I_j)$ be a real inter-image distance between two images I_i and I_j. Let m be the number of images that are closed to the query Q that the user wants to retrieve such that $m < n$. This image retrieval can be defined as the efficient retrieval of the best of m images based on IDB from a database on n images.

Image retrieval algorithms roughly belong to three categories: text-based approaches, content-based methods and object-based image retrieval.
The following subsection provides a brief discussion of these three categories.

10.2.3.1 Text-Based Approach

The text-based approach can be tracked back to 1970s. In such systems, the images are manually annotated by text descriptors, which are then used by a database management system (DBMS) to perform image retrieval [1]. The text-based approaches are based on the idea of storing set of keywords, or a textual description of the image content [1,5].This approach has two main disadvantages (1) A considerable level of human labour is required for manual annotation and (2) the annotation inaccuracy due to the subjectivity of human perception. To overcome the above disadvantages in text-based retrieval system, CBIR was introduced in the early 1980 [22].

10.2.3.2 Content-Based Approach

CBIR is an important alternative and complement to traditional text-based image searching and can greatly enhance the accuracy of the information being returned. It aims to develop an efficient visual-content-based technique to search, browse and retrieve relevant images from large-scale digital image collections. Most proposed CBIR techniques automatically extract low-level features, for example: color, texture, shapes and layout of objects, to measure the similarities among images by comparing the feature [1,22]. CBIR retrieved images based on information automatically extracted from pixels. A major problem in this area is computer perception. It is very difficult to make a computer think like a human being to sense what the image or object within the image is. In other words, there remains a big gap between low-level features like shape, color, texture and spatial relations and high-level features like table, car, etc. [20,21].

10.2.3.3 Object-Based Approach

The evolution of OBIR systems attempt to overcome the semantic and sensory gap where loss of information from an image to a representation by features is called the semantic gap and sensory gap that describes the loss between the actual structure and the representation in a digital image [18,19,20,21], by representing images as collections of regions that may correspond to objects such as flowers, trees, skies, and mountains.

OBIR systems retrieve images from a database based on the appearance of physical objects in those images. Examples of objects are elephants, stop sign, helicopters, buildings, faces, or any other object that the user wishes to find [15,20,21]. One common way to search database for certain image object is to segment the images in the database, then compare each segmented region against a region in a query image presented by the user. Such image retrieval systems are generally successful for objects that can be easily separated from the background. This chapter focuses on object- based techniques that allow the user to specify a particular region of an image and request the database to retrieve images that contain similar regions.

10.3 Object-Based Image Retrieval System

In this work, we assume that user is only interested in one semantic region of the query image. The main objective is to retrieve images that contain similar semantic regions. In the initial query, user identifies a query image with no information is provided as to which specific part of the image is of the users interest. Therefore, for all the images in the image database, we compute the similarity between the regions of these images and the regions of the query image.

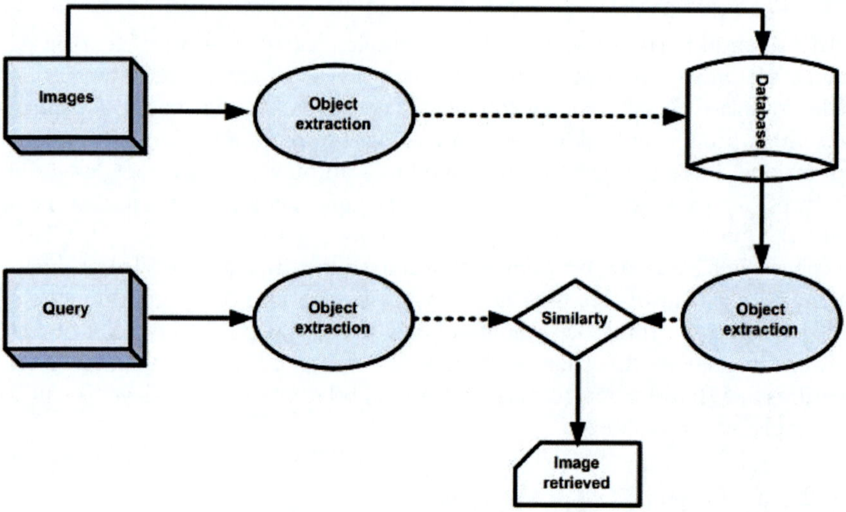

Fig. 10.2 Object-based image retrieval system

The similarity between the query image and an image in the image database is represented by the proposed value *Object Similarity Ratio (OSR)*; as described in Algorithm 2. These *OSR* are then sorted in an ascending order and the top n images are returned to the user for feedback.

A block diagram represents our proposed approach to OBIR system is shown in Figure (10.2). In this proposed system, the relevance between a query and any target image is ranked according to a similarity measure. The similarity comparison is performed based on object that appears in the images.

10.3.1 Pre-processing: Segmentation and Feature Extraction

Segmentation of an image to obtain different objects (regions), followed by the analysis of their structural information to recognize the desired model. The quality of segmentation directly affects the performance of an object retrieval system, which tends to classify image similarity on the basis of similar objects.

Let us describe a method of object segmentation or extraction based on the principle of minimizing the roughness of both object and background regions (i.e. minimizing RM_T for different granule size). Minimizing of RM_T tends to minimize the uncertainty arising from vagueness of the boundary region of the object. Therefore, for a given granule size, the threshold for object-background classification can be obtained through minimizing RM_T.

By computing the RM_T of the image for every gray level T, such that $(0 \ldots T)$ and $(T+1 ... L-1)$, represent the background and object regions and select $T^* = RM_T$ as the optimum threshold to provide the object-background segmentation. Minimize the rough mean to get the required threshold basically implies minimizing both the object roughness and background roughness.

Definition 2. *(Image object) Image object is a set of regions located near the center of the image, which has significant distribution compared with its surrounding (or background) region. Objects formed by grouping pixels according to certain of homogeneity by image segmentation are the basic processing units in object oriented image analysis approach.*

We have to note that it is difficult to describe the objects of interest using only the low-level image features like colour, texture, shape or spatial location. The concept of OBIR is based on implementing image segmentation, then, low-level features can be extracted from the segmented regions. Similarity between two images is defined based on region features [4,14].

A measure is defined, by which we can compute an optimal threshold for image segmentation. In this chapter, we adapt a measure called Mean Roughness Measure (RM_T) for an image. It denotes the average of roughness of object and background at a certain threshold T, defined in the following equation:

$$RM_T = \frac{R_{O_T} + R_{B_T}}{2} \tag{10.3}$$

We can deduce that the value of RM_T lies between 0 and 1 because $0 \leq R_{O_T} \leq 1$ and $0 \leq R_{B_T} \geq 1$ thus RM_T has a maximum value of unity when R_{O_T} and R_{B_T} equal one, and minimum value of zero when R_{O_T} and R_{B_T} equal zero. Similarly the value of RM_T determines the roughness of the region determined.

Algorithm 1 shows the best selection of T^* thresholding, where max_{gray} is the maximum gray level values of the image A, and min_{gray} minimum gray level values of the image A. A window of m×n pixels represents a granule and the total number of granules is given by $total_{no-granule}$.

10.3.2 Similarity and Retrieval System

In this stage we demonstrate an application of algorithm 2. The value of OSR takes the values $0 < p \leq 1$, where p equal one if two compared images are identical. Consequently, the proposed system retrieves the images that have OSR value approximately equals 1.

Algorithm 1. Selection of optimal threshold T^*

Input: Gray-scaled image.
Output: Segmented image according to optimal threshold T^*
Processing:

1: Divide image A into granules, each granule (G_i) represented by one column in matrix B.
2: Initialize: Four integer arrays namely $OL = object_{lower}$, $OU = object_{upper}$, $BL = background_{lower}$, and $BU = background_{upper}$, Let j, k, s and c are parameter with initial value equal one.
3: **for** i =1 to $total_{no-granule}$ **do**
4:   **if** $B(i) > min_{gray} + 1$ **then**
5:     Set OL(j) = B(i)
6:   **else**
7:     Set BU(k) = B(i)
8:   **end if**
9:   **if** $B(i) \leq min_{gray} + 1$ **then**
10:     Set BL(s) =B(i)
11:   **else**
12:     Set OU(c)= B(i)
13:   **end if**
14: **end for**
15: Calculate roughness of object(RoO) set using

$$RoO = 1 - \frac{| OL |}{| OU |} \tag{10.4}$$

16: Calculate roughness of background (RoB) set using the following equation

$$RoB = 1 - \frac{| BL |}{| BU |} \tag{10.5}$$

17: Calculate the mean roughness (MR) as follows:

$$MR = \frac{object_{roughness} + background_{roughness}}{2} \tag{10.6}$$

{*Suppose mean roughness at* $T = min_{gray} + 1$ *is minimum mean roughness measure.*}
18: ***for*** *Every gray level value* ***do***
19:   *Repeat step 2*
20: ***end for***
21: Find optimal threshold T^* *corresponding to minimum mean roughness using:*

$$Optimal_{Threshold}(T*) = MR_T \tag{10.7}$$

Algorithm 2. Object retrieval algorithm

Input: Gray scaled image with one centered object.
Output: Optimal similar image
Processing:

1: Construct image Database
2: Store the images in the database
3: Store $object_{lower}$ approximation array according to optimal-threshold T^* for each image in database
4: Calculate $object_{lower}$ approximation array according to optimal-threshold T^* for query image that selected by user
5: Compare the $object_{lower}$ approximation array of query image with $object_{lower}$ approximation array of each image in database table
6: Find the Object Similarity Ratio for each image in image database by using the following equation:

$$ObjectSimilarityRatio(R) = \frac{N}{T} \tag{10.8}$$

Where N is the number of pixels which have the same value in two $object_{lower}$ approximation array of images (query image and database image) and T is the size of $object_{lower}$ approximation array.
7: The images with largest Object Similarity Ratio are retrieved from database; the number of retrieved images is specified by user

10.4 Experimental Results and Discussion

In order to retrieve the similar images to the user query, at first user enter the query image. Then its object is extracted, by computing $object_{lower}$ approximation array according to optimal threshold T^* using the algorithm 1. To match between two objects in two different images we compare between $object_{lower}$ approximation array of the two images. In the proposed method we compare between $object_{lower}$ approximation array of the query image, and $object_{lower}$ approximation array of each image in the input image database. Compute the number N of the pixels that has the similar values in the two arrays. Then the Object Similarity ratio is calculated for each image in the database.

Object Similarity Ratio takes the values $0 < R \leq 1$, where R equals one if two compared images are the same. Consequently, the proposed system retrieves the images that have object similarity ratio value approximately equals 1. The Object Similarity Ratio for all database images are sorted in descending order and the first four images are retrieved according to user's choice.

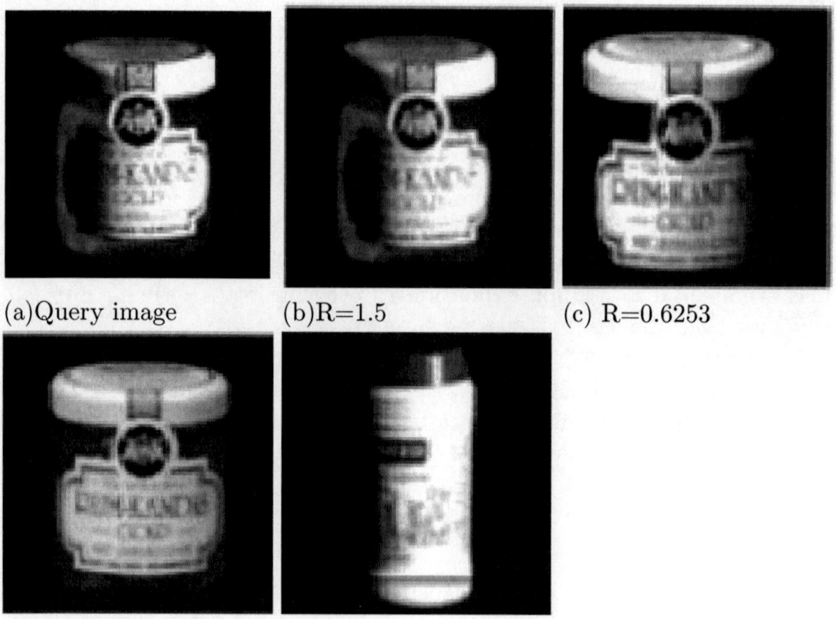

(a)Query image (b)R=1.5 (c) R=0.6253

(d)R=0.64831 (e) R=0.6525

Fig. 10.3 The sorted retrieved images according to Object Similarity Ratio (R)

Figures (10.3,10.4) show the sorted retrieved images according to Object Similarity Ratio resulted from a case study of searching the shown image within an image database consisting of 51 objects.

Usually precision and recall are used in retrieval system to measure retrieval performance. Precision Pr is defined as the ratio of the number of relevant images retrieved N_r to the number of total retrieved images K. Recall Re is defined as the number of retrieved relevant images N_r over the total number of relevant images available in the database N_t. Precision and recall are defined using the following equations:

$$Re = \frac{Nr}{Nt} \tag{10.9}$$

$$Pr = \frac{Nr}{K} \tag{10.10}$$

Table (1) shows results for retrieval images with different quires measured in terms of recall R_e, precision P_r and retrieval time (Ret-time). It shows that the retrieval time is very small for all quires. For all experiments, the total time required for a query did not exceed 5s.

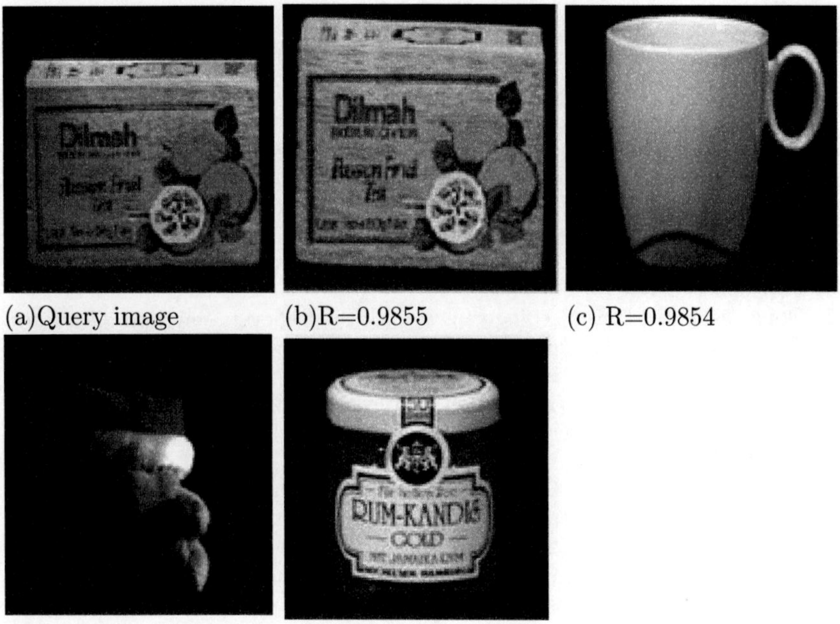

(a)Query image (b)R=0.9855 (c) R=0.9854

(d)R=0.9852 (e) R=0.9852

Fig. 10.4 Another sorted retrieved images according to Object Similarity Ratio (R)

Table 10.1 Retrieval images with different quires measured in terms of recall, precision and retrieval time

Qimage	(N_t, K, N_r)	Re	Pr	ave.	ret-time
Q1	51,4,4	0.08	1.0	0.540	5s
Q2	51,4,1	0.02	0.25	0.135	4s
Q3	51,4,1	0.02	0.25	0.135	4s
Q4	51,4,1	0.02	0.25	0.135	4s
Q5	51,4,3	0.06	0.75	0.405	4

10.5 Conclusion

We have presented a rough based approach to object-based image retrieval using rough set theory. The system incorporates two major modules: Preprocessing and Object-based image retrieval. In preprocessing, an image-based object segmentation algorithm in the context of the rough set theory is used to segment the images into meaningful semantic regions. A new object similarity measure is proposed and the performance is evaluated on a large image database and experimental results show that the proposed system performs well in terms of speed and accuracy.

References

1. Amato, A., Lecce, V.D.: A knowledge based approach for a fast image retrieval system. Image and Vision Computing 26, 1466–1480 (2008)
2. Hirano, S., Tsumoto, S.: Rough representation of a region of interest in medical images. Int. J. of Approximate Reasoning 40, 23–34 (2005)
3. Chen, Y., Li, J., Wang, J.Z.: Machine learning and statistical modeling approaches to image retrieval. Kluwer Academic Publishers (2004)
4. Deb, S.: Multimedia Systems and Content-Based Image Retrieval. Idea Group Publishing (2004)
5. Dong-cheng, S., Lan, X., Ling-yan, H.: Image Retrieval using both Color and Texture Features. The Journal of China Universities of Posts and Telecommunications 14(1), 94–99 (2007)
6. Goodrum, A.A.: Image Information Retrieval: An overview of current research. Informing Science, Special Issue on Information Science Research 3(2), 63–67 (2000)
7. Wan-Ting, S., Ju-Chin, C., Jenn-Jier, L.: Region-based image retrieval system with heuristic pre-clustering relevance feedback. Expert Systems with Applications 37(7), 4984–4998 (2010)
8. Liu, Y., Zhang, D., Lu, G., Ma, W.: Asurvey of content-based image retrieval with high-level semantics. Pattern Recognition 40(1), 262–282 (2007)
9. Müller, H., Michoux, N., Bandon, D., Geissbuhler, A.: A Review of content-based image retrieval systems in medical applications-clinical benefits and future directions. International Journal of Medical Informatics 73(1), 1–23 (2004)
10. Shankar, P.K., Uma, S.B., Mitra, P.: Granular Computing, Rough Entropy and Object Extraction. Pattern Recognition Letters 26, 2509–2517 (2005)
11. Pawlak, Z.: On Rough Sets. Bulletin of the European Association for Theoretical Computer Science (24), 94–109 (1984)
12. Pawlak, Z.: Rough Sets. International Journal of Computer and Information Scieviolnces 11(5) (1982)
13. Pawlak, Z.: Some Remarks about Rough Sets. In: ICS PAS, vol. 456 (1982)
14. Wang, Y., Ding, M., Zhou, C., Hu, Y.: Interactive relevance feedback mechanism for image retrieval using rough set. Knowledge-Based Systems 19(8), 696–703 (2006)
15. Yan, G.: Pixel Based and Object Oriented Image Analysis for Coal Fire Research, Master thesis, Enschede, The Netherlands, International Institute for Geoinformation Science and Earth Observation (ITC), Netherlands (2003)
16. Im, Y.H., Oh, S.G., Chung, M.J., Yu, J.H., Lee, H.S., Chang, J.K., Park, D.H.: A KFD web database system with an object-based image retrieval for family art therapy assessments. The Arts in Psychotherapy 37(3), 163–171 (2010)
17. Alajlan, N., Kamel, M., Freeman, G.: Multi-object image retrieval based on shape and topology. Signal Processing: Image Communication 21(10), 904–918 (2006)
18. Shao, L., Brady, M.: Specific object retrieval based on salient regions. Pattern Recognition 39(10), 1932–1948 (2006)
19. Liu, D., Chen, T.: Video retrieval based on object discovery. Computer Vision and Image Understanding 113(3), 397–404 (2009)
20. Xie, Z., Roberts, C., Johnson, B.: Object-based target search using remotely sensed data: A case study in detecting invasive exotic Australian Pine in South Florida. ISPRS Journal of Photogrammetry and Remote Sensing 63(6), 647–660 (2008)

21. Carson, C., Thomas, M., Belongie, S., Hellerstein, J.M., Malik, J.: Blobworld: A system for region-based image indexing and retrieval. In: Third International Conference on Visual Information Systems, Amsterdam, Netherland, pp. 509–516 (1999)
22. Graham, M.E.: Enhancing visual resources for searching and retrieval-Is content based image retrieval solution? Literary and Linguistic Computing 19(3), 321–333 (2004)
23. Viitaniemi, V., Laaksonen, J.: Techniques for Still Image Scene Classification and Object Detection. In: Kollias, S.D., Stafylopatis, A., Duch, W., Oja, E. (eds.) ICANN 2006. LNCS, vol. 4132, pp. 35–44. Springer, Heidelberg (2006)
24. Abraham, A., Hassanien, A., Carvalho, A.: Foundations of Computational Intelligence: Bio-Inspired Data Mining. Studies in Computational Intelligence, vol. 4. Springer, Germany (2009) ISBN: 978-3-642-01087-3
25. Hassanien, A.: Intelligent techniques for prostate ultrasound image analysis. Int. Jour. of Hybrid Intelligent Systems 6, 155–167 (2009), doi:10.3233/HIS-2009-0092
26. Hassanien, A., Abraham, A.: Rough morphology hybrid approach for mammography image classification and prediction. Int. Jour. of Computational Intelligence and Applications 7(1), 17–42 (2008)
27. Hassanien, A.: Pulse coupled neural network for detection of masses in digital mammogram. Neural Network World Journal 2(6), 129–141 (2006)
28. Mohabey, A., Ray, A.K.: Rough set theory based segmentation of color images. In: Proc. of NAFIPS 19th Int. Conf., pp. 338–342 (2000)
29. Sinha, D., Laplante, P.: A rough set-based approach to handling spatial uncertainty in binary images. Eng. Appl. Artif. Intell. 17, 97–110 (2004)
30. Pal, S.K., Shankar, U., Mitra, P.: Granular computing, rough entropy and object extraction. Pattern Recognition Letters 26(16), 2509–2517 (2005)
31. Hassanien, A., Abraham, A., Peters, J.F., Schaefer, G., Henry, C.: Rough sets and near sets in medical imaging: A review. IEEE Trans. Info. Tech. in Biomedicine (2009), doi:10.1109/TITB.2009.2017017

Chapter 11
Paraconsistent Artificial Neural Networks and Delta, Theta, Alpha, and Beta Bands Detection

Jair Minoro Abe[1,2], Helder F.S. Lopes[2], and Kazumi Nakamatsu[3]

[1] Graduate Program in Production Engineering, ICET - Paulista University
R. Dr. Bacelar, 1212, CEP 04026-002 São Paulo – SP – Brazil
[2] Institute For Advanced Studies – University of São Paulo, Brazil
jairabe@uol.com.br, helder.mobile@gmail.com
[3] School of Human Science and Environment/H.S.E. – University of Hyogo – Japan
nakamatu@shse.u-hyogo.ac.jp

Abstract. In this work we present a study of brain EEG waves – delta, theta, alpha, and gamma bands employing a new ANN based on Paraconsistent Annotated Evidential Logic Eτ which is capable of manipulating concepts like impreciseness, inconsistency, and paracompleteness in a nontrivial manner. We present the Paraconsistent Artificial Neural Network – PANN with some detail and discuss some applications.

Keywords: Artificial neural network, paraconsistent logics, EEG analysis, pattern recognition, Dyslexia.

11.1 Introduction

Generally speaking, Artificial Neural Network (ANN) can be described as a computational system consisting of a set of highly interconnected processing elements, called artificial neurons, which process information as a response to external stimuli. An artificial neuron is a simplistic representation that emulates the signal integration and threshold firing behavior of biological neurons by means of mathematical structures. ANNs are well suited to tackle problems that human beings are good at solving, like prediction and pattern recognition. ANNs have been applied within several branches, among them, in the medical domain for clinical diagnosis, image analysis and interpretation signal analysis and interpretation, and drug development.

So, ANN constitutes an interesting tool for electroencephalogram (EEG) qualitative analysis. On the other hand, in EEG analysis we are faced with

R. Kountchev et al. (Eds.): Advances in Reasoning-Based Image Processing, ISRL 29, pp. 331–364.
springerlink.com

imprecise, inconsistent and paracomplete data. In order to manipulate them directly, some interesting theories have been proposed recently: Fuzzy sets, Rough sets, among others.

In this work we employ a new kind of ANN based on paraconsistent annotated evidential logic Eτ, which is capable of manipulating imprecise, inconsistent and paracomplete data in order to make a first study of the recognition of EEG standards.

The EEG is a brain electric signal activity register, resultant of the space-time representation of synchronic postsynaptic potentials. The most probable is that the main generating sources of these electric fields are perpendicularly guided regarding to the cortical surface, as the cortical pyramidal neurons.

The graphic registration of the sign of EEG can be interpreted as voltage flotation with mixture of rhythms, being frequently sinusoidal, ranging from 1 to 70 Hz. In the clinical-physiological practice, such frequencies are grouped in frequency bands (Fig. 1.1): delta (0.5 to 4.0 Hz), theta (4.1 to 8.0 Hz), alpha (8.1 to 12.5 Hz), and beta (> 13.0 Hz). During the relaxed awake, normal EEG in adults is predominantly composed by alpha band frequency, which is generated by interactions of the slum-cortical and thalamocortical systems.

The traditional method of interpreting an EEG examination consists of a visual analysis, which is called domain analysis of time, for this analysis studies the morphology of the waves as they were captured. Such analysis is important to distinguish between brain activity and artifacts. Artifacts have features that can be relatively easy to identify, such as eye movements, muscle movements and even by external electromagnetic interference (Aminoff, 2003; Anghinah et al., 2006). However, new concepts such as quantitative EEG (qEEG), aid in the quantification of EEG tracings, with the objective to streamline and increase the accuracy of the analysis of EEG interpretations. The qEEG can be considered an accessory to the

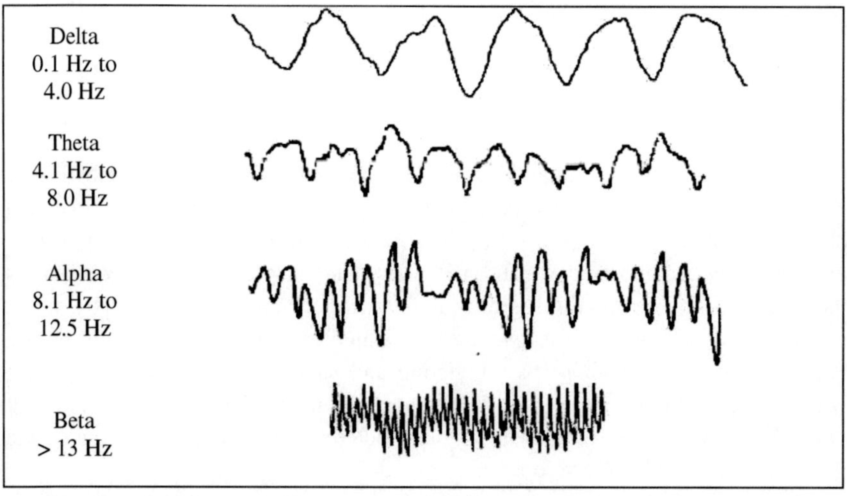

Fig. 11.1 Frequency bands clinically established and usually found in EEG [23].

EEG, because it is used in the EEG information and transform them into quantitative data through mathematical algorithms, such as the FFT - Fast Fourier transform. Currently the qEEG is used to create a topographical map showing the distribution and intensity of brain electrical activity as compared to a normal pattern, it can detect anomalies and interprets them as electroencephalographic findings, i.e., features in the distribution of activity brain that indicate some pathology (AANACNS, 1997; Anghinah, 2003; Chabot, 2005). Several studies demonstrate that the qEEG is an efficient tool and its use in clinical protocols has been gradually employed.

EEG analysis, as well as any other measurements devices, is limited and subjected to the inherent imprecision of the several sources involved: equipment, movement of the patient, electric registers and individual variability of physician visual analysis. Such imprecision can often include conflicting information or paracomplete data. The majority of theories and techniques available are based on classical logic and so they cannot handle adequately such set of information, at least directly. Although several theories have been developed in order to overcome such limitations, v.g. Fuzzy set theory, Rough set theory, non-monotonic reasoning, among others, they cannot manipulate inconsistencies and paracompleteness directly. So, we need a new kind of logic to deal with uncertainty, inconsistent and paracomplete data. Thus, we employ in this task the paraconsistent annotated evidential logic Eτ.

11.2 Background

Paraconsistent Artificial Neural Network (PANN) is a new artificial neural network introduced in [8]. Its basis leans on paraconsistent annotated logic *E*τ [1]. Let us present it briefly.

The atomic formulas of the logic *E*τ are of the type $p_{(\mu, \lambda)}$, where $(\mu, \lambda) \in [0, 1]^2$ and [0, 1] is the real unitary interval (*p* denotes a propositional variable). $p_{(\mu, \lambda)}$ can be intuitively read: "It is assumed that *p*'s favorable evidence is μ and contrary evidence is λ." Thus:

$p_{(1.0,\, 0.0)}$ can be read as a true proposition.

- $p_{(0.0,\, 1.0)}$ can be read as a false proposition.
- $p_{(1.0,\, 1.0)}$ can be read as an inconsistent proposition.
- $p_{(0.0,\, 0.0)}$ can be read as a paracomplete (unknown) proposition.
- $p_{(0.5,\, 0.5)}$ can be read as an indefinite proposition.

We introduce the following concepts (all considerations are taken with $0 \le \mu, \lambda \le 1$):

- Uncertainty degree (Eq. 11.2.1);
- Certainty degree (Eq. 11.2.2);

$$G_{un}(\mu, \lambda) = \mu + \lambda - 1 \tag{11.2.1}$$

$$G_{ce}(\mu, \lambda) = \mu - \lambda \tag{11.2.2}$$

An order relation is defined on $[0, 1]^2$: $(\mu_1, \lambda_1) \le (\mu_2, \lambda_2) \Leftrightarrow \mu_1 \le \mu_2$ and $\lambda_1 \le \lambda_2$, constituting a lattice that will be symbolized by τ.

With the uncertainty and certainty degrees we can get the following 12 output states (table 11.2.1): *extreme states*, and *non-extreme states*.

Table 11.2.1 Extreme and Non-extreme states

Extreme states	Symbol	Non-extreme states	Symbol
True	V	Quasi-true tending to Inconsistent	QV→T
False	F	Quasi-true tending to Paracomplete	QV→⊥
Inconsistent	T	Quasi-false tending to Inconsistent	QF→T
Paracomplete	⊥	Quasi-false tending to Paracomplete	Qf→⊥
		Quasi-inconsistent tending to True	QT→V
		Quasi-inconsistent tending to False	QT→F
		Quasi-paracomplete tending to True	Q⊥→V
		Quasi-paracomplete tending to False	Q⊥→F

Some additional control values are:

- V_{scct} = maximum value of uncertainty control = Ft_{un}
- V_{scc} = maximum value of certainty control = Ft_{ce}
- V_{icct} = minimum value of uncertainty control = $-Ft_{un}$
- V_{icc} = minimum value of certainty control = $-Ft_{ce}$

All states are represented in the next Figure (Fig. 11.2.1).

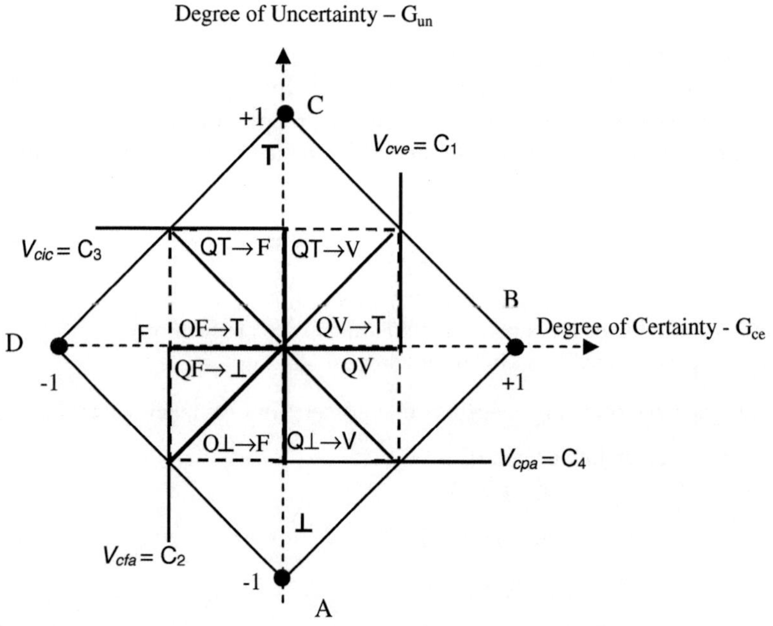

Fig. 11.2.1 Extreme and Non-extreme states.

11.3 The Main Artificial Neural Cells

In the PANN, the certainty degree G_{ce} indicates the 'measure' falsity or truth degree. The uncertainty degree G_{un} indicates the 'measure' of the inconsistency or paracompleteness. If the certainty degree in module is low or the uncertainty degree in module is high, it generates an indefinition.

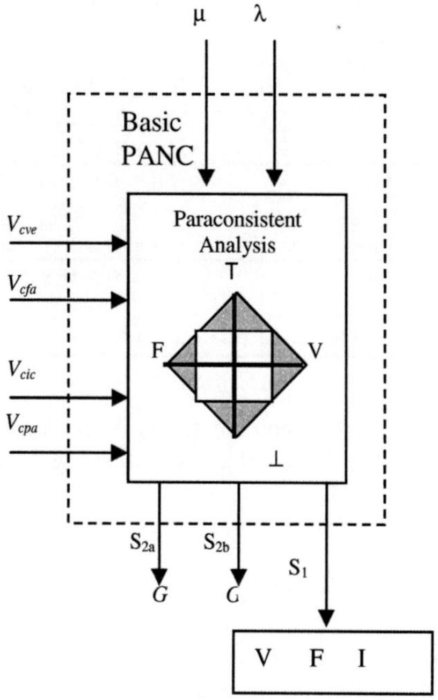

Fig. 11.3.1 Basic cell of PANN

The resulting certainty degree G_{ce} is obtained as follows:

- If: $V_{cfa} = G_{ce} = V_{cve}$ or $-Ft_{ce} = G_{ce} = Ft_{ce} \Rightarrow G_{ce} =$ Indefinition
- For: $V_{cpa} = G_{un} = V_{cic}$ or $-Ft_{un} = G_{un} = Ft_{un}$
- If: $G_{ce} = V_{cfa} = -Ft_{ce} \Rightarrow G_{ce} =$ False with degree G_{un}
- If: $Ft_{ce} = V_{cve} = G_{ce} \Rightarrow G_{ce} =$ True with degree G_{un}

A Paraconsistent Artificial Neural Cell – PANC – is called *basic* PANC (Fig. 11.3.1) when given a pair (μ, λ) is used as input and resulting as output:

- $S_{2a} = G_{un}$ = resulting uncertainty degree
- $S_{2b} = G_{ce}$ = resulting certainty degree
- $S_1 = X$ = constant of Indefinition.

Using the concepts of *basic* Paraconsistent Artificial Neural Cell , we can obtain the family of PANC considered in this work: Analytic connection (PANCac), Maximization (PANCmax), and Minimization (PANCmin) as described in Table 11.3.1 below:

Table 11.3.1 Paraconsistent Artificial Neural Cells

PANC	Inputs	Calculations	Output
Analytic connection: PANCac	μ λ Ft_{un} Ft_{un}	$\lambda_c = 1 - \lambda$ $G_{un}\ G_{ce}$, $\mu_r = (G_{ce} + 1)/2$	If $\vert G_{ce}\vert > Ft_{ce}$ then $S_1 = \mu_r$ and $S_2 = 0$ If $\vert G_{un}\vert > Ft_{ct}$ and $\vert G_{un}\vert > \vert G_{ce}\vert$ then $S_1 = \mu_r$ and $S_2 = \vert G_{un}\vert$ if not $S_1 = ½$ and $S_2 = 0$
Maximization: PANCmax	μ λ	G_{ce} $\mu_r = (G_{ce} + 1)/2$	If $\mu_r > 0.5$, then $S_1 = \mu$ If not $S_1 = \lambda$
Minimization: PANCmin	μ λ	G_{ce} $\mu_r = (G_{ce} + 1)/2$	If $\mu_r < 0.5$, then $S_1 = \mu$ if not $S_1 = \lambda$

To facilitate understanding on the implementation of the algorithms of PANC, we use a programming language Object Pascal, following logic of procedural programming in all samples.

11.3.1 Paraconsistent Artificial Neural Cell of Analytic Connection – PANCac

The paraconsistent artificial neural cell of analytic connection cell (PANCac) is the principal cell of all PANN, obtaining the certainty degree (G_{ce}) and the uncertainty degree (G_{un}) from the inputs and the tolerance factors.

This cell is the link which allows different regions of PANN perform signal processing in distributed and through many parallel connections [8].

The different tolerance factors certainty (or contradiction) acts as inhibitors of signals, controlling the passage of signals to other regions of the PANN, according to the characteristics of the architecture developed.

In table 11.3.2, we have a sample of implementation made in Object Pascal.

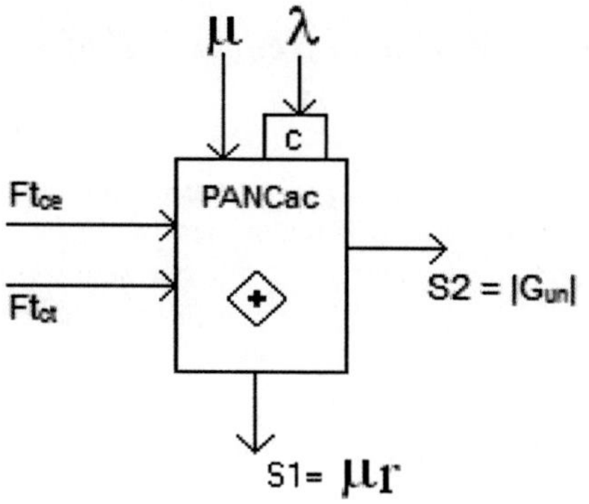

Fig. 11.3.2 Graphic representation of PANCac.

Table 11.3.2 PANCac implementation.

```
function TFaPANN.PANCAC(mi, lambda, Ftce, Ftct: real; output:
integer): real;
var
  Gce: real;
  Gun: real;
  lambdacp: real;
  mir: real;
  S1, S2: real;
begin
  lambdacp := 1 - lambda;
  Gce := mi - lambdacp;
  Gun := mi + lambdacp - 1;
  mir := (Gce + 1) / 2;

  if (abs(Gce) > Ftce) then
  begin
    S1 := mir;
    S2 := 0;
  end
  else
  begin
    if (abs(Gun) > Ftct) and (abs(Gun) > abs(Gce)) then
    begin
      S1 := mir;
      S2 := abs(Gun);
    end
    else
    begin
      S1 := 0.5;
      S2 := 0;
    end;
  end;
  if output = 1 then result := S1 else result := S2;
end;
```

11.3.2 Paraconsistent Artificial Neural Cell of Maximization– PANCmax

The paraconsistent artificial neural cell of maximization cell (PANCmax) allows selection of the maximum value among the entries.

Such cells operate as logical connectives OR between input signals. For this is made a simple analysis, through the equation of the Degree of Evidence (Table 11.3.1) which thus will tell which of the two input signals is of greater value, thus establishing the output signal [8].

In table 11.3.3, we have a sample of implementation made in Object Pascal.

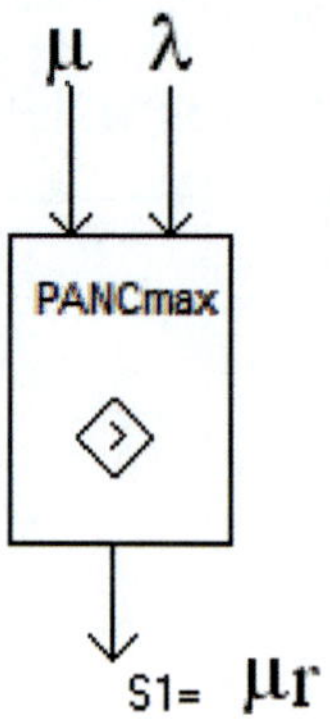

Fig. 11.3.3 Graphic representation of PANCmax.

Table 11.3.3 PANCmax implementation.

```
Function TFaPANN.PANCMAX(mi, lambda: real): real;
var
  mir: real;
begin
  mir := ((mi - lambda) + 1) / 2;

  if (mir > 0.5) then
     result := mi
  else
     result := lambda;
end;
```

11.3.3 Paraconsistent Artificial Neural Cell of Minimization– PANCmin

The paraconsistent artificial neural cell of maximization cell (PANCmin) allows selection of the minimum value among the entries.

Such cells operate as logical connectives AND between input signals. For this is made a simple analysis, through the equation of the Degree of Evidence (Table 11.3.1) which thus will tell which of the two input signals is of smaller value, thus establishing the output signal [8].

In table 11.3.4, we have a sample of implementation made in Object Pascal.

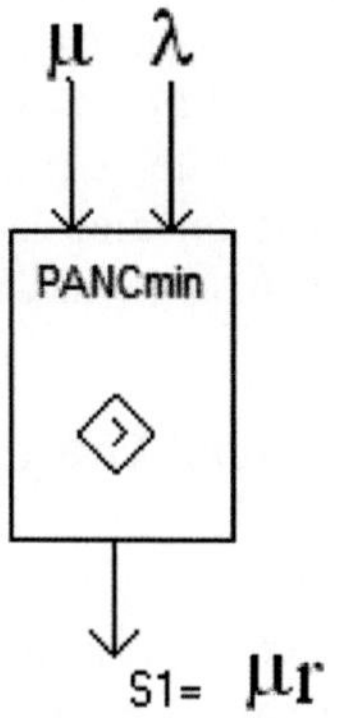

Fig. 11.3.4 Graphic representation of PANCmin.

Table 11.3.4 PANCmin implementation.

```
Function TFaPANN.PANCMIN(mi, lambda: real): real;
var
  mir: real;
begin
  mir := ((mi - lambda) + 1) / 2;

  if (mir < 0.5) then
     result := mi
  else
     result := lambda;
end;
```

11.3.4 Paraconsistent Artificial Neural Unit

A Paraconsistent Artificial Neural Unit (PANU) is characterized by the association ordered PANC, targeting a goal, such as decision making, selection, learning, or some other type of processing.

When creating a PANU, one obtains a data processing component capable of simulating the operation of a biologic neuron.

11.3.5 Paraconsistent Artificial Neural System

Classical systems based on binary logic are difficult to process data or information from uncertain knowledge. These data are captured or received information from multiple experts usually comes in the form of evidences.

Paraconsistent Artificial Neural Systems (PANS) modules are configured and built exclusively by PANU, whose function is to provide the signal processing 'similar' to processing that occurs in the human brain.

11.4 PANN for Morphological Analysis

The process of morphological analysis of a wave is performed by comparing with a certain set of wave patterns (stored in the control database). A wave is associated with a vector (finite sequence of natural numbers) through digital sampling. This vector characterizes a wave pattern and is registered by PANN. Thus, new waves are compared, allowing their recognition or otherwise.

Each wave of the survey examined the EEG corresponds to a portion of 1 second examination. Every second of the exam contains 256 positions.

The wave that has the highest favorable evidence and lowest contrary evidence is chosen as the more similar wave to the analyzed wave.

A control database is composed by waves presenting 256 positions with perfect sinusoidal morphology, with 0,5 Hz of variance, so taking into account Delta, Theta, Alpha and Beta (of 0.5 Hz to 30.0 Hz) wave groups.

In other words, morphological analysis checks the similarity of the passage of the examination of EEG in a reference database that represents a wave pattern.

11.4.1 Data Preparation

The process of wave analysis by PANN consists previously of data capturing, adaptation of the values for screen examination, elimination of the negative cycle and normalization of the values for PANN analysis.

As the actual EEG examination values can vary highly, in module, something 10 μV to 1500 μV, we make a normalization of the values between 100μV and -100 μV by a simple linear conversion, to facilitate the manipulation the data:

$$x = \frac{100.a}{m} \quad (11.4.1)$$

Where: m is the maximum value of the exam.
a is the current value of the exam.
x is the current normalized value.

The minimum value of the exam is taken as zero value and the remaining values are translated proportionally.

It is worth to observe that the process above does not allow the loss of any wave essential characteristics for our analysis.

11.4.2 The PANN Architecture

The architecture of the PANN used in decision making is based on the architecture of Paraconsistent Artificial Neural System for Treatment of Contradictions.

Such a system promotes the treatment of contradictions, so continuous between information signals, which receives three input signals and presents as a result, a value that represents the consensus between the three information. The contradictions between the two values are added to the third value, so that the output, the value proposed by the dominant majority. The analysis is instantly carrying all processing in real time, similar to the functioning of biological neurons.

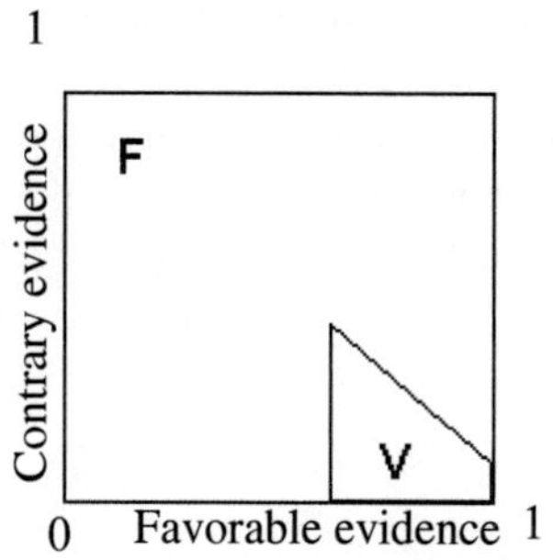

Fig. 11.4.1 Lattice for decision-making used in morphological analysis F: logical state false (it is interpreted as wave not similar); V: logical state true (it is interpreted as wave similar).

Table 11.4.1 Lattice for decision-making used in the morphological analysis (Fig. 11.4.1).

Limits of areas of lattice	
True	Fe > 0.61 Ce < 0.40 G_{ce} > 0,22
False	Fe < 0.61 Ce > 0.40 G_{ce} <= 0,23

Ce: contrary evidence; Fe: favorable evidence; G_{ce}: certainty degree;

This method is used primarily for PANN (fig. 11.4.2) to balance the data received from expert systems. After this process uses a decision-making lattice to determine the soundness of the recognition (fig. 11.4.1).

A sample of morphological analysis implementation using Object Pascal is show in table 11.4.2.

The definition of regions of the lattice decision-making was done through double-blind trials, i.e., each battery of tests, a validator checked the results and returning only the percentage of correct answers. After testing several different configurations, set the configuration of the lattice regions whose decision-making had a better percentage of success.

For an adequate PANN wave analysis, it is necessary that each input of PANN is properly calculated. These input variables are called expert systems, as they are specific routines for extracting information.

In analyzing EEG signals, one important aspect to take into account is the morphological aspect. To perform this task, it is valuable to build a very simple Expert System, which allows analyses the signal behavior verifying which band it belongs to (delta, theta, alpha and beta).

The method of morphological analysis has three expert systems that are responsible for feeding the inputs of PANN with information relevant to the wave being analyzed: number of peaks, similar points and different points.

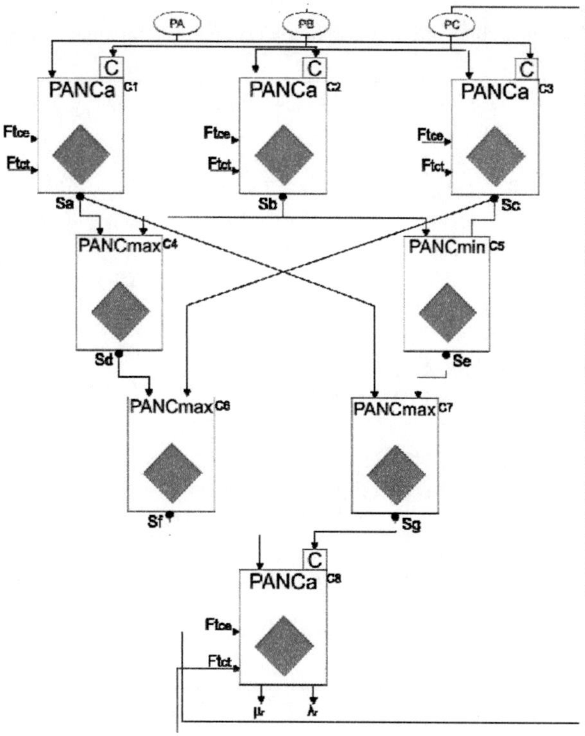

Fig. 11.4.2 The architecture for morphological analysis. Three expert systems operate: PA, for check the number of wave peaks; PB, for checking similar points, and PC, for checking different points.

The 1st layer of the architecture: C1–PANC which processes input data of PA and PB; C2–PANC which processes input data of PB and PC; C3–PANC which processes input data of PC and PA.

The 2nd layer of the architecture: C4–PANC which calculates the maximum evidence value between cells C1 and C2; C5–PANC which calculates the minimum evidence value between cells C2 and C3; The 3rd layer of the architecture: C6–PANC which calculates the maximum evidence value between cells C4 and C3; C7–PANC which calculates the minimum evidence value between cells C1 and C5.

The 4th layer of the architecture: C8 analyzes the experts PA, PB, and PC and gives the resulting decision value.

PANC A = Paraconsistent artificial neural cell of analytic connection.

$PANCLs_{Max}$ = Paraconsistent artificial neural cell of simple logic connection of maximization.

$PANCLs_{Min}$ = Paraconsistent artificial neural cell of simple logic connection of minimization.

Ft_{ce} = Certainty tolerance factor; Ft_{un} = Uncertainty tolerance factor.

S_a = Output of C1 cell; S_b = Output of C2 cell; S_c = Output of C3 cell; S_d = Output of C4 cell; Se = Output of C5 cell; Sf = Output of C6 cell; Sg = Output of C7 cell.

C = Complemented value of input;

μ_r = Value of output of PANN; λ_r = Value of output of PANN;

Table 11.4.2 The architecture for morphological analysis implementation (Fig. 11.4.2).

```
function   Tf_pann.Morphological_analysis(PA,   PB,   PC:
real; tipo: integer): real;
var
   C1, C2, C3, C4, C5, C6, C7: real;
begin
   C1 := FaPANN.PANCAC(PA, PB, 0, 0, 1);
   C2 := FaPANN.PANCAC(PC, PB, 0, 0, 1);
   C3 := FaPANN.PANCAC(PC, PA, 0, 0, 1);

   C4 := FaPANN.PANCMAX(C1, C2);
   C6 := FaPANN.PANCMAX(C4, C3);

   C5 := FaPANN.PANCMIN(C2, C3);
   C7 := FaPANN.PANCMIN(C1, C5);

   if tipo = 1 then
      result := FaPANN.CNAPCA(C6, C7, PC, PB, 1)
   else
      result := FaPANN.CNAPCA(C6, C7, PC, PB, 2);
end;
```

11.4.3 Expert System 1 – Checking the Number of Wave Peaks

The aim of the ***expert system 1*** is to compare the waves and analyze their differences regarding the number of peaks.

In practical terms, one can say that when we analyzed the wave peaks, we are analyzing the resulting frequency of wave (so well rudimentary).

It is worth remembering that, because it is biological signal, we should not work with absolute quantification due to the variability characteristic of this type of signal. Therefore, one should always take into consideration a tolerance factor.

A sample of checking the number of wave peaks function implementation using Object Pascal is show in table 11.4.3.

$$Se_1 = 1 - \left(\frac{(|bd - vt|)}{(bd + vt)} \right) \tag{11.4.2}$$

Where: vt is the number of peaks of the wave.

bd is the number of peaks of the wave stored in the database.

Se_1 is the value resulting from the calculation.

Table 11.4.3 Checking the number of wave peaks function implementation.

```
function Tf_pann.f_ EstimatedAveragePeak(vv: array of real;
total_elements: integer; Ftr: real): real;
  var
     last_larger_point, mean_peaks: real;
     peak_check: boolean;
     v_vector_aux_larger_value: real;
     a: integer;
     peaks: integer;

  begin
     last_larger_point := 0;
     peak_check := false;
     v_vector_aux_larger_value := 0;
     peaks := 0;
     mean_peaks := 0;

     for a := 1 to total_elements do
     begin
        if abs(vv[a - 1]) > v_vector_aux_larger_value then
           v_vector_aux_larger_value := vv[a - 1];

        if vv[a - 1] >= last_larger_point then
        begin
           last_larger_point := vv[a - 1];
           peak_check := false;
        end
        else
        begin
           if (peak_check = true) and
              (vv[a - 1] > vv[a - 2]) then
              last_larger_point := vv[a - 1];
           if abs(last_larger_point - vv[a - 1]) >= ((Ftr /
100) * last_larger_point) then
           begin
              if peak_check = false then
              begin
                 peaks := peaks + 1;
                 mean_peaks := mean_peaks + last_larger_point;
                 peak_check := true;
              end;
           end;
        end;

   end;

  result := mean_peaks / picos;

  end;
```

11.4.4 Expert System 2 – Checking Similar Points

The aim of the ***expert system 2*** is to compare the waves and analyze their differences regarding to similar points.

When we analyze the similar points, it means that we are analyzing how one approaches the other point.

It is worth remembering that, because it is biological signal, we should not work with absolute quantification due to the variability characteristic of this type of signal. Therefore, one should always take into consideration a tolerance factor.

A sample of checking similar points function implementation using Object Pascal is show in table 11.4.4.

$$Se_2 = \frac{\sum_{j=1}^{n} (x_j)}{n} \tag{11.4.3}$$

Where: n is the total number of elements.

x is the element of the current position.

j is the current position.

Se_2 is the value resulting from the calculation.

Table 11.4.4 Checking similar points function implementation

```
Function  Tf_pann.f_SimilarPoints(vv,  vb:  array  of  real;
total_elements:   integer;   Ftr:   real;   max_value:real;
lager_field_value:real): real;
var
     a: integer;
     fieldx_bd: real;
     q: real;
begin
      q:=0;
      for a := 1 to total_elements do
      bcgin
         fieldx_bd := vb[a - 1];
         fieldx_bd   :=   ((max_value   *   fieldx_bd)   /
lager_field_value);
        if  abs(fieldx_bd  -  vv[a  -  1])  <=  ((Ftr  /  100)  *
max_value) then
         begin
            q := q + 1;
         end;
      end;
  result   :=   1   -   (strtofloat(floattostrf(((q   /
total_elementos)), ffnumber, 18, 2)));
end;
```

11.4.5 Expert System 3 – Checking Different Points

The aim of the ***expert system 3*** is to compare the waves and analyze their differences regarding of different points.

When we analyze the different points, it means that we are analyzing how a point more distant from each other, so the factor of tolerance should also be considered.

A sample of checking different points function implementation using Object Pascal is show in table 11.4.5.

$$Se_3 = 1 - \left(\frac{\sum_{j=1}^{n} \left(\frac{|x_j - y_j|}{a} \right)}{n} \right) \tag{11.4.4}$$

Where: n is the total number of elements.
a is the maximum amount allowed.
j is the current position.
x is the value of wave 1.
y is the value of wave 2.
Se_3 is the value resulting from the calculation.

Table 11.4.5 Checking different points function implementation.

```
function Tf_pann.f_DifferentPoints(vv, vb: array of real;
total_elements: integer; Ftr, max_value,
lager_field_value:real): real;
var
     a: integer;
     fieldx_bd, q: real;
begin
      q:=0;
      for a := 1 to total_elements do
      begin
         fieldx_bd := vb[a - 1];
         fieldx_bd := ((max_value * fieldx_bd) /
lager_field_value);
         if abs(fieldx_bd - vv[a - 1]) > ((Ftr / 100) *
max_value) then
         begin
            q := q + (abs(fieldx_bd - vv[a - 1]) /
max_value);
         end;
      end;

  result := 1 - (strtofloat(floattostrf(((q /
total_elementos)), ffnumber, 18, 2)));

end;
```

11.5 A Didactic Sample

The following is an example that shows the operation of the methodology of morphologic analysis. In this example it will be considered three waves (Fig. 11.5.1) of 20 elements, with maximum amplitude of 11 points (from 0 to 10), and hypothetical values (Table 11.5.1).

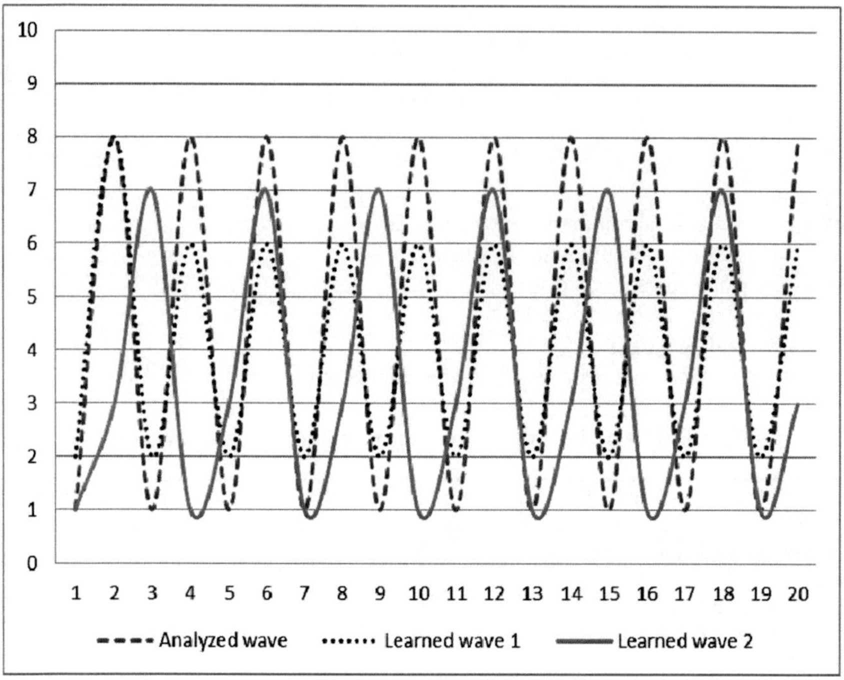

Fig. 11.5.1 Visual waves representation used in morphological analysis. The values of these waves are shown in Table 11.5.1.

Table 11.5.1 Values of the waves recognition. Their shapes can be seen in Fig. 11.5.1.

Wave	Values																			
Analyzed wave	1	8	1	8	1	8	1	8	1	8	1	8	1	8	1	8	1	8	1	8
Learned wave 1	2	8	2	6	2	6	2	6	2	6	2	6	2	6	2	6	2	6	2	6
Learned wave 2	1	3	7	1	3	7	1	3	7	1	3	7	1	3	7	1	3	7	1	3

The *analyzed wave* is the wave that will be submitted to the RNAP recognition. The *Learned wave 1* and *Learned wave 2* are two waves that were previously stored in the database control (normal).

Performing the comparison enters the waves to analyze, using experts systems we have, respectively: Expert system 1 (table 11.5.2), expert system 2 (table 11.5.3 and 11.5.4) and expert system 3 (table 11.5.5 and 11.5.6).

Table 11.5.2 Expert system 1 – Checking the number of wave peaks.

Waves	**Analyzed wave**	**Learned wave 1**	**Learned wave 2**
Peaks	9	9	6
Difference between the number of peaks, normalized by the sum of the number of peaks:		0	0,2
Expert system 1 (Se_1):		1	0,8

Table 11.5.3 Expert system 2 – Checking similar points. Comparison between the *analyzed wave* and the *learned wave 1.*

Analyzed wave	**Learned wave 1**	**Number of points equal**	**Analyzed wave**	**Learned wave 1**	**Number of points equal**
1	2	0	1	2	0
8	8	1	8	6	0
1	2	0	1	2	0
8	6	0	8	6	0
1	2	0	1	2	0
8	6	0	8	6	0
1	2	0	1	2	0
8	6	0	8	6	0
1	2	0	1	2	0
8	6	0	8	6	0
Sum of normalized differences: 1					
Expert system 2 (Se_2) - Normalized by the total element: 0,05					

Table 11.5.4 Expert system 2 – Checking similar points. Comparison between the *analyzed wave* and the *learned wave 2.*

Analyzed wave	**Learned wave 1**	**Number of points equal**	**Analyzed wave**	**Learned wave 1**	**Number of points equal**
1	2	0	1	2	0
8	8	1	8	6	0
1	2	0	1	2	0
8	6	0	8	6	0
1	2	0	1	2	0
8	6	0	8	6	0
1	2	0	1	2	0
8	6	0	8	6	0
1	2	0	1	2	0
8	6	0	8	6	0
Sum of normalized differences: 1					
Expert system 2 (Se_2) - Normalized by the total element: 0,05					

Table 11.5.5 Expert system 3 – *Analyzed wave* and the *learned wave 1*.

Analyzed wave	Learned wave 1	Difference (in module)	Normalization of the maximum amplitude difference
1	2	1	0,1
8	8	0	0
1	2	1	0,1
8	6	2	0,2
1	2	1	0,1
8	6	2	0,2
1	2	1	0,1
8	6	2	0,2
1	2	1	0,1
8	6	2	0,2
1	2	1	0,1
8	6	2	0,2
1	2	1	0,1
8	6	2	0,2
1	2	1	0,1
8	6	2	0,2
1	2	1	0,1
8	6	2	0,2
1	2	1	0,1
8	6	2	0,2
Sum of normalized differences: 2,8			
Expert system 3 (Se_3) - Normalized by the total element: 0,14 -) - Supplemented: 0,86			

Table 11.5.6 Expert system 3 – Checking different points. Comparison between the *analyzed wave* and the *learned wave 2*.

Analyzed wave	Learned wave 1	Difference (in module)	Normalization of the maximum amplitude difference
1	2	1	0,1
8	8	5	0,5
1	2	5	0,5
8	6	5	0,5
1	2	1	0,1
8	6	1	0,1
1	2	1	0,1
8	6	3	0,3
1	2	5	0,5
8	6	5	0,5
1	2	1	0,1
8	6	1	0,1
1	2	1	0,1
8	6	3	0,3
1	2	5	0,5
8	6	5	0,5
1	2	1	0,1
8	6	1	0,1
1	2	1	0,1
8	6	3	0,3
Sum of normalized differences: 5,4			
Expert system 3 (Se_3) - Normalized by the total element: 0,27 - Supplemented: 0,73			

The following (table 11.5.7) there are the values of each expert system that will be used as input values to PANN (fig. 11.4.1). After processing, the PANN has the resulting output values (table 11.5.8).

Table 11.5.7 Expert systems values.

Case	Expert system 1 (Se_1)	Expert system 2 (Se_2)	Expert system 3 (Se_3)
Analyzed wave × Learned wave 1	1,00	0,05	0,86
Analyzed wave × Learned wave 2	0,80	0,05	0,73

Table 11.5.8 Contrary evidence and favorable evidence resulting.

Case	Favorable evidence	Contrary evidence
Analyzed wave × Learned wave 1	0,69	0,48
Analyzed wave × Learned wave 2	0,58	0,38

According to Table 11.5.8, we see that the wave with the greatest evidence for is the *learned wave 1*, in other words, this is the wave more similar to the *analyzed wave*. In case of a draw between the values of favorable evidence will be used to wave with the slightest evidence to the contrary.

11.6 Experimental Procedures – Attention-Deficit / Hyperactivity Disorder

Recent researches reveal that 10% of the world population in school age suffer of learning and/or behavioral disorders caused by neurological problems, such as ADHD, dyslexia, and dyscalculia, with predictable consequences in those students' insufficient performance in the school [5], [6], [10], [11], [21], [22].

Concisely, a child without intellectual lowering is characterized as bearer of Attention-deficit/hyperactivity disorder (ADHD) when it presents signs of:

Inattention: difficulty in maintaining attention in tasks or games; the child seems not to hear what is spoken; difficulty in organizing tasks or activities; the child loses things; the child becomes distracted with any incentive, etc.

Hyperactivity: frequently the child leaves the class room; the child is always inconveniencing friends; the child runs and climbs in trees, pieces of furniture, etc; the child speaks a lot, etc.

Impulsiveness: the child interrupts the activities of colleagues; the child doesn't wait his time; aggressiveness crises; etc.

Dyslexia: when the child begins to present difficulties to recognize letters or to read them and to write them, although the child has not a disturbed intelligence, that is, a normal IQ;

Dyscalculia: when the child presents difficulties to recognize amounts or numbers and/or to figure out arithmetic calculations.

A child can present any combination among the disturbances above. All those disturbances have their origin in a cerebral dysfunction that can have multiple causes, many times showing a hereditary tendency. Since from the first discoveries made by [8], those disturbances have been associated to cortical diffuse lesions and/or more specific, temporal-parietal areas lesions in the case of dyslexia and dyscalculia [5], [11], [22]. The disturbances of ADHD disorder seem to be associated to an alteration of the dopaminergic system, that is, it is involved with mechanisms of attention and they seem to involve a frontal-lobe dysfunction and basal ganglia areas [6], [22].

EEG alterations seem to be associated those disturbances. Thus, some authors have proposed that there is an increase of the delta activity in EEG in those tasks that demand a larger attention to the internal processes. Other authors [16], [20] have described alterations of the delta activity in dyslexia and dyscalculia children sufferers. [12] has proposed that a phase of the EEG component would be associated to the action of the memory work. More recently, [14] has showed delta activity is reduced in occipitals areas, but not in frontals, when dyslexic's children were compared with normal ones.

In this way, the study of the delta and theta bands becomes important in the context of the analysis of learning disturbances.

So, in this paper we've studied two types of waves, specifically delta and theta waves band, where the size of frequency established clinically ranges from 1.0 Hz to 3.5 Hz and 4.0 Hz to 7.5 Hz respectively.

Seven exams of different EEG were analyzed, being two exams belonging to adults without any learning disturbance and five exams belonging to children with learning disturbances (exams and respective diagnoses given by ENSCER - Teaching the Brain, EINA - Studies in Natural Intelligence and Artificial Ltda).

Each analysis was divided in three rehearsals, each rehearsal consisted of 10 seconds of the analyzed, free from visual analysis of spikes and artifacts regarding to the channels T3 and T4.

In the first battery it was used of a filter for recognition of waves belonging to the Delta band. In the second battery it was used a filter for recognition of waves belonging to the Theta band. In the third battery it was not used any filters for recognition, i.e., the system was free to recognize any wave type.

At the end of the battery of tests, we obtained the following results (tables 11.6.1 to 11.6.6):

Table 11.6.1 Contingency table.

		Visual Analysis					
		Delta	Theta	Alpha	Beta	Unrecognized	Total
PANN Analysis	Delta	31	3	0	0	0	34
	Theta	15	88	1	1	0	105
	Alpha	0	5	22	0	0	27
	Beta	0	0	1	3	0	4
	N/D	7	2	1	0	0	10
	Total	53	98	25	4	0	180

Index Kappa = 0.80.

Table 11.6.2 Statistical results - sensitivity and specificity: Delta waves.

		Visual analysis		
		Delta	Not Delta	Total
PANN Analysis	True	31	124	155
	False	22	3	25
	Total	53	127	180

Sensitivity = 58%; Specificity = 97%.

Table 11.6.3 Statistical results - sensitivity and specificity: Theta waves.

		Visual analysis		
		Theta	Not Theta	Total
PANN Analysis	True	88	65	153
	False	10	17	27
	Total	98	82	180

Sensitivity = 89%; Specificity = 79%.

Table 11.6.4 Statistical results - sensitivity and specificity: Alpha waves.

		Visual analysis		
		Alpha	Not Alpha	Total
PANN Analysis	True	22	150	172
	False	3	5	8
	Total	25	155	180

Sensitivity = 88%; Specificity = 96%.

Table 11.6.5 Statistical results - sensitivity and specificity: Beta waves.

		Visual analysis		
		Beta	Not Beta	Total
PANN Analysis	True	3	175	178
	False	1	1	2
	Total	4	176	180

Sensitivity = 75%; Specificity = 99%.

Table 11.6.6 Statistical results - sensitivity and specificity: Unrecognized waves.

		Visual analysis		
		Unrecognized	Recognized	Total
PANN Analysis	True	0	170	170
	False	0	10	10
	Total	0	180	180

Sensitivity = 100%; Specificity = 94%.

Table 11.6.7 Result os tests performed.

Test	Visual	PANN	Test	Visual	PANN	Test	Visual	PANN	Test	Visual	PANN
1	D	D	46	T	T	91	A	-	136	D	D
2	T	T	47	D	D	92	T	T	137	D	D
3	D	D	48	A	A	93	T	T	138	D	-
4	D	T	49	T	T	94	T	T	139	T	T
5	T	T	50	T	T	95	T	T	140	T	T
6	T	T	51	T	T	96	A	A	141	D	D
7	D	D	52	T	T	97	T	T	142	T	T
8	A	A	53	T	T	98	T	T	143	T	T
9	T	T	54	T	T	99	T	T	144	D	-
10	T	T	55	T	T	100	T	T	145	T	T
11	T	T	56	T	D	101	T	T	146	D	D
12	T	T	57	T	T	102	T	A	147	D	D
13	T	T	58	A	A	103	T	T	148	T	T
14	T	T	59	A	B	104	D	T	149	A	A
15	T	T	60	D	T	105	T	T	150	A	A
16	T	T	61	T	A	106	A	A	151	D	D
17	T	T	62	T	D	107	T	T	152	D	T
18	A	A	63	B	T	108	T	T	153	D	D
19	B	B	64	T	A	109	T	T	154	T	T
20	D	T	65	D	D	110	D	-	155	D	-
21	D	D	66	T	A	111	A	A	156	D	-
22	T	T	67	T	-	112	T	A	157	A	A
23	D	D	68	T	T	113	T	T	158	A	A
24	D	T	69	D	D	114	T	T	159	D	D
25	T	T	70	T	T	115	T	T	160	T	T
26	T	T	71	D	-	116	D	D	161	D	D
27	D	D	72	D	T	117	D	T	162	D	D
28	A	A	73	T	T	118	A	A	163	D	T
29	T	T	74	T	T	119	T	T	164	D	D
30	T	T	75	T	T	120	D	D	165	T	T
31	T	T	76	A	A	121	A	A	166	D	T
32	T	T	77	T	T	122	T	T	167	D	D
33	T	T	78	D	T	123	T	T	168	T	T
34	T	T	79	T	T	124	T	T	169	T	T
35	T	T	80	T	T	125	T	T	170	T	T
36	D	T	81	A	A	126	D	D	171	D	D
37	D	T	82	T	D	127	T	T	172	D	-
38	A	A	83	B	B	128	T	T	173	T	T
39	B	B	84	A	A	129	A	A	174	A	A
40	T	T	85	D	D	130	T	T	175	A	T
41	D	D	86	A	A	131	T	T	176	T	T
42	D	D	87	T	-	132	T	T	177	A	A
43	D	D	88	T	T	133	D	T	178	T	T
44	D	T	89	D	D	134	T	T	179	T	T
45	T	T	90	T	T	135	D	D	180	A	A

Test: Number of test 10; **Visual**: Frequency band found by visual analysis; **PANN**: Frequency band found by PANN analysis (**-**: unrecognized; **D**: delta recognition; **T**: theta recognition; **A**: alpha recognition; **B**: beta recognition).

11.7 Experimental Procedures – Applying in Alzheimer Disease

Alzheimer's disease (AD) is a brain disorder characterized by cognitive impairment, which leads to a progressive dementia occurring in middle age or senectude (McKhann at al., 1984).

The AD corresponds roughly fifty percent of all cases of dementia (Berger et al., 1994; Herrera et al., 1998) and shows an increase in prevalence with advancing age. It is more prevalent among women (Fratiglioni et al., 1997) and the population aged 65 to 85 years. The incidence is approximately fourteen times higher in people aged 85 compared to 65 years (Herbert et al., 1995).

The definitive diagnosis of AD has not be established without histology of the brain (biopsy or autopsy), in which there is a specific degeneration in brain tissue, especially in pyramidal neurons, with a marked presence of intracellular neurofibrillary tangles and senile plaques in extra-cellular, accompanied by other structural changes, such as granulovacuolar degeneration, dendritic atrophy and loss of neuronal synapses (Terry, 1994).

The study model of the state follows the sequence-specific neuropathological findings in the course of cognitive decline in AD, which begins with changes in long-term memory, especially episodic visuospatial functions and attention. The following are changes in verbal and memory functions of short-term (Almkvist et al., 1993).

Because it is a disease where neurons are affected, a study may show differences in EEG brainwave patterns.

During relaxed wakefulness, the EEG in normal adults is predominantly composed of frequencies belonging to the alpha band, which are generated by interactions of the cortico-cortical and thalamocortical (Steriade et al., 1990; Lopes da Silva, 1991).

Many studies have shown that the visual analysis of EEG patterns may be useful in aiding the diagnosis of AD, and indicated in some clinical protocols for diagnosing the disease (Claus et al., 1999; Crevel et al., 1999).

In the tables of AD, the most common findings on visual analysis of EEG patterns are slowing of brain electrical activity based on predominance of delta and theta rhythms and decrease or absence of alpha rhythm. However, these findings are more common and evident in patients in moderate or advanced stages of disease (Silva et al., 1995; Alexander et al., 2006; Kwak, 2006).

In this context, the use of morphological analysis shows a promising tool, because with your application, we can quantify the concentration wave of bands of an examination and cross this information with the electroencephalographic findings, and since so, perform a diagnostic examination.

In this study we sixty-seven Analyzed EEG records, thirty-four normals and thirty-three probable AD (Table 11.7.2) during the awake state at rest (ie, eyes closed).

All tests were subjected to morphological analysis methodology for measuring the concentration of waves. Later this information is submitted to an artificial neural Paraconsistent another unit responsible for assessing the data and arriving at a classification of the examination in Normal or probable AD (Fig. 11.7.1).

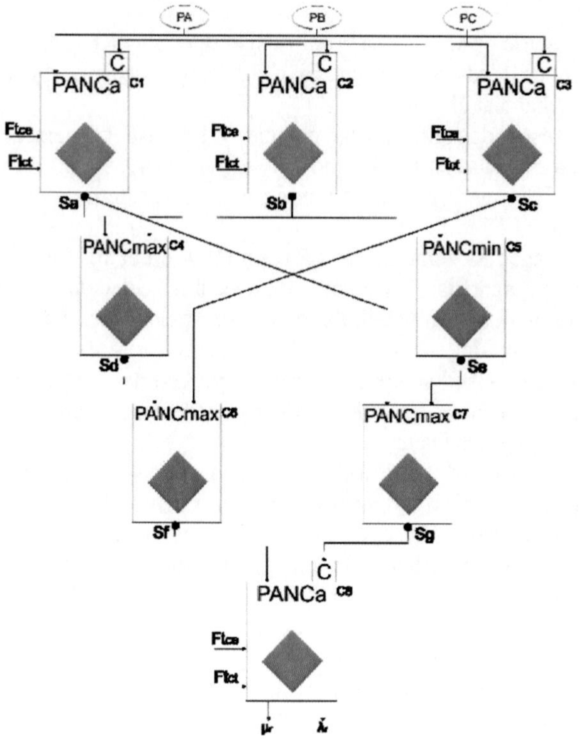

Fig. 11.7.1 The architecture for diagnostic analysis. Three expert systems operate: PA, for check the number of wave peaks; PB, for checking similar points, and PC, for checking different points:

The 1st layer of the architecture: C1–PANC which processes input data of PA and PB; C2–PANC which processes input data of PB and PC; C3–PANC which processes input data of PC and PA.

The 3rd layer of the architecture: C4–PANC which calculates the maximum evidence value between cells C1 and C2; C5–PANC which calculates the minimum evidence value between cells C2 and C3; C4 and C5 constitute the 2nd layer of the architecture; C6–PANC which calculates the maximum evidence value between cells C4 and C3; C7–PANC which calculates the minimum evidence value between cells C1 and C5.

The 4th layer of the architecture: C8 analyzes the experts PA, PB, and PC and gives the resulting decision value.

PANC A = Paraconsistent artificial neural cell of analytic connection.

$PANCLs_{Max}$ = Paraconsistent artificial neural cell of simple logic connection of maximization.

$PANCLs_{Min}$ = Paraconsistent artificial neural cell of simple logic connection of minimization.

Ft_{ce} = Certainty tolerance factor; Ft_{ct} = Contradiction tolerance factor.

S_a = Output of C1 cell; S_b = Output of C2 cell; S_c = Output of C3 cell; S_d = Output of C4 cell; Se = Output of C5 cell; Sf = Output of C6 cell; Sg = Output of C7 cell.

C = Complemented value of input;

μ_r = Value of output of PANN; λ_r = Value of output of PANN;

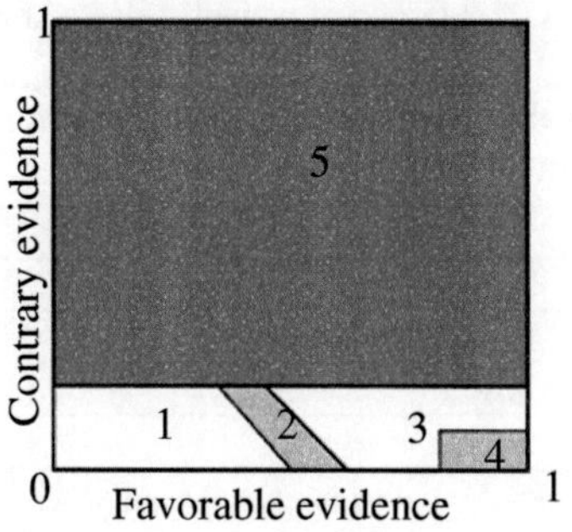

Fig. 11.7.2 Lattice for decision-making used in diagnostic analysis used after making PANN analisys (Fig. 11.7.1). Area 1: State logical False (AD likely below average population), 2: State logical Near-real (AD likely than average population); Area 3: State-Almost logical false (Normal below average population); Area 4: State logical True (Normal above average population); Area 5: logical state of uncertainty (not used in the study area).

Table 11.7.1 Lattice for decision-making used in diagnostic analysis used after making PANN analisys (Fig. 11.7.1).

Limits of areas of lattice	
Area 1	$G_{ce} <= 0{,}1999$ $G_{ce} >= 0{,}5600$ $\mid G_{un} \mid < 0{,}3999$ $\mid G_{un} \mid >= 0{,}4501$
Area 2	$0{,}2799 < G_{ce} < 0{,}5600$ $0{,}3099 <= \mid G_{un} \mid < 0{,}3999$ Fe $< 0{,}5000$
Area 3	$0{,}1999 < G_{ce} < 0{,}5600$ $0{,}3999 <= \mid G_{un} \mid < 0{,}4501$ Fe $> 0{,}5000$
Area 4	$G_{ce} > 0{,}7999$ $\mid G_{un} \mid < 0{,}2000$

Ce: contrary evidence; Fe: favorable evidence; G_{ce}: certainty degree; G_{un}: uncertainty degree;

Table 11.7.2 Group of patients selected for the study.

	Normal Individuals Control Group	Probable AD Individuals AD Group
Male	8	6
Female	26	27
Mean	61,38	68
Schooling	8,12	6,21
MEEM	24,53	20,58

Male: Male patients; Female: Female patients; Mean: Mean age of patients; Schooling: Mean of years of studies of patients; MMSE: Mean score of the mini mental state exam; p = 0.8496.

Table 11.7.3 The architecture for AD diagnostic analysis implementation (Fig. 11.7.1).

```
function  Tf_pann.Ad_diagnostic_analysis(PA,  PB,  PC:
real; tipo: integer): real;
var
   C1, C2, C3, C4, C5, C6, C7: real;
begin
   C1 := FaPANN.PANCAC(PA, PB, 0, 0, 1);
   C2 := FaPANN.PANCAC(PC, PB, 0, 0, 1);
   C3 := FaPANN.PANCAC(PC, PA, 0, 0, 1);

   C4 := FaPANN.PANCMAX(C1, C2);
   C6 := FaPANN.PANCMAX(C4, C3);

   C5 := FaPANN.PANCMIN(C2, C3);
   C7 := FaPANN.PANCMIN(C1, C5);

   if tipo = 1 then
      result := FaPANN.CNAPCA(C6, C7, 0, 0, 1)
   else
      result := FaPANN.CNAPCA(C6, C7, 0, 0, 2);
end;
```

11.7.1 Expert System 1 – Detecting the Diminishing Average Frequency Level

The aim of the ***expert system 1*** is An expert system verifies the average frequency level of Alpha waves and compares them with a fixed external one (external parameter wave).

Such external parameter can be, for instance, the average frequency of a population or the average frequency of the last exam of the patient. This system also generates two outputs: favorable evidence μ (normalized values ranging from

0 (corresponds to 100% – or greater frequency loss) to 1 (which corresponds to 0% of frequency loss) and contrary evidence λ (Eq. 11.7.1).

The average frequency of population pattern used in this work is 10 Hz.

$$\lambda = 1 - \mu \tag{11.7.1}$$

Table 11.7.4 Detecting the diminishing average frequency level function implementation.

```
function Tf_pann.CompareAveragePeak(fm, Freq_Avr_pop: real):
real;
var
  aux: real;
begin
  aux := ((100 * fm) / Freq_Avr_pop) / 100;
  if aux > 1 then aux := 1;
  if aux < 0 then aux := 0;
  result := aux;
end;
```

11.7.2 Expert System 2 – High Frequency Band Concentration

The aim of the ***expert system 2*** is the expert system is utilized for alpha band concentration in the exam. For this, we consider the quotient of the sum of fast alpha and beta waves over slow delta and theta waves (Eq. 11.7.2) as first output value (favorable evidence μ). For the second output value (contrary evidence λ) is used Eq. 11.7.1.

$$\mu = \left(\frac{(A+B)}{(D+T)} \right) \tag{11.7.2}$$

Where: A is the alpha band concentration.
B is the beta band concentration.
D is the delta band concentration.
T is the theta band concentration.
μ is the value resulting from the calculation.

Table 11.7.5 High frequency band concentration function implementation.

```
function Tf_pann.ChecksAlphaConcentration(Aband, Bband, Dband,
Tband: real): real;
var
  aux: real;
begin
  aux := (Aband + Bband) / (Dband + Tband);
  result := aux;
end;
```

11.7.3 Expert System 3 – Low Frequency Band Concentration

The aim of the ***expert system 3*** is the expert system is utilized for tetha band concentration in the exam. For this, we consider the quotient of the sum of slow delta and theta waves over fast alpha and beta waves (Eq. 11.7.3) as first output value (favorable evidence μ). For the second output value (contrary evidence λ) is used Eq. 11.7.1.

$$\mu = \left(\frac{(D+T)}{(A+B)} \right) \quad (11.7.3)$$

Where: A is the alpha band concentration.
B is the beta band concentration.
D is the delta band concentration.
T is the theta band concentration.
μ is the value resulting from the calculation.

Table 11.7.6 Low frequency band concentration function implementation.

```
function Tf_pann.ChecksThetaConcentration(Aband, Bband, Dband,
Tband: real): real;
var
  aux: real;
begin
  aux := (Dband + Tband) / (Aband + Bband);
  result := aux;
end;
```

11.7.4 Results

The results obtained in the study using the casuistry shown in Table 11.7.1 showed a promising performance, as shown in table 11.7.7.

Table 11.7.7 Diagnosis – Normal x Probable AD patients

		Gold Standard		
		AD patient	Normal patient	Total
PANN	AD patient	35.82%	14.93%	50.75%
	Normal patient	8.96%	40.30%	49.25%
	Total	44.78%	55.22%	100.00%

Sensitivity = 80%; Specificity = 73%; Index of coincidence (Kappa): 76%

11.8 Discussion

We believe that a process of the examination analysis using a PANN attached to EEG findings, such as relations between frequency bandwidth and inter hemispheric coherences, can create computational methodologies that allow the automation of analysis and diagnosis.

These methodologies could be employed as tools to aid in the diagnosis of diseases such as dyslexia or Alzheimer, provided they have defined electroencephalographic findings.

In the case of Alzheimer's disease, for example, in studies carried out previously shown satisfactory results [12] (but still far from being a tool to aid clinical) that demonstrated the computational efficiency of the methodology using a simple morphological analysis (only paraconsistent annotated logic $E\tau$). These results encouraged us to improve the morphological analysis of the waves and try to apply the method in other diseases besides Alzheimer's disease.

With the process of morphological analysis using the PANN, it becomes possible to quantify the frequency average of the individual without losing its temporal reference. This feature becomes a differential, compared to traditional analysis of quantification of frequencies, such as FFT (Fast Fourier Transform), aiming at a future application in real-time analysis, i.e. at the time of acquisition of the EEG exams.

For this future application, it must be assumed that the automatic detection of spikes and artifacts are important functions that should be aggregated for analysis, thus creating variations in morphology specialized detection devices, for example.

It is noteworthy that by treating the PANN a relatively new theory and extend the operation of classical PAN is justified to use different approaches (as discussed in this work) to know the full potential of the theory applied to the specific and real needs.

11.9 Conclusions

These findings suggest that the sensitivity with respect to the Delta waves is 58%. This is an indication that there must be improvements in the detection of peaks in the band Delta. We believe that such improvements are possible to be made in this direction. The sensitivities of the theta, alpha and beta waves are reasonable, but that improvements can be tried.

Regarding the specificity, the method showed more reliable results. Taking into account an overall assessment in the sense we take the arithmetic mean of sensitivity (75.50%) and specificity (92.75%), we find reasonable results that encourage us to seek improvements in this study.

Regarding the results in table 11.17 show that the data RNAP always performs some sort of recognition. The concept of non-recognition used here means that the degree of similarity found by RNAP was lower than that stipulated by the decision making lattice (Fig. 6). Thus, further studies must also be done in lattice configuration decision and the database of waves learned in order to refine them.

Even finding a low sensitivity in the recognition of delta waves, the methodology of pattern recognition using morphological analysis showed to be effective, achieving recognize patterns of waves similar to patterns stored in the database, allowing quantifications and qualifications of the examination of EEG data to be used by PANN in their process analysis of examination.

References

1. Abe, J.M.: Foundations of Annotated Logics, PhD thesis University of São Paulo, Brazil (1992) (in Portuguese)
2. Abe, J.M.: Some Aspects of Paraconsistent Systems and Applications. Logique et Analyse 157, 83–96 (1997)
3. Abe, J.M., Lopes, H.F.S., Anghinah, R.: Paraconsistent Artificial Neural Network and Alzheimer Disease: A Preliminary Study. Dementia & Neuropsychologia 3, 241–247 (2007)
4. Anghinah, R.: Estudo da densidade espectral e da coerência do eletrencefalograma em indivíduos adultos normais e com doença de Alzheimer provável, PhD thesis, Faculdade de Medicina da Universidade de São Paulo, São Paulo (2003) (in Portuguese)
5. Ansari, D., Karmiloff-Smith, A.: Atypical trajectories of number development: a neuroconstructivist perspective. Trends In Cognitive Sciences 12, 511–516 (2002)
6. Blonds, T.A.: Attention-Deficit Disorders and Hyperactivity. Developmental Disabilities in Infancy and Ramus, F., Developmental dyslexia: specific phonological deficit or general sensorimotor dysfunction? Current Opinion in Neurobiology 13, 1–7 (2003)
7. Da Silva Filho, J.I.: Métodos de interpretação da Lógica Paraconsistente Anotada com anotação com dois valores LPA2v com construção de Algoritmo e implementação de Circuitos Eletrônicos, EPUSP, PhD thesis, São Paulo (1999) (in Portuguese)
8. Da Silva Filho, J.I., Abe, J.M., Torres, G.L.: Inteligência Artificial com as Redes de Análises Paraconsistentes. LTC-Livros Técnicos e Científicos Editora S.A., São Paulo, p. 313 (2008) (in Portuguese)
9. Gallarburda, A.M., Sherman, G.F., Rosen, G.G., Aboitiz, F., Genschiwind, N.: Developmental dyslexia: four consecutive patients with cortical anomalies. Ann. Neurology 18, 2122–2333 (1985)
10. Hynd, G.W., Hooper, R., Takahashi, T.: Dyslexia and Language-Based disabilities. In: Coffey, Brumbak (eds.) Text Book of Pediatric Neuropsychiatry, pp. 691–718. American Psychiatric Press (1985)
11. Lindsay, R.L.: Dyscalculia. In: Capute, Accardo (eds.) Developmental Disabilities in Infancy and Childhood, pp. 405–415. Paul Brookes Publishing Co., Baltimore (1996)
12. Lopes, H.F.S.: Aplicação de redes neurais artificiais paraconsistentes como método de auxílio no diagnóstico da doença de Alzheimer, MSc Dissertation, Faculdade de Medicina-USP, São Paulo, p. 473 (2009) (in Portuguese)

13. Klimeshc, W.: EEG alpha and theta oscillations reflect cognitive and memory performance: a review and analysis. Brain Res. Ver. 29, 169–195 (1999)
14. Klimesch, W., Doppelmayr, H., Wimmer, J., Schwaiger, D., Rôhm, D., Bruber, W., Hutzler, F.: Theta band power changes in normal and dyslexic children. Clinical Neurophysiology 113, 1174–1185 (2001)
15. Kocyigit, Y., Alkan, A., Erol, H.: Classification of EEG Recordings by Using Fast Independent Component Analysis and Artificial Neural Network. Journal of Medical Systems 32(1), 17–20 (2008)
16. Niedermeyer, E., da Silva, F.L.: Electroencephalography, 5th edn. Lippincott Williams & Wilkins (2005)
17. Rocha, A.F., Massad, E.: How the human brain is endowed for mathematical reasoning. Mathematics Today 39, 81–84 (2003)
18. Rocha, A.F., Massad, E., Pereira Jr., A.: The Brain: From Fuzzy Arithmetic to Quantum Computing, pp. 1434–9922. Springer, Heidelberg (2005)
19. Temple, E.: Brain mechanisms in normal and dyslexic readers. Current Opinion in Neurobiology 12, 178–183 (2002)
20. Voeller, K.K.S.: Attention-Deficit / Hyperactivity: Neurobiological and clinical aspects of attention and disorders of attention. In: Coffey, Brumbak (eds.) Text Book of Pediatric Neuropsychiatry, pp. 691–718. American Psychiatric Press (1998)
21. Montenegro, M.A., Cendes, F., Guerreiro, N.M., Guerreiro, C.A.M.: EEG na prática clínica. Lemos Editorial, Brasil (2001)
22. McKhann, G., Drachman, D., Folstein, M., Katzman, R., Price, D., Stadlan, E.M.: Clinical diagnosis of AD: report of the NINCDS-ADRDA work group under the auspices of Deparment of health and human services task force on AD. Neurology 34, 939–944 (1984)
23. Berger, L., Morris, J.C.: Diagnosis in Alzheimer Diasease. In: Terry, R.D., Katzman, R., Bick, K.L. (eds.), pp. 9–25. Reaven Press, Ltd., New York (1994)
24. Herrera, J.E., Camarelli, P., Nitrini, R.: Estudo epidemológico populacional de demência na cidade de Catanduva, estado de São Paulo. Brasil. Rev. Psiquiatria Clínica 25, 70–73 (1998)
25. Fratiglioni, L., Viitanen, M., von Strauss, E., Tontodonai, V., Wimblad, H.A.: Very old women at highest risk of demencia and AD: Incidence data from Kungsholmen project, Stockhom. Neurology 48, 132–138 (1997)
26. Herbert, L.E., Scherr, P.A., Beckett, L.: Age-specific incidence of AD in a community population. JAMA 273, 1359 (1995)
27. Terry, R.D.: Neuropathological changes in AD. In: Svennerholm, L. (ed.) Progress in Brain Research. ch. 29, vol. 101, pp. 383–390. Elservier Science BV (1994)
28. Almkvist, O., Backman, L.: Detection and staging of early clinical dementia. Acta. Neurol. Scand. 88 (1993)
29. Steriade, M., Gloor, P., Llinás, R., Lopes da Silva, F., Mesulan, M.: Basic mechanisms of cerebral rhytmic activities. Electroencephalogr. Clin. Neurophysiol. 76, 481–508 (1990)
30. Lopes da Silva, F.: Neural mechanisms underlying brain waves: from neural membranes to network. Electroencephalogr. Clin. Neutophysiol. 79, 81–93 (1991)
31. Claus, J.J., Strijers, R.L.M., Jonkman, E.J., Ongerboer De Visser, B.W., Jonker, C., Walstra, G.J.M., Scheltens, P., Gool, W.: The diagnostic value of EEG in mild senile Alzheimer´s disease. Clin. Neurophysiol. 18, 15–23 (1999)

32. Crevel, H., Gool, W.A., Walstra, G.: Early diagnosis of dementia: Which tests are indicated? What are their costs: J. Neurol. 246, 73–78 (1999)
33. Silva, D.F., Lima, M.M., Anghinah, R., Lima, J.: Mapeamento cerebral. Rev. Neurociências 3, 11–18 (1995)
34. Alexander, D.M., Arns, M.W., Paul, R.H., Rowe, D.L., Cooper, N., Esser, A.H., Fallahpour, H., Stephan, B.C.M., Heesen, E., Breteler, R., Williams, L.M., Gordon, E.E.: markers for cognitive decline in elderly subjects with subjective memory complaints. Journal of Integrative Neuroscience 5(1), 49–74 (2006)
35. Kwak, Y.T.: Quantitative EEG findings in different stages of Alzheimer's disease. J. Clin. Neurophysiol. (Journal of clinical neurophysiology: official publication of the American Electroencephalographic Society) 23(5), 456–461 (2006)

Chapter 12
Paraconsistent Artificial Neural Networks and Pattern Recognition: Speech Production Recognition and Cephalometric Analysis

Jair Minoro Abe[1,2] and Kazumi Nakamatsu[3]

[1] Graduate Program in Production Engineering, ICET - Paulista University
R. Dr. Bacelar, 1212, CEP 04026-002 São Paulo – SP – Brazil
[2] Institute For Advanced Studies – University of São Paulo, Brazil
jairabe@uol.com.br
[3] School of Human Science and Environment/H.S.E. – University of Hyogo – Japan
nakamatu@shse.u-hyogo.ac.jp

Abstract. In this expository work we sketch a theory of artificial neural network, based on a paraconsistent annotated evidential logic $E\tau$. Such theory, called Paraconsistent Artificial Neural Network - PANN - is built from the algorithm Para-analyzer and has as characteristics the capability of manipulating uncertainty, inconsistent and paracomplete concepts. Some applications are presented in speech production analysis and cephalometrich variables analysis.

Keywords: Artificial neural network, paraconsistent logics, annotated logics, pattern recognition, speech disfluence, cephalometric variables.

12.1 Introduction

Many pattern recognition applications use statistical models with a large number of parameters, although the amount of available training data is often insufficient for robust parameter estimation. In order to overcome these aspects, a common technique to reduce the effect of data sparseness is the divide-and-conquer approach, which decomposes a problem into a number of smaller subproblems, each of which can be handled by a more specialized and potentially more robust model. This principle can be applied to a variety of problems in speech and language processing: the general procedure is to adopt a feature-based representation for the objects to be modeled (such as phones or words), learn statistical models describing the features of the object rather than the object itself, and recombine these partial probability estimates. Although this enables a more efficient use of data, other interesting techniques have been employed for the task. One of the most successful theories is the so-called artificial neural networks - ANN.

R. Kountchev et al. (Eds.): Advances in Reasoning-Based Image Processing, ISRL 29, pp. 365–382.
springerlink.com

In this work we are concerned in applying a particular ANN, namely the paraconsistent artificial neural network – PANN, introduced in [2] which is based on paraconsistent annotated evidential logic $E\tau$ [1] to speech signals recognition by using phonic traces signals. The PANN is capable of manipulating concepts like uncertainty, inconsistency and paracompleteness in its interior. To test the theory presented here, we develop one computer system with the capability of capturing and converting a speech signal as a vector. After this, we analyze the percentage recognizing results shown below. With these studies we point out some of the most important features of PANN: firstly, the PANN recognition becomes 'better' in every new recognition step, so it is a consequence of discarding contradicting signals and recognizing them by proximity, without trivializing the results. Finally, the performance and efficiency of the PANN is enough to recognize in real time, any speech signal.

Now we show how the PANN was efficient in formants recognition. The tests were made in Portuguese and 3 pairs of syllables were chosen 'FA-VA', 'PA-BA', 'CA-GA' presenting one articulation and differences in sonority (see table 12.2). The speaker is an adult, masculine sex, 42 years old, Brazilian, from São Paulo city.

After the sixth speech step, the PANN was able to recognize efficiently every signal with recognizing factor higher than 88%. Every signal lower than this factor can be considered as unrecognized. When the PANN was learned the 'FA' syllable (10 times) and it is asked the 'VA' syllable for recognizing, the recognizing factor is never higher than 72%. For the remaining pairs of syllables, this factor showed lower.

Cephalometrics is the most useful tool for orthodontic diagnosis, since assess craniofacial skeletal and dental discrepancies. However, conventional cephalometrics holds important limitations, mostly due to the fact that the cephalometric variables are not assessed under a contextualized scope and carry on important variation when compared to samples norms. Because of that, its clinical application is relative, subjective, and routinely less effective than the expected. In addition, discordance between orthodontists about diagnosis and treatments it is not uncommon, due to the inevitable uncertainties involved in the cephalometrics variables. In our point of view, this is a perfect scenario to evaluate the paraconsistent neural network capacity to perform with uncertainties, inconsistencies, and paracompleteness in a practical problem.

In this work an expert system to support orthodontic diagnosis was developed based on the paraconsistent approach. In the structure proposed the inferences were based upon the degrees of evidence (favorable and unfavorable) of abnormality for cephalometrics variables, which may have infinite values between "0" and "1". Therefore, the system may be refined with more or less outputs, depending upon the need. Such flexibility allows that the system can be modeled in different ways, allowing a finer adjusting.

In order to evaluate the practical aspects of this paraconsistent neural network we analyzed the concordance with the system and an expert opinion about 40 real cases. As preliminary results, the degrees of evidence of abnormality were tested for the three Units. Kappa values, comparing the software and the opinion of the expert were: Unit 1 = 0.485; Unit 2 =0.463 and Unit 3 = 0.496 (upper incisors), = 0.420 (lower incisors) and = 0.681 (upper combined with lower incisors). The strength of agreement is at least moderate. It is important to highlight that the initial data, used for the classification of each group presented significant variation and the opinions of the specialist about particular problems hold important subjective weight.

Finally, although the system needs more accurate validation, the preliminary results are encouraged and show in a doubtless way that paraconsistent neural networks may contribute for the development of expert systems taking to account uncertainties and contradictions, presented in the most real problems, particularly heath areas, opening a new promising tool of research.

12.2 Background

As mentioned in the previous paragraph, Paraconsistent Artificial Neural Network – PANN is based on the paraconsistent annotated evidential logic $E\tau$. Let us present the main ideas underlying its studies.

The atomic formulas of the paraconsistent annotated logic $E\tau$ is of the type $p_{(\mu, \lambda)}$, where $(\mu, \lambda) \in [0, 1]^2$ and [0, 1] is the real unitary interval (*p* denotes a propositional variable). An order relation is defined on $[0, 1]^2$: $(\mu_1, \lambda_1) \leq (\mu_2, \lambda_2) \Leftrightarrow \mu_1 \leq \mu_2$ and $\lambda_1 \leq \lambda_2$, constituting a lattice that will be symbolized by τ. A detailed account of annotated logics is to be found in [1]. $p_{(\mu, \lambda)}$ can be intuitively read: "It is assumed that *p*'s favorable evidence (or belief degree) is μ and contrary evidence (or disbelief degree) is λ." Thus, (1.0, 0.0) intuitively indicates total favorable evidence, (0.0, 1.0) indicates total contrary evidence, (1.0, 1.0) indicates total inconsistency, and (0.0, 0.0) indicates total paracompleteness. The operator $\sim : |\tau| \rightarrow |\tau|$ defined in the lattice, $\sim [(\mu, \lambda)] = (\lambda, \mu)$, works as the "meaning" of the logical negation of $E\tau$.

We can consider several important concepts (all considerations are taken with $0 \leq \mu, \lambda \leq 1$):
Segment *DB* - segment perfectly defined: $\mu + \lambda - 1 = 0$
Segment *AC* - segment perfectly undefined: $\mu - \lambda = 0$
Uncertainty Degree: $G_{un}(\mu, \lambda) = \mu + \lambda - 1$; Certainty Degree: $G_{ce}(\mu, \lambda) = \mu - \lambda$;

With the uncertainty and certainty degrees we can get the following 12 regions of output: *extreme states* that are, False, True, Inconsistent and Paracomplete, and *non-extreme states*. All the states are represented in the lattice of the Figure 12.1: such lattice τ can be represented by the usual Cartesian system.

These states can be described with the values of the certainty degree and uncertainty degree by means of suitable equations. In this work we have chosen the resolution 12 (number of the regions considered according to the Figure 12.1), but the resolution is totally dependent on the precision of the analysis required in the output and it can be externally adapted according to the applications.

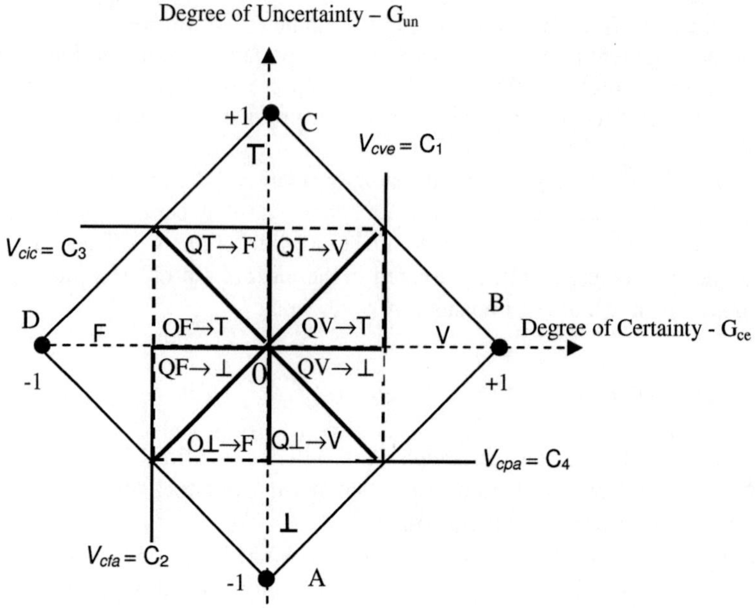

Fig. 12.1 Representation of the certainty and uncertainty degrees

So, such limit values called Control Values are:

V_{cic} = maximum value of uncertainty control = C_3
V_{cve} = maximum value of certainty control = C_1
V_{cpa} = minimum value of uncertainty control = C_4
V_{cfa} = minimum value of certainty control = C_2

For the discussion in the present work we have used: $C_1 = C_3 = ½$ and $C_2 = C_4 = -½$.

Table 12.1 Extreme and Non-extreme states

Extreme States	Symbol	Non-extreme states	Symbol
True	V	Quasi-true tending to Inconsistent	QV→T
False	F	Quasi-true tending to Paracomplete	QV→⊥
Inconsistent	T	Quasi-false tending to Inconsistent	QF→T
Paracomplete	⊥	Quasi-false tending to Paracomplete	QF→⊥
		Quasi-inconsistent tending to True	QT→V
		Quasi-inconsistent tending to False	QT→F
		Quasi-paracomplete tending to True	Q⊥→V
		Quasi-paracomplete tending to False	Q⊥→F

12.3 The Paraconsistent Artificial Neural Cells - PANC

In the paraconsistent analysis the main aim is to know how to measure or to determine the certainty degree concerning a proposition, if it is False or True. Therefore, for this, we take into account only the certainty degree G_{ce}. The uncertainty degree G_{un} indicates the measure of the inconsistency or paracompleteness. If the certainty degree is low or the uncertainty degree is high in mudule, it generates an indefinition.

The resulting certainty degree G_{ce} is obtained as follows:

If: $V_{cfa} = G_{ce} = V_{cve}$ or $C_2 = G_{ce} = C_1 \Rightarrow$ Gce = Indefinition
For: $V_{cpa} \leq G_{un} \leq V_{cic}$
If: $C_1 = V_{cve} = G_{ce} \Rightarrow G_{ce}$ = True with degree G_{un}
$V_{cic} \leq G_{un} \Rightarrow G_{ce}$ = True with degree G_{un}

The algorithm that expresses a basic Paraconsistent Artificial Neural Cell - PANC - is:

```
* /Definition of the adjustable values * /
   Vcve = C1 * maximum value of certainty control * /
Vcfa =C2 * / minimum value of certainty control * /
   Vcic =C3 * maximum value of uncertainty control * /
   Vcpa =C4 * minimum value of uncertainty control* /
      * Input /Variables * /
    μ, λ
     * Output /Variables *
   Digital output = S1
   Analog output = S2a
   Analog output = S2b
* /Mathematical expressions * /
begin:
0≤ μ ≤ 1 e 0≤ λ≤ 1
Gun = μ + λ - 1
Gce = μ - λ
           * / determination of the extreme states * /
if Gce ≥ C1 then S1 = V
if Gce ≥ C2 then S1 = F
if Gun ≥ C3 then S1 = T
if Gun ≤ C4 then S1 = ⊥
If not: S1 = I - Indetermination
   Gun = S2a
   Gce = S2b
```

A PANC is called *basic* PANC when given a pair (μ, λ) is used as input and resulting as output: G_{un} = resulting uncertainty degree, G_{ce} = resulting certainty degree, and X = constant of Indefinition, calculated by the equations $G_{un} = \mu + \lambda - 1$ and $G_{ce} = \mu - \lambda$.

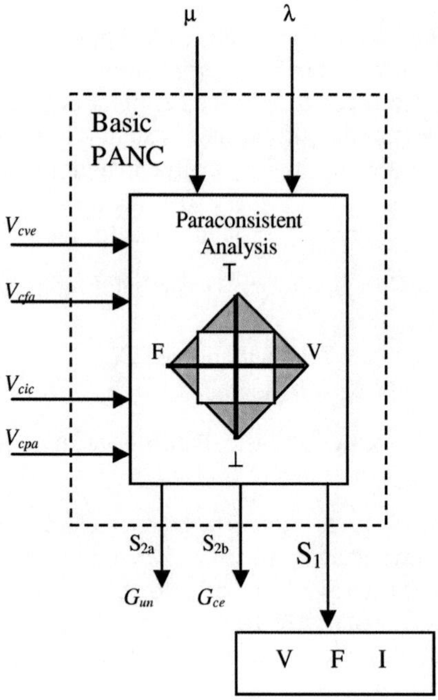

Fig. 12.2 The Basic Paraconsistent Artificial Neural Cell

12.4 The Paraconsistent Artificial Neural Cell of Learning - PANC-L

A Paraconsistent Artificial Neural Cell of Learning – PANC-l is obtained from a basic PANC.

In this learning Cell, sometimes we need the action of the operator Not in the training process. Its function is to do the logical negation in the resulting output sign. For a training process, we consider initially a PANC of Analytic Connection the one not undergoing any learning process.

According to the paraconsistent analysis, a cell in these conditions has two inputs with an Indefinite value ½.

So, the basic structural equation yields the same value ½ as output, having as result an indefinition. For a detailed account see [4].

The learning cells can be used in the PANN as memory units and pattern sensors in primary layers. For instance, a PANC-l can be trained to learn a pattern by using an algorithm. For the training of a cell we can use as pattern real values between 0 and 1. The cells can also be trained to recognize values between 0 and 1.

The learning of the cells with extreme values 0 or 1 composes the primary sensorial cells. Thus, the primary sensorial cells consider as pattern a binary digit where the value 1 is equivalent to the logical state True and the value 0 is equivalent to the logical state False. The first feeding of the cell is made by the input $\mu_{1,1} = \mu_r(0)$, resulting into the output $\mu_r(1)$. This one will be the second input $\mu_{1,2} = \mu_r(1)$, that, in a feed-back process, will result in the output $\mu_r(2)$ and so successively. In short, the input $\mu_r(k)$ produces the output $\mu_r(k+1)$.

The occurrence of the input $\mu_{1,k+1} = \mu_r(k) = 0$ repeated times means that the resulting favorable evidence degree is going to increase gradually in the output reaching the value 1. In these conditions we say that the cell has learned the falsehood pattern.

The same procedure is adopted when the value 1 is applied to the input repeated times. When the resulting of favorable evidence degree in the output reaches the value $\mu_r(k+1) = 1$, we say that the cell has learned the truth pattern. Therefore a PANC can learn two types of patterns: the truth pattern or the falsity pattern. In the learning process of a PANC, a learning factor can be introduced (LF) that is externally adjusted. Depending on the value of LF, it gives the cell a faster or slower learning. In the learning process, given an initial belief degree $\mu_r(k)$, we use the following equation to reach $\mu_r(k) = 1$, for some k.

So, for truth pattern we have

$$\mu_r(k+1) = \frac{(\mu_1 - \mu_r(k)_c)LF + 1}{2}$$

where $\mu_r(k)_c = 1 - \mu_r(k)$, and $0 \le LF \le 1$. For falsity pattern, we have

$$\mu_r(k+1) = \frac{(\mu_{1c} - \mu_r(k)_c)LF + 1}{2}$$

where $\mu_r(k)_c = 1 - \mu_r(k)$, $\mu_{1c} = 1 - \mu_1$, and $0 \le LF \le 1$

So we can say that the cell is completely learned when $\mu_r(k+1) = 1$.

If LF = 1, we say that the cell has a natural capacity of learning. Such capacity decreases as LF approaches 0. When LF = 0, the cell loses the learning capacity and the resulting belief degree will always have the indefinition value ½ .

12.5 Unlearning of a PANC-l

Even after having a cell trained to recognize a certain pattern, if insistently the input receives a value $\mu_{1,k+1} = \mu_r(k)$ totally different, the high uncertainty makes the cell unlearn the pattern gradually. The repetition of the new values implies in a decreasing of the resulting belief degree. Then, the analysis has reached an indefinition. By repeating this value, the resulting favorable evidence degree reaches 0 meaning that the cell is giving the null favorable evidence degree to the former proposition to be learned. This is equivalent to saying that the cell is giving the maximum value to the negation of the proposition, so the new pattern must be confirmed. Algorithmically, this is showed when the certainty degree G_{ce} reaches the value –1. In this condition the negation of the proposition is confirmed. This is obtained by applying the operator Not to the cell. It inverts the resulting belief degree in the output. From this moment on the PANC considers as a new pattern the new value that appeared repeatedly and unlearning the pattern learned previously.

By considering two factors, LF – learning factor and UF – unlearning factor, the cell can learn or unlearn faster or slower according the application. These factors are important giving the PANN a more dynamic process.

The graphic below presents the result of the learning PANC using the learning algorithm seen in an application of a pattern to the sinusoid form.

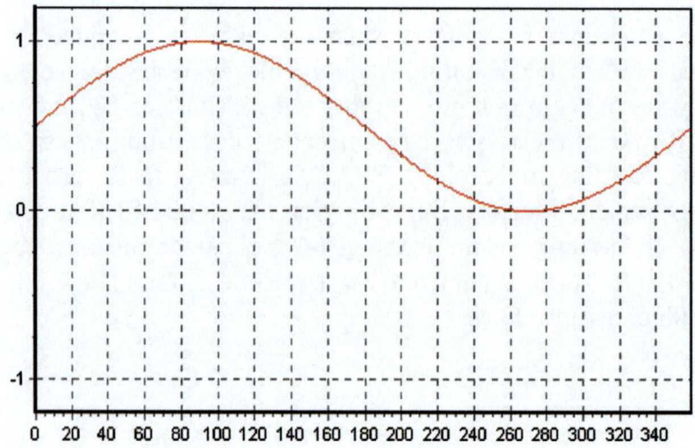

Fig. 12.3 Pattern versus number of steps

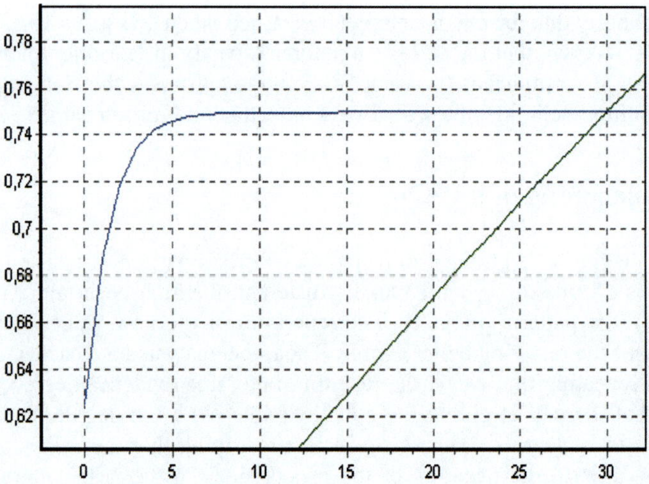

Fig. 12.4 Learning cell behavior

The figure 12.3 displays pattern versus number of steps by applying the equation and we have: sign [k] = (Sin ((k x Pk) / 180) + 1) / 2

The figure 12.4 displays the pattern versus number of steps for learning, showing that the cell has learned the applied function as input pattern after 30 steps.

12.6 Using PANN in Speech Production Recognition

Through a microphone hardwired to a computer, a sonorous signal can be caught and transformed to a vector (finite sequence of natural numbers x_i) through a digital sampling. This vector characterizes a sonorous pattern and it is registered by the PANN. So, new signals are compared, allowing their recognition or not. For the sake of completeness, we show some basic aspects of how PANN operates. Let us take three vectors: $V_1 = (2, 1, 2, 7, 2)$; $V_2 = (2, 1, 3, 6, 2)$; $V_3 = (2, 1, 1, 5, 2)$. The favorable evidence is calculated as follows: given a pair of vectors, we take '1' for equal elements and '0' to the different elements, and we figure out its percentage.

Comparing V_2 with V_1: 1 + 1 + 0 + 0 + 1 = 3; in percentage: (3/5)*100 = 60%
Comparing V_3 with V_1: 1 + 1 + 0 + 0 + 1 = 3; in percentage: (3/5)*100 = 60%

The contrary evidence is the weighted addition of the differences between the different elements, in module:

Comparing V_2 with V_1 = 0 + 0 + 1/8 + 1/8 + 0 = (2/8)/5 = 5%
Comparing V_3 with V_1 = 0 + 0 + 1/8 + 2/8 + 0 = (3/8)/5 = 7.5%

Therefore, we can say that V_2 is 'closer' to V_1 than V_3. We use a PANN to recognize this technical system.

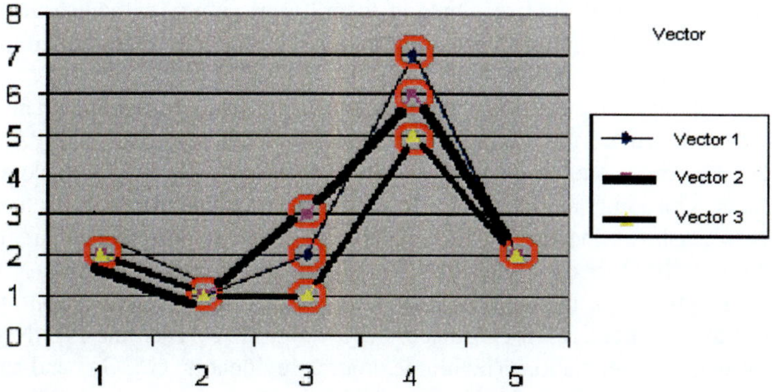

Fig. 12.5 Vector's representation

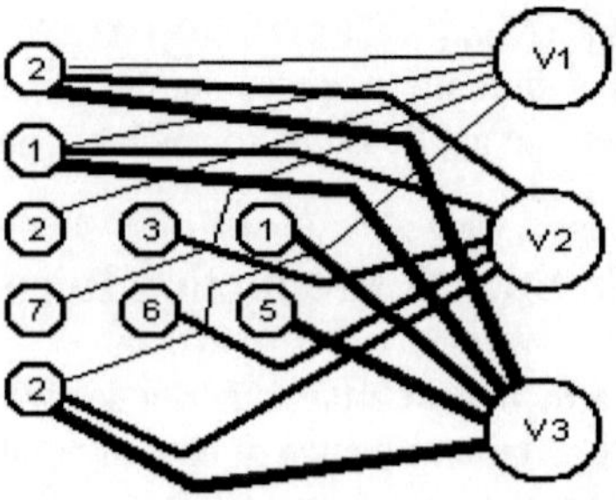

Fig. 12.6 PANN and layers

We can improve this technical method by adding more capabilities to the PANN, like 'proximity' concept and 'recognizing' level.

Also, the PANN has the capability of adjusting its own recognizing factor through the recognizing factor internal to the Neural Cell that can be propagated to higher neural levels. Thus, the PANN can improve its capability in each recognizing speech.

Another important functionality aspect of the PANN is the processing velocity, so we can work in real time producing and identifying speech.

12.7 Practical Results

To test the theory presented here, we develop one computer system with the capability of capturing and converting a speech signal as a vector. After this, we analyze the percentage recognizing results shown below. With these studies we point out some of the most important features of PANN: firstly, the PANN recognition becomes 'better' in every new recognition step, so it is a consequence of discarding contradicting signals and recognizing them by proximity, without trivializing the results. Finally, the performance and efficiency of the PANN is enough to recognize in real time, any speech signal.

Now we show how the PANN was efficient in formants recognition. The tests were made in Portuguese and 3 pairs of syllables were chosen 'FA-VA', 'PA-BA', 'CA-GA' presenting one articulation and differences in sonority (see table 12.2). The speaker is an adult, masculine sex, 42 years old, Brazilian, from São Paulo city.

Table 12.2 shows the recognizing capability. The recognizing percent in the first column is 100% because the PANN is empty and the syllables are just being learned. The process of recognition is made in the computer system as follows: in the step 2 the speaker says, for instance, the syllable 'FA'. Then the PANN gives an output with the calculations (favorable/contrary evidences, G_{ce}, G_{un}) and asks to the speaker (operator) if the data is acceptable or not. If the answer is 'Yes', the PANN keep the parameters for the next recognition. If the answer is 'Not", the PANN recalculate the parameters in order to criticize the next recognition, till such data becomes belongs to False state (fig. 12.1), preparing for the next step to

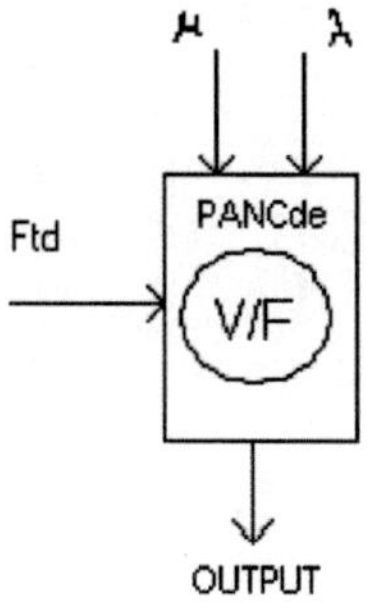

Fig. 12.7 PANCde

repeat the process (in this way, improves the recognition). This is performed by the neural cell PNACde (see table 12.2):

Table 12.2 Syllable recognition

Steps											
Syllable	**1**	**2**	**3**	**4**	**5**	**6**	**7**	**8**	**9**	**10**	**Average**
FA	100%	87%	88%	91%	90%	92%	95%	94%	94%	95%	**91,78%**
VA	100%	82%	85%	87%	88%	90%	94%	92%	96%	95%	**89,89%**
PA	100%	83%	86%	90%	88%	95%	89%	94%	95%	95%	**90,56%**
BA	100%	85%	82%	87%	90%	89%	92%	95%	95%	97%	**90,22%**
CA	100%	82%	87%	89%	88%	92%	90%	94%	92%	95%	**89,89%**
GA	100%	84%	88%	92%	90%	89%	95%	95%	95%	92%	**91,11%**

These adjustments are made automatically by the PANN except only the learning steps, which is made the intervention of the operator to feed the Yes/No data. More details can be found in [4]. Thus, from the second column on, PANN is recognizing and learning adjusts simultaneously, as well as adapting such improvements; this is the reason that the recognizing factor increases. In the example, we can see that after the sixth speech step, the PANN is able to recognize efficiently every signal with recognizing factor higher than 88%. Every signal lower than this factor can be considered as unrecognized.

Table 12.3 shows the recognition factor percentage when PANN analyzes a syllable with one different speech articulation. As we can see, when the PANN was learned the 'FA' syllable (10 times) and it is asked the 'VA' syllable for recognizing, the recognizing factor is never higher than 72%. For the remaining pairs of syllables, this factor showed lower.

Table 12.3 Recognition of pairs of syllables with one different speech articulation

Steps											
Pairs	**1**	**2**	**3**	**4**	**5**	**6**	**7**	**8**	**9**	**10**	**Average**
FA・VA	70%	67%	72%	59%	65%	71%	64%	69%	66%	63%	66,60%
PA・BA	51%	59%	49%	53%	48%	52%	46%	47%	52%	48%	**50,50%**
CA・GA	62%	59%	61%	62%	58%	59%	60%	49%	63%	57%	**59,00%**

12.8 Cephalometric Variables

Craniofacial discrepancies, either skeletal or dental, are assessed in lateral cephalograms by cephalometric analyses. In Orthodontics, such quantitative analysis compares an individual with a sample of a population, matched by gender and age. However, cephalometric measurements hold significant degrees of insufficiency and inconsistency, making its clinical application less effective than ideal.

A conventional cephalometric analysis compares individual measurements to a pattern, i.e., a norm for that variable, assessed in a sample of patients which has the same age and gender. Such piece of information is, in the best scenario, a suggestion of the degree of deviation from the norm, for such particular variable. A better scenario would be to knowing how much the value of a variable of a certain patient is deviated from its norm. It would be better if we could quantify the "noise" carried by a cephalometric value and filter its potential damage to the contextual result. In this sense, developing a mathematical structure able to provide quantitative information and modeling the inconsistencies, contradictions and evidences of abnormality of these variables is relevant and useful.

In order to analyze skeletal and dental changes we selected a set of cephalometric variables based on an expert knowledge. Figures 12.1 and 12.2 show these variables and analysis. In this work we propose an expert system able to assess the degree of evidence of abnormality of each variable, suggesting a diagnosis for the case and, consequently, an adequate treatment plan. The expectance is that system increases the potential clinical application of the cephalometric analysis, potentially better addressing a more efficient therapy.

1. Basion
2. Sella
3. Nasion
4. Posterior Nasal Spine
5. Anterior Nasal Spine
6. Inter-Molars
7. Inter-Incisors
8. Gonion
9. Menton

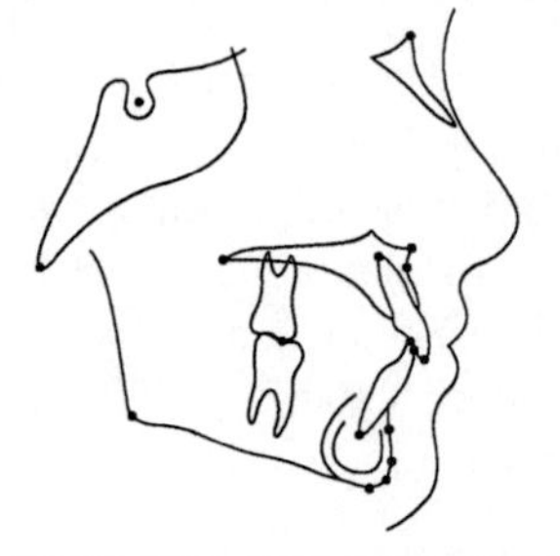

10. Gnathion
11. A Point
12. B Point
13. Pogonion
14. Incisal Edge - Upper Incisor
15. Apex - Upper Incisor
16. Incisal Edge - Lower Incisor
17. Apex - Lower Incisor

Fig. 12.8 Cephalometric Variables

1. Anterior Cranial Base
2. Palatal Plane (PP)
3. Oclusal Plane (OP)
4. Mandibular Plane (MP)
5. Cranial Base
6. Y Axis
7. Posterior Facial Height
8. Anterior Facial Height - Median Third

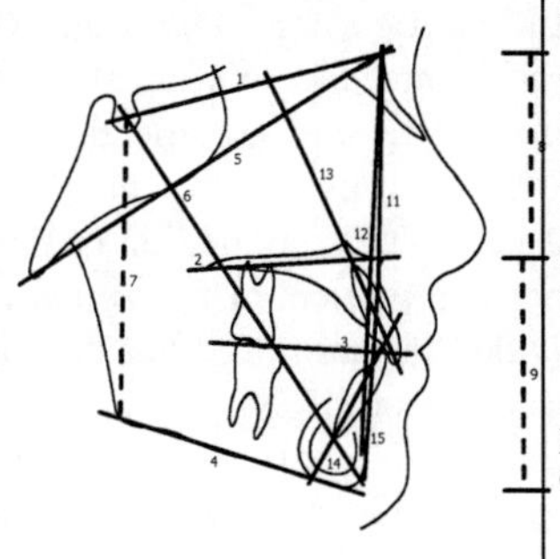

9. Anterior Facial Height - Lower Third
10. Anterior Facial Height
11. SNA
12. SNB
13. Long Axis - Upper Incisor
14. Long Axis - Lower Incisor
15. A Point - Pogonion Line

Wits: distance between the projections of the A and B Points on the occlusal plane.

Fig. 12.9 Proposed Cephalometric Analysis

12.9 Architecture of the Paraconsistent Artificial Neural Network

The selected cephalometric variables are inserted in the paraconsistent network in the following three units: Unit I, considering the antero-posterior discrepancy, Unit II, considering vertical discrepancy, and Unit III, taking into account dental discrepancy (see Fig. 12.10).

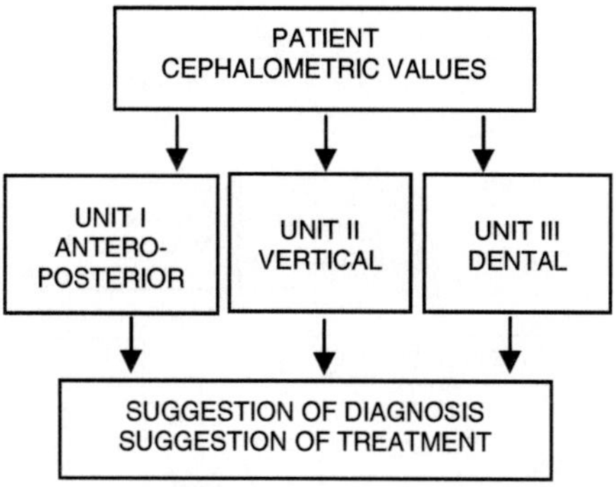

Fig. 12.10 Functional macro view of the neural architecture

Unit I is made by 2 levels. The first one involves the ANB and Wits variables. At the second level, there is a combination of the result of the level 1 and the variables SNA and SNB. The output of the second level is regard to the position of the maxilla and the mandible, classifying it as: well positioned, protruded, retruded, tending to protruded and tending to retruded. The classes protruded and retruded coming with their respective degrees of evidence of abnormality. Moreover, the classes "tending to" suggest the assessment of the outputs of the Units II and III. The variables pertaining to the Unit II are divided in three different groups: Group I: Se-Go/Na-Me proportion. The value of this proportion may result in normal, vertical or horizontal face. Group II: Y Axis. The value of this angle may also result in normal, vertical or horizontal face. Group III: angles Se-Na/PP, SeNa/OP and SeNa/MP. Each one of the three angles may also result in normal, vertical or horizontal face. The combination of the output form Groups I, II and III will also result in normal, vertical or horizontal face. The variables pertaining to the Unit III are divided in three different groups: Group I: U1.PP and U1.SN angles and linear measurement U1-NA, taking in account the SNA angle (Unit I). The upper incisors may be in a normal position, proclined or retroclined. Group II: L1.APg, L1. NB and L1.GoMe angles and linear measurements L1-APg, L1-NB, taking in account the SNB angle. The lower incisors may be in a normal position, proclined or retroclined. Group III: angle U1.L1. This value results in three possible positions: normal, proclined and retroclined. The combination of the outputs of the Groups I, II and III will result in normal, proclined, retroclined, tending to proclined and tending to retroclined. Each unit has the following components, represented in the Fig. 12.11:

a) Standardization: the difference between the data from the patient radiographs and its relative norm, by age and gender;
b) Data modeling: matrices with possible degrees of evidence of abnormality.

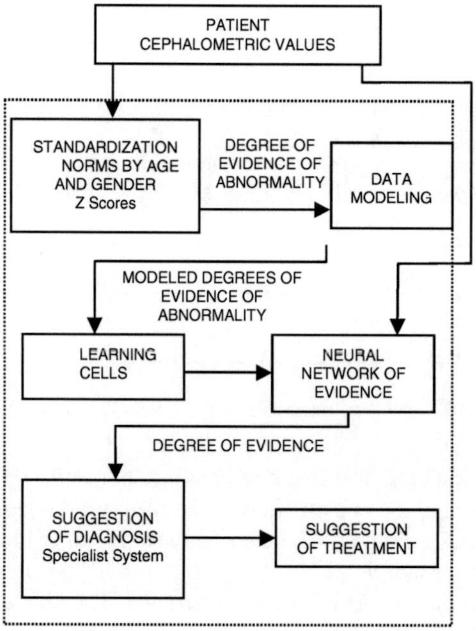

Fig. 12.11 Functional micro view of the structure of each unit of the Fig 5.

Modeling Tool: matrices with all possible values of evidence of abnormality related to the variables of the patients, using the standard deviation values and providing the degree of evidence of abnormality.

Learning Cell: "learn" the functions of the matrices which contain the degrees of evidence of abnormality of the variables.

Neural Network of Evidence: file the learned information and return with the degree of evidence according to the standard deviation of each variable in a contextualized form, i.e., considering the degrees of evidence of abnormality of the other variables.

Expert System: provides the diagnosis, inferred from the neural degrees of evidence of abnormality of each unit.

Treatment Plan: based upon the specialist system provides a suggestion of diagnosis for that specific orthodontic case.

The system works with four craniofacial measurements in the Unit I, resulting in 46 inferences, giving 33 outputs of types of malocclusion (diagnosis). In the Unit II, 5 craniofacial measurements allow 90 inferences and 4 outputs of diagnosis. In the Unit III, 9 cephalometric measurements were assessed, giving 87 inferences and 12 outputs of diagnosis. Suggestions of treatment are proposed for all diagnosis pointed. At total, the expert system is based upon 18 craniofacial cephalometric measurements, 223 inferences and 49 outputs of diagnosis and suggestions of treatment.

12.10 Results

In order to evaluate the system performance 120 orthodontic protocols with the cephalometric information contained all variables values (only measurements) considered in the model were analyzed by the model and three orthodontic experts. Kappa agreement concordance was evaluated comparing the diagnostic proposal of the experts and the system (Siegel and Castellan, 1988; Fleiss, 1981).

The analyzed data is composed by 44,17% of males and 55,83% of females, with age raging between 6 to 53; 18.33% are more than 18 years old (corresponding to 22 in 120), which were treated by the model as 18 years old for simplicity. The sample is composed for the most part by a white group. However, as ethnical definition in Brazil is a complex matter, such group must be regarded with restriction.

Table 12.4 Kappa values of agreement where E_1, E_2, and E_3 indicates the experts:

	E1 & model	E2 & model	E3 & model	E1&E2&E3
Anteroposterior region	0.343	0.289	0.369	0.487
Mandible position	0.296	0.245	0.306	0.404
Maxilla position	0.343	0.289	0.369	0.421
Vertical discrepancy	0.75	0.372	0.67	0.534
Superior incisive	0.443	0.216	0.454	0.468
Inferior incisive	0.451	0.084	0.457	0.418
Lip position	0.924	0.849	0.885	0.838

12.11 Discussion

Cephalometrics is the most useful tool for orthodontic diagnosis, since assess craniofacial skeletal and dental discrepancies. However, conventional cephalometrics holds important limitations, mostly due to the fact that the cephalometric variables are not assessed under a contextualized scope and carry on important variation when compared to samples norms. Because of that, its clinical application is relative, subjective, and routinely less effective than the expected. In addition, discordance between orthodontists about diagnosis and treatments it is not uncommon, due to the inevitable uncertainties involved in the cephalometrics variables. In our point of view, this is a perfect scenario to evaluate the paraconsistent neural network capacity to perform with uncertainties, inconsistencies, and paracompleteness in a practical problem.

In this work an expert system to support orthodontic diagnosis was developed based on the paraconsistent approach. In the structure proposed the inferences were based upon the degrees of evidence (favorable and unfavorable) of abnormality for cephalometrics variables, which may have infinite values between "0" and "1". Therefore, the system may be refined with more or less outputs, depending upon the need. Such flexibility allows that the system can be modeled in different ways, allowing a more fine adjusting.

The system require measurements taken from the head lateral radiography of the patient that will be assessed. The precision of the system increase as much as data is added, enriching the learning cells for that specific situation. On the other hand, if the radiographic information provided was insufficient, the system give back an Undefined (*Un*) output, state that is also predictable in the paraconsistent logic. Therefore, possible "noise" as consequence of the lack of data does not prevent that the initial goals of neural network can be achieved.

In order to evaluate the practical aspects of this paraconsistent neural network we analyzed the concordance with the system and an expert opinion about 40 real cases. As preliminary results, the degrees of evidence of abnormality were tested for the three Units. Kappa values, comparing the software and the opinion of the expert were: Unit 1 = 0.485; Unit 2 =0.463 and Unit 3 = 0.496 (upper incisors), = 0.420 (lower incisors) and = 0.681 (upper combined with lower incisors). The strength of agreement is at least moderate. It is important to highlight that the initial data, used for the classification of each group presented significant variation and the opinions of the specialist about particular problems hold important subjective weight.

Finally, although the system needs more accurate validation, the preliminary results are encouraged and show in a doubtless way that paraconsistent neural networks may contribute for the development of expert systems taking to account uncertainties and contradictions, presented in the most real problems, particularly heath areas, opening a new promising tool of research.

12.12 Conclusions

The variation of the analyzed values are interpreted by the PANN and adjusted automatically by the system. Due the PANN structural construction, the network is able to identify small variations between the pairs of syllables chosen. One central reason is its capability of proximity recognition and discarding contradictory data without trivialization. In the examples above, we can define as recognized if the factor is higher than 88%, and non-recognized, if the factor is lower than 72%. The difference of 16% (between recognition and non-recognition) is enough to avoid mistakes in the interpretation of the results. Thus, PANN shows itself as a superior system in being capable to manipulate the factors described showing high accuracy in data analysis.

The results presented in this work show that PANN can be a very efficient structure for speech analysis. Of course, new concepts are necessary for a more complete study of speech production, but this is in course. We hope to say more in forthcoming works.

References

[1] Abe, J.M.: Fundamentos da Lógica Anotada (Foundations of Annotated Logics), in Portuguese, Ph. D. Thesis, University of São Paulo, São Paulo (1992)

[2] Da Silva Filho, J.I., Abe, J.M.: Para-Analyzer and Inconsistencies in Control Systems. In: Proceedings of the IASTED International Conference on Artificial Intelligence and Soft Computing (ASC 1999), Honolulu, Hawaii, USA, August 9-12, pp. 78–85 (1999)

[3] Da Silva Filho, J.I., Abe, J.M.: Paraconsistent analyzer module. International Journal of Computing Anticipatory Systems 9 (2001); ISSN: 1373-5411, ISBN: 2-9600262-1-7, 346-352

[4] Da Silva Filho, J.I., Abe, J.M.: Fundamentos das Redes Neurais Paraconsistentes – Destacando Aplicações em Neurocomputação. In: Portuguese, Editôra Arte & Ciência, 247 (2001)

[5] Dempster, A.P.: Generalization of Bayesian inference. Journal of the Royal Statistical Society Series B-30, 205–247 (1968)

[6] Hecht-Nielsen, R.: Neurocomputing. Addison Wesley Pub. Co., New York (1990)

[7] Kohonen, T.: Self-Organization and Associative Memory. Springer, Heidelberg (1984)

[8] Kosko, B.: Neural Networks for signal processing. Prentice-Hall, USA (1992)

[9] Sylvan, R., Abe, J.M.: On general annotated logics, with an introduction to full accounting logics. Bulletin of Symbolic Logic 2, 118–119 (1996)

[10] Fausett, L.: Fundamentals of Neural Networks Architectures, Algorithms and Applications. Prentice-Hall, Englewood Cliffs (1994)

[11] Stephens, C., Mackin, N.: The validation of an orthodontic expert system rule-base for fixed appliance treatment planning. Eur. J. Orthod. 20, 569–578 (1998)

[12] Martins, D.R., Janson, G.R.P., Almeida, R.R., Pinzan, A., Henriques, J.F.C., Freitas, M.R.: Atlas de Crescimento Craniofacial, Santos, São Paulo, SP (1998) (in Portuguese)

[13] Mario, M.C.: Modelo de análise de variáveis craniométricas através das Redes Neurais Artificiais Paraconsistentes, Ph. D. Thesis, University of São Paulo (2006)

[14] Sorihashi, Y., Stephens, C.D., Takada, K.: An inference modeling of human visual judgment of sagittal jaw-base relationships based on cephalometry: Part II. J. Orthod. Dentofac. Orthop. 117, 303–311 (2000)

[15] Steiner, C.: Cephalometrics for you and me. Am. J. Orthod. 39, 729–755 (1953)

[16] Jacobson, A.: Wits appraisal of jaw disharmony. Am. J. Orthod. 67, 125–138 (1975)

[17] Jacobson, A.: The application of the Wits appraisal. Am. J. Orthod. 70, 179–189 (1971)

[18] Ricketts, R.M.: Cephalometric analysis and synthesis. Angle Orthod. 31, 141–156 (1961)

[19] Abe, J.M., Ortega, N., Mario, M.C., Del Santo Jr., M.: Paraconsistent Artificial Neural Network: an Application in Cephalometric Analysis. In: Khosla, R., Howlett, R.J., Jain, L.C. (eds.) KES 2005. LNCS (LNAI), vol. 3684, pp. 716–723. Springer, Heidelberg (2005)

[20] Weinstein, J., Kohn, K., Grever, M.: Neural Computing in Cancer Drug Development: Predicting Mechanism of Action. Science 258, 447–451 (1992)

[21] Baxt, W.J.: Application of Artificial Neural Network to Clinical Medicine. Lancet 346, 1135–1138 (1995)

[22] Subasi, A., Alkan, A., Koklukaya, E., Kiymik, M.K.: Wavelet neural network classification of EEG signals by using AR model with MLE preprocessing. Neural Networks 18(7), 985–997 (2005)

[23] Russell, M.J., Bilmes, J.A.: Introduction to the special issue on new computational paradigms for acoustic modeling in speech recognition. Computer Speech & Language 17(2-3), 107–112 (2003)

Chapter 13
On Creativity and Intelligence in Computational Systems

Stuart H. Rubin

SSC-Pacific, US Navy
53560 Hull St., San Diego, CA, USA 92152-5001
stuart.rubin@navy.mil

This chapter presents an investigation of the potential for creative and intelligent computing in the domain of machine vision. It addresses such interrelated issues as randomization, dimensionality reduction, incompleteness, heuristics, as well as various representational paradigms. In particular, randomization is shown to underpin creativity, heuristics are shown to serve as the basis for intelligence, and incompleteness implies the need for heuristics in any non trivial machine vision application, among others. Furthermore, the evolution of machine vision is seen to imply the evolution of heuristics. This conclusion follows from the examples supplied herein.

13.1 Introduction

In current and future operational environments, such as the Global War on Terrorism (GWOT) and Maritime Domain Awareness (MDA), war fighters require technologies evolved to support information needs regardless of location and consistent with the user's level of command or responsibility and operational situation. This chapter addresses current shortfalls in C2 systems resulting from limitations in image recognition technologies and associated uncertainties. According to DoD definitions, Command and Control (C2) Systems are "the facilities, equipment, communications, procedures, and personnel essential to a commander for planning, directing, and controlling operations of assigned forces pursuant to the missions assigned." This chapter develops a technology that (1) provides an automated approach for real-time image processing and analysis; (2) identifies and integrates informational content from multiple information sources; and (3) provides automatic correlation, fusion, and insight to support user-cognitive processes. In short, this chapter provides a novel approach for automated image understanding and automated processes for recognizing target feature information.

Images are important forms of information for understanding the battle-space situation. Automated image/scene understanding provides the battlefield commander

R. Kountchev et al. (Eds.): Advances in Reasoning-Based Image Processing, ISRL 29, pp. 383–421.
springerlink.com

with enhanced situational awareness by fully exploiting the capabilities of reconnaissance/surveillance platforms. As sensor systems are upgraded, sensors are added, and new systems constructed, more images will be collected than can be manually screened due to the limited number of analysts available. The current processes for analyzing images, in advance of military operations, are mostly manual and time consuming and otherwise not of sufficient complexity. Time critical and autonomous operations using autonomous platforms equipped with image sensors demand automated approaches for real-time image processing and analysis for (a) extraction of relevant features, (b) association of extracted features from different sensor modalities, (c) efficient representation of associated features, and (d) understanding of images/scenes. This chapter provides for:

- Automatic extraction of relevant features
- Association of extracted features from different sensor modalities
- Recognition of objects
- Tracking of objects in different modalities
- Activity/change detection
- Behavior analysis based on detected activities
- Threat assessment based on behavior analysis
- Image/scene understanding.

The Navy is likely to be interested in adapting the contents of this chapter for the recognition of all manner of objects – including human faces, terrain, hyperspectral imagery for use in automating robots, including robotic vehicles or UAVs (Fig. 13.1). Such robots could be trained to autonomously survey dangerous terrain using infrared, microwaves, visible light, and/or neutron back scatter. The technology could also be applied to scan suitcases in airports, vehicles passing through checkpoints, and a plethora of similar applications. Data fusion technology is inherent to its operation. This means that given sufficient, albeit attainable processing power, war fighters may be gradually removed from hazardous duties – being replaced by ever-more capable UAVs.

The Joint Services have a vested interest in computer vision solutions for surveillance, reconnaissance, automated weaponry, treaty monitoring, and a plethora of related needs.[1] The problem is that so far no system has come even close to mimicking the visual capabilities of a carrier pigeon. These birds play a primary role in naval rescue operations by spotting humans at sea from the vantage point of a helicopter bay.

Successful approaches for the construction of visual systems include edge detection and extraction, texture analysis, feature recognition, neural analysis and classification, inverse computer graphics techniques (pioneered in France), pyramid decomposition (pioneered in Bulgaria), fractals, rotation, reflection, and so on.[2] Each of these approaches has inherent limitations; where, these limitations differ from system to system.

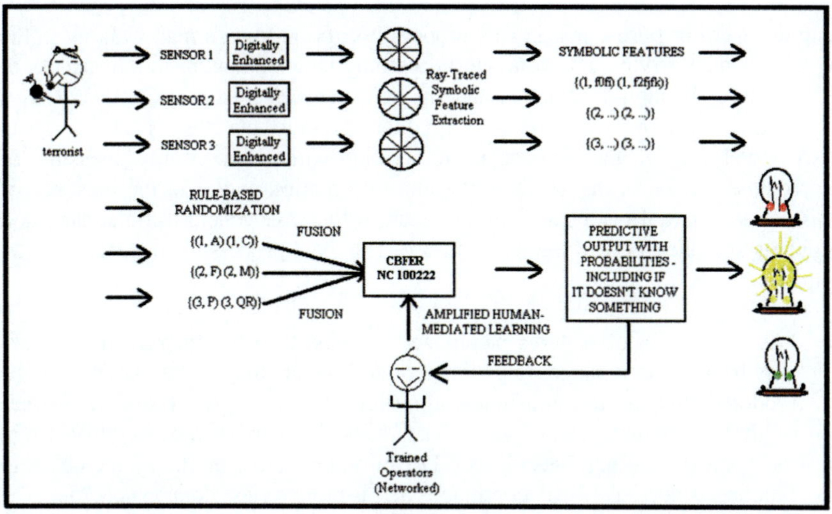

Fig. 13.1 Block diagram of a system and method for symbolic vision

13.2 On the Use of Ray Tracing for Visual Recognition

The concept of symbolic vision emphasizes the reuse of information as a hedge to insure tractability of the algorithm. Basically, ray tracing is used to count the number of edges, color and/or frequency changes (e.g., in conjunction with Doppler radar) among a hierarchy of symmetric line segments. Each image thus contains an (overlapping) hierarchy of circular rings (Fig. 13.2), where each ring has its symmetric line counts mapped to an object in a database. The areas to be fused are geometrically proximal because effective vision tends to depend on continuity. Notice that the exponents shown in Fig. 13.2 double with each successive level. Given about a kilobyte (2^{10} bytes) per ring at the lowest level, it follows that a total of three levels will suffice for most applications (top level having 2^{40} or about a trillion bytes) – see Fig. 13.3. This system learns to map each ring to the object described by it (e.g., a house, tree, John Doe, etc.). Rings are designed to cover an area no smaller than that which can report back a feature of some sort on average. We find that the square root of the image size, measured in square inches, provides a good estimate of the number of rings of minimal size that are needed at the primitive level. This system will operate correctly despite the presence of limited error. This is an on-going process with machine learning central to its success.

13.2.1 Case Generalization for Ray Tracing

The (overlapping) hierarchical rings are replaced by the symbolic representations, which they define. Naval patent disclosure NC 100222, "Adaptive case-based reasoning system using dynamic method for knowledge acquisition," is used to map the word or phrase matrix to a proper description (e.g., a man walking in the park, a woman driving a Toyota, etc.). Learning is again central to the success of this method. It should be noted that naval patent disclosure NC 100222 uses Computing with Words to achieve qualitative fuzzy matches. That is, it can associatively recall the nearest match, along with a possibility measure of correctness – even in the absence of some information and/or in the presence of noise and/or partial occlusion. An approach to the case generalization/adaptation problem suitable for ray tracing follows. The post-processor backend may be found in NC 100222.

1. Cases are a set of features paired with a consequent(s). In practice, features may have numeric ranges (e.g., temperature), which may or may not need to be rendered Boolean for an efficient algorithm. For example, 70 and 72 degrees might both be described as warm; but, 31 and 33 degrees are vastly different in the context of water freezing at STP. In other words, the fuzzy membership function is not necessarily continuous and is more or less domain specific.
2. Boolean questions provide an environmental context for the proper interpretation of the continuous ray-degree signature class mappings. This is superior to a Markov chain (e.g., used in island-driven speech recognition algorithms) because it is both context-sensitive and symbolic (e.g., unlike Hearsay and Harpy).
3. Cases can introduce new features and expunge the least relevant ones.
4. Cases serve as analogical explanations (e.g., a similar image and/or class of images). That is, the current image is most similar to this most-recent image(s), which will all be in the same class if reliable and not otherwise.
5. Exercised cases are moved to the head of the list. Expunge the LFU cases as memory limitations necessitate.
6. User (vision program) supplies a contextual case.
7. In vision, a case consists of ray tracings that count the number of B&W and/or other previously mentioned transitions in their paths. Even color pixels have exactly two possible states – on or off for each filtered color (RGB). This scheme can also represent stereopticons, as necessary.
8. Features may be entered in order from highest to lowest weighting, where the most-likely to be true are presented first (i.e., to effect an associative recall).
9. The essential idea is to find a weight vector that maps an arbitrary context into the best-matching antecedent *well* associated with the proper class consequent.
10. The present method can take advantage of massive parallelism and allows for greater accuracy than does neural networks alone, which do not allow for the storage and comparison of cases. It also allows for heuristic analogical explanations ascribed with probabilistic and possibilistic metrics. Note that in accordance with Lin and Vitter,[3] if a conventional neural network has at least one hidden layer, its runtime will be *NP-hard*. Clearly, our algorithm can perform no worse temporally and the spatial allowance enables the predicted improved performance. We have already built and successfully tested a better neural network based on well formation (reference naval patent disclosure NC 98330) – even without the presented capability for feature evolution.

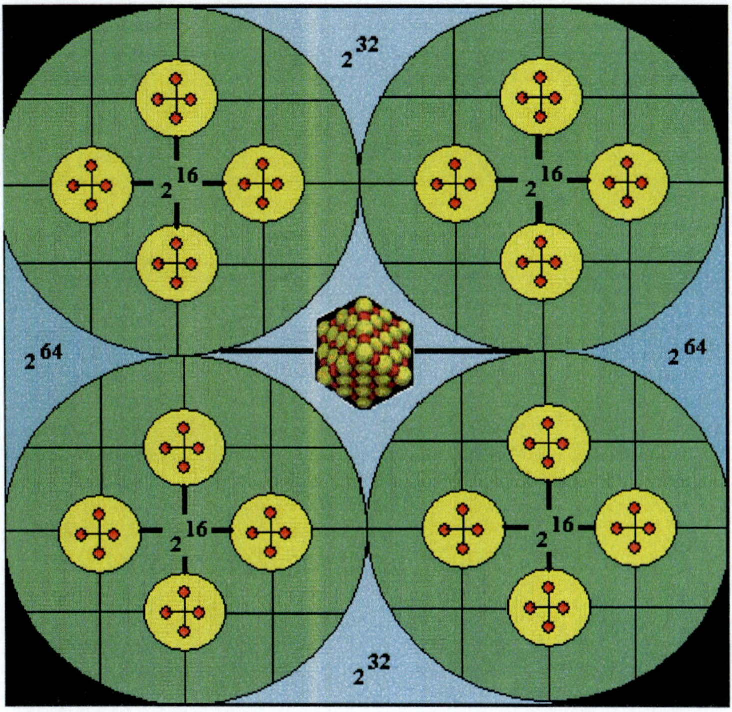

Fig. 13.2 Hierarchy of ray-traced symbols for extraction of macro-features

11. Thus far, the algorithm has replaced the innermost red circles in Fig. 13.2 with symbolic descriptions of recognized objects (and their possibilities). Not all such objects (at each level of recursion) are necessarily correct. It is almost trivial to provide feedback at all levels of recursion to correct a recognized object so as to conform to reality (i.e., learning). What is not so trivial is the proper recognition of said objects in the presence of error at each level of recursion. Naval patent disclosure NC 100222 details an implemented and operational system for Computing with Words. It can associatively recall the proper semantics for a phrase or even sentence in the presence of syntactic error. For example, "The image is a bee with leaves." is properly recognized as, "The image is a tree with leaves." Similarly, the successively complex sentential output from successively higher levels as shown in Fig. 13.2 can be fused to arrive at a correct description of the image. While the implementation of NC 100222 has been made with due regard to efficiency, far greater speed and memory capabilities are available through its parallelization (see Fig. 13.3).

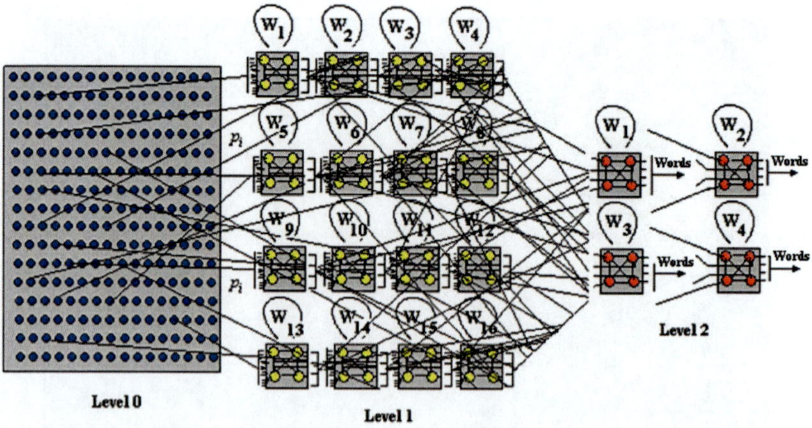

Fig. 13.3 Hierarchy of 2-level ray-traced symbols with chip realization (W_i are co-processed parallel Cbfer-word NC 100222 computations)

13.2.2 The Case Generalization for Ray Tracing Algorithm

A more-detailed description of the algorithm follows.

1) Consider just three cases, where W_i are evolved weights for each Boolean feature, f_i, (four features in the illustration below). The number of features (or consequents) per case need not be uniform (e.g., if a question(s) cannot be answered, or if a feature(s) was not available at the time of case creation). The context will not present values for features having less than a defined minimal weighting, which is determined as a function of memory size, processor speed, and the algorithm itself. The same features and associated weights will be evaluated for each case; although, individual cases can be missing as well as have extraneous features relative to the contextual set. The more significant features a context and case antecedent have in common, the more reliable the case(s) score. Note that a "hidden layer" would tend to use the match information for all features in determining the utility of any one feature, which requires on the order of the square of the number of weights otherwise used. That will be too many weights to be tractable in the case of vision applications. If one could solve for an ideal set of weights to map every (out-of-sample) context to its proper class antecedent (i.e., without the use of case wells); then, one would not need the cases in the first place, which is a contradiction on the utility of the presented method. Furthermore, experiments completed in the prior year have demonstrated that it would take an exponential increase in neural network training to substitute for case wells as defined herein. Three normalized case weight vectors follow.

w_0	w_1	w_2	w_3...	
.5	0	.5	0	→ A
.33	.33	.33	0	→ A
.33	.33	0	.33	→ B

2) Each case defines a *well* and implicitly fuses the column features. A supplied context is compared to get a metric match for each case in the base. Expunged features have their associated case-column set to "—" in preparation for reassignment.

Example 1

feature:	$\underline{f_0\ f_1\ f_2\ f_3}$	(the four features having the greatest weights)
context:	1 1 1 1	
case:	1 0 1 0 → A	
match:	1 -1 1 -1	1 = match; -1 = not match;
W_0:.	25 .25 .25 .25	score = .25(1) + .25(-1) + .25(1) + .25(-1) = 0;
W_1:.	20 .20 .40 .20	score = .2(1) + .2(-1) + .4(1) + .2(-1) = 0.2 (better);

Example 2

feature:	$\underline{f_0\ f_1\ f_2\ f_3\ f_4}$	
context:	1 1 1 1 1	
case:	1 0 1 0 -- → A	(Cases may be lacking one or more features)
match:	1 -1 1 -1 0	1 = match; -1 = not match; 0 = omitted;
W_0:.	20 .20 .20 .20 .20	score = .2(1) + .2(-1) + .2(1) + .2(-1) + .2(0) = 0;
W_1:.	17 .17 .33 .17 .17	score = .17(1) + .17(-1) + .33(1) + .17(-1) + .17(0) = 0.167 (better);

Example 3

feature:	$\underline{f_0\ f_1\ f_2\ f_3\ f_4}$	
context:	1 1 1 1 --	(Contextual terms may be unknown.)
case:	1 0 1 0 1 → A	
match:	1 -1 1 -1 0	1 = match; -1 = not match; 0 = omitted;
W_0:.	20 .20 .20 .20 .20	score = .2(1) + .2(-1) + .2(1) + .2(-1) + .2(0) = 0;
W_1:.	17 .17 .33 .17 .17	score = .17(1) + .17(-1) + .33(1) + .17(-1) + .17(0) = 0.167 (better);

…

3) Summing W_i going across yields unity. The weights are normalized to enable comparisons using differing numbers of weights. The most-recently acquired/fired case (i.e., the one towards the head of the list) receiving the highest score (not necessarily the maximum possible) provides the consequent class. Individual weights, w_i, are in the range [0, n], where n is the number of features taken (e.g., 4 or 5 above). The weight vector is then normalized as shown for the results above. This best-matching case(s), or *well*, in the same class also serves as an analogical explanation(s).
4) Adjust the weight vectors, W_i, using evolutionary programming (EP) so that each case – including singletons – evaluates to the same proper consequent(s)

when multiplied by the weight vector and scored. Actually, we want to minimize the error map over the entire case base in minimal time. To do this, first evaluate the top ceil (SQRT r) rows using the previous best W_i and the candidate W_i. The candidate W_i must produce at least as good a metric, or another will be selected. Otherwise, proceed to evaluate the W_i across all rows (i.e., up to the user-defined cutoff limit). If the summed metrics for the r rows is better, then the candidate W_i becomes the new best weight vector. Otherwise, another W_i will be selected. Notice that this method favors the most-recently acquired cases for the speedup, but does not sacrifice the optimization of the overall metric (i.e., up to the user-defined cutoff limit). The square root was taken because it relates the size of b to A in Ax = b, where A represents the case base, x the weight vector space, and b the best weight vector. Again, we use insert-at-the-head and move-to-the-head operations. Perfectly matched cases having the wrong consequent are reacquired at the head using the correct consequent (expunging the incorrect case). Otherwise, incorrectly mapped cases are acquired at the head with the correct consequent(s) as a new case. Correctly mapped cases are just moved to the head. The LFU or least-frequently used (bottom) cases are expunged whenever the memory limit is reached to make room for insertion at the head.

5) When evaluating a W_i, each row in the range, where there must be at least two rows in the range having the same consequent, will in turn have its antecedent, $a_{i,j}$, serve as a context, c_j. This context will be compared against every row excepting that from which it was derived. The score of the *i*th row is given by $\sum_{j=1}^{n} w_j(c_j - a_{i,j})$, where $\forall_j \mid w_j \geq 0$ and $\sum w_j = 1$. In the case of Boolean functions, define $(c_j - a_{i,j}) = \begin{cases} +1, & a_{i,j} = c_j; \\ -1, & a_{i,j} = \overline{c_j}; \\ 0, & otherwise. \end{cases}$. Here, if the row having the *maximum* score has the correct consequent, award +1; otherwise, -1. In the case of real-valued functions, define $(c_j - a_{i,j}) = \lvert c_j - a_{i,j} \rvert$, which is always defined. Here, if the row having the *minimum* score has the correct consequent, award +1; otherwise, -1. Thus, the higher the score, the better the W_i, where a perfect score is defined to be the number of rows in the range – the number of singleton classes there.
6) Questions may be presented in order of decreasing feature weights (which are presented) to insure that the user is aware of their relative importance. Questions are ordered to present the most-likely to be true (and otherwise highest-weighted) question next on the basis of matching on-going answers (probabilities presented), which provides a context. Such an associative Markov-recall mechanism facilitates human-information system interaction and is *purpose-driven*. Questions may attach functions, which have Boolean evaluations. Some questions are of the type, "Is S a person?" Given a reply

here, one can infer the answer to such complimentary questions as, "Is S an inanimate object?" automatically. Indeed, complementary questions are redundant and should not be recorded. The *user* might scan the list of current questions to be sure that any new questions are not equivalent or complementary. The user might also benefit from a capability to expunge any features found to be equivalent or complementary at the time of query. It is not necessary that relations more complex than equivalent or complementary be captured for the same reason that one does not optimize Karnough maps anymore – the net gain does not justify the cost. Order of presentation is irrelevant to the quantitative vision component.

7) The evolutionary program for weight mutation (*vibrational annealing*, or *dreaming*), which respects randomness and symmetry in keeping with Rubin[4] follows. Note that the random and symmetric steps can be dove-tailed, or preferably run on parallel processors:

   a. An upper limit is set by the user (e.g., default = 100) such that any metric may be evaluated on less, but never on more than this number of rows. The remaining rows (i.e., case repository) are heuristically assumed to follow suit. This limit is needed to provide a hedge against an $O(m^2)$ slowdown. Remaining rows are not included in the evaluation because of temporal locality. However, all the rows in the base will be searched in linear time for a simple best match.
   b. $Soln_1 \leftarrow W_0$; {e.g., 0, 0, 0, 0};
   c. $Soln_2 \leftarrow W_1$; {e.g., 4, 4, 4, 4}; {Without loss of generality, assume $Soln_2$ is at least as good as $Soln_1$. Here, n = 4.}
   d. Symmetric_Step: Form the symmetric range from the union of the elements in $Soln_1$ and $Soln_2$: {e.g., [0, 4], [0, 4], [0, 4], [0, 4], which in this initial instance is the same as random} Use the Mersenne Twister algorithm to uniformly vibrate each un-normalized weight within the range defined by its interval to select individual weights, $Soln_i$.
   e. Normalize W_i;
   f. Evaluate W_i on first ceil (SQRT r) rows.
   g. If the metric is at least as good as $Soln_1$ for these rows (a new row may have been added or an old one replaced), then evaluate W_i by computing the metric as the sum over all rows
   h. Else go to Random_Step; {symmetry fails}
   i. If the metric is at least as good as $Soln_1$ for all rows, then $Soln_1 \leftarrow Soln_i$
   j. Else go to Random_Step; {symmetry fails}
   k. If $Soln_1$ is better than $Soln_2$, swap.
   l. Go to Symmetric_Step; {symmetry succeeds}
   m. Random_Step: Use the Mersenne Twister algorithm to uniformly *vibrate* each un-normalized weight within the range [0, n] to select individual weights in the vector, $Soln_i$.
   n. Normalize W_i;
   o. Evaluate W_i on first ceil (SQRT r) rows.

p. If the metric is at least as good as $Soln_1$ for these rows (a new row may have been added or an old one replaced), then evaluate W_i by computing the metric as the sum over all rows
q. Else go to Symmetric_Step; {random fails}
r. If the metric is at least as good as $Soln_1$ for all rows, then $Soln_1 \leftarrow Soln_i$
s. Else go to Symmetric_Step; {random fails}
t. If $Soln_1$ is better than $Soln_2$, swap.
u. Go to Random_Step; {random succeeds}
v. Exit by interrupt, quantum expiration, when the number of superior solution vectors, $Soln_i$, discovered turns convex as a function of the elapsed time (e.g., Regula Falsi method), or a perfect match for all cases is found (i.e., captured by the prior test).

8) The lowest-weighted feature(s) are iteratively replaced with new feature(s) whenever space is at a premium. The number of features so replaced depends on the number of new ones available. Revert to the previous feature set just in case the new set results in a worse (weighted) global metric. All feature weights should be co-evolved (i.e., evolved at the same time) for maximal efficiency. The lowest-weighted feature(s) are iteratively expunged upon the conclusion of the evolutionary step until such deletion results in a worse (weighted) global metric.
9) The best contemporary CCD cameras capture over 12 mega-pixels per highest-quality image. Surely, one mega-pixel per image (combined with a consequent-directed fovea having digital zoom) is sufficient for all robotic reconnaissance. That allows for on the order of eight thousand basis images per gigabyte of RAM, which is sufficient and readily realizable.
10) Features may be manually selected, or in the case of computer vision, pixels may be automatically selected up to the defined memory bounds. However, pixels are voluminous and sensitive to orientation. Rays (i.e., line segments, which pass through the center of an image) can count the number of pixel transitions from on to off in their path. Thus, they are minimally sensitive to images being "off-center". However, the Rubin and Kountcheva algorithm should be used to normalize intensities a priori.[5] There will be 180 degrees of ray tracings as a consequence of path symmetry (see Figure 13.2a). Thus, one can trace 180*n rays for n = 1, 2, … (though real spacing may also be used). One may increase or decrease the resolution obtained by increasing or decreasing, respectively, the number of equally-spaced ray tracings in a ray-degree signature (RDS). The number of rays may not exceed the square-root of the number of pixels (i.e., the optimal number, based on the number of rows and columns in a square matrix) to eliminate redundancy. Too few rays will limit the resolution of the system. An RDS consists of 180*n non-negative integers – each of which serves as a context-sensitive feature. Color vision can be represented through the fusion of RBG filters (i.e., three distinct filtered runs). Stereo vision (or echolocation) can be conveniently represented by case antecedents consisting of a pair of RDSs. Example 2 is reworked below to incorporate arbitrary numbers of pixel transitions. By design, there will never be any omitted features here.

Example 4:

feature:	$\underline{f_0\ f_1\ f_2\ f_3\ f_4}$	
context:	1 1 2 4 1	(can be rays, track prediction, etc.)
case:	3 1 3 3 1 → A	
Δ:	2 0 1 1 0	perfect match = 0;
W_0:.	33 .17 .17 .17 .17	score = .33(2) + .17(0) + .17(1) + .17(1) + .17(0) = 1.0 (can be larger);
W_1:.	20 .20 .20 .20 .20	score = .2(2) + .2(0) + .2(1) + .2(1) + .2(0) = 0.8 (better);

11) The evolved patterns of pixels or rays may be expected to be similar to that found by Hubert and Weisel in cat visual cortices after being exposed to vertical stripes during neurogenesis.[6] Case consequents, output from naval patent disclosure NC 100222, which computes with words, directs the motion of the CCD matrices so as to center the object(s) of interest in the visual field as well as to normalize the digital zoom ratio so as to present a normalized image (sub-image). Images can be further normalized by rotation and/or reflection for specialized tasks, which is not anticipated to be necessary in the case of robotic vision; although, it is not too difficult to add this capability using simple optical transforms. Similarly, images may be preprocessed using edge detection, texture recognition, 3-D affine transforms, and motion estimation algorithms as provided by Kountcheva.[5]
12) It is useful to provide an estimate of the possibility and/or probability of a predicted class membership – something that too cannot be done using neural networks. The relative likelihood of a fired class is a function of the mean distance between the context and members of the fired class and that between the context and members of the remaining classes.

   a. Weighted scores will be in the range [-1.0, +1.0] with plus one the score for a perfect match and minus one the score for a perfect mismatch using Boolean functions. (RDS-weighted scores will be in the range [0, ∞) with zero the score for a perfect match and the larger the number, the worse the match.) Note that in either case, the perfect scores are independent of the chosen weights.
   b. Take the mean score for each class (e.g., D_u is the mean score for class D), where $D_u \in$ [-1.0, +1.0] for Boolean functions and $D_u \in$ [0, top] for RDS-weighted scores, where top is the worst-matching score.
   c. For example, let, A_u = -1.0; B_u = 0; and, C_u = +1.0. Then, the relative likelihoods are defined in Table 13.1. Relative possibilities are analogous to certainty factors (CF) in expert systems. Relative possibilities are defined on a scale from 0 to 100 and as shown below, C_u has twice the possibility of B_u and both are infinitely more (i.e., asymptotically) possible than A_u. Normalizing the relative possibilities yields probabilities, which shows that A_u has no (i.e., asymptotically) chance of being correct, B_u a 33 percent chance of being correct, and C_u a 67

percent chance of being the true class membership. Probabilities are to be preferred here because they can be compared across runs as a result of normalization. Note that the best-matched case and the best-matched class mean can differ significantly. That is why the relative possibilities for the best-matched case and the best-matched class (which is never higher) are presented to the user or learning algorithm in addition to the normalized probabilities for all the classes. If the class of the best-matched case differs from the class having the highest probability, then the predicted class assignment is said to be "unknown" because there is likely to be no correlation of case consequents with case antecedents. Reliability and the best-matching case are thus both accounted for. RDS-weighted scores are handled similarly, where again the worst-matching score is taken as the upper bound.

Table 13.1 Calculated relative possibilities and normalized probabilities.

A_u	B_u	C_u	Class Means
-1.0	0	+1.0	Weighted Scores
0	50	100	Relative Possibility
0%	33%	67%	Normalized Probability

13) The Boolean environmental queries are fused with the RDS class predictions to provide Markov-like visual recognition. It is clear that humans do this too. Note that acoustic data can be fused with visual data using the same paradigm previously outlined for stereopticon recognition. Here, symbolic abstract sensory information is to be incorporated. For example, a robotic sentry is more likely to recognize a snowman than a light-colored scarecrow say, if a temperature sensor reports that the ambient temperature is at or below freezing (crops don't grow at these temperatures). The case for multiple sensors (e.g., including the season and geo-position) has been made above. For example, consequent-directed sub-image gathering can be so fused to "make sense" of the picture. We do this when we visually "look around" an image prior to deciding on a course of action.
14) CCD (or superconducting SQUID) hardware will serve as an ideal visual sensor here. We will come to know that our algorithm works on the basis of the rate it learns to map previously unknown images to their proper class (case) as can be best ascertained by visual inspection and/or robotic performance. We have already outperformed the best neural networks for this task so there is nothing else to be used to draw the desired comparison – assuming that we outperform prior results, which is a fair assumption. That is, we have already succeeded by default.

15) One of the major unsolved problems in case-based reasoning (CBR) pertains to how to automatically adapt a retrieved case to not only match the current context, but to adapt its consequent to be a proper match as well. Naval patent disclosure, NC 100222, takes the view that:

    1. A Computing with Words (CW) methodology[7,8,9] will best enable matching the context to a case
    2. Granular computing will then be applied to selecting the best case
    3. Case adaptation, where appropriate, is an emergent property of fired case interactions and is not to be accomplished through case parameterization.

16) Super-linear learning occurs if an acquired corrective case can be or is properly matched by more than one distinct context. Produced consequents can request further information, be fused with other produced consequents, or invoke various procedure calls to produce the results of a complex query process. Networking allows answers supplied by different segments to be fused in one or more segments for iterative processing. It follows from the Incompleteness Theorem[10] that no matter how strong the super-linear learning, there will be potentially a countably infinite number of cases for an arbitrary (i.e., non trivial) domain. This also serves to underscore the importance of representation and generalization towards minimizing the requisite number of cases and thus making an improved machine vision a reality.
17) The output from all rings, at every level, is read in a fixed sequence (see Fig. 13.3). The goal performed by naval patent disclosure NC 100222 is to find and map that maximal subsequence for each ring at each level to the best matching case as stored in local memory. Cbfer is currently operational for up to about 100,000 local cases per processor; although, there is no fundamental reason why this number could not be increased. Cbfer cases serve in the same role as the brain's associative memory. Thus, errors propagate out at each level and images are recognized after having been trained on only a basis set of them. The following major points should by now be clear.

    - Ray tracing (preprocessed and/or feature-based) images assists in removing variance from the input stream
    - Computing with Words is fundamental to the tractable operation of any non-trivial vision system or System of Systems
    - The brain may or may not represent images using natural language labels, but primitive images are likely represented in iconic form for processing as indicated herein
    - Fine-grained parallel processing is necessary for real-time vision.

13.3 On Unmanned Autonomous Vehicles (UAVs)

There have been planes constructed, which can stay aloft indefinitely on account of being solar powered. Such planes and others would be quite suitable for photo reconnaissance missions save for the fact that they require a human operator to direct them at all times. The present section provides for the automation of this task by way of learning from the remote pilot what differentiates an interesting scene from a mundane one and then being capable of acting upon it; that is, *learning by doing*. The scene may be photographed under unobstructed daylight or starlight (using a Starlight scope). There are several degrees of freedom for this operation as follows. First, the UAV (Fig. 13.4) may autonomously steer right or left. Next, it may ascend or descend in altitude. Finally, it may navigate a circular pattern of arbitrary radius (i.e., to make another pass). The camera also has a few degrees of freedom. In particular, the lens may zoom in or out with loss or gain in field of view, respectively. Gyroscopic stabilization and automatic centering of the image in the field of view are separately maintained. A distinct fail-safe system prevents imminent crashes with the ground, trees, buildings, or other aircraft.

13.4 Overview

The central concept throughout this section is machine learning. In its simplest form, an object-oriented database records and pairs images and flight instructions (e.g., zoom in, veer left, descend, *et al.*). A handshake architecture allows for asynchronous control. Images are matched using rotation and scaling. The problem with this simple approach is that (a) the method is quite storage intensive and (b) if an image does not exactly match a stored one, it cannot be processed. Therein lies the crux of this disclosure; namely, (1) how to generalize the image base, (2) how not to over-generalize this base, (3) how to acquire and time stamp images, (4) how to evolve figures of merit to evaluate the likelihood of a match, and (5) how to evolve heuristics, which serve to speedup the matching process, among related lesser details. Matched objects are always paired with some desired operator action, which is replayed in case of a match. Such actions might bring the UAV in for a closer look in a process that is akin to context-directed translation. These actions may optionally be smoothed so as to render all changes in the UAV gradual and continuous. The methodology developed below is also compatible with hyperspectral imagery and can autonomously fuse the various sensor modalities employed. This generally yields a more accurate sensor description than is otherwise possible.

Version spaces[11] have shown that the number of possible generalizations, known as the generalization space, grows exponentially in the number of predicate features. A version space saves all most general and most specific productions and converges as the most general become specialized and the most specific become generalized. The idea is that the version space will collapse to context-sensitive production(s) – representing the true correction – when the true production(s) have been learned.[11] The implied need then is for knowledge-based translation, where the knowledge takes the form of heuristics.

Conceptually, what is it that makes an image interesting – is it the mix of colors, and/or its smokestack, and/or its silhouette, etc? Clearly, the number of possibilities

Fig. 13.4 A smart photo reconnaissance UAV

is exponential in the number of features. The goal here is to develop an algorithm, which ties in with the UAV control system and learns to extract reason for its human-supplied direction – much as a co-pilot would without being explicitly told because there is no practical way to accomplish that. In this manner, the system evolves a true understanding of what is interesting in an image (or sequence thereof). For all engineering intents and purposes, the system will evolve the instincts of its trainer(s). Not only does such a system serve naval *Information, Surveillance, and Reconnaissance* (ISR) needs, but from a purely scientific perspective, it can be said to conceptualize causality, or equivalently converge on image semantics. This then is the critical missing step in enabling our UAVs and *unmanned underwater vehicles* (UUVs) to become truly autonomous. The US military's un-crewed aerial vehicles are a critical component of its search-and-destroy missions in warring regions. Michael Goodrich, Lanny Lin, and colleagues at Brigham Young University in Provo, UT adapted a small propeller-driven plane to fly *search and rescue* (SAR) missions – even in perilous weather conditions that can ground helicopter-led rescue missions.[12] Topographical and environmental factors play a big role in determining where someone ends up. At the present time, the UAV needs to work with a human in the loop to analyze images. Moreover, the Air Force Research Laboratory at Ohio's Wright-Patterson Air Force Base said it will soon solicit engineers to design an algorithm to allow drones to integrate seamlessly with piloted planes for takeoff and landing.[13] These are areas where the subject of this chapter is relevant.

13.5 Alternate Approaches

The concept is to employ downward-facing image sensors in a UAV, which are architecturally designed to be a hierarchical vision cone. One should use as few pixels, for identification, as possible – including most importantly reference to the proper segment for the next level of identification. This dictum serves to minimize processing demands. Next, image sensors will not be practical if they only focus at the level of the individual pixel – great for photography, but unacceptable for automated image recognition. Rather, an image is defined by the features that it embodies and thus triggers.

Features can be static as well as dynamic. For example, a static feature could be the outline of a ship against the horizon; whereas, a dynamic feature could be the same ship whose gun turrets differ more or less in position from that of the last captured image.

Features are a product of evolution (e.g., click-detectors in cats and the ability to zero in on a fly in frogs and some Amazonian fish). Quinlan reported that the manual extraction of features in chess took upwards of one person-month per feature.[14,15]

Consider a Tic-Tac-Toe board, where each of nine cells may assume one of three possible values; namely, blank, X, or O. This allows for the formation of 3^9 or 19,683 features (boards). Of course, playing this game using features would be enormously wasteful of space and utterly impossible for more complex games like chess, where algorithms are more appropriate as randomizations.[4] Consider next a context-free grammatical (CFG) representation, which it will be noted, captures the inherent hierarchy in the game of Tic-Tac-Toe. Fig. 13.5 gives a partially completed grammar for the game:

S → Loss | Win | Draw
Loss → O O O
Win → X X X
Draw → A Y | Y B
Y → X | O
A → X - | O –
B → - X | - O

Fig. 13.5 The partial grammar for Tic-Tac-Toe

Notice that the Draw feature is recursively decomposed into its four constituent patterns; namely, X-X, X-O, O-X, and O-O. The extraction of sub-patterns facilitates the recognition process in the large. For example, instead of learning - - X, we may learn – B, where B is a previously acquired sub-pattern. Such randomization[16] allows for the definition of features. Here, B is a feature, where features may be recursively defined, which is in keeping with the concept of a hierarchical architecture.

In a B&W image of only 262,144 pixels, the feature space is $2^{262,144}$ – which is clearly intractable. A perfect hierarchical approach would have $\log_2 2^{262,144}$ = 262,144 features, which while enormously better, does not go far enough in distinguishing good from bad features. Rather, the search for features must be directed by acquired heuristics. Therein lies the classic *generalization problem.* The solution requires a generalization language (e.g., A, B, Y, ...) and an iterative search process for randomization. Moreover, it is well-known from the various *pumping lemmas* that many problems cannot be represented in a context-free language (e.g., languages of the form, $a^n b^n c^n$).[16] The most general representation language is the Type 0 (i.e., phrase structure or contracting) grammar, which is best exemplified by natural language. Thus, it is suggested that images must hierarchically be converted to English, using a contracting grammar, for the most general pattern recognition to be attainable. The problem with this approach is that English is not interpretable in the same way that a context-free computer language is. While possible in theory, it too is impractical for this reason. Thus, while two-level, or *w-grammars* can attain the complexity of natural language, they lack a mechanics to ascribe a proper semantics to transformational images of productions.

The solution is to store all images in randomized form, where subsequent to expansion a greater randomization is attempted, which is concomitant with knowledge acquired in the interim. Now randomization does not necessarily imply lossless compression. There needs to be an allowance for imprecision – not only in sensor measurement, but in conceptual definition. The degree of this allowance is acquired as context-sensitive transformation rules by the learning algorithm. In any case, a basis is to be acquired and when generalization of that basis allows inclusions, which should not be admitted, another basis function must be acquired to exclude them. Clearly, the larger the base of like entities, the greater the degree of possible randomizations and vice versa.

The method of this section, which we will term, *event-driven randomization*, works well with the aforementioned hierarchical approach. Here, one starts with say the pixel image in full detail and iteratively averages neighboring pixels over a progressively coarser mesh until such generalization would allow an improper categorical inclusion. Indeed, the human brain may work in this manner, which is suggested by the fact that we can readily mentally map cartoon characters onto their male or female counterparts. Improperly mapped concepts will be acquired at a higher level of resolution. Then, in order to prevent an unknown object from improperly mapping onto an object, which is too general, it will need to be compared with the objects stored at the next more-specific level, if possible. If no objects have been stored at that level, or if the current level is at the level of the individual pixel, then the current level of generalization is used for conceptual definition. However, if a more-specific level exists and is populated with a match having a higher figure of merit, then the more-specific match is taken.

Here, the feature space is not mathematically defined by the intractable number of possible feature combinations. Rather, it is defined by the tractable number of experienced categories of generalization. What makes this methodology desirable is that it only makes use of structural generalization, where the attached semantics evolve from the replication of similar structures whose apparent complexity is offset by randomization. These descriptions will be codified by the algorithm that follows below.

13.5.1 Theory

The multisensory hyperspectral recognition and fusion of "images" is broadly defined by the Theory of Randomization.[4] Here, it is developed for minimal storage, retrieval, and learning time, and maximal accuracy.

Definition 5. The fundamental unit of 'mass' storage, but by no means the only one, is the *image* or slide. Then, the representation for a dynamic set of such images is to be randomized.

Definition 6. An image is said to be at Level n if it contains $2^n x 2^n$ pixels.

Definition 7. An image may be strictly converted from Level n to Level n-1 as follows.

$$\bigcup_{i=0}^{2^{n-1}-1} \bigcup_{j=0}^{2^{n-1}-1} a_{i,j} = \begin{cases} 1, & \delta_{i,j} > \text{threshold} \\ 0, & \text{otherwise;} \end{cases}$$
$$where,\ \delta_{i,j} = \frac{a_{2i+1,2j+1} + a_{2i+2,2j+1} + a_{2i+1,2j+2} + a_{2i+2,2j+2}}{4}. \quad (13.1)$$

Here, threshold is a constant that determines if a pixel will be black or white (colored or not).

Definition 8. A color image may be captured using three filters – red, green, and blue (RGB). Hence, a color image is defined by $\bigcup\{A_r, A_g, A_b\}$. Similarly, an auditory image may be captured using the three filters – low pass (L), medium pass (M), and high pass (H). Hence, an auditory image is defined by $\bigcup\{A_L, A_M, A_H\}$.

Definition 9. A color image at Level n is defined by 2^n non-symmetric ray traces (see Fig. 13.7). A ray trace counts the number of changes in pixel brightness – either moving from a dark to a light pixel or vice versa adds one to the ray trace count. The union of these counts, in any fixed order, comprises a vector signature, which defines the image at that level. Then, a color vector signature is comprised of a vector of ordered triples, representing the RGB scans for the same ray. An auditory image is similar.

Theorem 10. (*The tracking theorem*). An object(s) at the center of attention may be tracked under the assumption of continuous movement using ray tracing.

Proof: Assume high enough speed vector matching for the evaluation of an outer ring of concentric images, which may overlap and containing the maximum degree of such overlap as is practical. Then, it follows that the ring containing the closest vector signature using the 2-norm will designate the track. If this were not the case then either the image is not unique or the system is not sufficiently trained. The proof follows. □

Theorem 11. (*The hyperspectral fusion theorem*). Any vector, or vector of vectors, may be integrated into a systemic vector signature. This provides for multisensory fusion.

Proof. It suffices to show that any number of vectors of vectors can be fused into a single integer because this integer may be mapped to countably infinite semantics. Then, it needs to be shown that every semantics (integer) has an inverse mapping, which may be a "do nothing" mapping. That a vector of vectors may be mapped to a single integer follows from the definition of pairing functions.[16] Then, projection functions exist,[16] which can reverse map a paired function back to its constituent elements, where integers that are not mapped by pairing functions are reverse mapped to "do nothing". □

Example 12. Let k be the temporal sampling differential, which may be thought of as the appropriate interval for whatever is being observed – from a snail to a bullet, etc. Let the vector signature at time t be given by V_t and the next vector signature be given by V_{t+k}. Then, velocity is defined by $V_{t+k} - V_t$ and acceleration is defined by $(V_{t+2k} - V_{t+k}) - (V_{t+k} - V_t) = V_{t+2k} - 2\ V_{t+k} + V_t$, which is used in the numerical solution of PDEs. For example, let $V_t = (1, 2, 4, 6, 5, 4, 4, 5)$; $V_{t+k} = (1, 2, 5, 8, 6, 4, 4, 5)$; and $V_{t+2k} = (1, 3, 7, 11, 8, 5, 4, 5)$. Then, $V_{t+k} - V_t = (0, 0, 1, 2, 1, 0, 0, 0)$ and $V_{t+2k} - 2\ V_{t+k} + V_t = (0, 1, 1, 1, 1, 1, 0, 0)$. Vector elements may be negative because differences, rather than pixel crossings, are being measured. Here, the non-zero elements reflect velocity and acceleration, respectively. These two vectors may be fused to allow for motion recognition. This result has been presented for B&W, but may be adapted, using a vector of vectors, to accommodate color and/or sound, etc. too.

Comment 13. It follows from the hyperspectral fusion theorem that each vector signature should independently map to a semantics and the collection of these (or collection of collections) should map to a fused semantics. This is how pairing functions work. The output semantics are then the fused semantics and may be identical to or distinct from the pre-fused semantics at each lower stage. A union vector signature is said to be matched if and only if each member of said union is matched.

Comment 14. Visual vector signatures may be permuted (i.e., iteratively move the first element to the position of the last element) to account for rotation of the image without the need to reclassify it (and cost more storage space). This is *normalization*. Other sensory modalities may also be normalized, where the mechanics of so doing must remain domain specific. For example, acoustic sensors may be normalized for relative motion, ambient noise, altitude, etc. Preprocessing functions (e.g., edge detectors, texture analysis, etc.) also fall into this domain-specific category. The proof of the need for domain specificity follows from the need for different algorithms for normalizing image rotation and Doppler shifting, which may be found in the algorithm for image randomization below. A pair of exceptions such as this is enough to insure the existence of a countably infinite number of them and would lead us all the way into evolutionary biology to justify their existence. Machine learning[17] plays a major role because some connections are just too dynamic to be efficiently hardwired. Higher cognitive functions (knowledge bases expressed in a natural language equivalent) evolved to satisfy a need.[18] Wernicke's area of the brain is but one human codification of that evolution.[6]

Theorem 15. (*The randomization theorem*)**.** The following stipulations define a unified multi-level storage methodology. The proofs are by induction.

a. Every signature vector is to be paired with a semantics, M:1, which is defined in natural language. Two or more signature vectors may be recursively fused to yield one semantics, M:1. The requirement for a unique semantics only holds for a given level. That is, distinct levels may hold distinct semantics; albeit, only one. We may express this as M:1/i, where i is the level number. Every syntax may be paired with exactly one semantics, M:1/i. However, in practice distinct levels may be expected to hold the same semantics as defined by the same syntax.

   One practical way to randomize the storage, decrease recognition time, and increase storage capacity, but at the cost of less verbal discriminatory power is to employ object-oriented class semantics and either single or multiple inheritance. The former is most economical, while the latter allows for the development of greater verbal discriminatory power.

b. Two signature vectors, on the same level, are said to be *equivalent* if they are equal or they share a common permutation – the definition of which is necessarily domain specific. Two unions of signature vectors, on the same level, are said to be equivalent if they are equal or they can be put in bijective correspondence (i.e., 1:1 and onto), where each such constituent pairing is equivalent.

c. Two signature vectors or unions of signature vectors, if equivalent, must point to the same semantics, in which case they are redundant. Otherwise, non determinism is created, which is obviously not permissible.

d. If a signature vector or unions of signature vectors has no equivalent at a certain level, then it may be saved at the most-general level where this occurs – regardless of the semantics it is paired with. Conversely, if any member of the union fails to be matched at Level i, the next more-specific level is computed, if any.
e. In case of non determinism, more detail is sought to resolve it. Thus, (1) if a more-specific level exists, the non deterministic pair of signature vectors or unions of signature vectors are removed from the present level and recomputed using the resolution available at the next more-specific level until acquired or not acquired after having reached the most-specific level. If the original pair is deterministic at some more-specific level, but induces a non determinism with one or two signature vectors or unions of signature vectors (two is the maximum number), then the deterministic pairing is saved (expunging the *least-recently used* (LRU) member at Level n as needed to free space, where Level n is the most-specific and thus the LRU level, which holds the greatest amount of storage), the non deterministic pairing(s) is removed, stacked, and recursively processed as indicated herein. Any deterministic pairs may be saved at the level they were found at. If either member of a non deterministic pairing is not available for more resolute computing, then it is expunged and forgotten. If (2), the non deterministic pair of signature vectors or unions of signature vectors is not acquired after having reached the most-specific level or if deterministic and insufficient space for the storage of both, then the most recently acquired member of the pairing is saved (i.e., temporal locality) and the LRU member at that level is expunged as needed to free space.

   Furthermore, such movement is self-limiting because there are going to be far fewer incidents of non determinism at the more-specific levels. This is because the combinatorics of ray tracing (or equivalent) grow exponentially; whereas, the number of pixels on successive levels only grows quadratically. Also, it is not appropriate to allow for inexact matching at any level, since that is already accounted for by the representational formalism at the next more-general level.
f. Finally, if space becomes a premium, the LRU most-specific unions of signature vectors or signature vectors are expunged in that order. This minimizes the functional number of lost memories, while maximizing the reclaimed space. In effect, the system cannot learn complex new patterns, but does not forget simple old ones either. Of course, given today's pentabytes of hand-held on-line storage (i.e., several days worth of streaming video), this need not be a problem for a large class of applications.

Figs. 13.6a, 13.6b, and 13.6c below present a flowchart for a smart photo-reconnaissance UAV/UUV. This flowchart is expanded into algorithmic form in what follows. Here, dotted lines are indications for expansions along domain-specific paths. Both the flowchart and its algorithmic expansion derive from the theory above.

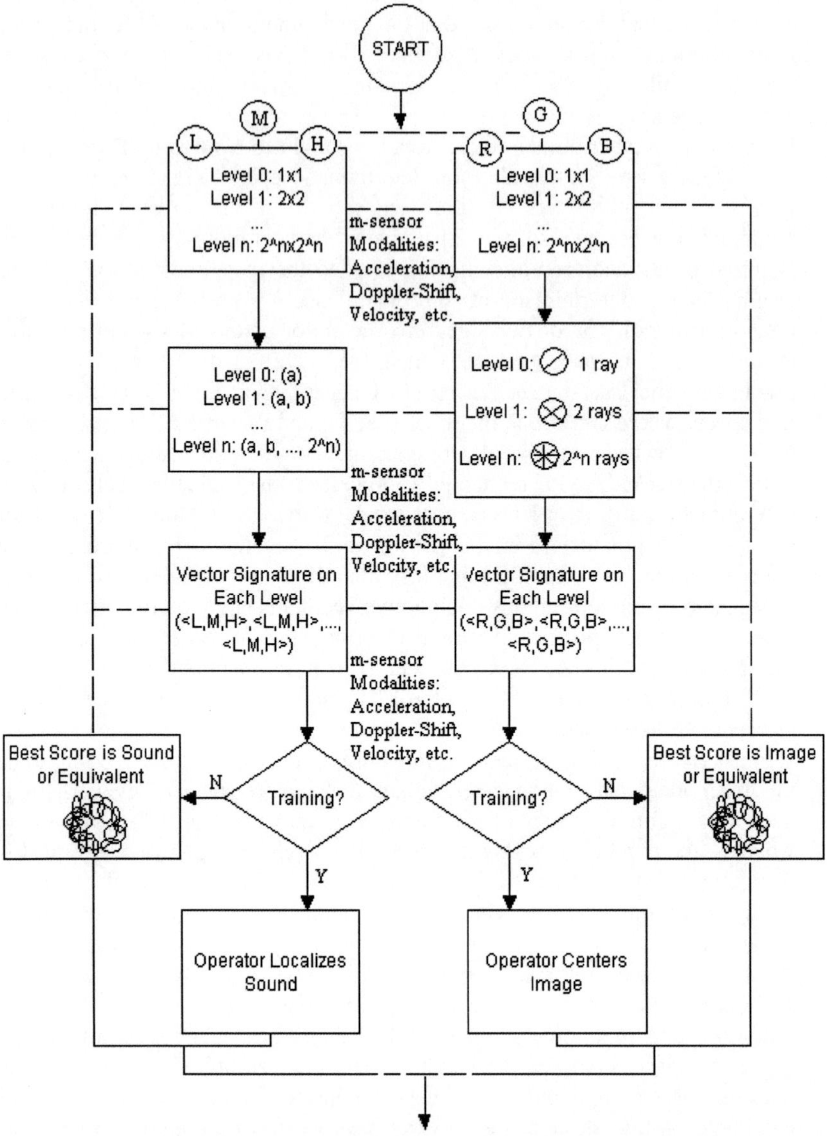

Fig. 13.6a Block diagram for a smart photo-reconnaissance UAV/UUV

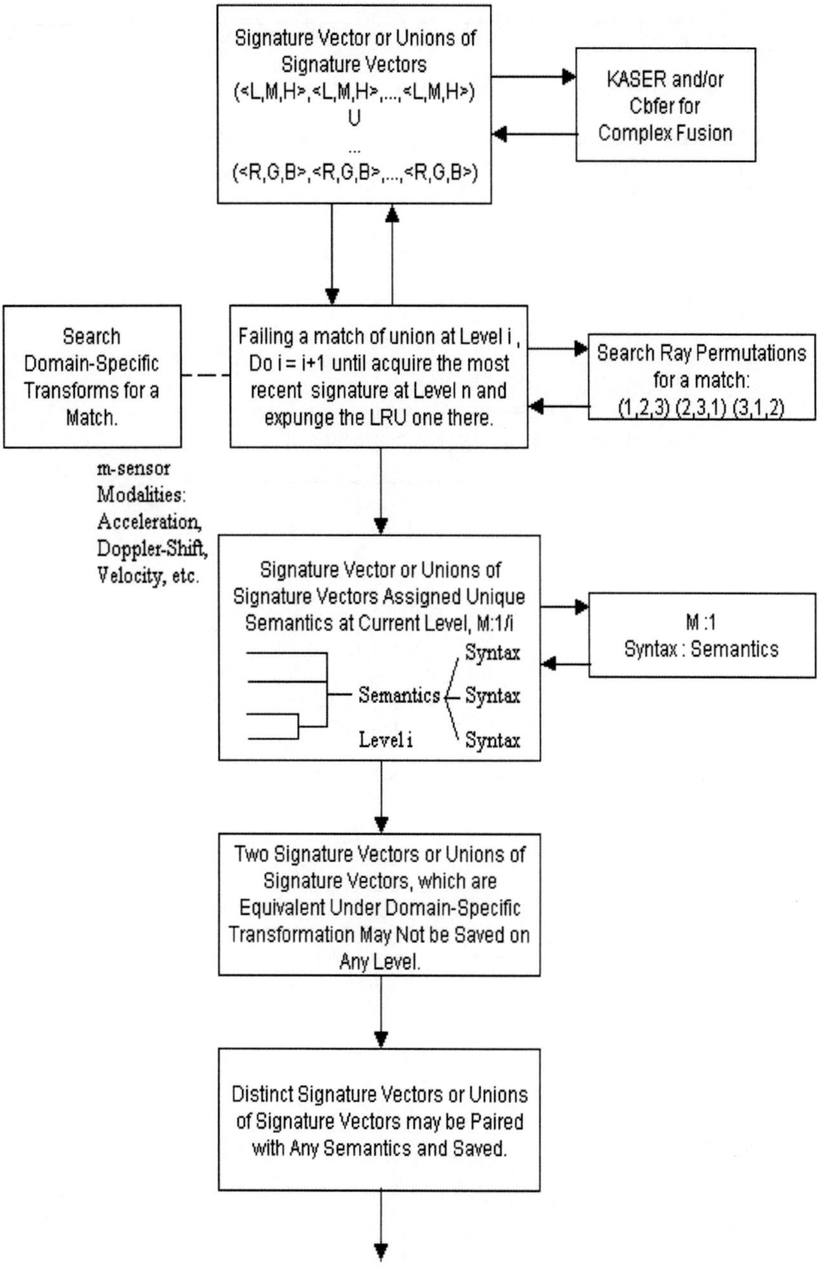

Fig. 13.6b Block diagram for a smart photo-reconnaissance UAV/UUV

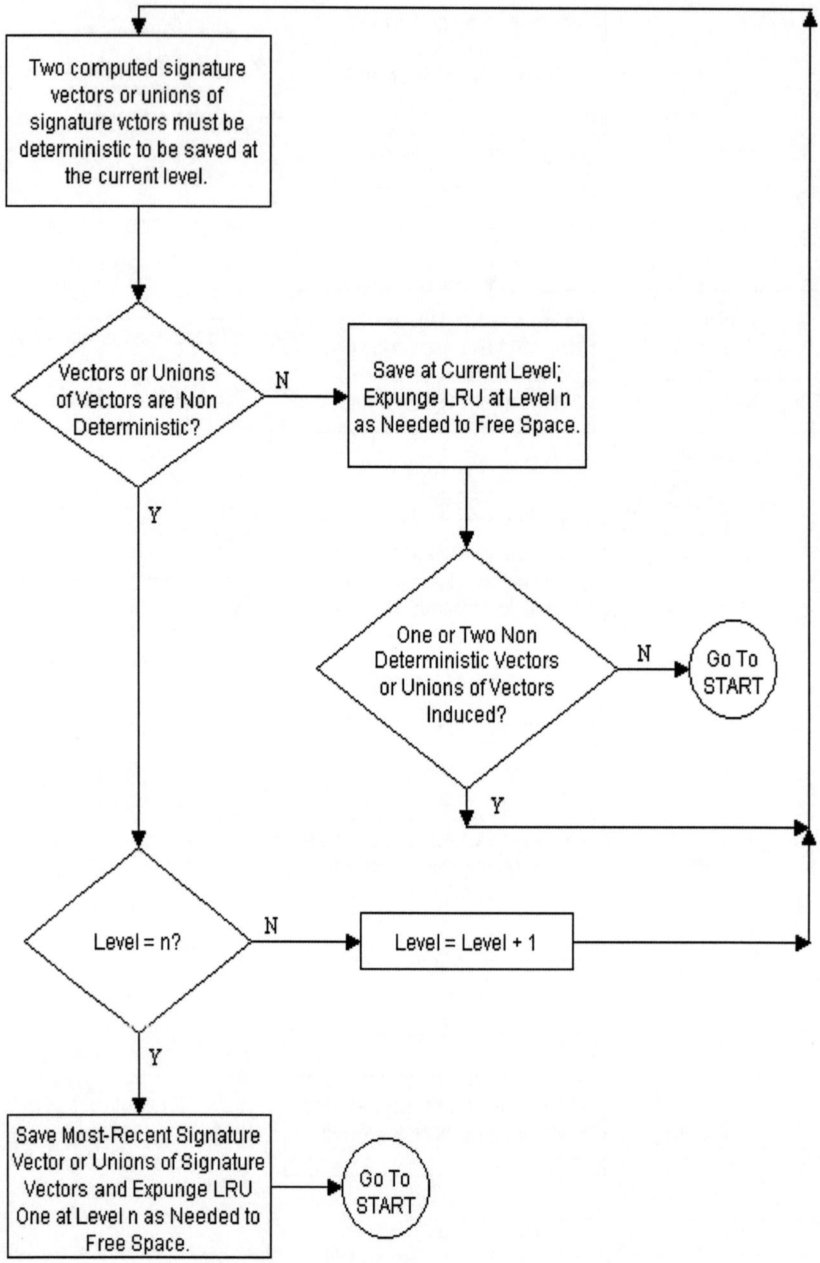

Fig. 13.6c Block diagram for a smart photo-reconnaissance UAV/UUV

13.6 Algorithm for Image Randomization

The following algorithm presents a methodology for the randomization of images (see Definition 13.5) as well as the inverse methodology for their recognition and acquisition, where appropriate. In particular, rotational and positional invariance is required to minimize the number of instances of a particular image (i.e., fundamental memories) that need be stored.

1. START
2. *Repeat*
3. Define Level n of an image to be the full resolution, which without loss of generality may be assumed to be a power of 2 and square in the layout of its pixels (e.g., 1x1, 2x2, 4x4, 8x8,..., 2^nx2^n). Hence, Level 0 consists of one pixel, Level 1, 4 pixels, Level 2, 16 pixels, …, and Level n, 2^{2n} pixels. At this rate, a megapixel image is encoded at Level 10. It is simple to convert an image from Level n to Level n-1. Simply convert every 2x2 contiguous grouping to a 1x1 (i.e., a single pixel). Notice that there is no overlap among 2x2 groups. Average the shading of the four contiguous pixels to find that for their reduction (e.g., black or white depending on the threshold). The same extends to the use of pixels as measured through red, green, and blue (RGB) color filters. An auditory image is similar.
4. The number of rays in the tracing should bear proportion to the number of pixels in the image. In practice, this number must be more constrained due to tractability and constraints on time and space. Here, the number of rays is defined by the square root of the number of pixels, which follows from the definition of level. This methodology compresses an image into effectively less than the square root of the number of pixels in weights (i.e., given the advantages provided by multiple levels), which can not only outperform neural networks in the required storage space for the weights (i.e., the number of fundamental memories that can be saved), but does not incur the need for iterative retraining. Thus, while neural networks having a hidden layer are *NP-hard* in their training time,[3] the present method operates in polynomial time. That difference is profound. It is also amenable to fine-grained parallel realization to boot.
5. Consider an image of a fork. At least two images will need to be captured for its recognition – one top-down and one sideways. If the image is far from centered, then it will need to be recaptured as part of the image may have been truncated. In addition, the image may not be angularly aligned. Also, the fork may contain a bit of tarnish, but this is not sufficient to cause it not to be a fork. Thus, a hierarchical visual recognition approach is appropriate. The essential ray-tracing methodology is presented in Fig. 13.7:

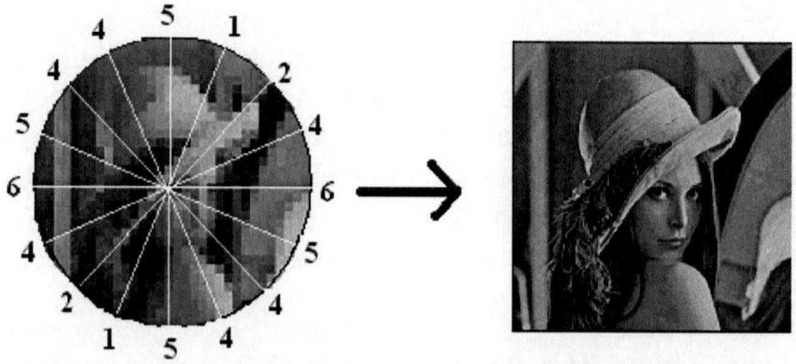

a. Unknown Ray-Traced Image b. Best-Matched Image is in "Lena" Class

Fig. 13.7 Ray tracing approach to image invariance

6. Fig. 13.7 shows that the generalized image of Lena has the following vector signature (1, 2, 4, 6, 5, 4, 4, 5). These integers represent the number of threshold changes incurred by the ray as it traverses the diameter of the image having a central focus. Again, color images can be so processed as the fusion of the same using distinct red, green, and blue filters. Here, the vector signature would become that of ordered triples instead of singletons (e.g., (<1, 1, 3>, <1, 2, 3>, <1, 4, 7>, <6, 4, 2>, <3, 5, 4>, <5, 4, 3>, <4, 4, 4>, <5, 5, 0>). The processing of hyperspectral images is similar. Ray tracing has an advantage in that it is somewhat insensitive to the requirement for exact image centering.
7. In training, it can be assumed that the operator will center the image. In recognition, we assume that multiple images are snapped using a circular pattern around the presumed center. Then, all images, save the one having the best recognition score, will be freed. Here, the most-specific image having the highest figure of merit becomes the focus of attention. Clearly, this response is dynamic with machine learning (as well as view). This then is perhaps the most practical method for image centering in the field (including zoom-in). It may make use of parallel processors, which are sure to be part of the most cost-effective solution in such applications as these.
8. In particular, if the circular pattern can be followed in real time, then once a center of attention is achieved, it may be followed and even reacquired (e.g., visually following a car, which passes under an overpass and thus is obscured for a second or two) by maintaining the best-matched vector signature(s) (i.e., as the exact one will likely change over time) in a cache as the frame of reference.
9. Next, we consider the question of how to produce a vector signature using acoustic, radar, SAR, ISAR, sonar, and/or more sophisticated sensors. Such vectors may be fused by taking their union. 3D imaging however is not recommended for inclusion because the registration of camera separation precludes the use of the more general levels, which would find no difference in vector signatures. Consider the formation of an acoustic vector signature

without loss of generality. As before, each successively more-specific level has double the number of vectorized elements. These elements uniformly sweep the desired frequency bands at time t (e.g., (500 Htz, 1000 Htz, 1,500 Htz, 2,000 Htz)). However, what is being sensed are interval frequencies (e.g., ([250 Htz – 749 Htz], [750 Htz – 1,249 Htz], [1,250 Htz – 1,749 Htz], [1,750 Htz – 2,249 Htz])) – with a noted allowance for fuzzy intervals as well. Then, the vector signature records relative amplitudes supplied on a scale (e.g., decibels), which makes sense for the application domain (e.g., (10, 40, 40, 10)). Notice that more-specific levels are associated with a decreased width in frequency response – allowing for the better characterization of nuances as desired.

10. Just as permutations capture rotations of visual images in the plane (see below); Doppler shifts capture the motion of acoustical contacts towards or away from the observer. A Doppler shift is confirmed by computing the permutations in an attempt to obtain a match. For example, a centroid of (500 Htz, 1000 Htz, 1,500 Htz, 2,000 Htz), or (500, 1,000, 1,500, 2,000) is perceived to be rapidly approaching if it is received as (1,000, 1,500, 2,000, or 2,500) say. Such up or down shifts of one or more displacements are easily computed and matched. The more-specific the level, the greater the sensitivity to Doppler-shifting motion relative to the observer. Moreover, the recognition of lightning is often a fusion of an immediate flash-bulb effect followed by a delayed cannon-like thunder. Fusion here requires searching all auditory feedback over an interval of k seconds, where $0 \le k \le c$ and c may be fuzzy in an attempt to satisfy, $\bigcup(\text{flash}_t, \text{thunder}_{t+k}) \rightarrow$ lightning. Simple fusion, such as typified here, may be hardwired or processed through a knowledge base (e.g., KASER,[17] CBFER[18]) if more complex. Similarly, Area 17 of the human visual cortex may pass on more complex recognition tasks to higher areas of the brain.[6]
11. Next, consider the task of motion recognition. Typically, such motion is based on velocity, but may include acceleration too. Higher derivatives may remain academic curiosities. Example 12 shows how to compute velocity and acceleration vectors. Some vision systems may only perceive objects that move and/or accelerate their motion. (While humans don't strictly perceive their world in this manner, it is known that the dinosaur, TRex did.)
12. In recognizing an image, shape, color, velocity, acceleration, etc. can all play a fundamental role. Here, the union of these vectors is used to best characterize the image. The multiple vectors (at each level), produced by the union, is then the vector signature (reference Theorem 11). It follows from Comment 13 that the semantics of the vector signature is that of the union of the semantics of each of multiple vectors. For example, lightning has the same image signature of a flash bulb and thunder presumably has the same auditory signature as a cannon. But, $\bigcup(\text{flash}_t, \text{thunder}_{t+k}) \rightarrow$ lightning, where $0 \le k \le c$ and c may be fuzzy. If any member of the union fails to be matched at Level i, the next more-specific level is computed, where it exists.

13. Images may be normalized by way of permuting their signature vectors until a best-matching vector is found (see Comment 14). Here, the permutations are given by (1, 2, 4, 6, 5, 4, 4, 5), (2, 4, 6, 5, 4, 4, 5, 1), (4, 6, 5, 4, 4, 5, 1, 2), (6, 5, 4, 4, 5, 1, 2, 4), (5, 4, 4, 5, 1, 2, 4, 6), (4, 4, 5, 1, 2, 4, 6, 5), (4, 5, 1, 2, 4, 6, 5, 4), and (5, 1, 2, 4, 6, 5, 4, 4). Note that a megapixel image can be covered by 1,000 ray traces, for which the permutations are deemed to be tractable. The greater the number of vectors, the more times that the same vector will be saved at each level (e.g., due to differing accelerations), but the vectors are implicitly fuzzified through the use of the level concept, which tends to minimize the number of such savings. That is, the method implicitly and maximally avoids redundancy by only storing a more-specific vector when a more-general one (in union) won't do (as described below). Most significantly, the union of multiple signature vectors is processed piecewise independently and then fused for a combined semantics. Note that if say A and B fuse to give C, then they may not give anything but C at said level. However, they may fuse to give C or a distinct D on a distinct level, since consistency need only be maintained intra-level (reference Theorem 15a). The nature and number of such signature vectors is necessarily domain-specific (reference Comment 14).
14. Each level stores the ray vector for each image (or equivalent, or in union) at that level. Each image (or equivalent) is ascribed one or more syntactic labels having a common semantics. This mimicks Broca's area for speech in humans.[6] Only one semantics is ascribed to any image (or equivalent) at each level. The levels are visited in order from most-general (level 0) to most-specific (level n), which can serve to speed up the recognition process as well. Images (or equivalent) are only saved to a level of specificity associated with their actual use.
15. An image (or equivalent) is saved at the current level if its semantics is distinct from all other images at that level and the new image does not share a common permutation (or equivalent) with an existing image, at that level, lest non determinism be enabled (reference Theorem 15c). In what follows, images are again meant to include other modalities or unions thereof. Images sharing a common semantics and a common permutation are not allowed due to redundancy. This is easy to effect by not acquiring such a semantics and associated vector. Images sharing a common semantics or not, but without a common permutation are saved. Images sharing a common permutation, but having distinct semantics are indicative of failure of the number of pixels at the current level to discriminate the images as evidenced by the non determinism. Here, both images (or equivalent) are removed from the current level and storage is attempted at the next more-specific level (i.e., using the computed enhanced resolution) until the pair is saved or the most-specific level is attained. Even if the original pair is successfully saved at some level (expunging the least-recently used (LRU) member at Level n as needed to free space), a check needs to be performed at that level for induced non determinism in up to two pairs. If such new non determinism is found where the original pairing is now deterministic, then the new non deterministic

pair(s) are removed, stacked, and recursively processed as before. Any deterministic pairs may be saved at the level they were found at. If either member of a non deterministic pairing is not available for more resolute computing, then it is expunged and forgotten. If the storage attempt fails at the most-specific level, the most-recent image (or equivalent) is saved and the LRU member at that level is expunged if needed to free space (reference Theorem 15e).

16. *forever*
17. END.

13.7 A Theory for Machine Learning

There are certain things that all predictive methods will have in common:

1. **Representation** – a characteristic framework, which may be dynamic, depending on the problem domain
2. **Problem Reduction** – determining those features and combinations of features, along with methods for recombination, which can serve as most-accurate predictors
3. **Generalization** – increasing the applicability of a case or segment of knowledge by decreasing its required context

The theory to be developed below is a generalized theory of machine learning. It need capture evolutionary or so-called spiral development because any body of knowledge, sufficiently complex to be capable of self-reference, is necessarily incomplete.[16] First, rule or case-based knowledge is *universal* in that it is theoretically sufficient to represent and solve any problem. The difficulty addressed by this theory is three-fold; namely,

1. the acquisition of ever-better representations of knowledge;
2. the acquisition of salient rules or cases; and,
3. the evolution of tractable search methodologies to retrieve said knowledge.

13.7.1 Case vs. Rule-Based Learning

The difference between a rule and a case is that the latter may contain extraneous antecedent predicates. This allowance is to be preferred because such an excess permits the evolution of the most salient predicates. Next, a case may embody one or more consequents, where a single consequent suffices through the use of pairing and projection functions.[16] Thus, define a case (equivalently knowledge) base as follows. Let,

$$KB_i = \begin{Bmatrix} s_{1,1}, s_{1,2}, s_{1,3}, ..., s_{1,n}; f_{1,1}, f_{1,2}, f_{1,3}, ..., f_{1,r} \\ s_{2,1}, s_{2,2}, s_{2,3}, ..., s_{2,n}; f_{2,1}, f_{2,2}, f_{2,3}, ..., f_{2,r} \\ s_{3,1}, s_{3,2}, s_{3,3}, ..., s_{3,n}; f_{3,1}, f_{3,2}, f_{3,3}, ..., f_{3,r} \\ ... \\ s_{m,1}, s_{m,2}, s_{m,3}, ..., s_{m,n}; f_{m,1}, f_{m,2}, f_{m,3}, ..., f_{m,r} \end{Bmatrix} \rightarrow \begin{Bmatrix} d_1 \\ d_2 \\ d_3 \\ ... \\ d_m \end{Bmatrix} \tag{13.2}$$

This defines the ith among a countably infinite number of knowledge bases, where s defines sensors (i.e., unprocessed inputs), f defines features (i.e., permutations of sensors and/or features as defined by an algorithm), and d is an implied dependency. The n sensors, r features, and m dependencies are dynamic in magnitude so long as they are all computable by some finite state machine. We may equivalently write, $KB_i = S_i F_i D_i$, where SF is a set – not a sequence.

Dependencies are arranged into pragmatic groups. Too many groups and learning by the system will occur with less rapidity. Too few groups and learning by the system, while rapid, will not serve to distinguish situations of practical concern. In general, dependencies are procedural. They may also communicate by way of posting to/erasing from/updating a blackboard, or global database.

A *context* is defined to be the left-hand side of the implication ($\rightarrow$). It is said to be *conformal* just in case the context carries n' sensors and r' features such that $n' \geq n$ and $r' \geq r$. If $n' < n$ or $r' < r$, the context is said to be *non conformal*, in which case it cannot be matched. Sensors and features not contained in SF are simply discarded in making a local match. Again, a context must be conformal to be matched against a KB.

The independent (antecedent) portion of the ith knowledge base is weighted as the following formalism makes clear. The two horizontal bars over the KB designate the inclusion of the weights, where $\sum_{i=1}^{n+r} w_i = 1$, which is to say that the weights are *normalized*.

$$\overline{\overline{KB_i}}^{a} = \begin{Bmatrix} w_1, w_2, w_3, ..., w_n; w_{n+1}, w_{n+2}, w_{n+3}, ..., w_{n+r} \\ s_{1,1}, s_{1,2}, s_{1,3}, ..., s_{1,n}; f_{1,1}, f_{1,2}, f_{1,3}, ..., f_{1,r} \\ s_{2,1}, s_{2,2}, s_{2,3}, ..., s_{2,n}; f_{2,1}, f_{2,2}, f_{2,3}, ..., f_{2,r} \\ s_{3,1}, s_{3,2}, s_{3,3}, ..., s_{3,n}; f_{3,1}, f_{3,2}, f_{3,3}, ..., f_{3,r} \\ ... \\ s_{m,1}, s_{m,2}, s_{m,3}, ..., s_{m,n}; f_{m,1}, f_{m,2}, f_{m,3}, ..., f_{m,r} \end{Bmatrix} \tag{13.3}$$

13.7.2 The Inference Engine

Next, we need a formalism to find the best matching rows, given a conformal context, and an approximation of the errors in match for use in segregating the matched rows from the mismatched rows. Moreover, many KBs may need to be searched for a best match; whereas, not all such KBs can be searched due to contention on the busses, limited frequency discrimination using RF, or equivalent. Thus, a heuristic mechanism is needed to map (or learn to map) the conformal context to a finite subset of KBs. The *inference engine* has the task of matching a context against a knowledge or case base to find those cases having minimal difference with the context.

A "software" mapping process (using one base to map to another) is not practical because a higher-level map need not be conformal – reducing the efficiency of the mapping process, which was the initial concern. Thus, a physical heuristic is sought; namely, one based on Euclidean Geometry. Here, each knowledge base has storage coordinates in 3-space. Such factors as contention on the bus, memory, processor(s) speed, and time for data transfer determine how many knowledge bases may be referenced by a given one. References typically occur for sub-domain knowledge. For example, in deciding what crop to plant, a knowledge base will reference another specializing in weather prediction, and so on in an acyclic architecture. Given a uniform distribution and the physical size of said knowledge bases, one can specify a radius, χ, the interior of which is used to specify that finite subset of KBs, which are local to a given one. Clearly, knowledge bases to the corners, edges, or walls may have larger radii; but, there is no great loss if this is not addressed due to the fact that the interior is larger. Here, $\chi_{i,j} = \sqrt{(x_j - x_i)^2 + (y_j - y_i)^2 + (z_j - z_i)^2}$, which gives the radius from the ith knowledge base to the jth knowledge base.

Non-local sensor data should be taken for the current time period. Knowledge will iteratively propagate, across these radii, from the outermost to the innermost knowledge base and vice versa. All knowledge bases may be simultaneously searched by the inference engine.

Assume a conformal context, *SF*, $s_1, s_2, s_3, \ldots, s_n; f_1, f_2, f_3, \ldots, f_r$, and a weighted case antecedent matrix, $\overline{\overline{KB_i}}^{a}$, equation (13.2). Given the geometric mapping heuristic previously described, the 2-norm is most-closely aligned with determining the matching error. It is determined by,

$$\alpha_j = \sqrt{\sum_{i=1}^{n} w_i (s_i - s_{j,i})^2 + \sum_{i=n+1}^{n+r} w_i (f_{i-n} - f_{j,i-n})^2}\text{, where } j = 1, m \qquad (13.4)$$

A dynamic positive constant, δ, is set such that if, $\alpha_j < \delta$, the jth dependency in KB_i is included as a potential response to *SF*. If $\alpha_j = 0$, then a perfect match occurs. However, that does not necessarily mean that the jth dependency is the correct answer. Rather, the learning methodology is based on statistical mechanics and thus it is the most frequently occurring dependency (i.e., the one with greatest multiplicity), which provides the initial response. Such predictions are more likely to be valid because they have multiple supports. Note that many applications do not have numerical dependencies, in which case it makes no sense to average them. Thus, $d_j \mid \forall i, i \neq j, count(d_i) \leq count(d_j)$. If $count(d_i) = count(d_j)$, then resolve ties in favor of that predictive group having the closest to a zero-error sum, or $d_j \mid \forall i, i \neq j, sum(d_i) \geq sum(d_j)$, where *sum* is given by α_k and any further ties are broken arbitrarily. Alternatively, in lieu of a closest to a zero-error sum, ties may be broken in favor of the most-frequently used (MFU) group. Here, fired rows in KB_i are logically transposed with the one above them, if any. Then, the MFU group averages a position closest to the top. The choice between methods for breaking ties is made arbitrarily.

The initial response vector is saved and all of the rows in this vector are set to the correct response when it is forthcoming. In this manner, the system learns on the basis of first principles of statistical mechanics. There can be no one exact design here so long as it adheres to the aforementioned principles. No prediction is issued if no rows are returned as being within δ error (i.e., the response vector is empty). This may be an indication that δ needs to be larger. Conversely, if δ is set too large, then accuracy will fall off due to decreasing dependency on the values specified within a context.

Next, we address the issue as to how to set δ. This depends in part on the problems incurred in the event of a non prediction. If such an event can be tolerated, then the target is to set δ so as to return one row; otherwise, it is set so as to return up to $\left\lfloor \frac{m}{2} \right\rfloor$ rows. The value for δ can be set algorithmically as follows, where $\varepsilon \in \left[1, \left\lfloor \frac{m}{2} \right\rfloor\right]$ and may be set by one or more fired dependencies in KB_i.

```
If the magnitude of the response vector < ε
    δ ← 2δ
Else
    If the magnitude of the response vector > ε
        δ ← δ/2
// Otherwise, don't change δ.
```

Fig. 13.8 Algorithmic setting of δ using ε

Each KB_i has an ideal size, which enables the most rapid searching of their collective. It can be shown that this ideal size is a square matrix. This occurs where $m = n + r$, where $n + r$ should be set in accordance with the number of sensors and features allowed to be resident at any one time. Ideally, each KB_i will be assigned its own asynchronous processor (e.g., using a data flow architecture).

Whenever no prediction is issued, an approximate square matrix is maintained by expunging that least-frequently used (LFU) row and logically acquiring the new case, once known, at the head of the matrix. In every case where learning occurs, the weights need to be initialized and updated.

The $n+r$ weights in $\overline{\overline{KB_i}}^{a}$ are initialized to one and normalized to yield an average value of $1/_{n+r}$. Normalization occurs immediately subsequent to deletion, where the number of sensors plus features is bounded above by β (assuming a homogeneous processor architecture), which serves as an upper bound for $n + r$. Normalization is defined by, $w_j \leftarrow \frac{w_j}{\sum_{j=1}^{n+r} w_j}$. New features, f_j, arrive by two routes; that is, by being synthesized in the same knowledge base, or by being imported from a different local knowledge base within a radius, χ. If the former, $w_{n+r} \leftarrow \frac{\sum_{j=1}^{n+r-1} w_j}{n+r-1}$. Alternatively, if imported, since the number of sensors plus features is approximately the same in all KB_i, or bounded above by β, w_{n+r} is imported from KB_j, where $j \neq i$. In either case, the resulting weight vector is normalized. Clearly, the greater the number of asynchronous processors (i.e., one per knowledge base) the higher will be the quality of performance per unit of time.

13.7.3 On Making Predictions

Next, consider the members of the response vector (Fig. 13.8), which is designated by, $\underline{D}$. The subsequently found actual(s) are designated by, $\overline{D}$. A correct prediction is said to occur where $\underline{d}_j = \overline{d}_j$. These dependencies were correct predictors. We want to increase the value of the weights for the antecedents associated with these dependencies, which most contributed to the correct prediction in context and decrease the value of the weights otherwise. Thus, $H_i = \sum_{j=1}^{m} a_{j,i}$, s.t. $\underline{d}_j = \overline{d}_j$, where H_i sums the ith column for each correctly predicting row, if any and a represents a sensor or a function as appropriate.

Next, we partition the weights into two classes – those for whom $H_i \leq \mu H$ and the remainder. Here, $\mu H = \frac{\sum_{i=1}^{n+r} H_i}{n+r}$. If $H_i \leq \mu H$ then $w_i \leftarrow 2w_i$. These weights are doubled because the context of these variables serve best as predictors. Also, if approximately half of the weights are doubled, then heuristic learning across similar problems will be maximized, which can be shown not to be the case with multipliers other than two.

H_i must be less than or *equal* to μH to properly handle the situation where exactly one row is within δ error. Then, the difference would need to be greater than or *equal* to μH for those weights to be halved – resulting in a contradiction, where the difference is exactly μH. It follows that only one side of the mean is to be handled as previously described.

Finally, renormalize all columns. Notice that the resulting weights have increased as well as decreased in concert in proportion to their most-recently experienced utilities.

In order to prevent the number of sensors plus features from exceeding the upper bound of β in all KB_i, it becomes necessary to replace that sensor or feature having the least weight in its local knowledge base, where a sensor or feature may not be replaced if it is used to define at least one resident feature. A normalized weight may only be replaced by a greater normalized weight if imported. The new feature may not be redundant with the local set of features. Finally, where the weights are exactly the same, select the LFU weight and its associated sensor/feature for deletion because it was most recently created and thus its weight is most subject to error. Similarly, select the MFU weight and its associated sensor/feature for export because it was least recently created and thus its weight is least subject to error. A practical solution is to initially non-redundantly populate all β sensors/features, in all KB_i, at the outset. Again, whenever a column is lost or acquired, renormalization must immediately follow.

A feature may not be expunged from a knowledge base until at least one update cycle of its weights has completed, since the last feature replacement. This constraint is necessary to prevent the rapid overwriting of the feature set(s). If $H_i > \mu H$, then the ith features (not necessarily sensors) are simultaneously replaced. The goal here is to avoid the so-called mesa phenomenon,[19] whereby global optimality can never be assured if less than all features are simultaneously evolved (though it can get arbitrarily close). Given distinct contexts over time, this set of least desirable features will not only be replaced (evolved), but it can include any subset selected from the ith knowledge base set, $\{f_1, f_2, \ldots, f_r\}$. It may also import one non-redundant feature per cycle, from the global knowledge base set, having the greatest normal weight. The imported feature must come from the same schema or a sub-schema so as to make sense (see below). Co-evolution is thus facilitated with regard to the mesa phenomenon as well as for saving the more desirable features and their associated weights. Imports take advantage of proven features, while evolutionary processes seek to synthesize new ones.

Values for the new feature are recomputed. This may introduce limited non determinism in the local cases. However, such limited non determinism will have no deleterious effects on performance. This follows because typically, $|\underline{D}|>1$, where the response favors the dependency having the greatest multiplicity for a locally best-matched context. That is, the provision for limited non determinism will only improve the accuracy of response. Notice too that the set, which is reset will change on most iterations of the method – insuring diversity in the voting process.

The magnitude of the response vector may be zero because δ and/or ε is too small. In this case again, no dependency is associated and the LFU row in equation (13.1) is expunged to make room for the mapping of the conformal context to the actual value for the dependency, $SF \rightarrow \overline{d}_j$. This mapping is logically acquired at the head of its resident knowledge base. All rows in equation (1) are logically updated using the method of transposition, where a fired case is logically transposed with the one above it, where possible. A local case is said to have been fired where, $\underline{d}_j = \overline{d}_j$, j = 1, m and in that order.

Evolutionary feature definition requires the manual specification of schema for evolutionary instantiation. The value of the computer is that it searches out the best instances as well as the best instances in combination with other instances for capturing the defining details of any among a set of dependencies. A balance between the human (Yin) and the machine (Yang) is sought. The same functionality is implemented for all knowledge bases in the system, with one important addition. Here, each distinct processor, which is instantiating the same schema (or a sub-schema thereof) is also able to export better instances, whenever found. This is symmetric evolution;[20] whereas, finding local instances is random evolution,[20] which is conditioned by the choice of enabled operands and operators.

13.7.4 On Feature Induction

At this point, we address the construction and instantiation of feature schemata. The first principle is that as many schema as practical should have the same set definition or sub-definition to enable symmetric evolution and the attendant greater use of distributed processors.

Features are based on a composition of functions, which are subject to realization on parallel/distributed processors. Each function is said to be immutable because it may embody the full range of features in the programming language used for realization; yet, once specified, it exists in a library of functions and may only be subject to manual update. Such functions may be iteratively composed using one or more pairs of nested braces, { }. An example or two will serve to clarify the design and instantiation of a schema through their use. Consider:

```
Current_Altitude_Sea_Level?
FN1{Barometer_Falling?,
        Wind_Changing_Direction?,
        Clear_Sky?}
FN0{Temperature_Below_Freezing?,
    Temperature_Above_Freezing?}
FN0{Local_Sea_Level_Cities?,
        Local_Mountain_Cities?}
```

Fig. 13.9 A weather prediction feature-definition schema for snow

In Fig. 13.9, the knowledge engineer has attempted to write schema having a minimal search space and a maximal penetrance to capture the conditions that might lead to a dependency of snow (in combination with other sensors/features). FN0 is a special function and disappears after the single argument is selected. FN1 counts the number of formal parameters and randomly selects exactly enough actual parameters to fill these slots. If too few actual parameters are supplied, then an error message will be generated. Notice the composition of functions in the implied sequence. Chance evolution would lead to the following correct instance.

```
Current_Altitude_Sea_Level?
FN1 (Barometer_Falling?,
        Wind_Changing_Direction?)
Temperature_Below_Freezing?
Local_Sea_Level_Cities?
```

Fig. 13.10 A weather prediction feature-definition instance for snow

Functions may return a Boolean result, as in Fig. 13.10. They may also return words or numbers, where they would otherwise return TRUE. Parameters, if included, follow the rules of the local programming language.

Clearly, the search space for FN0 is the product of the number of arguments to each function – 1. It tends to be the same for higher-level functions (i.e., FN1, FN2, …) as well. Selections are made at uniform chance. The search space here can rapidly grow to be intractable if certain precautions are not taken. The complexity of definition here usually results in features being ascribed a heuristic definition. Notice how symmetry provides direct support for heuristic programming.

Fig. 13.11 presents a schema having a search space of $\prod_{i=1}^{m}(X_{FN1_i}-1)\prod_{i=1}^{n}(X_{FN2_i}-1)$. However, by careful pruning, this search space is reduced to $\sum_{i=1}^{m}(X_{FN1_i}-1)(X_{FN2_i}-1)$, where $m = n$ as shown in Fig. 13.12. Here,

$\sum_{i=1}^{m}(X_{FN1_i}-1)(X_{FN2_i}-1) << \prod_{i=1}^{m}(X_{FN1_i}-1)\prod_{i=1}^{n}(X_{FN2_i}-1)$ with scale. Thus, knowledge should be added in this manner whenever possible to minimize the implicit search space. Here, optimizations may be realized by an optimizing compiler; although, the continual evolution of the features makes this less necessary than with conventional programs, since instancing the schemas will be the main time sink. These schemas may be manually optimized by making use of the *triangle inequality* as demonstrated in Figs. 13.11 and 13.2.

```
DE FNi (X1?, X2?, …, XFN1?, XFN2?):
FN11 (X1?, X2?, …, XFN1,1?)
FN21 (X1?, X2?, …, XFN2,1?)
…
FN1m (X1?, X2?, …, XFN1,m?)
FN2n (X1?, X2?, …, XFN2,n?)
END
```

Fig. 13.11 An abstract feature-definition schema

```
DE FNj (X1?, X2?, …, XFN1?, XFN2?):
(FN1 (X1?, X2?, …, XFN1,1?)
FN1 (X1?, X2?, …, XFN2,1?))
…
(FN1m (X1?, X2?, …, XFN1,m?)
FN2n (X1?, X2?, …, XFN2,n?))
END
```

Fig. 13.12 A refined abstract feature-definition schema

The triangle inequality tells us that simply put, the less that is left to chance the more efficient the model will be at producing viable features. The process of manually and iteratively removing chance is referred to as the triangle inequality (e.g., $|A| + |B| \geq |A + B|$). This means that it is better to have several program schemata with few articulation points than to place all of the articulation points in one schema, where the resultant complexity can easily overwhelm at least a low-end computer. A key concept in writing the features is that instead of the user searching for the correct code construct or function at various points in the program, the user specifies a space of alternative constructs at a limited number of articulation points over a maximal number of schemata. In any case, the user specifies reasonable alternatives, which are captured in the form of a set – possibly tagged with a pneumonic id. Here, the user need not contemplate the details – details that would detract from his/her capability to be an efficient programmer/debugger. The computer will find for programs that satisfy all of the test vectors, if possible.

13.8 Conclusions and Outlook

Creativity and intelligence are generally ascribed as human attributes when people make correct decisions in the presence of uncertainty. This chapter has provided a few demonstrations of machine intelligence in the context of machine vision. It has argued in favor of the mechanization of these attributes.

A central theme underpinning creativity and intelligence is randomization,[20] or the discovery and extraction of syntactic and semantic patterns in a learning environment. Randomized patterns can be put in isomorphism and this chapter has provided evidence that this is the basis for intelligence and creativity given suitable representation(s) of knowledge. Indeed, any mechanics for selecting a representational formalism may be guided by randomized patterns. This means that randomization may be self-referential and thus mathematically non trivial.

Essential incompleteness can only be avoided through the use of heuristic search, which comes with the attendant advantage in that it can always enable tractable search. Heuristics too must be learned and are inherently part of non trivial randomization. Fast forwarding a few steps, it follows that evolution synthesizes paradigms for learning to evolve and computer vision must follow suit if it is to be deemed mathematically non trivial.

The outlook for this chapter is that it may be possible to evolve heuristics and heuristic transforms with ever greater speed and reliability, which serve the domain of computer vision among others. However, this is best done from a standpoint of novel basic theory and far less likely from a requirements base of application needs. After all, such an approach itself defines the mechanics of the most general, or Type 0 grammar.

Acknowledgments. The author thanks the Space and Naval Warfare Systems Center, San Diego, California for financial support. He also thanks their Office of Patent Council for their assistance in filing disclosures with the U.S. Patent Office on much of the work described herein. This work was produced by a U.S. government employee as part of his official duties and no copyright subsists therein. It is approved for public release with an unlimited distribution.

References

1. USN PEO for C4I and Space and USAF Electronic Systems Center. Net-Centric Enterprise Solutions for Interoperability, Net-Centric Implementation, Part 2: ASD (NII) Checklist Guidance 1.2 (20) (2005)
2. Rubin, S.H.: On the Auto-Randomization of Knowledge. In: Proc. IEEE Intern. Conf. Info. Reuse and Integration, Las Vegas, NV, pp. 308–313 (2004)
3. Lin, J.H., Vitter, J.S.: Complexity Results on Learning by Neural Nets. Mach. Learn. 6(3), 211–230 (1991)
4. Rubin, S.H.: On Randomization and Discovery. Info. Sciences 177(1), 170–191 (2007)

5. Rubin, S.H., Kountchev, R., Todorov, V., Kountcheva, R.: Contrast Enhancement with Histogram-Adaptive Image Segmentation. In: Proc. IEEE Intern. Conf. Info. Reuse and Integration, Waikaloa, HI, pp. 602–607 (2006)
6. Eccles, J.C.: Understanding of the Brain, 2nd edn. McGraw-Hill Co. (1976)
7. Rubin, S.H.: Computing with Words. IEEE Trans. Syst. Man, Cybern. 29(4), 518–524 (1999)
8. Zadeh, L.A.: From Computing with Numbers to Computing with Words – from Manipulation of Measurements to Manipulation of Perceptions. IEEE Trans. Ckt. Syst. 45, 105–119 (1999)
9. Pedrycz, W., Rubin, S.H.: Data Compactification and Computing with Words. Intern. J. Engineering Applications of Artificial Intelligence 23, 346–356 (2010)
10. Uspenskii, V.A.: Gödel's Incompleteness Theorem, Translated from Russian. Ves Mir Publishers (1987)
11. Mitchell, T.M.: Version Spaces: A Candidate Elimination Approach to Rule Learning, Ph.D. Thesis, Stanford University (1979)
12. Duncan, G.R.: Cheap Drones Could Replace Search-And-Rescue Choppers. New Scientist, `http://www.newscientist.com/article/mg20727696.000-cheap-drones-could-replace-searchandrescue-choppers.html`
13. Ackerman, S.: Air Force Wants Drones to Sense Other Planes' 'Intent', `http://www.wired.com/dangerroom/2010/07/air-force-wants-drones-to-sense-other-planes-intent/`
14. Quinlan, J.R.: C4.5: Programs for machine learning (September 1997)
15. Quinlan, J.R.: Bagging, Boosting, and C4.5. In: Proc. of the Thirteenth National Conference on Artificial Intelligence, pp. 725–730. AAAI Press, MIT Press, Cambridge, MA (1996)
16. Kfoury, A.J., Moll, R.N., Arbib, M.A.: A Programming Approach to Computability. Springer, Heidelberg (1982)
17. Rubin, S.H.: On Knowledge Amplification by Structured Expert Randomization (KASER), U.S. Patent No. 7,047,226. Space and Naval Warfare Systems Center, San Diego Biennial Review (2001)
18. Rubin, S.H.: CBFER: Case-Based Field-Effect Reasoning. USN Patent Pending, NC 100222 (2009)
19. Feigenbaum, E.A., Feldman, J. (eds.): Computers and Thought. McGraw-Hill Inc. (1963)
20. Chaitin, G.J.: Randomness and Mathematical Proof. Sci. Amer. 232(5), 47–52 (1975)

Chapter 14
Method for Intelligent Representation of Research Activities of an Organization over a Taxonomy of Its Field

Boris Mirkin[1], Susana Nascimento[2], and Luís Moniz Pereira[2]

[1] Department of Computer Science, Birkbeck University of London, London, UK, and School of Applied Mathematics and Informatics, Higher School of Economics, Moscow, RF
mirkin@dcs.bbk.ac.uk

[2] Department of Computer Science and Centre for Artificial Intelligence (CENTRIA), Faculdade de Ciências e Tecnologia, Universidade Nova de Lisboa, 2829-516 Caparica, Portugal
{snt,lmp}@di.fct.unl.pt

Abstract. We describe a novel method for the analysis of research activities of an organization by mapping that to a taxonomy tree of the field. The method constructs fuzzy membership profiles of the organization members or teams in terms of the taxonomy's leaves (research topics), and then it generalizes them in two steps. These steps are: (i) fuzzy clustering research topics according to their thematic similarities in the department, ignoring the topology of the taxonomy, and (ii) optimally lifting clusters mapped to the taxonomy tree to higher ranked categories by ignoring "small" discrepancies. We illustrate the method by applying it to data collected by using an in-house e-survey tool from a university department and from a university research center. The method can be considered for knowledge generalization over any taxonomy tree.

14.1 Introduction

14.1.1 Motivation

Our subject should be counted as what can be referred to as organizational knowledge management. We represent activity of an organization in a novel way by mapping it to an ontology of the field. Our method involves three stages: (i) data integration, (ii) ontology usage, and (iii) activity visualization.

To give an intuitive idea of the method, let us first consider three similar stages of data representation for operative control, such as that by a company delivering electricity to homes in a town zone. Fig. 14.1, taken from [2], represents an energy network over a map of the corresponding district on which the topography and the network data are integrated in such a way that gives the company "an unprecedented

R. Kountchev et al. (Eds.): Advances in Reasoning-Based Image Processing, ISRL 29, pp. 423–454.
springerlink.com

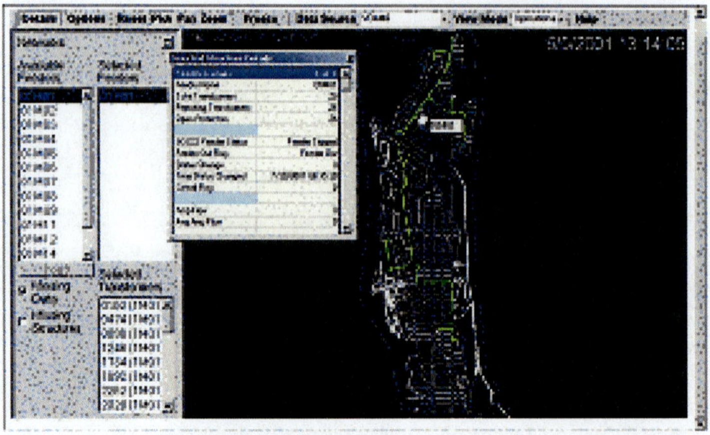

Fig. 14.1 Energy network of Con Edison Company on Manhattan New-York USA visualized by Advanced Visual Systems [2].

ability" to control the flow of energy by following all the maintenance and repair issues on-line in a real time framework.

What we are concerned with is whether a similar mapping is possible for a long-term analysis of an organization whose activity is much less tangible, such as a university research department. There are three major ingredients that allow for a successful representation of the energy network:

(1) map of the district,
(2) the energy network units, and
(3) mapping (2) at (1).

Moreover, one could imagine an extension of this mapping to other infrastructure items, such as the water supply, sewage type and transports, so that the map could be used for more long-term city planning tasks such as development of leisure or residential areas and the like. This allows us to take, for a research department, the following analogues to the elements of the mapping in Fig. 14.1:

(1') an ontology of Computer Science (CS),
(2') the set of CS research subjects being developed by members of the department, and
(3') representation of the research on the ontology.

Why would one want to do that? There can be various management goals such as, for example:

– Positioning of the research organization within the taxonomy;
– Analyzing and planning the structure of research conducted in the organization;

- Finding nodes of excellence, nodes of failure and nodes needing improvement for the organization;
- Discovering research elements that poorly match the structure of the taxonomy;
- Planning of research and investment;
- Integrating data of different research organizations in a region for the purposes of regional planning and management.

Before moving further, let us take a look at the structure of Classification of Computer Subjects developed by the Association for Computing Machinery (ACM-CCS) [1] which is the heart of the Computer Science ontology. Its highest layer is presented in Fig.14.2 that shows the whole of Computer Science as divided into eleven first-layer subjects:

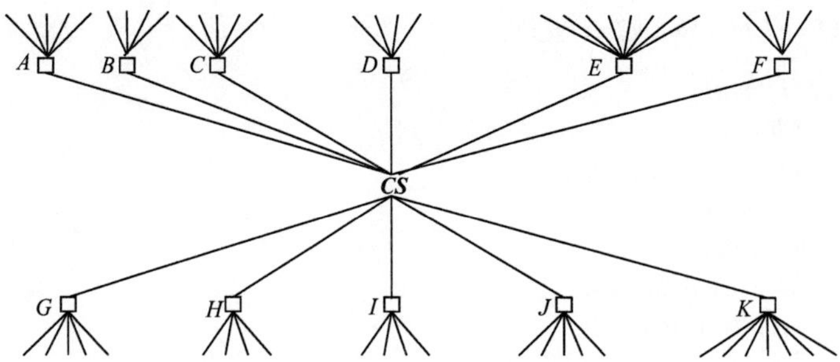

Fig. 14.2 The higher level of ACM-CCS Taxonomy.

A. General Literature
B. Hardware
C. Computer Systems Organization
D. Software
E. Data
F. Theory of Computation
G. Mathematics of Computing
H. Information Systems
I. Computing Methodologies
J. Computer Applications
K. Computing Milieux

Each of the eleven subjects is further subdivided into subjects of the second layer. For example, the subject of our current interest, *'I. Computing Methodologies'*, consists of seven specific subjects plus two of general interest, *'0. GENERAL'* and *'m. MISCELLANEOUS'* these two are present in every division of ACM-CCS:

I. Computing Methodologies

I.0 GENERAL
I.1 SYMBOLIC AND ALGEBRAIC MANIPULATION
I.2 ARTIFICIAL INTELLIGENCE
I.3 COMPUTER GRAPHICS
I.4 IMAGE PROCESSING AND COMPUTER VISION
I.5 PATTERN RECOGNITION
I.6 SIMULATION AND MODELING (G.3)
I.7 DOCUMENT AND TEXT PROCESSING (H.4, H.5)
I.m MISCELLANEOUS

Many of these are further subdivided into subjects of the third-layer such as, for instance,

I.5 PATTERN RECOGNITION

I.5.0 General
I.5.1 Models
I.5.2 Design Methodology
I.5.3 Clustering
Algorithms
Similarity measures
I.5.4 Applications
I.5.5 Implementation (C.3)
I.5.m Miscellaneous

There can be also units of the fourth layer that are not indexed and used mainly as indications of the contents of the third layer subjects, such as those two shown for topic *'I.5.3 Clustering'*. One can also see a number of collateral links between topics both on the second and the third layers - they are in the parentheses at the end of some topics such as *G.3*, referred to at *I.6*.

At first glance, mapping of subjects under development in a department to the taxonomy is a rather straightforward exercise. For example, a survey found that 25 of the total of 81 meaningful subjects of the second layer are being developed in a research department[1]. After these 25 subjects are mapped to the taxonomy, one can see them visualized by black boxed nodes in Fig. 14.3. This portrayal is not entirely unhelpful - the visualization does provide a useful information of the coverage of the taxonomy subjects by the research. Yet this representation gives no hints of the structure of the research: the subjects are presented with no indication of relation between them according to the working of the department. The taxonomy reflects only those relations between the subjects that have been specified in it according to the working of the entire community of computer scientists. The structure of research projects in a department may impose a different taxonomy of the subjects. This "local" taxonomy would reflect the relations between subjects according to research

[1] Survey conducted in the CS department of Faculdade de Ciências e Tecnologia, Universidade Nova de Lisboa (DI-FCT-UNL) in 2007 [33].

projects worked on in the organization: the larger the number and weight of research projects that involve two taxonomy subjects, the greater association between the subjects in the department. The local taxonomy may not necessarily follow the structure of the global taxonomy, and it would be of interest to map the local taxonomy to the global one. Although we are going to explore this problem in the future, this paper concerns a less challenging undertaking. Instead of presenting the "local" structure of research as a hierarchical taxonomic structure, we present it as a "flat" set of not necessarily disjoint clusters of ACM-CCS subjects in such a way that the clusters reflect the "local" associations between the subjects - the greater the weight and number of the projects at which two subjects are involved the greater the association between those subjects and greater the chance that the two subjects belong to the same cluster.

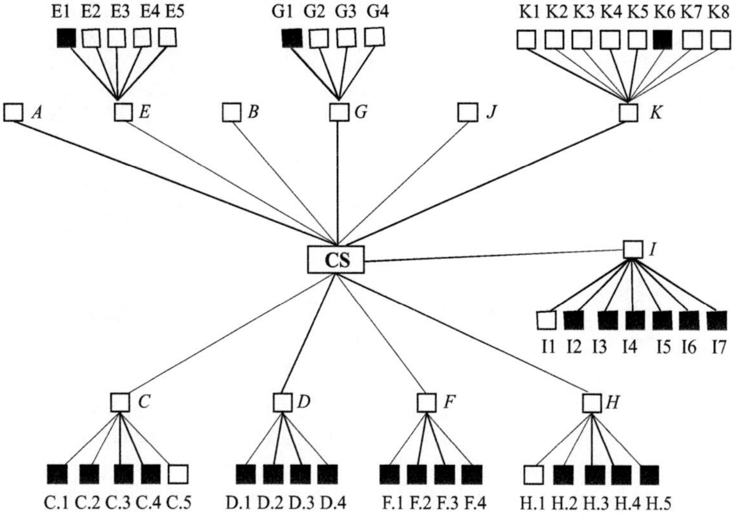

Fig. 14.3 Twenty five ACM-CCS subjects of a University department mapped to the ACM-CCS taxonomy.

Returning to the 25 ACM-CCS subjects, we have found that they can be reasonably divided into 6 clusters which are mapped to the ACM-CCS taxonomy separately to produce the following portrayal (Fig. 14.4) [33]. The mapping involves the concept of "head subject" that can be defined as the highest rank node(s) covering, in general, the cluster. Since the coverage is not necessarily exact (see Sect. 14.4 for definitions and method) two more types of elements emerge. These are: "gap", that is a node covered by a cluster head subject but not belonging to the cluster, and "offshoot", that is a node belonging to the cluster but not covered by its head subject node(s). These are illustrated in Fig. 14.4 with different graphic elements. Among interesting features of the research conducted in the department, three clusters have

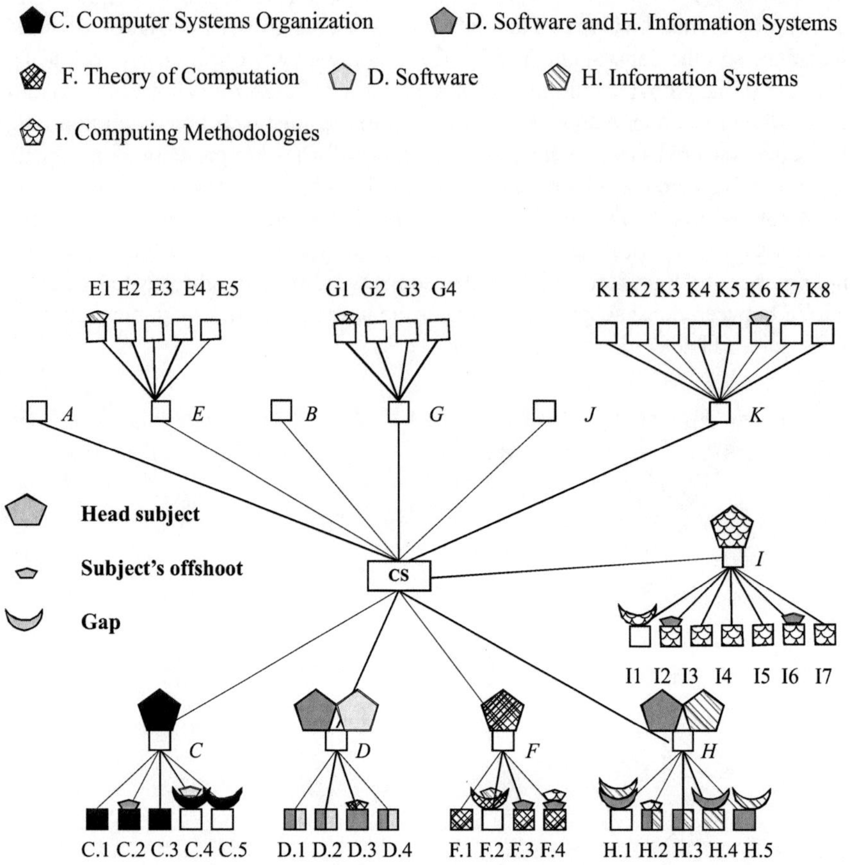

Fig. 14.4 Six clusters visualized on the ACM-CCS taxonomy by larger pentagrams of "head subjects" differently colored. Now we can see more structure in the organization's research as described in the text.

been found to relate to head subjects *'D.Software'* and *'H. Information Systems'* so that two of the clusters have only one of them as their respective head subject, whereas the third one received both of the nodes as its head subjects, thus suggesting an integrating development that has been underway in the department. This integrating development can be attributed to establishment in recent years of *'D.2 Software Engineering'* as a major subject in Computer Science which is yet to be reflected by raising the node in the ACM-CCS taxonomy.

Our method for finding clusters of research subjects according to the workings of the organization involves the following steps:

1. defining research units representing the activities;
2. determining research profiles of the units, that is, crisp or fuzzy sets of the taxonomy subjects to represent every unit;

3. integrating the profiles in a matrix of similarity scores between the taxonomy subjects which are worked on in the organization; and
4. finding clusters of taxonomy subjects representing the similarity matrix.

The cluster finding completes the first stage of the approach, (i) the organization data integration. The second stage, (ii) the ontology usage, works as follows: a thematic cluster found at stage (i) would be considered as a query to the ontology requesting to find a node or two in the taxonomy, the head subject(s), as high as possible, to cover all the nodes in the query in such a way that the gaps and offshoots emerging at the high rank head subject would be not too extensive, or too expensive in the penalty function defined for any set of head subjects. The third stage, (iii) the activity visualization, presents the results over the ontology in the manner of Fig. 14.4. The rules for interpretation of the results are yet to be produced, though some simple observations like those above can be recommended already.

This three-stage approach has been sketched out in our previous paper [33]. The current paper outlines the current state of the approach. Specifically, the following novel features are described here. First, for stage (i), instead of a crisp clustering approach, we developed a genuine fuzzy clustering method based on an additive model of the subject-to-subject similarity matrix and involving the spectral clustering approach [35]. Second, for stage (ii), instead of some informal considerations described in [33], we developed an optimal lifting method based on minimization of a penalty function embracing all the elements of the lift, the head subjects, the gaps and the offshoots [34]. Using the software developed, the visualization stage (iii), now can be conducted automatically, not manually, which raises some new possibilities in manipulating the relative weights of lifting elements.

14.1.2 Background

Since the stages of our approach involve relatively different techniques, the background will be described in the following subsections along the separate lines of development in the literature: Fuzzy clustering background; Ontology usage background; Activity visualization background.

14.1.2.1 Fuzzy Clustering Background

The major fuzzy clustering algorithms, such as c-means, an extension of the popular k-means approach, work on data in the entity-to-feature format [5]. Yet the result of our first stage is a square subject-to-subject similarity matrix. Thus, we are concerned here mainly with the so-called relational fuzzy clustering, an activity of deriving fuzzy clusters from a relation, that is, a matrix of a similarity or dissimilarity index. The published work on this can be divided in two major streams: one utilizing the fuzzy logics operations such as minimum or plus but no operation of division, and the other involving all the numeric operations, including division. The former is rather thin and less developed (see, for instance, [52] and [20]). We adhere to the latter stream, which can be traced to papers [40] and [49] that utilized, essentially, the sum $\sum_{k=1}^{K}\sum_{t,t'} u_{tk}^2 u_{t'k}^2 d(t,t')$ as the criterion to minimize over unknown membership vectors $\mathbf{u}_k = (u_{tk}), k = 1,...,K$ where t,t denote ontology subjects. A similar criterion, proven to be equivalent to the criterion of popular fuzzy c-means method [5], was utilized by Hathaway, Davenport and Bezdek [21] to derive their RFCM algorithm, that works in two-phase iterations similar to c-means, including a relational analogue to the concept of cluster centroid. Specifying the so-called "fuzzifying" constant at the level of 2, the RFCM criterion is the sum over $k = 1,...,K$ of items $\sum_{t,t'} u_{tk}^2 u_{t'k}^2 d(t,t')/\sum_t u_{tk}^2$ where $d(t,t')$ is the squared Euclidean distance at different d RFCM may lead to negative memberships. But even in this format, RFCM appears to be superior to Windham's assignment-prototype algorithm [4]. Later this restriction was relaxed, initially, by modifying RFCM into NERFCM algorithm to include the addition of a positive number to all off diagonal distances [22] and, more recently, by directly imposing the non-negativity constraint for memberships [9]. The latter paper also extended the concept of fuzzy clustering to include the so-called "noise" cluster to hold the bulk of membership for entities that are far away from the K clusters being built. Brouwer [6] makes use of a two-stage procedure in which the first stage supplies the entities with a few distance-approximating features so that the second stage utilizes a conventional algorithm such as fuzzy c-means for building fuzzy clusters in the feature space. This approach proved superior to the others in experiments reported in [6].

Yet there are a number of issues related to these approaches that are not quite satisfactory:

1. The cluster memberships form what is called a fuzzy partition so that each entity shares its full membership among the clusters. This does not allow an entity to belong to no cluster or partly belong to the clusters.
2. The clusters do not feed back to the similarity data - they do not represent the data as a function of clusters, which would be a desirable option when modelling research activities. A nice additive clustering model of similarity data has been introduced, in English, by Shepard and Arabie [44] for crisp clusters. The model proposes that each cluster is characterized by a positive constant, intensity, that adds up to the similarity between entities in the cluster. Paper [31] referred to earlier publications, in Russian, and proposed an iterative crisp cluster extraction framework in that setting. However, the additive clustering model had not been extended to relational fuzzy clustering until a simplified version of the model, involving constant, not cluster-specific, intensity weights, was considered in [41] citing no specific applications and using Newton's descent method for fitting the model. This method involves many initialization parameters that need to be prespecified, which is not what an innocent user would be willing to do. In this paper, we would like to use a proper extension of the additive model to fuzzy clustering such as introduced by Mirkin and Nascimento [35].

3. There is no explicit instruction for the choice of number of clusters in the fuzzy clustering models. Typically, the number of clusters is considered to be determined based on some post-processing considerations related to stability of the clusters regarding some random changes in either the initialization or the data themselves [29]. In this regard, the sequential manner of the method proposed in [35] can be considered an advantage because it allows choosing the option of stopping computations after any number of clusters.

14.1.2.2 Ontology Usage Background

The concept of ontology as a computationally feasible environment for knowledge maintenance has recently emerged to comprise, first of all, the set of concepts and relations between them pertaining to the knowledge of the domain. The initial attempts have been concentrated at automatically generating ontologies from web and other document resources; a review of the efforts to about 2000 can be found in [10]. Meanwhile, it became clear that currently a relevant ontology can be produced only manually, and big ontologies are being built, first of all, in medicine (see SNOMED [46]) and bioinformatics (GO [18]). Currently, most research efforts by computer scientists, beyond developing platforms and languages for ontology representation (see, for example, developments of OWL language (e.g. [39]), are concentrated on computational methods for (a) integrating ontologies and (b) using them for various purposes.

The issue of (a) integration of different ontologies requires developing a common ground for representing different elements of ontologies as well as methods for mapping the elements of the same type between ontologies. Examples of the former can be found in [48, 15]. Examples of the latter can be found in [7, 19]. Both of these are in rather initial states of development, which is supported by the findings of [19]: this shows that a simple text matching method outperforms those involving the ontologies.

The issue of (b) usage of ontologies is of especial importance to us because our paper should be counted in this category. Most efforts here are devoted to building rules for ontological reasoning and querying utilizing the inheritance relation supplied by the ontology's taxonomy in the presence of different data models ([8, 3, 47]). These do not attempt approximate representations but just utilize additional possibilities supplied by the ontology relations. Another type of ontology usage is in using its taxonomy nodes for interpretation of data mining results such as association rules [27] and clusters [11, 14]. Our approach naturally falls within this category. What we want is to generalize the query set within the taxonomy in a 'soft' manner by allowing some "non-costly" discrepancies between the set and the subtree rooted at the generalized concept, which differs from the other work on queries to ontologies that strictly conform to the crisp meanings [8, 3, 47].

14.1.2.3 Visualization Background

The subject of visualization attracts increasing attention of computer scientists. In this regard, usually the visualization of activities does not much differ from visualization of any other concept, of which many papers and websites inform (see, for a recent reference on visualization [28]). Some aspects of activities have been covered such as, for instance, web related activities [12], and, more recently, modelling activity in general is being considered as well [42, 17]. Our case, as illustrated in Fig.14.4, is very much clear cut: the organization's activity is represented by a set of clusters that are supplied to a taxonomy as query sets to be lifted to head subjects expressing the general tendencies of the activities, not without some gap or offshoot pitfalls. This visualization attends here to not just cognitive abilities of the user but to more tangible issues related to the analysis of matches and mismatches between the query and taxonomy, that could be interpreted, in respect, as points of strength or weakness, and give rise to questions of their meaning in the context of the taxonomy to be used in planning of the organization changes and investment policies. Potentially, after integration of activities of a number of organizations in the taxonomy, one could use the discrepancies to feedback to the taxonomy itself, as the points requiring a most urgent taxonomy updating.

14.2 Taxonomy-Based Profiles

In the case of investigation of activities of a university department or research center, a research team's profile can be defined as a fuzzy membership function on the set of leaf-nodes of the taxonomy under consideration so that the memberships reflect the extent of the team's effort put into corresponding research topics.

14.2.1 E-Screen Survey Tool

Fuzzy membership profiles are derived from either automatic analysis of documents posted on the web by the teams or by explicitly surveying the members of the department. The latter option is especially convenient in situations in which the web contents do not properly reflect the developments, for example, in non-English speaking countries with relatively underdeveloped internet infrastructures for the maintenance of research results. We have developed an interactive survey tool that provides two types of functionality: i) collection of data about ACM-CCS based research profiles of individual members; ii) statistical analysis and visualization of the data and results of the survey on the level of a department. The respondent is asked to select up to six topics among the leaf nodes of the ACM-CCS tree and assign each with a percentage expressing the proportion of the topic in the total of the respondent's research activity for, say, the past four years. This describes the respondent's activity fuzzy membership profile. Fig. 14.5 shows a screenshot of the baseline interface for a respondent who has chosen six ACM-CCS topics during his/her survey session.

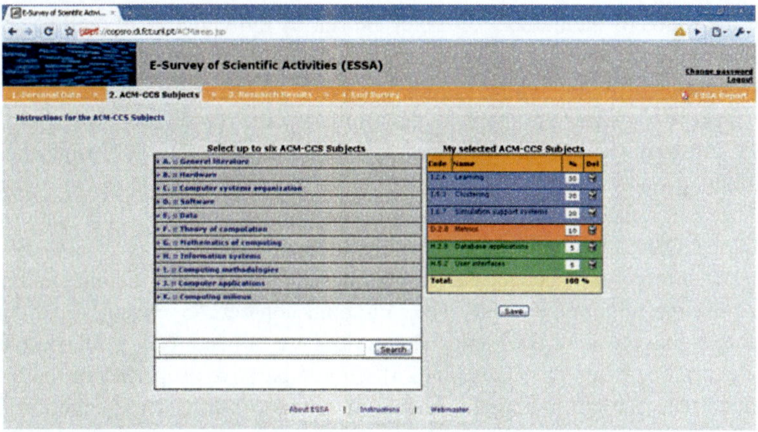

Fig. 14.5 Screenshot of the interface survey tool for selection of ACM-CCS topics.

The set of profiles supplied by respondents forms an $N \times M$ matrix F where N is the number of ACM-CCS topics involved in the profiles and M the number of respondents. Each column of F is a fuzzy membership function, rather sharply delineated because only six topics maximum may have acknowledged membership in each of the columns.

14.3 Representing Research Organization by Fuzzy Clusters of ACM-CCS Topics

14.3.1 Deriving Similarity between ACM-CCS Research Topics

We represent a research organization by clusters of ACM-CCS topics to reflect thematic communalities between activities of members or teams working on these topics. The clusters are found by analyzing similarities between topics according to their appearances in the profiles. The more profiles contain a pair of topics i and j and the greater the memberships of these topics, the greater is the similarity score for the pair.

Consider a set of V individuals ($v = 1, 2, \cdots, V$), engaged in research over some topics $t \in T$ where T is a pre-specified set of scientific subjects. The level of research effort by individual v in developing topic t is evaluated by the membership f_{tv} in profile f_v ($v = 1, 2, \cdots, V$).

Then the similarity $w_{tt'}$ between topics t and t' is defined as

$$w_{tt'} = \sum_{v=1}^{V} \frac{n_v}{n_{max}} f_{tv} f_{t'v}, \qquad (14.1)$$

where the ratios of the number of topics chosen by individual v, n_v, and n_{max}, the maximum n_v over all $v = 1, 2, \cdots, V$, are introduced to balance the scores of individuals bearing different numbers of topics.

To make the cluster structure in the similarity matrix sharper, we apply the spectral clustering approach to pre-process the similarity matrix W using the so-called Laplacian transformation [26]. First, an $N \times N$ diagonal matrix D is defined, with (t,t) entry equal to $d_t = \sum_{t' \in T} w_{tt'}$, the sum of t's row of W. Then unnormalized Laplacian and normalized Laplacian are defined by equations $L = D - W$ and $L_n = D^{-1/2} L D^{-1/2}$, respectively. Both matrices are semipositive definite and have zero as the minimum eigenvalue. The minimum non-zero eigenvalues and corresponding eigenvectors of the Laplacian matrices are utilized then as relaxations of combinatorial partition problems [45, 26]. Of comparative properties of these two normalizations, the normalized Laplacian, in general, is considered superior [26]. Since the additive clustering approach described in the next section relies on maximum rather than minimum eigenvalues, we use the Laplacian PseudoINverse transformation, Lapin for short, defined by

$$L_n^+(W) = \tilde{Z} \tilde{\Lambda}^{-1} \tilde{Z}'$$

where $\tilde{\Lambda}$ and $\tilde{Z}$ are defined by the spectral decomposition $L_n = Z \Lambda Z'$ of matrix $L_n = D^{-1/2}(D - W) D^{-1/2}$. To specify these matrices, first, set T' of indices of elements corresponding to non-zero elements of Λ is determined, after which the matrices are taken as $\tilde{\Lambda} = \Lambda(T', T')$ and $\tilde{Z} = Z(:, T')$. The choice of the Lapin transformation can be explained by the fact that it leaves the eigenvectors of L_n unchanged while inverting the non-zero eigenvalues $\lambda \neq 0$ to those $1/\lambda$ of L_n^+. Then the maximum eigenvalue of L_n^+ is the inverse of the minimum non-zero eigenvalue λ_1 of L_n, corresponding to the same eigenvector.

14.3.2 Fuzzy Additive-Spectral Clustering

In spite of the fact that many fuzzy clustering algorithms have been developed already [5, 25], most of them are ad hoc and, moreover, they all involve manually specified parameters such as the number of clusters or threshold of similarity without providing any guidance for choosing them. We apply a model-based approach of additive clustering, combined with the spectral clustering approach, to develop a fuzzy clustering method that is both adequate and supplied with model-based parameters helping to choose the right number of clusters.

Thematic similarities $a_{tt'}$ between topics are but manifested expressions of some hidden patterns within the organization which can be represented by fuzzy clusters in exactly the same manner as the manifested scores in the definition of the similarity $w_{tt'}$ (14.1). We propose to formalize a thematic fuzzy cluster as represented by two items: (i) a membership vector $\mathbf{u} = (u_t)$, $t \in T$, such that $0 \leq u_t \leq 1$ for all $t \in T$, and (ii) an intensity $\mu > 0$ that expresses the extent of significance of the pattern corresponding to the cluster, within the organization under consideration. With the introduction of the intensity, applied as a scaling factor to $\mathbf{u}$, it is the product $\mu\mathbf{u}$ that is a solution rather than its individual co-factors. Given a value of the product μu_t, it is impossible to tell which part of it is μ and which u_t. To resolve this, we follow a conventional scheme: let us constrain the scale of the membership vector $\mathbf{u}$ on a constant level, for example, by a condition such as $\sum_t u_t = 1$ or $\sum_t u_t^2 = 1$, then the remaining factor will define the value of μ. The latter normalization better suits the criterion implied by our fuzzy clustering method and, thus, is accepted further on.

Our additive fuzzy clustering model follows that of [44, 31, 41] and involves K fuzzy clusters that reproduce the pseudo-inverted Laplacian similarities $a_{tt'}$ up to additive errors according to the following equations [35]:

$$a_{tt'} = \sum_{k=1}^{K} \mu_k^2 u_{kt} u_{kt'} + e_{tt'}, \qquad (14.2)$$

where $\mathbf{u}_k = (u_{kt})$ is the membership vector of cluster k, and μ_k its intensity.

The item $\mu_k^2 u_{kt} u_{kt'}$ is the product of $\mu_k u_{kt}$ and $\mu_k u_{kt'}$ expressing participation of t and t', respectively, in cluster k. This value adds up to the others to form the similarity $a_{tt'}$ between topics t and t'. The value μ_k^2 summarizes the contribution of the intensity and will be referred to as the cluster's weight.

To fit the model in (14.2), we apply the least-squares approach, thus minimizing the sum of all $e_{tt'}^2$. Since A is definite semi-positive, its first K eigenvalues and corresponding eigenvectors form a solution to this if no constraints on vectors $\mathbf{u}_k$ are imposed. Additionally, we apply the one-by-one principal component analysis strategy for finding one cluster at a time this makes the computation feasible and is crucial for determining the number of clusters. Specifically, at each step, we consider the problem of minimization of a reduced to one fuzzy cluster least-squares criterion

$$E = \sum_{t,t' \in T} (b_{tt'} - \xi u_t u_{t'})^2 \qquad (14.3)$$

with respect to unknown positive ξ weight (so that the intensity μ is the square root of ξ) and fuzzy membership vector $\mathbf{u} = (u_t)$, given similarity matrix $B = (b_{tt'})$.

At the first step, B is taken to be equal to A. Each found cluster changes B by subtracting the contribution of the found cluster (which is additive according to model (14.2)), so that the residual similarity matrix for obtaining the next cluster is equal to $B - \mu^2 \mathbf{u}\mathbf{u}'$ where μ and $\mathbf{u}$ are the intensity and membership vector of the found cluster. In this way, A indeed is additively decomposed according to formula (14.2) and the number of clusters K can be determined in the process.

Let us specify an arbitrary membership vector $\mathbf{u}$ and find the value of ξ minimizing criterion (14.3) at this $\mathbf{u}$ by using the first-order optimality condition:

$$\xi = \frac{\sum_{t,t' \in T} b_{tt'} u_t u_{t'}}{\sum_{t \in T} u_t^2 \sum_{t' \in T} u_{t'}^2},$$

so that the optimal ξ is

$$\xi = \frac{\mathbf{u}'B\mathbf{u}}{(\mathbf{u}'\mathbf{u})^2} \tag{14.4}$$

which is obviously non-negative if B is semi-positive definite.

By putting this ξ in equation (14.3), we arrive at

$$E = \sum_{t,t' \in T} b_{tt'}^2 - \xi^2 \sum_{t \in T} u_t^2 \sum_{t' \in T} u_{t'}^2 = S(B) - \xi^2 (\mathbf{u}'\mathbf{u})^2,$$

where $S(B) = \sum_{t,t' \in T} b_{tt'}^2$ is the similarity data scatter.

Let us denote the last item by

$$G(\mathbf{u}) = \xi^2 (\mathbf{u}'\mathbf{u})^2 = \left(\frac{\mathbf{u}'B\mathbf{u}}{\mathbf{u}'\mathbf{u}}\right)^2, \tag{14.5}$$

so that the similarity data scatter is the sum:

$$S(B) = G(\mathbf{u}) + E \tag{14.6}$$

of two parts, $G(\mathbf{u})$, which is explained by cluster $(\mu, \mathbf{u})$, and E, which remains unexplained.

An optimal cluster, according to (14.6), is to maximize the explained part $G(\mathbf{u})$ in (14.5) or its square root

$$g(\mathbf{u}) = \xi \mathbf{u}'\mathbf{u} = \frac{\mathbf{u}'B\mathbf{u}}{\mathbf{u}'\mathbf{u}}, \tag{14.7}$$

which is the celebrated Rayleigh quotient: its maximum value is the maximum eigenvalue of matrix B, which is reached at its corresponding eigenvector, in the unconstrained problem.

This shows that the spectral clustering approach is appropriate for our problem. According to this approach, one should find the maximum eigenvalue λ and corresponding normed eigenvector z for B, $[\lambda, z] = \Lambda(B)$, and take its projection to the set of admissible fuzzy membership vectors.

Our clustering approach involves a number of model-based criteria for halting the process of sequential extraction of fuzzy clusters. The process stops if either is true:

1. The optimal value of ξ (14.4) for the spectral fuzzy cluster becomes negative.
2. The contribution of a single extracted cluster to the data scatter becomes too low, less than a pre-specified $\tau > 0$ value.
3. The residual data scatter becomes smaller than a pre-specified ε value, say less than 5% of the original similarity data scatter.

The described one-by-one Fuzzy ADDItive-Spectral thematic cluster extraction algorithm is referred to as FADDI-S in [35]. It combines three different approaches: additive clustering [44, 31, 41], spectral clustering [45, 26, 55], and relational fuzzy clustering [5, 6] and adds an edge to each. In the context of additive clustering, fuzzy approaches were considered only by [41], yet in a very restricted setting: (a) the clusters intensities were assumed constant there, (b) the number of clusters was pre-specified, and (c) the fitting method was very local and computationally intensive - these all restrictions are overcome in FADDI-S. The spectral clustering approach is overtly heuristic, whereas FADDI-S is model-based. The criteria used in relational fuzzy clustering are ad hoc whereas that of FADDI-S is model-based, and, moreover, its combined belongingness function values $\mu\mathbf{u}$ are not constrained by unity as is the case in relational clustering, but rather follow the scales of the similarity relation under investigation, which is in line with the original approach by L. Zadeh [54].

14.3.3 Experimental Verification of FADDI-S

We describe here results of two of the experiments reported in [35].

14.3.3.1 Fuzzy Clustering Affinity Data

The affinity data is a relational similarity data obtained from a feature based dataset using a semi-positive definite kernel, usually the Gaussian one. Specifically, given an $N \times V$ matrix $Y = (y_{tv})$, $t \in T$ and $v = 1,2,...,V$, non-diagonal elements of the similarity matrix W are defined by equation

$$w_{tt'} = exp(-\frac{\sum_{v=1}^{V}(y_{tv} - y_{t'v})^2}{2\sigma^2}),$$

with the diagonal elements made equal to zero, starting from founding papers [45, 38]. The value $ss = 2\sigma^2$ is a user-defined parameter, that is pre-specified to make the resulting similarities $w_{tt'}$ spread over interval [0,1].

To compare our approach with other methods for fuzzy clustering of affinity data, we pick up an example from a recent paper by Brouwer [6]. This example concerns a two-dimensional data set, that we refer to as Bivariate4, comprising four clusters generated from bivariate spherical normal distributions with the same standard deviation 950 at centers (1000, 1000), (1000,4000), (4000, 1000), and (4000, 4000), respectively. The data forms a cloud presented in Fig. 14.6.

This data was analyzed in [6] by using the matrix D of Euclidean distances between the generated points. Five different fuzzy clustering methods have been compared, three of them relational, by Roubens [40], Windham [49] and NERFCFM [22], and two of fuzzy c-means (FCM) with different preliminary pre-processing options of the similarity data into the entity-to-feature format, FastMap and SMACOF [6]. Of these five different fuzzy clustering methods, by far the best results have been obtained with method FCM applied to a five-feature set extracted from D with FastMap method [6]. The Adjusted Rand index [24] of the correspondence between the generated clusters and those found with the FCM over FastMap method is equal on average, of 10 trials, 0.67 (no standard deviation is reported in [6]).

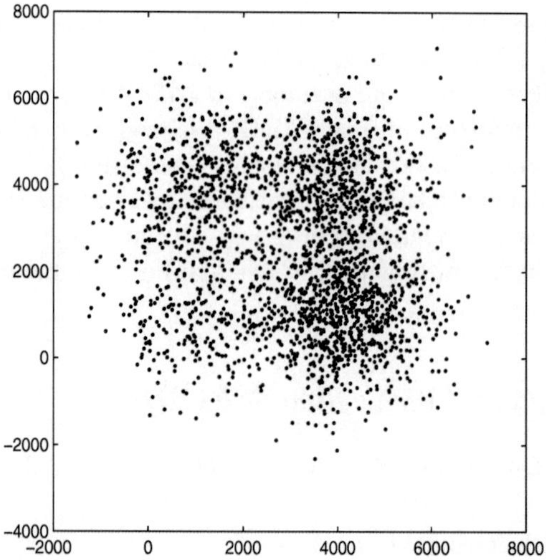

Fig. 14.6 Bivariate4: the data of four bivariate clusters generated from Gaussian distributions according to [6].

To compare FADDI-S with these, we apply Gaussian kernel to the data generated according to the Bivariate4 scheme and pre-processed by the z-score standardization so that similarities, after z-scoring, are defined as $a_{ij} = exp(-d^2(y_i, y_j)/0.5)$ where d is Euclidean distance. This matrix then is Lapin transformed to the matrix W to which FADDI-S is applied.

To be able to perform the computation using a PC MatLab, we reduce the respective sizes of the clusters, 500, 1000, 2000, and 1500 totaling to 5000 entities altogether in [6], tenfold to 50, 100, 200 and 150 totaling to 500 entities. The issue is of doing a full spectral analysis of the square similarity matrices of the entity set sizes, which we failed to do with our PC MatLab versions at a 5000 strong dataset. We also experimented with fivefold and twofold size reductions. This should not much change the results because of the properties of smoothness of the spectral decompositions [23].

Indeed, one may look at a 5000 strong random sample as a combination of two 2500 strong random samples from the same population. Consider a randomly generated $N \times 2$ data matrix X of N bivariate rows, thus leading to Lapin transformed $N \times N$ similarity matrix W. If one doubles the data matrix by replicating X as $XX = [X;X]$, in MatLab notation, which is just a $2N \times 2$ data matrix consisting of a replica of X under X, then its Lapin transformed similarity matrix will be obviously equal to

$$WW = \begin{bmatrix} W & W \\ W & W \end{bmatrix}$$

whose eigenvectors are just doubles (z,z) of eigenvectors z of W. If the second part of the double data matrix XX slightly differs from X, due to sampling errors, then the corresponding parts of the doubled similarity matrix and eigenvectors also will slightly differ from those of WW and (z,z). Therefore, the property of stability of spectral clustering results [23] will hold for thus changed parts. This argument equally applies to the case when the original sample is supplemented by four or nine samples from the same population.

In our computations, five consecutive FADDI-S clusters have been extracted for each of randomly generated ten Bivariate4 datasets. The very first cluster has been discarded as reflecting just the general connectivity information, and the remaining four were defuzzified into partitions so that every entity is assigned to its maximum membership class. The average values of the Adjusted Rand index, along with the standard deviations at Bivariate4 dataset versions of 500, 1000, and 2500 generated bivariate points are presented in Table 14.1 for FADDI-S. The results support our view that the data set size is not important if the proportions of the cluster structure are maintained. According to the table, FADDI-S method achieves better results than the ones obtained by the five fuzzy clustering methods reported in [6].

Table 14.1 Adjusted Rand Index values for FADDI-S at different sizes of Bivariate4 dataset

Size	Adjusted Rand Index	Standard deviation
500	0.70	0.04
1000	0.70	0.03
2500	0.73	0.01

A remark. The entity-to-feature format of the Bivariate4 data suggests that relational cluster analysis is not necessarily the best way to analyze it; a genuine data clustering method such as K-Means may bring better results. Indeed, an application of the "intelligent" K-Means method from [30] to the original data size of $N = 5000$ has brought results with the average adjusted Rand index of 0.75 (the standard deviation 0.045), which is both higher and more consistent than the relational methods applied here and in [6].

14.3.3.2 Finding Community Structure

The research in finding community structure in ordinary graphs has been a subject of intense research (see, for example, [37, 36, 50, 26]). The graph with a set of vertices T is represented by the similarity matrix $A = (a_{tt'})$ between graph vertices such that $a_{tt'} = 1$ if t and t' are connected by an edge, and $a_{tt'} = 0$, otherwise. Then matrix A is symmetrized by the transformation $(A + A')/2$ after which all diagonal elements are made zero, $a_{tt} = 0$ for all $t \in T$. We assume that the graph is connected; otherwise, its connected components are to be treated separately.

The spectral relaxation involves subtraction of the "background" random interactions from similarity matrix $A = (a_{tt'})$. The random interactions are defined with the same within-row summary values $d_t = \sum_{t' \in T} a_{tt'}$ as those used in Laplace matrices. The random interaction between t and t' is defined as the product $d_t d_{t'}$ divided by the total number of edges [36]. The modularity criterion is defined as a usual, non-normalized cut, that is the summary similarity between clusters to be minimized, with thus transformed similarity data [36]. The modularity criterion has proven good in crisp clustering. This approach was extended to fuzzy clustering in the space of the first eigenvectors in [55].

Our approach allows for a straightforward application of FADDI-S algorithm to the network similarity matrix A. It also involves a transformation of the similarity data which is akin to the subtraction of background interactions in the modularity criterion [36]. Indeed we find initially the eigenvector z_1 corresponding to the maximum eigenvalue λ_1 of A itself. As is well known, this vector is positive because the graph is connected. Thus z_1 forms a fuzzy cluster itself, because it is conventionally normed. We do not count it as part of the cluster solution, though, because it expresses just the fact that all the entities are part of the same network. Thus, we proceed to the residual matrix with elements $a_{tt'} - \lambda_1 z_{1t} z_{1t'}$. We expect the matrix A to be rather "thin" with respect to the number of positive eigenvalues, which should allow for a natural halting the cluster extracting process when there are no positive eigenvalues at the residual matrix W.

We apply the FADDI-S algorithm to Zachary karate club network data, which serves as a prime test bench for community finding algorithms. This ordinary graph consists of 34 vertices, corresponding to members of the club and 78 edges between them - the data and references can be found, for example, in [37, 55]. The members of the club are divided according to their loyalties toward the club's two prominent individuals: the administrator and instructor. Thus the network is claimed to consist of two communities, with 18 and 16 differently loyal members respectively.

Applied to this data, FADDI-S leads to three fuzzy clusters to be taken into account. Indeed, the fourth cluster accounts for just 2.4% of the data scatter, which is less than the inverse of the number of entities $\tau = 1/34$, reasonably suggested as a natural threshold value. Some characteristics of the found solution are presented in Table 14.2. All the membership values of the first cluster are positive - as mentioned above, this is just the first eigenvector; the positivity means that the network is well connected. The second and third FADDI-S clusters match the claimed structure of the network: they have 16 and 18 positive components, respectively, corresponding to the two observed groupings.

Table 14.2 Characteristics of Karate club clusters found with FADDI-S.

Cluster	Contribution, %	λ_1	Weight	Intensity
I	29.00	3.36	3.36	1.83
II	4.34	2.49	1.30	1.14
III	4.19	2.00	0.97	0.98

Let us compare our results with those of a recent spectral fuzzy clustering method developed in [55]. The latter method finds three fuzzy clusters, two of them representing the groupings, though with a substantial overlap between them, and the third, smaller cluster consisting of members 5,6,7,11,17 of just one of the groupings – see [55], p. 487. We think that this latter cluster may have come up from an eigenvector embracing the members with the largest numbers of connections in the network. It seems for certain that FADDI-S outperforms the method of [55] on Zachary club data.

14.4 Parsimonious Lifting Method

To generalize the contents of a thematic cluster, we propose a method for lifting it to higher ranks of the taxonomy so that if all or almost all children of a node in an upper layer belong to the cluster, then the node itself is taken to represent the cluster at this higher level of the ACM-CCS taxonomy (see Fig. 14.7). Depending on the extent of inconsistency between the cluster and the taxonomy, such lifting can be done differently, leading to different portrayals of the cluster on ACM-CCS tree depending on the relative weights of the events taken into account. A major event is the so-called "head subject", a taxonomy node covering (some of) leaves belonging to the cluster, so that the cluster is represented by a set of head subjects. The penalty of the representation to be minimized is proportional to the number of head subjects so that the smaller that number the better. Yet the head subjects cannot be lifted too high in the tree because of the penalties for associated events, the cluster "gaps" and "offshoots", where their number depends on the extent of inconsistency of the cluster versus the taxonomy.

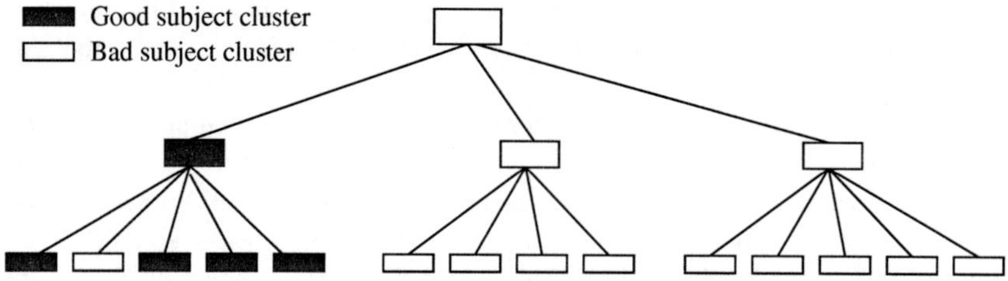

Fig. 14.7 Two clusters of second-layer topics, presented with checked and diagonal-lined boxes, respectively. The checked box cluster fits within one first-level category (with one gap only), whereas the diagonal line box cluster is dispersed among two categories on the right. The former fits the classification well; the latter does not.

The gaps are head subject's children topics that are not included in the cluster. An offshoot is a taxonomy leaf node that is a head subject (not lifted). It is not difficult to see that the gaps and offshoots are determined by the head subjects specified in a lifting (see Fig. 14.8).

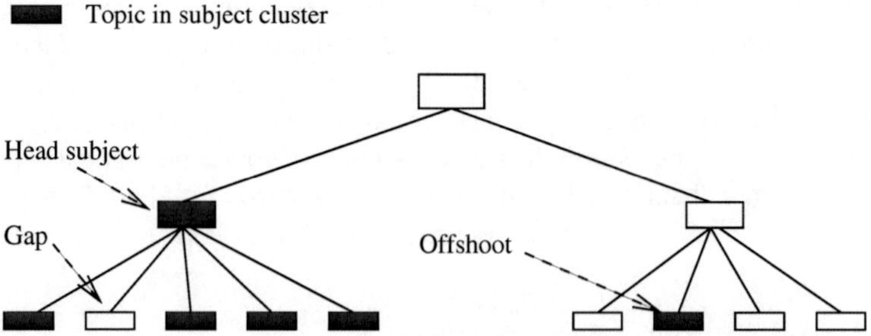

Fig. 14.8 Three types of features in lifting a subject cluster within taxonomy.

The total count of head subjects, gaps and offshoots, each weighted by both the penalties and leaf memberships, is used for scoring the extent of the cluster misfit needed for lifting a grouping of research topics over the classification tree. The smaller the score, the more parsimonious the lifting and the better the fit. Depending on the relative weighting of gaps, offshoots and multiple head subjects, different liftings can minimize the total misfit, as illustrated in Fig. 14.10 and Fig. 14.11 later.

Altogether, the set of topic clusters together with their optimal head subjects, offshoots and gaps constitute a parsimonious representation of the organization. Such a representation can be easily accessed and expressed. It can be further elaborated by highlighting those subjects in which members of the organization have been especially successful (i.e., publication in best journals or awards) or distinguished by a special feature (i.e., industrial use or inclusion in a teaching program). Multiple head subjects and offshoots, when they persist at subject clusters in different organizations, may show some tendencies in the development of the science, that the classification has not taken into account yet.

A parsimonious lift of a subject cluster can be achieved by recursively building a parsimonious representation for each node of the ACM-CCS tree based on parsimonious representations for its children as described in [34]. In this, we assume that any head subject is automatically present at each of the nodes it covers, unless they are gaps (as presented in Fig. 14.8). Our algorithm is set as a recursive procedure over the tree starting at leaf nodes.

The procedure [34] determines, at each node of the tree, sets of head subject gain and gap events to iteratively raise them to those of the parents, under each of two different assumptions that specify the situation at the parental node. One assumption is that the head subject has been inherited at the parental node from its own parent, and the second assumption is that it has not been inherited but gained in the node only. In the latter case the parental node is labeled as a head subject. Consider the parent-children system as shown in Fig. 14.9, with each node assigned with sets of gap and head subject gain events under the two inheritance of head subject assumptions.

Let us denote the total penalty, to be minimized, under the inheritance and non-inheritance assumptions by p_i and p_n, respectively. A lifting result at a given node is defined by a pair of sets (H, G), representing the tree nodes at which events of head subject gains and gaps, respectively, have occurred in the subtree rooted at the node. We use (H_i, G_i) and (H_n, G_n) to denote lifting results under the inheritance and non-inheritance assumptions, respectively. The algorithm computes parsimonious representations for parental nodes according to the topology of the tree, proceeding from the leaves to the root in the manner which is similar to that described in [32] for a mathematical problem in bioinformatics.

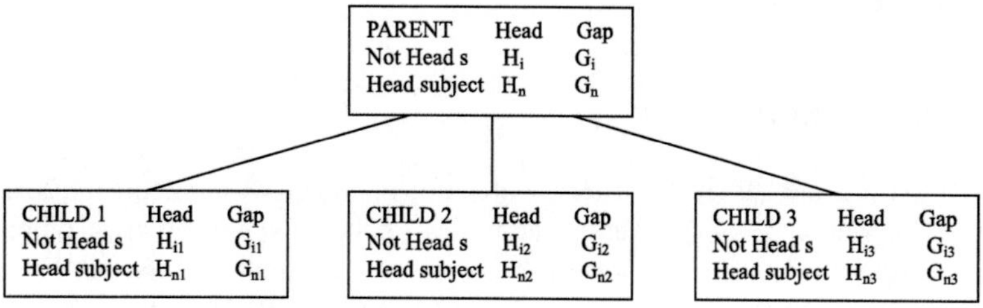

Fig. 14.9 Events in a parent-children system according to a parsimonious lift scenario.

We present here only a version of the algorithm for crisp clusters obtained by a defuzzification step. Given a crisp topic cluster S, and penalties h, o and g for being a head subject, offshoot and gap, respectively, the algorithm is initialized as follows.

At each leaf l of the tree, either $H_n = \{l\}$, if $l \in S$, or $G_i = \{l\}$, otherwise. The other three sets are empty. The penalties associated are $p_i = 0$, $p_n = o$ if H_n is not empty, that is, if $l \in S$, and $p_i = g$, $p_n = 0$, otherwise. This is obviously a parsimonious arrangement at the leaf level.

The recursive step applies to any node t whose children $v \in V$ have been assigned with the two couples of H and G sets already (see Fig. 14.9 at which V consists of three children): $(H_i(v), L_i(v);\ H_n(v), L_n(v))$ along with associated penalties $p_i(v)$ and $p_n(v)$.

(I) Deriving the pair $H_i(t)$ and $G_i(t)$, under the inheritance assumption, the one of the following two cases is to be chosen depending on the cost:

(a) The head subject has been lost at t, so that $H_i(t) = \cup_{v \in V} H_n(v)$ and $G_i(t) = \cup_{v \in V} G_n(v) \cup \{t\}$. (Note different indexes, i and n in the latter expression.) The penalty in this case is $p_i = \Sigma_{v \in V}\, p_n(v) + g$;

or

(b) The head subject has not been lost at t, so that $H_i(t) = \emptyset$ (under the assumption that no gain can happen after a loss) and $G_i = \cup_{v \in V} G_i(v)$ with $p_i = \Sigma_{v \in V}\, p_i(v)$.

The case that corresponds to the minimum of the two p_i values is returned then.

(II) Deriving the pair $H_n(t)$ and $G_n(t)$, under the non-inheritance assumption, the one of the following two cases is to be chosen that minimizes the penalty p_n:

(a) The head subject has been gained at t, so that $H_n(t) = \cup_{v \in V} H_i(v) \cup \{t\}$ and $G_n(t) = \cup_{v \in V} G_i(s)$ with $p_n = \Sigma_{v \in V} p_i(v) + h$;

or (b) The head subject has not been gained at t, so that $H_n(t) = \cup_{v \in V} H_n(v)$ and $G_n = \cup_{v \in V} G_n(v)$ with $p_n = \Sigma_{v \in V} p_n(v)$.

After all tree nodes t have been assigned with the two pairs of sets, accept the H_n, L_n and p_n at the root. This gives a full account of the events in the tree.

This algorithm leads indeed to an optimal representation; its extension to a fuzzy cluster is achieved through using the cluster memberships in computing the penalty values at tree nodes [34].

14.5 Case Study

In order to illustrate our cluster-lift&visualization approach we are going to use data from two surveys of research activities conducted in two Computer Science organizations: (A) the research Centre of Artificial Intelligence, Faculty of Science & Technology, New University of Lisboa and (B) Department of Computer Science and Information Systems, Birkbeck, University of London. The ESSA survey tool was applied for data collection and maintenance (see Sect. 14.2.1).

Because one of the organizations, A, is a research center whereas the other, B, is a university department, one should expect that the total number of research topics in A is smaller than that in B, and, similarly, the number of clusters in A should be less than that in B. In fact, research centers are usually created for a limited set of research goals, whereas university departments must cover a wide range of topics in teaching, which relates to research efforts. These appear to be true: the number of ACM-CCS third layer topics scored in A is 46 (out of 318) versus 54 in B. With the algorithm FADDI-S applied to the 46×46 and 54×54 topic-to-topic similarity matrices (see equation (14.1)), two fuzzy clusters (in case of center A) and four fuzzy clusters (in case of department B) have been sequentially extracted, after which the residual similarity matrix has become definite negative (stopping condition (1) of FADDI-S algorithm).

Let us focus our attention on the analysis of department B's research activities. On the clustering stage, as a result of the FADDI-S algorithm, four fuzzy clusters are obtained which are presented in Tables 14.3 and 14.4. Each of the topics in the tables is denoted by its ACM-CCS code and the corresponding string. The topics are sorted in the descending order of their cluster membership values (left columns of Tables 14.3 and 14.4). For each cluster, it is also presented its contribution to the data scatter, $G(\mathbf{u})$ (equation (14.5)), its intensity μ, and its weight ξ (equation (14.4)). Notice that the sum of clusters' contributions total to about 60%, which is a good result for clustering [2].

On the lifting stage, each of the found four clusters is mapped to and lifted in the ACM-CCS tree by applying the parsimonious lifting method with penalties for

[2] A 50% sum of clusters' contributions was obtained in the case of center A.

Table 14.3 Two clusters of research topics in department B

Cluster 1		
Contribution	26.7%	
Eigenvalue	37.44	
Intensity	5.26	
Weight	27.68	
Membership	**Code**	**Topic**
0.43055	K.2	HISTORY OF COMPUTING
0.39255	D.2.11	Software Architectures
0.35207	C.2.4	Distributed Systems
0.3412	I.2.11	Distributed Artificial Intelligence
0.3335	K.7.3	Testing, Certification, and Licensing
0.30491	D.2.1	Requirements/Specifications in D.2 Software Engineering
0.27437	D.2.2	Design Tools and Techniques in D.2 Software Engineering
0.24126	C.3	SPECIAL-PURPOSE AND APPLICATION-BASED SYSTEMS
0.19525	D.1.6	Logic Programming
0.19525	D.2.7	Distribution, Maintenance, and Enhancement in D.2 Software Engineering
Cluster 2		
Contribution	13.4%	
Eigenvalue	26.65	
Intensity	4.43	
Weight	19.60	
Membership	**Code**	**Topic**
0.66114	J.1	ADMINISTRATIVE DATA PROCESSING
0.29567	K.6.1	Project and People Management in K.6
0.29567	K.6.0	General in K.6 MANAGEMENT OF COMPUTING AND INF. SYSTEMS
0.29567	H.4.m	Miscellaneous in H.4 INF. SYSTEMS APPLICATIONS
0.29567	J.7	COMPUTERS IN OTHER SYSTEMS
0.2696	J.4	SOCIAL AND BEHAVIORAL SCIENCES
0.16271	J.3	LIFE AND MEDICAL SCIENCES
0.14985	G.2.2	Graph Theory
0.14593	I.5.3	Clustering
0.12307	I.6.4	Model Validation and Analysis
0.10485	I.6.5	Model Development

"head subjects" (h), "offshoots" (o) and "gaps" (g) of: $h = 1$, $o = 0.8$, and $g = 0.15$. We have chosen the gap penalty value considering that the numbers of children in ACM-CCS are typically around 10 so that two children belonging in the query would not be lifted to the parental node because the total gap penalty 8*0.15=1.2 would be greater than the decrease of head subject penalty 2-1=1. Yet if 3 of the children belong to the query, then it would be better to lift them to the parental node because the total gap penalty in this case, 7*0.15=1.05 would be smaller than the decrease of head subject penalty 3-1=2.

The parsimonious representation of the clusters in terms of the "head subjects", "offshoots", and "gaps" are described in Tables 14.5-14.8. Specifically, cluster 1 has

Table 14.4 Two other clusters of research topics in department B

Cluster 3		
Contribution	18.9%	
Eigenvalue	24.31	
Intensity	4.83	
Weight	23.31	
Membership	**Code**	**Topic**
0.613	E.2	DATA STORAGE REPRESENTATIONS
0.55728	I.0	GENERAL in I. Computing Methodologies
0.55728	H.0	GENERAL in H. Information Systems
Cluster 4		
Contribution	3.7%	
Eigenvalue	19.05	
Intensity	3.20	
Weight	10.26	
Membership	**Code**	**Topic**
0.35713	I.2.4	Knowledge Representation Formalisms and Methods
0.35636	F.4.1	Mathematical Logic
0.29495	F.2.0	General in F.2 ANAL. OF ALGORITHMS AND PROBLEM COMPLEXITY
0.28713	I.5.0	General in I.5 PATTERN RECOGNITION
0.28169	I.2.6	Learning
0.25649	K.3.1	Computer Uses in Education
0.24848	I.4.0	General in I.4 IMAGE PROCESSING AND COMPUTER VISION
0.24083	F.4.0	General in F.4 MATHEMATICAL LOGIC AND FORMAL LANGUAGES
0.18644	H.2.8	Database Applications
0.17707	H.2.1	Logical Design
0.17029	I.2.3	Deduction and Theorem Proving
0.15727	E.1	DATA STRUCTURES
0.15306	I.5.3	Clustering
0.14976	F.2.2	Nonnumerical Algorithms and Problems
0.14809	I.2.8	Problem Solving, Control Methods, and Search
0.14809	I.2.0	General in I.2 ARTIFICIAL INTELLIGENCE

as "head subject" *'D.2 SOFTWARE ENGINEERING'* with "offshoots" including *'C.2.4 Distributed Systems'*, *'D.1.6 Logic Programming'* and *'I.2.11 Distributed Artificial Intelligence'*. Cluster 2 is of *'J. Computer Applications'* with "offshoots" including *'G.2.2 Graph Theory'*, *'I.5.3 Clustering'*, *'K.6.0 General in K.6 - MANAGEMENT OF COMPUTING AND INFORMATION SYSTEMS'*, *'K.6.1 Project and People Management'*. Cluster 3 is described by the subjects (not lifted) *'E.2 DATA STORAGE REPRESENTATIONS'*, *'H.0 GENERAL in H. - Information Systems'*, *'I.0 GENERAL in I. - Computing Methodologies'*. Finally, cluster 4, with a more broad representation, has as "head subject" *'F. Theory of Computation'*, *'I.2 ARTIFICIAL INTELLIGENCE'*, and *'I.5 PATTERN RECOGNITION'*; its "offshoots" include *'E.1 DATA STRUCTURES'*, *'H.2.8 Database Applications'*, *'J.3 LIFE AND MEDICAL SCIENCES'* and *'K.3.1 Computer Uses in Education'*.

Let us illustrate the influence of the penalty parameters, more specifically the cost of gaps, g, on the parsimonious representation of cluster's research activities. Consider the scenario represented in Fig. 14.10 resulting from the lifting method with penalties of $h = 1$, $o = 0.8$, and $g = 0.3$. Due to the value of the gap penalty the cluster's topics (see Table 14.3) hold on as "leaf head subjects" as they are stated in the initialization of the lifting algorithm, being not lifted to higher ranks of the taxonomy (which would imply the appearance of some gaps). However, when decreasing the gap penalty from $g = 0.3$ to $g = 0.15$, it would lead to a different parsimonious generalization with subjects D.2.1, D.2.2, D.2.7 and D.2.11 generalized to "head subject" D.2, and the consequent assignment of the other subjects as "offshoots", as well as the occurrence of a set of gaps (i.e. the children of D.2 not present in the cluster). This scenario, described in Table 14.5, is visualized in Fig. 14.11.

Table 14.5 Parsimonious representation of department B cluster 1

	HEAD SUBJECT
D.2	SOFTWARE ENGINEERING
	OFFSHOTS
C.2.4	Distributed Systems
C.3	SPECIAL-PURPOSE AND APPLICATION-BASED SYSTEMS
D.1.6	Logic Programming
I.2.11	Distributed Artificial Intelligence
K.2	HISTORY OF COMPUTING
K.7.3	Testing, Certification, and Licensing
	GAPS
D.2.0	General in D.2 - SOFTWARE ENGINEERING
D.2.3	Coding Tools and Techniques
D.2.4	Software/Program Verification
D.2.5	Testing and Debugging
D.2.6	Programming Environments
D.2.8	Metrics
D.2.9	Management
D.2.10	Design
D.2.12	Interoperability
D.2.13	Reusable Software
D.2.m	Miscellaneous in D.2 - SOFTWARE ENGINEERING

Table 14.6 Parsimonious representation of department B cluster 2

	HEAD SUBJECT
J.	Computer Applications
	OFFSHOTS
G.2.2	Graph Theory
H.4.m	Miscellaneous in H.4 - INFORMATION SYSTEMS APPLICATIONS
I.5.3	Clustering
I.6.4	Model Validation and Analysis
I.6.5	Model Development
K.6.0	General in K.6 - MANAGEMENT OF COMPUTING AND INFORMATION SYSTEMS
K.6.1	Project and People Management
	GAPS
J.0	GENERAL in J. - Computer Applications
J.2	PHYSICAL SCIENCES AND ENGINEERING
J.5	ARTS AND HUMANITIES
J.6	COMPUTER-AIDED ENGINEERING
J.m	MISCELLANEOUS in J. - Computer Applications

Table 14.7 Parsimonious representation of department B cluster 3

	SUBJECTS
E.2	DATA STORAGE REPRESENTATIONS
H.0	GENERAL in H. - Information Systems
I.0 -	GENERAL in I. - Computing Methodologies

Additionally, Fig. 14.11 illustrates the present visualization stage of our approach. Each cluster is individually visualized on the ACM-CCS subtree that covers the clusters' topics, represented as a tree plot with nodes labeled with the corresponding ACM-CCS subjects's code. The "head subjects", "gaps" and "offshoots" are marked with distinct graphical symbols: black circle for "head subjects" (or leaf head subjects), open circle for "gaps", and dark grey square in case of "offshoots". Also, the children of an "head subjects" that were "head subjects" before the current lifting stage are marked with grey circle.

A similar analysis had been performed concerning the representation of research activities in center A. The parsimonious representations of the two clusters found correspond to cluster 1 having as "head subject" *'H. Information Systems'* and *'I.5*

Table 14.8 Parsimonious representation of department B cluster 4

	HEAD SUBJECTS
F.	Theory of Computation
I.2	ARTIFICIAL INTELLIGENCE
I.5	PATTERN RECOGNITION
	OFFSHOTS
D.2.8	Metrics
E.1	DATA STRUCTURES
G.2.2	Graph Theory
H.2.1	Logical Design
H.2.8	Database Applications
I.4.0	General in I.4 - IMAGE PROCESSING AND COMPUTER VISION
J.3	LIFE AND MEDICAL SCIENCES
K.3.1	Computer Uses in Education
	GAPS
F.0	GENERAL in F. - Theory of Computation
F.1	COMPUTATION BY ABSTRACT DEVICES
F.2.1	Numerical Algorithms and Problems
F.2.3	Tradeoffs between Complexity Measures
F.2.m	Miscellaneous in F.2 - ANAL. OF ALGORITHMS AND PROBLEM COMPLEXITY
F.3	LOGICS AND MEANINGS OF PROGRAMS
F.4.2	Grammars and Other Rewriting Systems
F.4.3	Formal Languages
F.4.m	Miscellaneous in F.4 - MATHEMATICAL LOGIC AND FORMAL LANGUAGES
F.m	MISCELLANEOUS in F. - Theory of Computation
I.2.1	Applications and Expert Systems
I.2.2	Automatic Programming
I.2.5	Programming Languages and Software
I.2.7	Natural Language Processing
I.2.9	Robotics
I.2.10	Vision and Scene Understanding
I.2.11	Distributed Artificial Intelligence
I.2.m	Miscellaneous in I.2 - ARTIFICIAL INTELLIGENCE
I.5.1	Models
I.5.4	Applications
I.5.5	Implementation
I.5.m	Miscellaneous in I.5 - PATTERN RECOGNITION

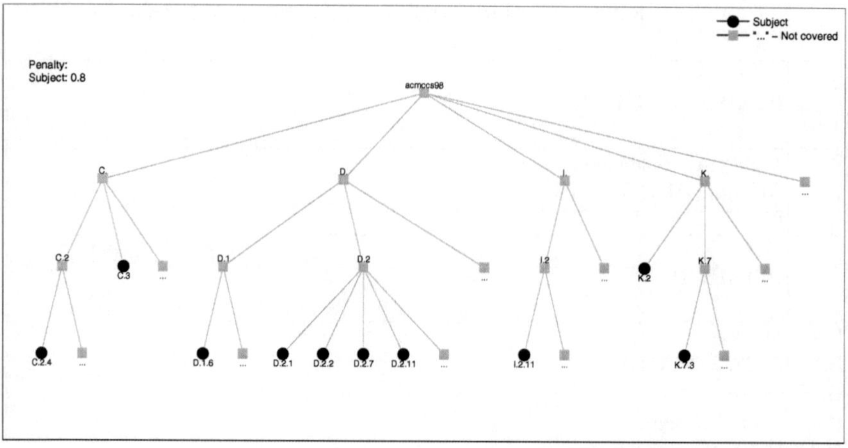

Fig. 14.10 Parsimonious representation lift of department B cluster 1 within the ACM-CCS tree with penalties of $h = 1$, $o = 0.8$, and $g = 0.3$

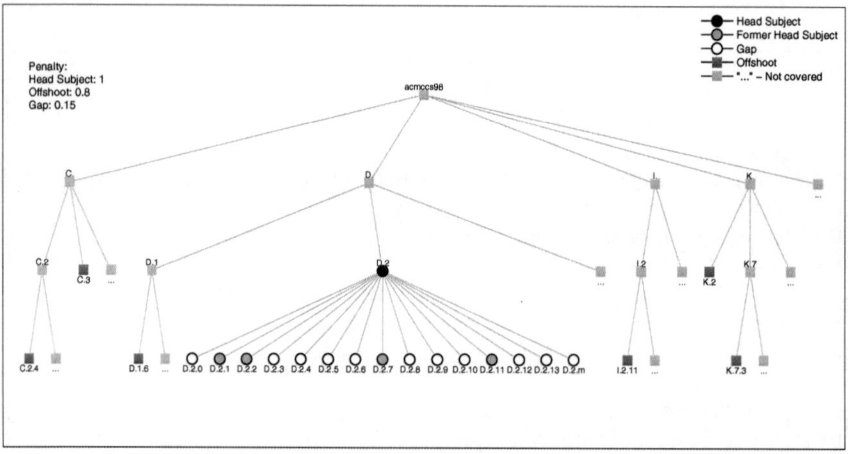

Fig. 14.11 Parsimonious representation lift of department B cluster 1 within the ACM-CCS tree with penalties of $h = 1$, $o = 0.8$, and $g = 0.15$

PATTERN RECOGNITION' with offshoots including *'I.2.6 Learning'*, *'I.2.6 Natural Language Processing'*, *'I.4.9 Applications'*, *'J.2 PHYSICAL SCIENCES AND ENGINEERING'*. Cluster 2 has as head subject *'G. Mathematics of Computing'* and its "offshoots" include *'F.4.1 Mathematical Logics'*, *'I.2.0 General in I.2 - ARTIFICIAL INTELLIGENCE'*, *'I.2.3 Deduction and Theorem Proving'* as well as *'J.3 LIFE AND MEDICAL SCIENCES'*.

Overall, the surveys' results analyzed in this study are consistent with the informal assessment of the research conducted in each of the research organizations. Moreover, the sets of research topics that have been chosen by individual members at the ESSA survey follow the cluster structure rather closely, falling mostly within one of them.

14.6 Conclusion

We have proposed a novel method for knowledge generalization that employs a taxonomy tree. The method constructs fuzzy membership profiles of the entities constituting the system under consideration in terms of the taxonomys leaves, and then it generalizes them in two steps. These steps are:

(i) fuzzy clustering research topics according to their thematic similarities, ignoring the topology of the taxonomy, and

(ii) lifting clusters mapped to the taxonomy to higher ranked categories in the tree.

These generalization steps thus cover both sides of the representation process: the empirical – related to the structure under consideration – and the conceptual – related to the taxonomy hierarchy.

Potentially, this approach could lead to a useful instrument for comprehensive visual representation of developments in any field of organized human activities.

However, there are a number of issues remaining to be tackled. They relate to all main aspects of the project: (a) data collection, (b) thematic clustering and (c) lifting. On the data collection side, the mainly manual e-survey ESSA tool should be supported by an automated analysis and rating of relevant research documents including those on the internet. The FADDI-S method, although already experimentally proven competitive to a number of existing methods, should be further explored and more thoroughly investigated. The issue of defining right penalty weights for parsimonious cluster lifting should be addressed.

Moreover, further investigation should be carried out with respect to the extension of this approach to more complex structures than the hierarchical tree taxonomy, ontology structures. Finally, there remains to be explored the usage of the cluster and head subjects information in query answering, and its visualization; as well as the updating of taxonomies (or other structures) on the basis of the empirical data found, possibly involving aggregating data from multiple organizations.

Acknowledgements. The authors are grateful to the staff members of the department of Computer Science, Faculdade de Ciências e Tecnologia, Universidade Nova de Lisboa, Centre for Artificial Intelligence, Universidade Nova de Lisboa, Lisboa, Portugal, and department of Computer Science, Birkbeck University of London, London, UK, that participated in the surveys. Igor Guerreiro is acknowledged for developing software for the ESSA tool. Rui Felizardo is acknowledged for developing software for the lifting algorithm with interface shown in Fig. 14.10 and Fig. 14.11. This work has been supported by grant

PTDC/EIA/69988/2006 from the Portuguese Foundation for Science & Technology. The partial financial support of the Laboratory of Choice and Analysis of Decisions at the State University – Higher School of Economics, Moscow RF, is also acknowledged.

References

1. ACM Computing Classification System (1998), http://www.acm.org/about/class/1998 (cited September 9, 2008)
2. Advanced Visual Systems (AVS), http://www.avs.com/solutions/avs-powerviz/utility_distribution.html (cited November 27, 2010)
3. Beneventano, D., Dahlem, N., El Haoum, S., Hahn, A., Montanari, D., Reinelt, M.: Ontology-driven semantic mapping. In: Enterprise Interoperability III, Part IV, pp. 329–341. Springer, Heidelberg (2008)
4. Bezdek, J., Hathaway, R.J., Windham, M.P.: Numerical comparisons of the RFCM and AP algorithms for clustering relational data. Pattern Recognition 24, 783–791 (1991)
5. Bezdek, J., Keller, J., Krishnapuram, R., Pal, T.: Fuzzy Models and Algorithms for Pattern Recognition and Image Processing. Kluwer Academic Publishers (1999)
6. Brouwer, R.: A method of relational fuzzy clustering based on producing feature vectors using FastMap. Information Sciences 179, 3561–3582 (2009)
7. Buche, P., Dibie-Barthelemy, J., Ibanescu, L.: Ontology mapping using fuzzy conceptual graphs and rules. In: ICCS Supplement, vol. 1724 (2008)
8. Cali, A., Gottlob, G., Pieris, A.: Advanced processing for ontological queries. Proceedings of the VLDB Endowment 3(1), 554–565 (2010)
9. Davé, R.N., Sen, S.: Robust fuzzy clustering of relational data. IEEE Transactions on Fuzzy Systems 10, 713–727 (2002)
10. Ding, Y., Foo, S.: Ontology research and development. Journal of Information Science 28(5), 375–388 (2002)
11. Dotan-Cohen, D., Kasif, S., Melkman, A.: Seeing the forest for the trees: using the gene ontology to restructure hierarchical clustering. Bioinformatics 25(14), 1789–1795 (2009)
12. Eick, S.G.: Visualizing online activity. Communications of the ACM 44(8), 45–50 (2001)
13. Feather, M., Menzies, T., Connelly, J.: Matching software practitioner needs to researcher activities. In: Proc. of the 10th Asia-Pacific Software Engineering Conference (APSEC 2003), vol. 6, IEEE (2003)
14. Freudenberg, J.M., Joshi, V.K., Hu, Z., Medvedovic, M.: CLEAN: CLustering Enrichment ANalysis. BMC Bioinformatics 10, 234 (2009)
15. Gahegan, M., Agrawal, R., Jaiswal, A., Luo, J., Soon, K.-H.: A platform for visualizing and experimenting with measures of semantic similarity in ontologies and concept maps. Transactions in GIS 12(6), 713–732 (2008)
16. Gaevic, D., Hatala, M.: Ontology mappings to improve learning resource search. British Journal of Educational Technology 37(3), 375–389 (2006)
17. Georgeon, O.L., Mille, A., Bellet, T., Mathern, B., Ritter, F.: Supporting activity modeling from activity traces. Expert Systems: The Journal of Knowledge Engineering (2010) (submitted)
18. The Gene Ontology Consortium. The Gene Ontology project in 2008. Nucleic Acids Research 36 (database issue), D4404 (2008); doi:10.1093/nar/gkm883, PMID 17984083
19. Ghazvinian, A., Noy, N., Musen, M.: Creating mappings for ontologies in Biomedicine: simple methods work. In: AMIA 2009 Symposium Proceedings, pp. 198–202 (2009)

20. Guh, Y.Y., Yang, M.S., Po, R.W., Lee, E.S.: Establishing performance evaluation structures by fuzzy relation-based cluster analysis. Computers and Mathematics with Applications 56, 572–582 (2008)
21. Hathaway, R.J., Davenport, J.W., Bezdek, J.C.: Relational duals of the c-means algorithms. Pattern Recognition 22, 205–212 (1989)
22. Hathaway, R.J., Bezdek, J.C.: NERF c-means: Non-Euclidean relational fuzzy clustering. Pattern Recognition 27, 429–437 (1994)
23. Huang, L., Yan, D., Jordan, M.I., Taft, N.: Spectral clustering with perturbed data. In: Koller, D., Schuurmans, D., Bengio, Y., Bottou, L. (eds.) Advances in Neural Information Processing Systems, Proceedings of the Twenty-Second Annual Conference on Neural Information Processing Systems, Vancouver, vol. 21, pp. 705–712. MIT Press (2009)
24. Hubert, L.J., Arabie, P.: Comparing partitions. Journal of Classification 2, 193–218 (1985)
25. Liu, J., Wang, W., Yang, J.: Gene ontology friendly biclustering of expression profiles. In: Proc. of the IEEE Computational Systems Bioinformatics Conference, pp. 436–447. IEEE (2004)
26. von Luxburg, U.: A tutorial on spectral clustering. Statistics and Computing 17, 395–416 (2007)
27. Marinica, C., Guillet, F.: Improving post-mining of association rules with ontologies. In: The XIII International Conference Applied Stochastic Models and Data Analysis (ASMDA), pp. 76–80 (2009); ISBN 978-9955-28-463-5
28. Mazza, R.: Introduction to Information Visualization, pp. 978–971. Springer, Heidelberg (2009); ISBN: 978-1-84800-218-0
29. McLachlan, G.J., Khan, N.: On a resampling approach for tests on the number of clusters with mixture model based clustering of tissue samples. J. Multivariate Anal. 90, 90–105 (2004)
30. Miralaei, S., Ghorbani, A.: Category-based similarity algorithm for semantic similarity in multi-agent information sharing systems. In: IEEE/WIC/ACM Int. Conf. on Intelligent Agent Technology, pp. 242–245 (2005)
31. Mirkin, B.: Additive clustering and qualitative factor analysis methods for similarity matrices. Journal of Classification 4(1), 7–31 (1987)
32. Mirkin, B., Fenner, T., Galperin, M., Koonin, E.: Algorithms for computing parsimonious evolutionary scenarios for genome evolution, the last universal common ancestor and dominance of horizontal gene transfer in the evolution of prokaryotes. BMC Evolutionary Biology 3(2) (2003)
33. Mirkin, B., Nascimento, S., Pereira, L.M.: Cluster-lift method for mapping research activities over a concept tree. In: Koronacki, J., Raś, Z.W., Wierzchoń, S.T., Kacprzyk, J. (eds.) Advances in Machine Learning II. SCI, vol. 263, pp. 245–258. Springer, Heidelberg (2010)
34. Mirkin, B., Nascimento, S., Fenner, T., Pereira, L.M.: Constructing and Mapping Fuzzy Thematic Clusters to Higher Ranks in a Taxonomy. In: Bi, Y., Williams, M.-A. (eds.) KSEM 2010. LNCS (LNAI), vol. 6291, pp. 329–340. Springer, Heidelberg (2010)
35. Mirkin, B., Nascimento, S.: Additive spectral method for fuzzy cluster analysis of similarity data including community structure and affinity matrices. Information Sciences 183, 16–34 (2012)
36. Newman, M.: Finding community structure in networks using the eigenvectors of matrices. Phys. Rev. E 74, 036104 (2006)
37. Newman, M., Girvan, M.: Finding and evaluating community structure in networks. Phys. Rev. E 69, 026113 (2004)

38. Ng, A., Jordan, M., Weiss, Y.: On spectral clustering: analysis and an algorithm. In: Ditterich, T.G., Becker, S., Ghahramani, Z. (eds.) Advances in Neural Information Processing Systems, vol. 14, pp. 849–856. MIT Press, Cambridge (2002)
39. OWL 2 Web Ontology Language Overview (2009), http://www.w3.org/TR/2009/RECowl2overview20091027/ (cited November 27, 2010)
40. Roubens, M.: Pattern classification problems and fuzzy sets. Fuzzy Sets and Systems 1, 239–253 (1978)
41. Sato, M., Sato, Y., Jain, L.C.: Fuzzy Clustering Models and Applications. Physica-Verlag, Heidelberg (1997)
42. Schattkowsky, T., Frster, A.: On the pitfalls of UML-2 activity modeling. In: International Workshop on Modeling in Software Engineering (MISE 2007), pp. 1–6 (2007)
43. Skarman, A., Jiang, L., Hornshoj, H., Buitenhuis, B., Hedegaard, J., Conley, L., Sorensen, P.: Gene set analysis methods applied to chicken microarray expression data. BMC Proceedings 3 (suppl. 4) (2009)
44. Shepard, R.N., Arabie, P.: Additive clustering: representation of similarities as combinations of overlapping properties. Psychological Review 86, 87–123 (1979)
45. Shi, J., Malik, J.: Normalized cuts and image segmentation. IEEE Transactions on Pattern Analysis and Machine Intelligence 22(8), 888–905 (2000)
46. SNOMED Clinical Terms (2010), http://www.nlm.nih.gov/research/umls/Snomed/snomed_main.html (cited November 27, 2010)
47. Sosnovsky, S., Mitrovic, A., Lee, D., Prusilovsky, P., Yudelson, M., Brusilovsky, V., Sharma, D.: Towards integration of adaptive educational systems: mapping domain models to ontologies. In: Dicheva, D., Harrer, A., Mizoguchi, R. (eds.), Procs. of 6th International Workshop on Ontologies and Semantic Web for ELearning (SWEL 2008) at ITS 2008 (2008), http://compsci.wssu.edu/iis/swel/SWEL08/Papers/Sosnovsky.pdf
48. Thomas, H., O'Sullivan, D., Brennan, R.: Evaluation of ontology mapping representation. In: Proceedings of the Workshop on Matching and Meaning, pp. 64–68 (2009)
49. Windham, M.P.: Numerical classification of proximity data with assignment measures. Journal of Classification 2, 157–172 (1985)
50. White, S., Smyth, P.: A spectral clustering approach to finding communities in graphs. In: SIAM International Conference on Data Mining (2005)
51. Thorne, C., Zhu, J., Uren, V.: Extracting domain ontologies with CORDER. Tech. Reportkmi-05-14. Open University, 1-15 (2005)
52. Yang, M.S., Shih, H.M.: Cluster analysis based on fuzzy relations. Fuzzy Sets and Systems 120, 197–212 (2001)
53. Yang, L., Ball, M., Bhavsar, V., Boley, H.: Weighted partonomy-taxonomy trees with local similarity measures for semantic buyer-seller match-making. Journal of Business and Technology 1(1), 42–52 (2005)
54. Zadeh, L.A.: Fuzzy sets. Information and Control 8, 338–353 (1965)
55. Zhang, S., Wang, R.-S., Zhang, X.-S.: Identification of overlapping community structure in complex networks using fuzzy c-means clustering. Physica A 374, 483–490 (2007)

Author Index

Abd-Elmonim, Wafaa G. 315
Abe, Jair Minoro 331, 365

Babel, Marie 91
Bédat, Laurent 91
Bekiarski, Alexander 173
Blestel, Médéric 91

Déforges, Olivier 91

Favorskaya, Margarita 211

Ghali, Neveen I. 315

Hassanien, Aboul Ella 315

Kountchev, Roumen 3, 35
Kountcheva, Roumiana 35
Kpalma, Kidiyo 255

Lopes, Helder F.S. 331

Mendi, Engin 147
Milanova, Mariofanna 147
Mirkin, Boris 423

Nakamatsu, Kazumi 3, 331, 365
Nakatani, Hiromasa 127
Nascimento, Susana 423

Pasteau, François 91
Pelcat, Maxime 91
Pereira, Luís Moniz 423

Rao, K.R. 9
Ravi, Aruna 9
Ronsin, Joseph 255
Rubin, Stuart H. 383

Strauss, Clément 91

Todorov, Vladimir 35
Tsagaan, Baigalmaa 127

Yang, Mingqiang 255